MCAT® Biochemistry

2025–2026 Edition: An Illustrated Guide

Copyright © 2024
On behalf of UWorld, LLC
Dallas, TX
USA

All rights reserved.
Printed in English, in the United States of America.

Reproduction or translation of any part of this work beyond that permitted by Sections 107 and 108 of the United States Copyright Act without the permission of the copyright owner is unlawful.

The Medical College Admission Test (MCAT®) and the United States Medical Licensing Examination (USMLE®) are registered trademarks of the Association of American Medical Colleges (AAMC®). The AAMC® neither sponsors nor endorses this UWorld product.

Facebook® and Instagram® are registered trademarks of Facebook, Inc. which neither sponsors nor endorses this UWorld product.

X is an unregistered mark used by X Corp, which neither sponsors nor endorses this UWorld product.

Acknowledgments for the 2025–2026 Edition

Ensuring that the course materials in this book are accurate and up to date would not have been possible without the multifaceted contributions from our team of content experts, editors, illustrators, software developers, and other amazing support staff. UWorld's passion for education continues to be the driving force behind all our products, along with our focus on quality and dedication to student success.

About the MCAT Exam

Taking the MCAT is a significant milestone on your path to a rewarding career in medicine. Scan the QR codes below to learn crucial information about this exam as you take your next step before medical school.

Basic MCAT Exam Information

Scores and Percentiles

MCAT Sections

Registration Guide

Preparing for the MCAT with UWorld

The MCAT is a grueling exam spanning seven subjects that is designed to test your aptitude in areas essential for success in medicine. Preparing for the exam can be intimidating—so much so that in post-MCAT questionnaires conducted by the AAMC®, a majority of students report not feeling confident about their MCAT performance.

In response, UWorld set out to create premier learning tools to teach students the entire MCAT syllabus, both efficiently and effectively. Taking what we learned from helping over 90% of medical students prepare for their medical board exams (USMLE®), we launched the UWorld MCAT Qbank in 2017 and the UWorld MCAT UBooks in 2024. The MCAT UBooks are meticulously written and designed to provide you with the knowledge and strategies you need to meet your MCAT goals with confidence and to secure your future in medical school.

Below, we explain how to use the MCAT UBooks and MCAT Qbank together for a streamlined learning experience. By strategically integrating both resources into your study plan, you will improve your understanding of key MCAT content as well as build critical reasoning skills, giving you the best chance at achieving your target score.

MCAT UBooks: Illustrated and Annotated Guides

The MCAT UBooks include not only the printed editions for each MCAT subject but also provide digital access to interactive versions of the same books. There are eight printed MCAT UBooks in all, six comprehensive review books covering the science subjects and two specialized books for the Critical Analysis and Reasoning Skills (CARS) section of the exam:

- Biology
- Biochemistry
- General Chemistry
- Organic Chemistry
- Physics
- Behavioral Sciences
- CARS (Annotated Practice Book)
- CARS Passage Booklet (Annotated)

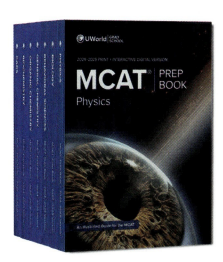

Each UBook is organized into Units, which are divided into Chapters. The Chapters are then split into Lessons, which are further subdivided into Concepts.

MCAT Sciences: Printed UBook Features

The MCAT UBooks bring difficult science concepts to life with thousands of engaging, high-impact visual aids that make topics easier to understand and retain. In addition, the printed UBooks present key terms in blue, indicating clickable illustration hyperlinks in the digital version that will help you learn more about a scientific concept.

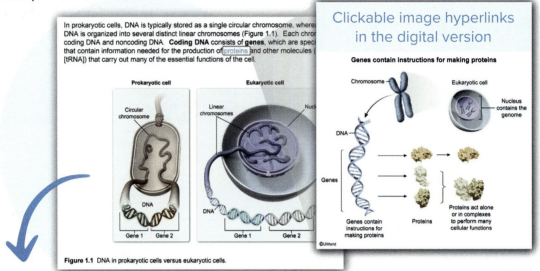

Thousands of educational illustrations in the print book

Test Your Basic Science Knowledge with Concept Check Questions

The printed UBooks also include 450 new questions—never before available in the UWorld Qbank—for Biology, General Chemistry, Organic Chemistry, Biochemistry, and Physics. These new questions, called Concept Checks, are interspersed throughout the entire book to enhance your learning experience. Concept Checks allow you to instantly test yourself on MCAT concepts you just learned from the UBook.

Short answers to the Concept Checks are found in the appendix at the end of each printed UBook. In addition, the digital version of the UBook provides an interactive learning experience by giving more detailed, illustrated, step-by-step explanations of each Concept Check. These enhanced explanations will help reinforce your learning and clarify any areas of uncertainty you may have.

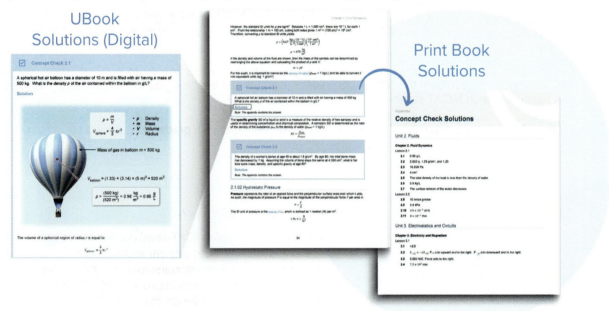

MCAT CARS Printed UBook Features

For CARS, the main book, or Annotated Practice Book, teaches you the specialized CARS skills and strategies you need to master and then follows up with multiple sets of MCAT-level practice questions.

Additionally, the CARS Passage Booklet includes annotated versions of the passages in the CARS Main Book. From these annotations, you will learn how to break down a CARS passage in a step-by-step manner to find the right answer to each CARS question.

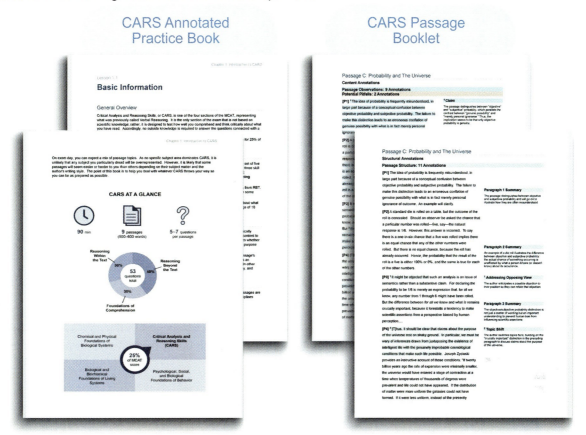

MCAT-Level Exam Practice with the UWorld Qbank

UWorld's MCAT UBooks and Qbank were designed to be used together for a comprehensive review experience. The UWorld Qbank provides an active learning approach to MCAT prep, with thousands of MCAT-level questions that align with each UBook.

The printed UBooks include a prompt at the end of each unit that explains how to access unit practice tests in the MCAT Qbank. In addition, the MCAT UBooks' digital platform enables you to easily create your own unit tests based on each MCAT subject.

To purchase MCAT Qbank access or to begin a free seven-day trial, visit gradschool.uworld.com/mcat.

Boost Your Score with the #1 MCAT Qbank

Scan for free trial

Why use the UWorld Qbank?

- Thousands of high-yield MCAT-level questions
- In-depth, visually engaging answer explanations
- Confidence-building user interface identical to the exam
- Data-driven performance and improvement tracking
- Fully featured mobile app for on-the-go review

Special Features Integrating Digital UBooks and the UWorld Qbank

The digital MCAT UBooks and the MCAT Qbank come with several integrated features that transform ordinary reading into an interactive study session. These time-saving tools enable you to personalize your MCAT test prep, get the most out of our detailed explanations, save valuable time, and know when you are ready for exam day.

My Notebook

My Notebook, a personalized note-taking tool, allows you to easily copy and organize content from the UBooks and the Qbank. Simplify your study routine by efficiently recording the MCAT content you will encounter in the exam, and streamline your review process by seamlessly retrieving high-yield concepts to boost your study performance—in less time.

Digital Flashcards

Our unique flashcard feature makes it easy for students to copy definitions and images from the MCAT UBooks and Qbank into digital flashcards. Each card makes use of spaced repetition, a research-supported learning methodology that improves information retention and recall. Based on how you rate your understanding of flashcard content, our algorithm will display the card more or less frequently.

Fully Featured Mobile App

Study for your MCAT exams anytime, anywhere, with our industry-leading mobile app that provides complete access to your MCAT prep materials and that syncs seamlessly across all devices. With the UWorld MCAT app, you can catch up on reading, flip through flashcards between classes, or take a practice quiz during lunch to make the most of your downtime and keep MCAT material top of mind.

Book and Qbank Progress Tracking

Track your progress while using the MCAT UBooks and Qbank, and review MCAT content at your own pace. Our learning tools are enhanced by advanced performance analytics that allow users to assess their preparedness over time. Hone in on specific subjects, foundations, and skills to iron out any weaknesses, and even compare your results with those of your peers.

Explore the Periodic Table

You will need to use the periodic table to answer questions on the MCAT for specific sections. Introductory general chemistry concepts constitute 30% of the material tested in the Chemical and Physical Foundations of Biological Systems section of the exam. In addition, General Chemistry constitutes 5% of the Biological and Biochemical Foundations of Living Systems section of the MCAT. Using and understanding the periodic table is a crucial skill needed for success in these sections.

1 H 1.0																	2 He 4.0
3 Li 6.9	4 Be 9.0											5 B 10.8	6 C 12.0	7 N 14.0	8 O 16.0	9 F 19.0	10 Ne 20.2
11 Na 23.0	12 Mg 24.3											13 Al 27.0	14 Si 28.1	15 P 31.0	16 S 32.1	17 Cl 35.5	18 Ar 39.9
19 K 39.1	20 Ca 40.1	21 Sc 45.0	22 Ti 47.9	23 V 50.9	24 Cr 52.0	25 Mn 54.9	26 Fe 55.8	27 Co 58.9	28 Ni 58.7	29 Cu 63.5	30 Zn 65.4	31 Ga 69.7	32 Ge 72.6	33 As 74.9	34 Se 79.0	35 Br 79.9	36 Kr 83.8
37 Rb 85.5	38 Sr 87.6	39 Y 88.9	40 Zr 91.2	41 Nb 92.9	42 Mo 95.9	43 Tc (98)	44 Ru 101.1	45 Rh 102.9	46 Pd 106.4	47 Ag 107.9	48 Cd 112.4	49 In 114.8	50 Sn 118.7	51 Sb 121.8	52 Te 127.6	53 I 126.9	54 Xe 131.3
55 Cs 132.9	56 Ba 137.3	57 La* 138.9	72 Hf 178.5	73 Ta 180.9	74 W 183.9	75 Re 186.2	76 Os 190.2	77 Ir 192.2	78 Pt 195.1	79 Au 197.0	80 Hg 200.6	81 Tl 204.4	82 Pb 207.2	83 Bi 209.0	84 Po (209)	85 At (210)	86 Rn (222)
87 Fr (223)	88 Ra (226)	89 Ac+ (227)	104 Rf (261)	105 Db (262)	106 Sg (266)	107 Bh (264)	108 Hs (277)	109 Mt (268)	110 Ds (281)	111 Rg (280)	112 Cn (285)	113 Uut (284)	114 Fl (289)	115 Uup (288)	116 Lv (293)	117 Uus (294)	118 Uuo (294)

*	58 Ce 140.1	59 Pr 140.9	60 Nd 144.2	61 Pm (145)	62 Sm 150.4	63 Eu 152.0	64 Gd 157.3	65 Tb 158.9	66 Dy 162.5	67 Ho 164.9	68 Er 167.3	69 Tm 168.9	70 Yb 173.0	71 Lu 175.0
+	90 Th 232.0	91 Pa (231)	92 U 238.0	93 Np (237)	94 Pu (244)	95 Am (243)	96 Cm (247)	97 Bk (247)	98 Cf (251)	99 Es (252)	100 Fm (257)	101 Md (258)	102 No (259)	103 Lr (260)

Table of Contents

UNIT 1 AMINO ACIDS AND PROTEINS

CHAPTER 1 AMINO ACIDS 1
- Lesson 1.1   Amino Acid Structure 3
- Lesson 1.2   Amino Acid Properties 15
- Lesson 1.3   Acid-Base Chemistry of Amino Acids 23

CHAPTER 2 PEPTIDES AND PROTEINS 37
- Lesson 2.1   Peptides 37
- Lesson 2.2   Levels of Protein Structure 47
- Lesson 2.3   Protein Folding and Stability 63
- Lesson 2.4   Protein Modifications 79

CHAPTER 3 NONENZYMATIC PROTEIN ACTIVITY 89
- Lesson 3.1   Protein-Ligand Interactions 89
- Lesson 3.2   Receptor Proteins 103
- Lesson 3.3   Membrane Transport Proteins 113

UNIT 2 ENZYMES

CHAPTER 4 ENZYME ACTIVITY 125
- Lesson 4.1   Enzymes are Catalysts 127
- Lesson 4.2   Enzyme Classification 139
- Lesson 4.3   Catalytic Mechanisms of Enzymes 167
- Lesson 4.4   Optimization of Enzyme Activity 187

CHAPTER 5 ENZYME KINETICS 199
- Lesson 5.1   The Michaelis-Menten Equation 199
- Lesson 5.2   Michaelis-Menten Parameters 209
- Lesson 5.3   Graphical Representations of Enzyme Kinetics 217
- Lesson 5.4   Enzyme Inhibitors 233

CHAPTER 6 ENZYME REGULATION 251
- Lesson 6.1   Feedback Regulation 251
- Lesson 6.2   Regulation by Covalent Modifications 263

UNIT 3 CARBOHYDRATES, NUCLEOTIDES, AND LIPIDS

CHAPTER 7 CARBOHYDRATES 271
- Lesson 7.1   Monosaccharides 273
- Lesson 7.2   Complex Carbohydrates 295

CHAPTER 8 NUCLEOTIDES AND NUCLEIC ACIDS 309
- Lesson 8.1   Nucleotides 309
- Lesson 8.2   Nucleic Acids 329

CHAPTER 9 LIPIDS 345
- Lesson 9.1   Energy Storage Lipids 345
- Lesson 9.2   Structural Lipids 353
- Lesson 9.3   Signaling Lipids 381

UNIT 4 METABOLIC REACTIONS

CHAPTER 10 CATABOLISM AND ANABOLISM 397
- Lesson 10.1   Catabolism 399
- Lesson 10.2   Anabolism 413

CHAPTER 11 CARBOHYDRATE METABOLISM 419
- Lesson 11.1   Glycolysis and Fermentation 419
- Lesson 11.2   Gluconeogenesis 435
- Lesson 11.3   The Pentose Phosphate Pathway 453
- Lesson 11.4   Glycogen Metabolism 465

CHAPTER 12 AEROBIC RESPIRATION 481
- Lesson 12.1   The Citric Acid Cycle 481
- Lesson 12.2   The Electron Transport Chain 495
- Lesson 12.3   Oxidative Phosphorylation 509

CHAPTER 13 NONCARBOHYDRATE METABOLISM.. 517
- Lesson 13.1   Fatty Acid Metabolism 517
- Lesson 13.2   Protein Catabolism 539

UNIT 5 BIOCHEMISTRY LAB TECHNIQUES

CHAPTER 14 BIOMOLECULE PURIFICATION AND CHARACTERIZATION 551
- Lesson 14.1   Gel Electrophoresis 553
- Lesson 14.2   Blotting Techniques 575
- Lesson 14.3   Chromatography 589
- Lesson 14.4   Additional Techniques 603

APPENDIX
CONCEPT CHECK SOLUTIONS 629
INDEX 637

Unit 1 Amino Acids and Proteins

Chapter 1 Amino Acids

1.1 Amino Acid Structure

- 1.1.01 Amino Acid Backbone
- 1.1.02 Variable Side Chains
- 1.1.03 Amino Acid Stereochemistry

1.2 Amino Acid Properties

- 1.2.01 Nonpolar Amino Acids
- 1.2.02 Polar Neutral Amino Acids
- 1.2.03 Charged Amino Acids

1.3 Acid-Base Chemistry of Amino Acids

- 1.3.01 Amino Acid Backbone Ionization
- 1.3.02 Ionizable Amino Acid Side Chains
- 1.3.03 Amino Acid Percent Ionization
- 1.3.04 Isoelectric Points of Amino Acids
- 1.3.05 Amino Acid Titration Curves
- 1.3.06 Nucleophilic Amino Acids

Chapter 2 Peptides and Proteins

2.1 Peptides

- 2.1.01 Peptide Bond Formation and Degradation
- 2.1.02 Peptide Bond Characteristics
- 2.1.03 Peptide Organization
- 2.1.04 Acid-Base Chemistry of Polypeptides

2.2 Levels of Protein Structure

- 2.2.01 Primary Protein Structure
- 2.2.02 Secondary Protein Structure
- 2.2.03 Tertiary Protein Structure
- 2.2.04 Quaternary Protein Structure

2.3 Protein Folding and Stability

- 2.3.01 The Hydrophobic Effect
- 2.3.02 Energy of Protein Folding
- 2.3.03 Conformational Changes
- 2.3.04 Protein Denaturation and Stability
- 2.3.05 Misfolding and Chaperones

2.4 Protein Modifications

- 2.4.01 Mutations
- 2.4.02 Post-translational Modifications
- 2.4.03 Cofactors

Chapter 3 Nonenzymatic Protein Activity

3.1 Protein-Ligand Interactions

- 3.1.01   Ligand Binding
- 3.1.02   The Dissociation Constant K_d
- 3.1.03   Binding Curves
- 3.1.04   Allostery and Cooperativity

3.2 Receptor Proteins

- 3.2.01   Ligand Binding by Receptor Proteins
- 3.2.02   Transmembrane Receptors
- 3.2.03   Cytosolic Receptors

3.3 Membrane Transport Proteins

- 3.3.01   Overview of Channels
- 3.3.02   Types of Channels
- 3.3.03   Carrier Proteins

Lesson 1.1
Amino Acid Structure

Introduction

Amino acids are the fundamental building blocks of proteins, which carry out nearly every necessary function of the cell. To fully understand how proteins form and how they interact with their environment, the properties of the amino acids that make up proteins must first be understood. The **proteinogenic amino acids**, the 20 amino acids that serve as building blocks for proteins in all organisms, each have a distinct structure. This structure determines the amino acid's properties and contributes to interactions between a protein and its environment. The 20 proteinogenic amino acids are shown in Figure 1.1.

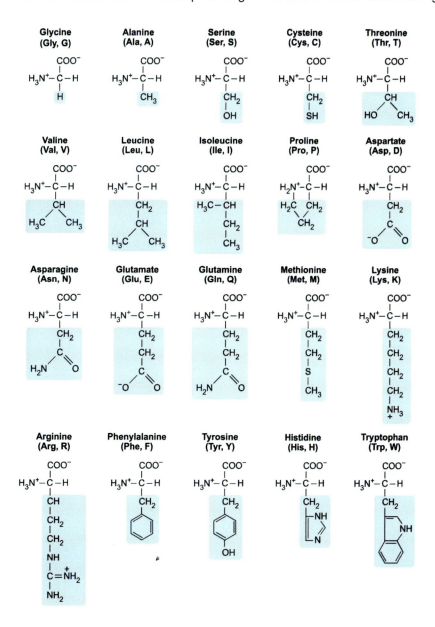

Figure 1.1 The 20 proteinogenic amino acids, arranged by increasing complexity.

This lesson provides an overview of the structural features of the amino acids and some important properties that arise from these features.

1.1.01 Amino Acid Backbone

The proteinogenic amino acids share a common structure called the backbone. The backbone consists of an amino group and a carboxylic acid group (also called a carboxyl group), giving rise to the name *amino acid*. In the 20 proteinogenic amino acids, the amino and carboxylic acid groups are both connected to a central carbon atom. Because this central carbon is directly adjacent, or *alpha*, to the carboxylic acid, the central carbon is also known as the alpha-carbon (α-carbon), and amino acids of this form are often called α-amino acids.

In contrast, some nonproteinogenic amino acids contain multiple carbons between the amino and carboxyl groups. These amino acids are not typically found in proteins, though they still play physiologically important roles. β-Alanine, for example, is a β-amino acid that has an amino group on its β-carbon and plays a structural role in the molecule coenzyme A. γ-Aminobutyric acid (GABA) is an amino acid with its amino group on the γ-carbon and is an important inhibitory neurotransmitter.

In addition to being bonded to the amino group and the carboxyl group, the α-carbon is also bonded to a hydrogen atom and to a variable group called the R-group or the side chain (Figure 1.2). The R-group determines the identity and unique properties of the amino acid.

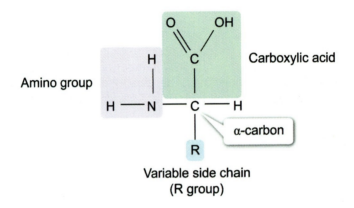

Figure 1.2 General structure of an α-amino acid.

Figure 1.2 depicts an amino acid with all atoms in an electrically neutral form (ie, with no formal charges). This form can be found in amino acids dissolved in aprotic solvents such as DMSO. However, amino acids in nature are typically found in the aqueous environment of a cell, usually at or near pH 7.4 (ie, mammalian physiological pH).

In this environment, the carboxyl group tends to lose a proton and the amino group tends to pick up a proton, resulting in a **zwitterion** (ie, a molecule that has a positive charge at one position canceled by a negative charge at another, resulting in a molecule that is electrically neutral overall). Figure 1.3 shows the zwitterionic form of a typical amino acid.

Chapter 1: Amino Acids

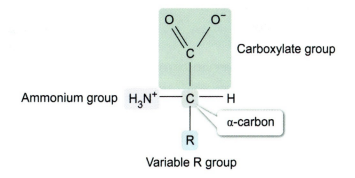

Figure 1.3 Zwitterionic form of an amino acid, as found in the conditions within a living cell.

1.1.02 Variable Side Chains

The unique side chain of each amino acid determines its identity and its properties, including how it behaves when it is part of a protein. The name of each amino acid can be abbreviated using either a three-letter code or a one-letter code as summarized in Figure 1.1.

For most amino acids, the three-letter code is the first three letters of its name, with a few exceptions detailed in the following list of amino acids. The one-letter code is the first letter of the amino acid name where possible, but in cases where multiple amino acids start with the same letter, some are assigned other one-letter codes.

For the exam, it is essential to memorize the structures of each amino acid along with the three- and one-letter codes. The following amino acids are presented in order of increasing structural complexity to allow for the memorization of the simple structures first. These can then be built upon to learn the more complex structures. The structures of the amino acids are shown in Figures 1.4 through 1.23.

Glycine

Glycine is the simplest of the amino acids and perhaps the easiest structure to remember. Glycine's side chain is a hydrogen atom. The three-letter code for glycine is Gly, and the one-letter code is G.

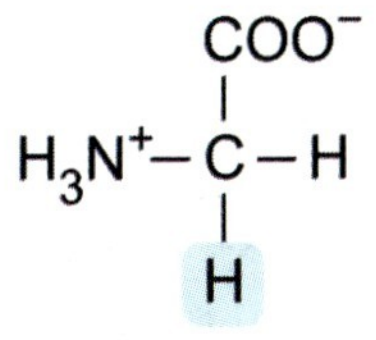

$$H_3N^+ - \underset{\underset{H}{|}}{\overset{\overset{COO^-}{|}}{C}} - H$$

Figure 1.4 Predominant structure of glycine (Gly, G) at physiological pH.

Alanine

The next simplest amino acid is alanine, which has a methyl (–CH3) group as its side chain. The three-letter code for alanine is Ala, and the one-letter code is A.

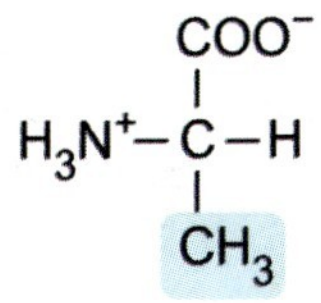

$$H_3N^+ - \underset{\underset{CH_3}{|}}{\overset{\overset{COO^-}{|}}{C}} - H$$

Figure 1.5 Predominant structure of alanine (Ala, A) at physiological pH.

Most amino acids are, in a sense, derived from alanine because they contain a methylene group (–CH$_2$–) where alanine has its methyl group.

Serine

The side chain of serine is a primary alcohol in the form of a methylene group linked to a hydroxyl (–OH) group. In other words, one of the H atoms in the –CH$_3$ group of alanine is replaced by an –OH group in serine. The three-letter code for serine is Ser, and the one-letter code is S.

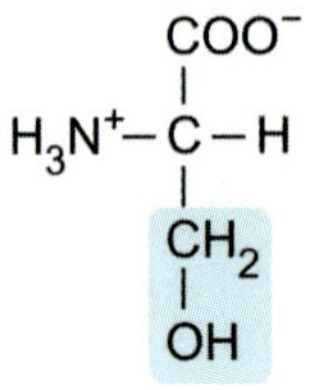

$$\begin{array}{c} COO^- \\ | \\ H_3N^+-C-H \\ | \\ CH_2 \\ | \\ OH \end{array}$$

Figure 1.6 Predominant structure of serine (Ser, S) at physiological pH.

Cysteine

The structures of serine and cysteine differ by one atom. Where serine has an oxygen atom in its hydroxyl group, cysteine has a sulfur atom, giving it a thiol (–SH) instead of a hydroxyl group. The three-letter code for cysteine is Cys, and the one-letter code is C.

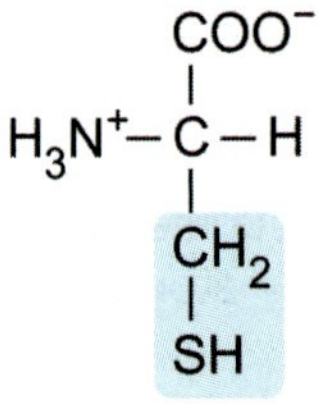

$$\begin{array}{c} COO^- \\ | \\ H_3N^+-C-H \\ | \\ CH_2 \\ | \\ SH \end{array}$$

Figure 1.7 Predominant structure of cysteine (Cys, C) at physiological pH.

Threonine

Threonine is nearly identical in structure to serine, but threonine contains an extra methyl group. Consequently, threonine contains a secondary alcohol. The three-letter code for threonine is Thr, and the one-letter code is T.

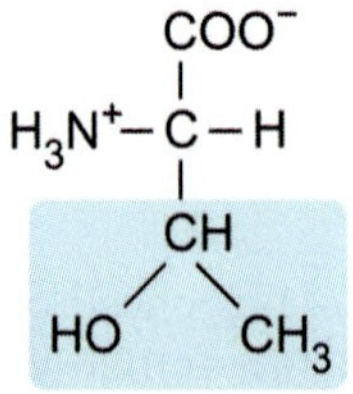

$$\begin{array}{c} COO^- \\ | \\ H_3N^+-C-H \\ | \\ CH \\ \diagup \quad \diagdown \\ HO \qquad CH_3 \end{array}$$

Figure 1.8 Predominant structure of threonine (Thr, T) at physiological pH.

Valine

Structurally, valine is nearly identical to threonine, except that valine has a methyl group where threonine has a hydroxyl group. The three-letter code for valine is Val, and the one-letter code is V. Note that the arrangement of the carbon atoms in the side chain of valine resembles the letter V.

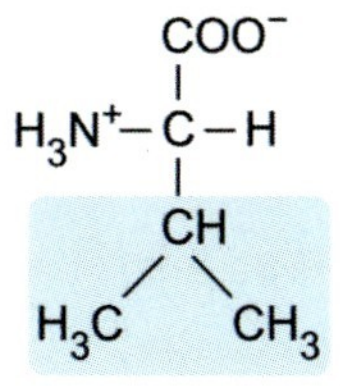

Chapter 1: Amino Acids

Figure 1.9 Predominant structure of valine (Val, V) at physiological pH.

Leucine

Leucine is similar to valine but has an extra methylene group connected to the α-carbon. The three-letter code for leucine is Leu, and the one-letter code is L.

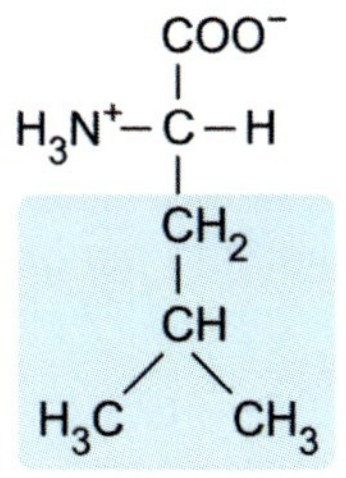

Figure 1.10 Predominant structure of leucine (Leu, L) at physiological pH.

Isoleucine

As its name implies, isoleucine is an isomer of leucine. In other words, isoleucine has the same molecular formula as leucine ($C_6H_{13}O_2N$), but the atoms are arranged differently. In leucine, both methyl groups are attached to the *same* carbon, whereas in isoleucine the methyl groups are attached to *different* carbons. The three-letter code for isoleucine is Ile (an exception to the rule of using the first three letters of the name), and the one-letter code is I.

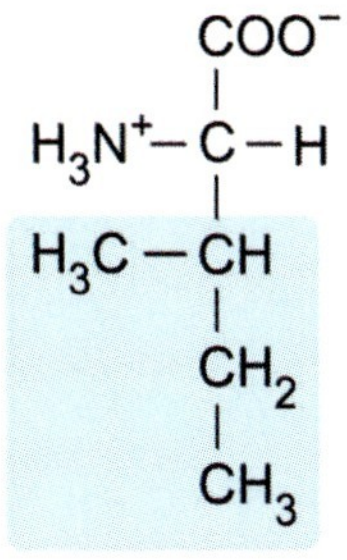

Figure 1.11 Predominant structure of isoleucine (Ile, I) at physiological pH.

Proline

Proline is unique among the amino acids because its side chain connects to the backbone in *two* places: the α-carbon (as in all amino acids) and the backbone nitrogen, forming a five-membered ring. Because of this feature, proline is the only proteinogenic amino acid that has a secondary amine instead of a primary amine. The three-letter code for proline is Pro, and the one-letter code is P.

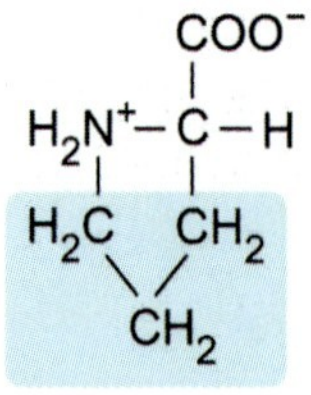

Figure 1.12 Predominant structure of proline (Pro, P) at physiological pH.

Aspartic Acid

Aspartic acid, with a carboxylic acid attached to the methylene group, forms a four-carbon dicarboxylic acid. Because carboxylic acids tend to lose protons in water, the side chain is typically deprotonated under physiological conditions and carries a negative charge. In recognition of the deprotonated side chain, the amino acid is often called aspartate instead of aspartic acid.

The three-letter code for aspartate (or aspartic acid) is Asp. Because A is already used as the one-letter code for alanine, a different one-letter code is needed. The one-letter code for aspartate (or aspartic acid) is D, which can be remembered as representing *asparDate* or *asparDic* acid.

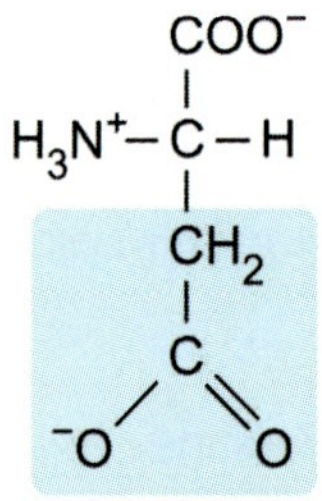

Figure 1.13 Predominant structure of aspartate (Asp, D) at physiological pH.

Asparagine

Asparagine is similar to aspartate, but where aspartate has a carboxyl group, asparagine has an amide. The side chain amide of asparagine does not ionize and remains neutral under physiological conditions. Asparagine and aspartate both start with the same three letters, so a different three-letter code was chosen for asparagine: Asn, which highlights the –NH$_2$ of the amide. The one-letter code is N, which can be remembered as the code for *asparagiNe*.

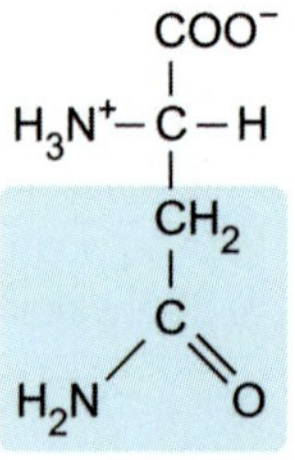

Figure 1.14 Predominant structure of asparagine (Asn, N) at physiological pH.

Glutamic Acid

Glutamic acid is nearly identical to aspartic acid, but glutamic acid has an extra methylene group (giving it five carbons in total). Like aspartic acid, glutamic acid tends to lose a proton from its side chain when in water and is commonly called glutamate.

The three-letter code for glutamate (or glutamic acid) is Glu. Because G is the one-letter code for glycine, E was chosen as the one-letter code for glutamate (or glutamic acid) because alphabetically it comes after D, the one-letter code for the similar aspartate molecule.

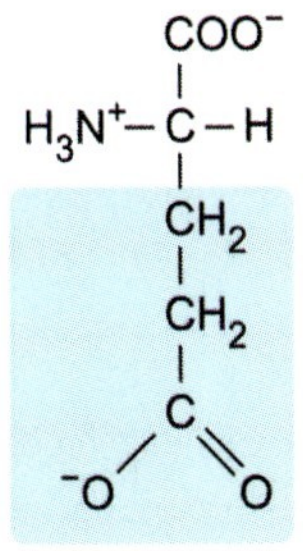

Figure 1.15 Predominant structure of glutamate (Glu, E) at physiological pH.

Glutamine

Like aspartate and asparagine, glutamate and glutamine differ only in that glutamine has an amide where glutamate has a carboxyl group. Glu is the three-letter code for glutamate, so a different code is needed for glutamine. Gln was chosen, which highlights the –NH₂ of the amide. The one-letter code for glutamine is Q, which can be remembered because glutamine sounds similar to *Q-tamine*.

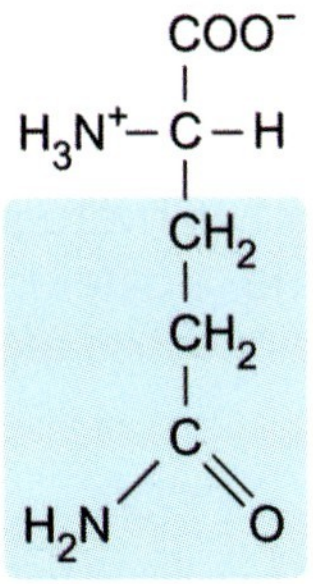

Figure 1.16 Predominant structure of glutamine (Gln, Q) at physiological pH.

Methionine

Like glutamate and glutamine, methionine has two methylene groups connected to the α-carbon. However, instead of a carboxyl group, the second methylene group is attached to a thioether—a sulfur atom that is in turn connected to an alkyl group. Specifically, **methio**nine contains a **me**thyl **thio**ether. The three-letter code for methionine is Met, and the one-letter code is M.

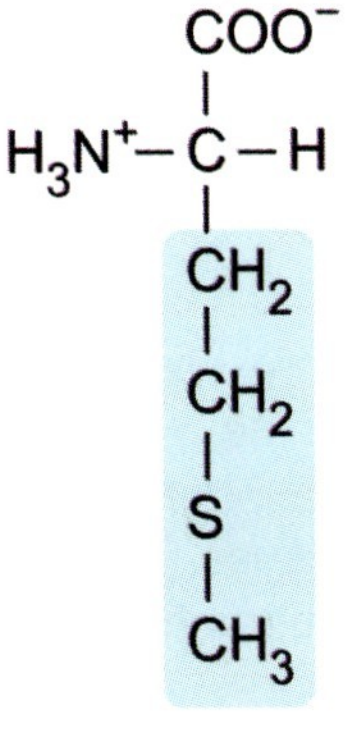

Figure 1.17 Predominant structure of methionine (Met, M) at physiological pH.

Lysine

Lysine contains a chain of four methylene groups attached to the α-carbon. At the end of this carbon chain is an amino group, which is protonated and carries a positive charge at physiological pH. The three-letter code for lysine is Lys. Because L is the one-letter code for leucine, a different letter is needed for lysine. K is adjacent to L in the alphabet and is the one-letter code for lysine.

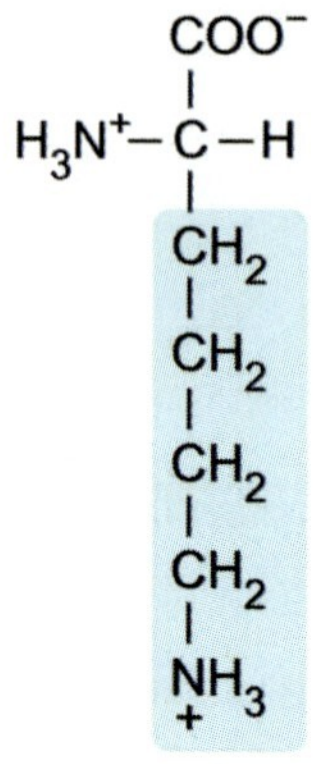

Figure 1.18 Predominant structure of lysine (Lys, K) at physiological pH.

Arginine

Arginine contains three methylene groups attached to a guanidinium group, which is protonated and carries a positive charge at physiological pH. The guanidinium group is an sp^2 carbon atom surrounded by three nitrogen atoms, each bonded to one or two hydrogen atoms. The three-letter code for arginine is Arg. Because A is the one-letter code for alanine, a different code is needed for arginine. R was chosen as the one-letter code and can be remembered because arginine sounds like *R-ginine*.

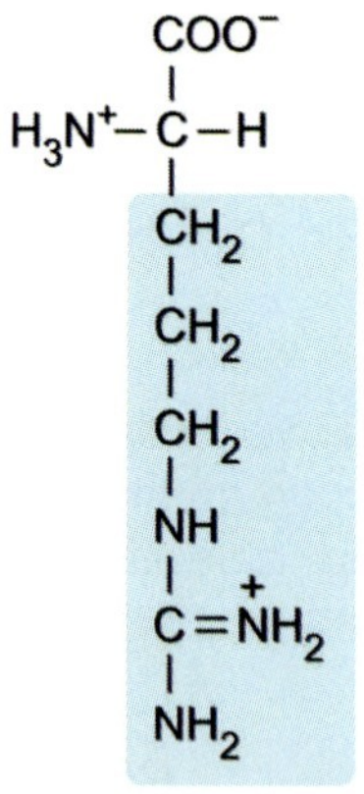

Figure 1.19 Predominant structure of arginine (Arg, R) at physiological pH.

Phenylalanine

The remaining amino acids contain aromatic rings connected to the methylene group of their side chains. The simplest of these amino acids is phenylalanine, which has a phenyl group (ie, a benzene ring) attached to its methylene group. In other words, phenylalanine is a phenyl-substituted alanine. The three-letter code for phenylalanine is Phe. Because P is the one-letter code for proline, a different letter is needed for phenylalanine. F was chosen because phenylalanine sounds like *Fenylalanine*.

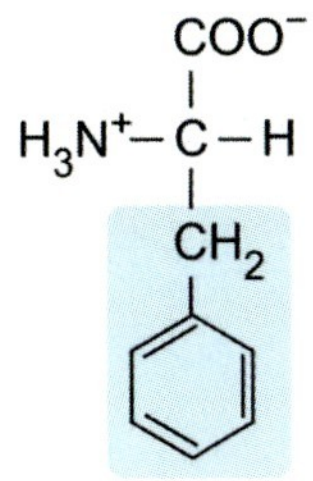

Figure 1.20 Predominant structure of phenylalanine (Phe, F) at physiological pH.

Tyrosine

Tyrosine is identical to phenylalanine except that it contains a hydroxyl group at the *para* position on the benzene ring (ie, it is a phenol group). The three-letter code for tyrosine is Tyr. Because T is the one-letter code for threonine, a different letter is needed for tyrosine. As the second letter in the name, Y was chosen for the one-letter code.

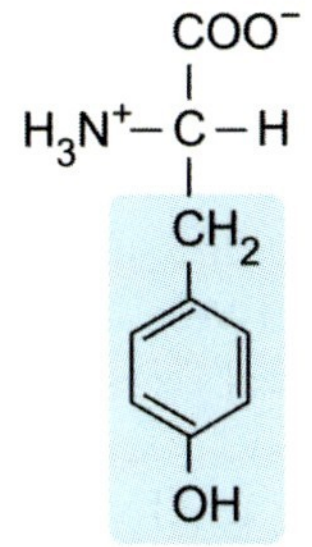

Figure 1.21 Predominant structure of tyrosine (Tyr, Y) at physiological pH.

Histidine

Histidine contains a five-membered aromatic ring called imidazole attached to its methylene group. The ring contains two nitrogen atoms and three carbon atoms. The nitrogen atoms are separated from each other by one of the three carbon atoms. The three-letter code for histidine is His, and the one-letter code is H.

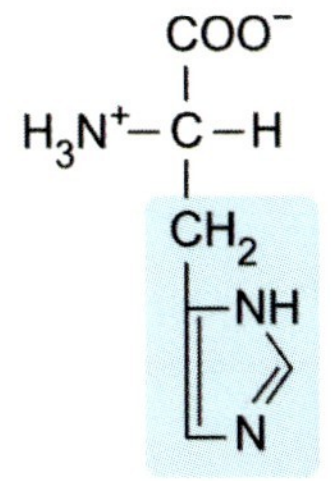

Figure 1.22 Predominant structure of histidine (His, H) at physiological pH.

Tryptophan

Tryptophan is the only amino acid with two aromatic rings (an indole group) in its side chain. A five-membered ring containing a nitrogen atom is connected to the methylene group, and a six-member ring is linked to the five-member ring. Two carbon atoms are shared by the rings. The three-letter code for tryptophan is Trp, a slight deviation from using the first three letters. Because T is the one-letter code for threonine, a different letter is needed for tryptophan. W was chosen for this purpose and can be remembered by thinking of tryptophan as *tWyptophan*.

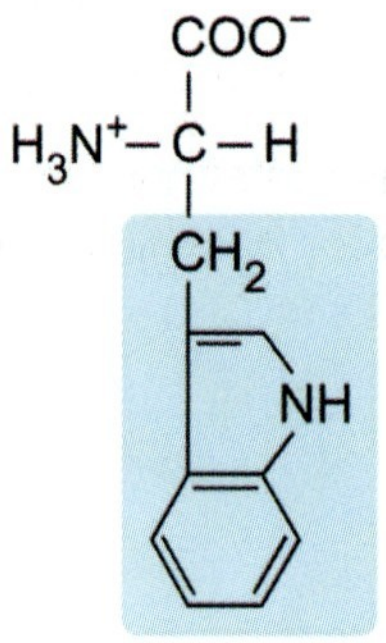

Figure 1.23 Predominant structure of tryptophan (Trp, W) at physiological pH.

1.1.03 Amino Acid Stereochemistry

The amino acids are often represented in two dimensions, causing them to appear flat. In reality, the amino acids are three-dimensional objects in which the α-carbon has tetrahedral geometry. Any tetrahedral atom that is bonded to four distinct groups is a chiral atom, meaning that its substituents (the chemical groups that surround it) can be arranged in two distinct, nonsuperimposable (and therefore nonidentical) configurations.

Apart from glycine, which is achiral, all proteinogenic amino acids have four distinct substituents around the α-carbon; therefore, these amino acids are chiral. In other words, there are at least two possible forms of each amino acid except glycine. Figure 1.24 shows different ways in which the constituents of an amino acid may be arranged.

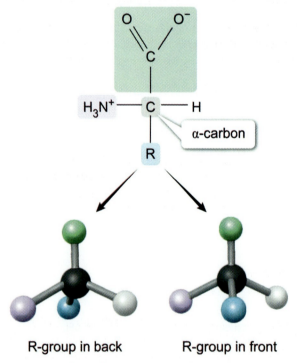

Figure 1.24 Two possible configurations of a typical amino acid.

Two conventions are commonly used to distinguish amino acid isomers. The first is the **L/D convention**, which names molecules based on their similarity to L- or D-glyceraldehyde (see Figure 1.25), respectively. L-Glyceraldehyde was given its designation because it is empirically observed to be levorotatory (ie, it rotates plane-polarized light counterclockwise), whereas D-glyceraldehyde is dextrorotatory (ie, it rotates plane-polarized light clockwise). Amino acids with α-carbon configurations similar to L-glyceraldehyde are L-amino acids, and amino acids that are similar to D-glyceraldehyde are D-amino acids.

In living cells, the chiral amino acids in proteins are almost exclusively L. A few rare exceptions exist (eg, certain bacteria convert L-amino acids to the D-form within specific proteins).

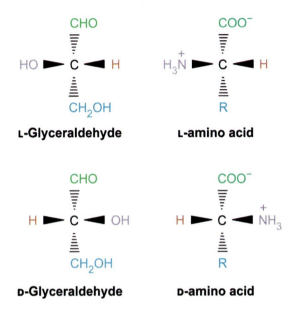

Figure 1.25 Comparison of the structures of L-and D-glyceraldehyde with L- and D-amino acids.

Note that designation of an amino acid as L- or D- does *not* indicate how the amino acid rotates plane-polarized light. Some L-amino acids rotate plane-polarized light clockwise, and some rotate it counterclockwise. The designation represents *only* the structural similarity to glyceraldehyde.

 Concept Check 1.1

Three different amino acids are analyzed by polarimetry. One amino acid rotates plane-polarized light clockwise, one rotates plane-polarized light counterclockwise, and one does not rotate plane-polarized-light. What conclusions can be drawn about the stereochemistry of these amino acids? Assume that solutions containing chiral amino acids have only the L-form or only the D-form (ie, the mixtures are not racemic).

Solution

Note: The appendix contains the answer.

The other convention for naming different forms of amino acids is the *R/S* system, which follows the Cahn-Ingold-Prelog priority rules. Under this system, all L-amino acids have *S*-configurations at the α-carbon except for cysteine. The sulfur atom in cysteine's side chain increases its priority, giving the α-carbon of L-cysteine an *R*-configuration. Note, however, that *R*-cysteine is still an L-amino acid.

In addition to stereochemistry at the α-carbon, two amino acids have chiral centers in their side chains. The chiral center in the side chain of L-threonine has an *R*-configuration (which can be remembered as L-*thR*eonine), whereas the chiral center in the side chain of L-isoleucine has an *S*-configuration (L-i*S*oleucine). Their diastereomers (allothreonine and alloisoleucine) are not prevalent in healthy humans but may accumulate in pathological conditions.

Lesson 1.2
Amino Acid Properties

Introduction

Each amino acid has distinct chemical properties. Because the backbones of the 20 proteinogenic amino acids are identical, the unique properties of an amino acid arise solely from its side chain. The unique properties of each amino acid allow for tremendous diversity in the ways different proteins behave and the functions they perform. For instance, proteins with an abundance of positively charged amino acid side chains are likely to interact with negatively charged molecules (and vice versa), and proteins with long hydrophobic regions are often found embedded in the hydrophobic environment of a cell membrane.

An understanding of the chemical properties of amino acids can facilitate predictions about interactions between proteins and their environments. Consequently, amino acids are commonly grouped according to their chemical properties. Broadly, amino acid side chains are classified as nonpolar, polar neutral, and charged. This lesson explains the categorizations of each amino acid.

1.2.01 Nonpolar Amino Acids

Most of the nonpolar amino acid side chains (Figure 1.26) contain only carbon and hydrogen atoms, which have similar electronegativities. These C–C and C–H bonds are nonpolar and typically do not give rise to a large net dipole moment.

The lack of polar bonds in many of these amino acid side chains means that they cannot form hydrogen bonds or dipole-dipole interactions with water. Consequently, the nonpolar amino acid side chains interact poorly with water and are considered **hydrophobic**.

However, two of the nonpolar amino acids contain heteroatoms (ie, atoms other than carbon or hydrogen) in their side chain: methionine and tryptophan. Methionine contains a sulfur atom in the form of a methyl thioether. Ethers are already relatively nonpolar *despite* the inclusion of an electronegative oxygen. Thioethers have a sulfur atom instead of oxygen, and sulfur has nearly the same electronegativity as carbon. Therefore, the thioether in methionine is even *less* polar than an ether, and methionine is nonpolar.

Tryptophan contains a large side chain with a single nitrogen atom, which is bonded to one hydrogen and two carbons. Nitrogen is significantly more electronegative than carbon or hydrogen, so this portion of the tryptophan side chain is polar and capable of acting as a hydrogen bond donor. However, the nonpolar character of the rest of the side chain overwhelms this effect, and tryptophan is considered nonpolar overall.

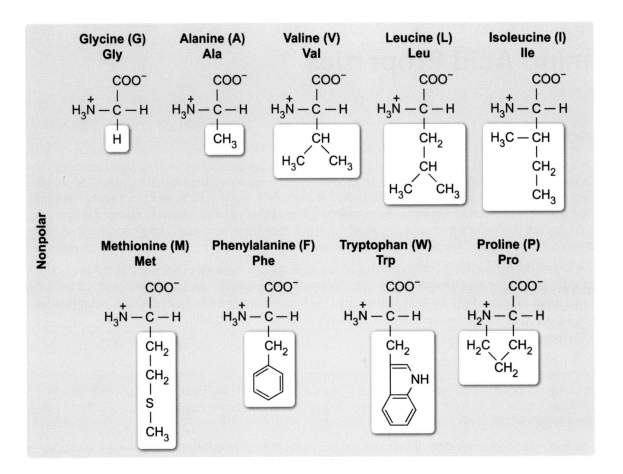

Figure 1.26 The nonpolar amino acids.

The nonpolar amino acids can be further subdivided into **aliphatic** and **aromatic** side chains.

Aliphatic Amino Acids

The aliphatic amino acids are nonpolar amino acids that are *not* aromatic (ie, they do not contain a fully conjugated, aromatic ring structure). This group includes glycine, alanine, methionine, valine, leucine, isoleucine, and proline (Figure 1.27). Although proline contains a ring, the ring is *not* aromatic, so proline is aliphatic.

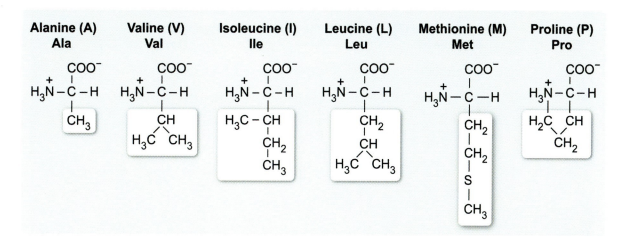

Figure 1.27 The aliphatic amino acids.

Among the aliphatic amino acids, **valine**, **leucine**, and **isoleucine** are commonly grouped as the **branched-chain amino acids** (BCAAs), shown in Figure 1.28. These three amino acids are grouped both because of their similar structures and because of their involvement in similar metabolic pathways (see Chapter 13) and pathological mechanisms.

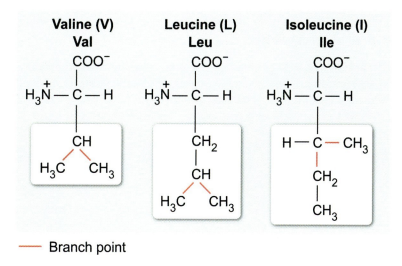

— Branch point

Figure 1.28 The branched-chain amino acids.

Note that although other amino acids such as threonine and arginine also have branch points, these branches link to nonalkyl groups (ie, groups of atoms other than carbon and hydrogen). The BCAAs include only those amino acids with branched *alkyl* chains; therefore, threonine and arginine are *not* included.

Aromatic Amino Acids

The remaining nonpolar amino acids (phenylalanine and tryptophan) contain aromatic rings in their side chains, and, consequently, they are grouped as aromatic amino acids (see Figure 1.29). Tyrosine is also included in this group. Although tyrosine has a hydroxyl group in its side chain and is often considered polar (see Concept 1.2.02 in this lesson), its large aromatic ring confers significant nonpolar character on tyrosine and allows it to be grouped with the other aromatic amino acids.

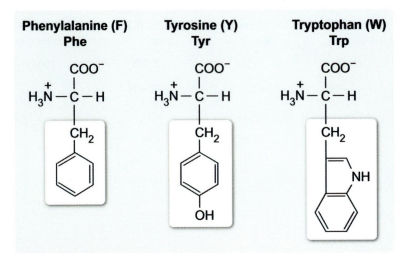

Figure 1.29 The aromatic amino acids.

Histidine also has an aromatic ring in its side chain, but it is often omitted from the aromatic amino acids because its smaller size and two-nitrogen side chain cause it to be much more polar than phenylalanine, tyrosine, or tryptophan; it can even carry a positive charge (see Lesson 1.3).

1.2.02 Polar Neutral Amino Acids

The polar neutral amino acids have side chains functional groups that have significant polar character (ie, a large net dipole moment) but do not predominantly carry an electric charge under physiological conditions (ie, the side chains are neutral). The side chains of these amino acids contain oxygen or nitrogen atoms or, in the case of cysteine, a sulfur atom. They are shown in Figure 1.30.

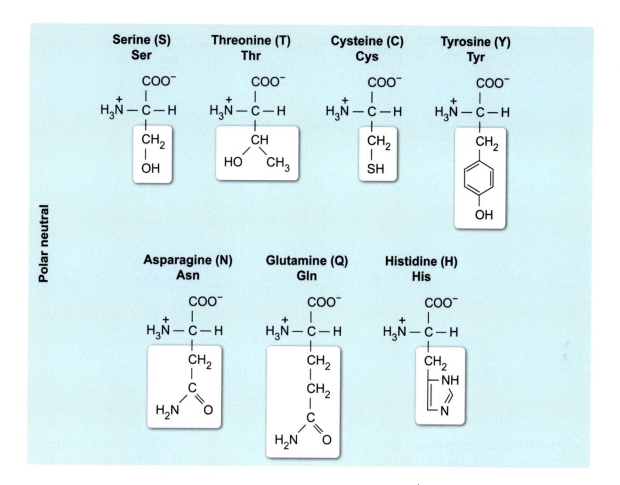

Figure 1.30 The polar neutral amino acids.

The oxygen and nitrogen atoms in these side chains are much more electronegative than the carbon and hydrogen atoms to which they are bonded. Consequently, these chemical groups have significant dipole moments and are capable of hydrogen bonding with water, making them generally **hydrophilic**.

The amino acids with amide and alcohol side chains (asparagine, glutamine, serine, and threonine) are, for all intents and purposes, truly neutral amino acids. Although it is possible to deprotonate alcohols to yield negative charges, this generally does not happen in the mild conditions found in living cells, except in specific contexts such as enzyme active sites (see Lesson 4.3). Amides require even harsher conditions than alcohols to ionize and are essentially never found to carry a positive or negative charge in biological contexts.

In contrast, tyrosine, cysteine, and histidine *can* carry charges on their side chains, as shown in Figure 1.31, but they are *predominantly* neutral under physiological conditions.

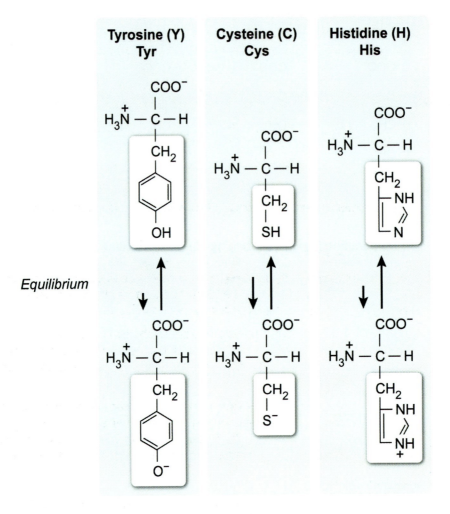

Figure 1.31 Relative equilibria of tyrosine, cysteine, and histidine neutral and ionized (ie, charged) forms at physiological pH.

The hydroxyl group of tyrosine is linked to a phenyl ring, making it a special type of alcohol called a phenol. The presence of the hydroxyl group outside of the ring allows this functional group to be much more accessible to other molecules and causes tyrosine to be more polar and slightly more hydrophilic than other aromatic amino acids, such as tryptophan. Because of resonance, phenols are much easier to deprotonate than other alcohols and, consequently, a small percentage of tyrosine side chains are deprotonated and negatively charged under physiological conditions.

Cysteine, unlike the other polar neutral amino acids, does *not* contain oxygen or nitrogen atoms in its side chain. Instead, it contains sulfur. The dipole moment of its thiol (–SH) group is weak, and cysteine cannot hydrogen bond with water. However, the bent geometry of the thiol group exposes the lone electron pairs on sulfur, and the small size of the hydrogen atom improves access to those lone pairs. This allows stronger dipole–induced dipole interactions between water and cysteine.

Because of cysteine's geometry, at physiological pH, cysteine is slightly more soluble in water than tyrosine is, allowing classification of cysteine as polar neutral. Although cysteine is predominantly neutral, a small but significant percentage of cysteine side chains are deprotonated and negatively charged at physiological pH.

Histidine is often excluded from the polar neutral amino acids and instead grouped with the basic (positively charged) amino acids. However, this characterization can be misleading. At physiological pH, *most* histidine side chains in solution are deprotonated and uncharged. Fewer than 10% of histidine side chains are protonated and positively charged at pH 7.4. The acid-base properties of several amino acids are discussed further in Lesson 1.3.

1.2.03 Charged Amino Acids

The remaining amino acids carry electric charges (either positive or negative) on their side chains under physiological conditions. These amino acids are arginine and lysine (positive charges) and aspartate and glutamate (negative charges). The charged amino acids are shown in Figure 1.32.

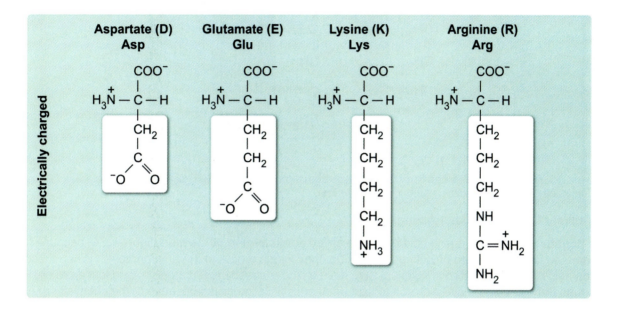

Figure 1.32 The charged amino acids.

Positively Charged (Basic) Amino Acids

The **positively charged** amino acids are also called **basic amino acids** (see Figure 1.33) because they become positively charged by acting as bases (ie, they accept a proton from water or other sources in their environment). Note that, as discussed for polar neutral amino acids, histidine is *capable* of becoming positively charged by accepting a proton and is often counted among the basic amino acids for this reason. However, under physiological conditions, most histidine side chains are deprotonated and neutral, whereas most lysine and arginine side chains are protonated and positively charged.

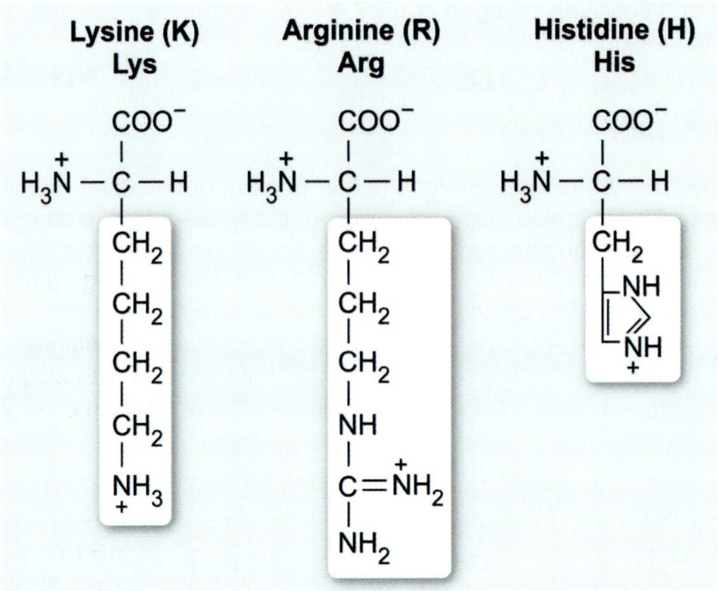

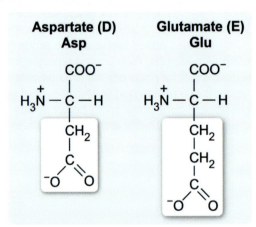

Figure 1.33 Basic amino acids can carry a positive charge at physiological pH. Lysine and arginine are predominantly charged; histidine is predominantly neutral.

Negatively Charged (Acidic) Amino Acids

The **negatively charged** amino acids are also called **acidic amino acids** (see Figure 1.34) because they become negatively charged by acting as acids (ie, they lose a proton to the environment). When these amino acids are encountered in solution, they have typically *already* acted as acids and lost a proton. Although they no longer have a proton to lose, they are still called acidic because of the chemistry they underwent (ie, loss of a proton) to gain the negative charge.

Figure 1.34 Acidic amino acids carry a negative charge under physiological conditions.

Although both tyrosine and cysteine are also able to lose a proton to become negatively charged, they are not usually counted among the acidic amino acids. For tyrosine, this is because the percentage of tyrosine molecules that lose a proton at physiological pH is small (~0.3%). For cysteine, it is because cysteine's most notable physiological role is to undergo a redox reaction to form or break a disulfide bond (see Lesson 2.2), rather than an acid-base reaction.

Lesson 1.3
Acid-Base Chemistry of Amino Acids

Introduction

Each of the amino acids has at least two functional groups that can exchange protons with their environment. In other words, amino acids can participate in acid-base chemistry. The extent to which a chemical group on an amino acid is protonated or deprotonated can be modulated by its local environment. This chapter examines the protonation states of various amino acid chemical groups under various conditions and explores the effect of protonation on the overall charges of amino acids.

1.3.01 Amino Acid Backbone Ionization

Chemical groups that participate in acid-base chemistry are said to be **ionizable** because they either become positively charged (ie, cations) by accepting protons, or they become negatively charged (ie, anions) by losing protons. All amino acids have at least two ionizable groups—the backbone amino group and the backbone carboxyl group. As discussed in Lesson 1 of this chapter, the backbone of an amino acid is a **zwitterion** at physiological pH. The amino group is protonated and positively charged, and the carboxyl group is deprotonated and negatively charged, resulting in a molecule with a net charge of 0.

However, the protonation states of the backbone groups can be altered by adjusting the pH, as shown in Figure 1.35. Decreasing the pH *increases* the number of protons available in the environment, making it easier for an amino acid carboxyl group to pick up one of these protons. Increasing the pH has the opposite effect: fewer free protons are available in the solution, so the amino group more easily *loses* protons.

Consequently, when the pH is low, the carboxyl and amino groups in an amino acid are more likely to be protonated, giving the backbone an overall positive charge. Similarly, at high pH, both the carboxyl and amino groups are likely to be deprotonated, giving the backbone an overall negative charge.

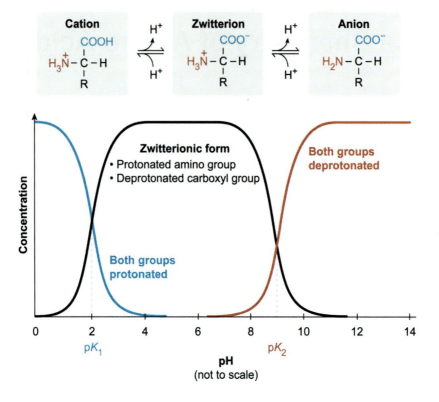

Figure 1.35 Amino acid backbones are cations (positive charge) at low pH, zwitterions (no net charge) at physiological pH, and anions (negative charge) at high pH.

1.3.02 Ionizable Amino Acid Side Chains

The side chains of most amino acids are electrically neutral at physiological pH, but some side chains can become protonated or deprotonated and carry a positive or negative charge. These amino acids—arginine, lysine, histidine, tyrosine, cysteine, glutamate, and aspartate—are collectively called the ionizable amino acids. They are shown in Figure 1.36.

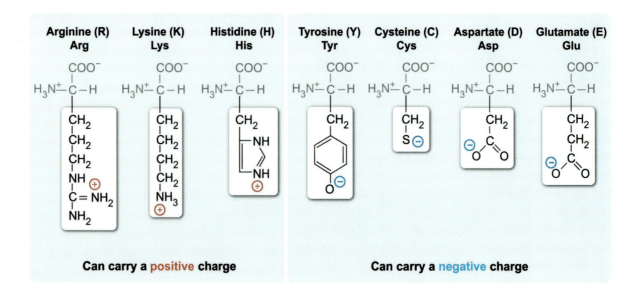

Figure 1.36 The seven ionizable amino acids shown in their ionized (ie, charged) forms.

The acidity or basicity of a chemical group (ie, its tendency to lose or pick up protons) can also be altered by its environment, as shown in Figure 1.37. For instance, carboxyl groups are normally acidic and are therefore deprotonated under physiological conditions. However, if a carboxyl group is placed near another negative charge, the carboxyl group becomes more likely to accept a proton, neutralizing it and reducing charge-charge repulsion. In other words, the carboxyl group becomes more basic in these conditions.

Similarly, if an alcohol (normally neutral) is placed near a positive charge, the alcohol becomes more likely to lose a proton (ie, it becomes more acidic). The resulting negative charge increases attractive interactions with the nearby positive charge.

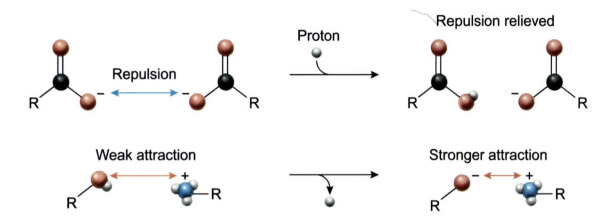

Figure 1.37 Acidity of chemical groups may be altered by their local environments to decrease repulsive forces and increase attractive forces.

Consequently, in addition to these seven amino acids, serine and threonine can be deprotonated and negatively charged in certain contexts, such as the active sites of some enzymes (see Chapter 4). However, because serine and threonine are difficult to deprotonate outside of these contexts, they are typically omitted from the ionizable amino acids.

Note that some of the ionizable amino acids (Arg, Lys, and His) are positively charged when protonated, whereas the others (Tyr, Cys, Glu, and Asp) are neutral when protonated. Similarly, Arg, Lys, and His are neutral when deprotonated whereas Tyr, Cys, Glu, and Asp are negatively charged when deprotonated. For the exam, it is sufficient to remember that the ionizable side chains that contain nitrogen can either be positive or neutral, and the ionizable side chains that contain oxygen or sulfur can either be neutral or negative. The charge associated with the protonation state of each side chain is summarized in Figure 1.38.

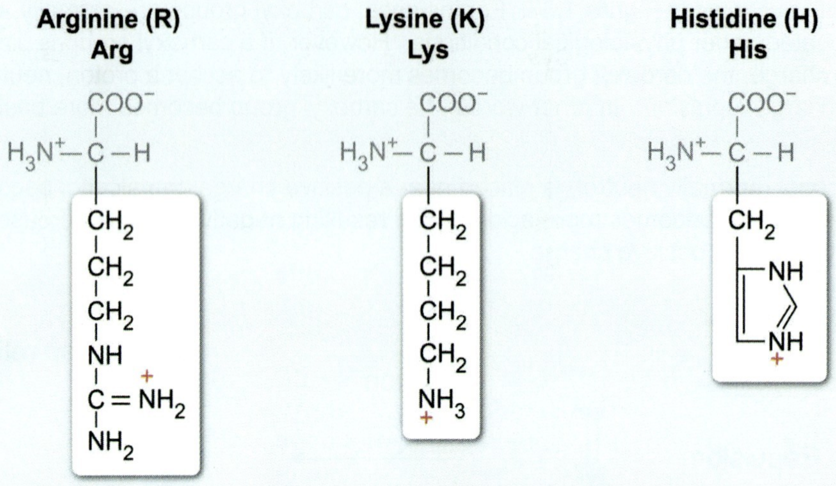

Chapter 1: Amino Acids

Arginine (R) Arg — COO⁻, H₃N⁺—C—H, CH₂-CH₂-CH₂-NH-C(=NH₂⁺)-NH₂

Lysine (K) Lys — COO⁻, H₃N⁺—C—H, CH₂-CH₂-CH₂-CH₂-NH₃⁺

Histidine (H) His — COO⁻, H₃N⁺—C—H, CH₂—imidazole ring (NH, NH⁺)

Positive when protonated

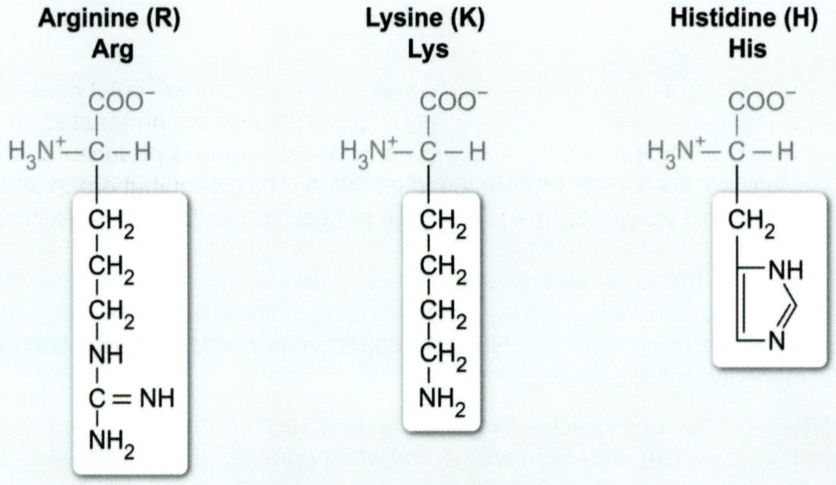

Arginine (R) Arg — COO⁻, H₃N⁺—C—H, CH₂-CH₂-CH₂-NH-C(=NH)-NH₂

Lysine (K) Lys — COO⁻, H₃N⁺—C—H, CH₂-CH₂-CH₂-CH₂-NH₂

Histidine (H) His — COO⁻, H₃N⁺—C—H, CH₂—imidazole ring (NH, N)

Neutral when deprotonated

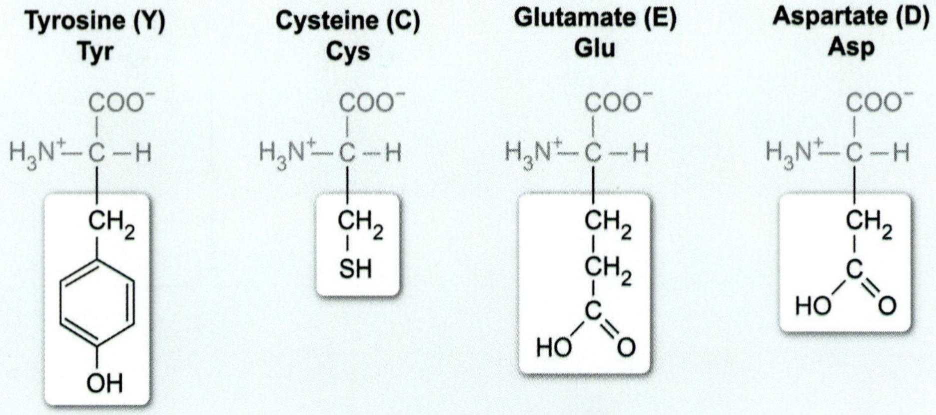

Tyrosine (Y) Tyr — COO⁻, H₃N⁺—C—H, CH₂—phenyl—OH

Cysteine (C) Cys — COO⁻, H₃N⁺—C—H, CH₂-SH

Glutamate (E) Glu — COO⁻, H₃N⁺—C—H, CH₂-CH₂-C(=O)-OH

Aspartate (D) Asp — COO⁻, H₃N⁺—C—H, CH₂-C(=O)-OH

Neutral when protonated

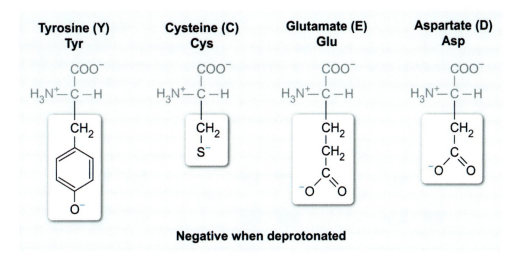

Figure 1.38 Some ionizable amino acids are neutral when deprotonated and positive when protonated; others are negative when deprotonated and neutral when protonated.

1.3.03 Amino Acid Percent Ionization

A single site on an ionizable group may either be protonated or deprotonated, but it cannot be partially protonated. However, in a *population* of ionizable groups, a percentage are protonated and the remainder are deprotonated. The probability that a given individual group is protonated, and the extent to which a population of ionizable groups is protonated, depends on the pK_a of that group and the pH to which it is exposed. As the pH increases, the percentage of deprotonated groups also increases (see Figure 1.39).

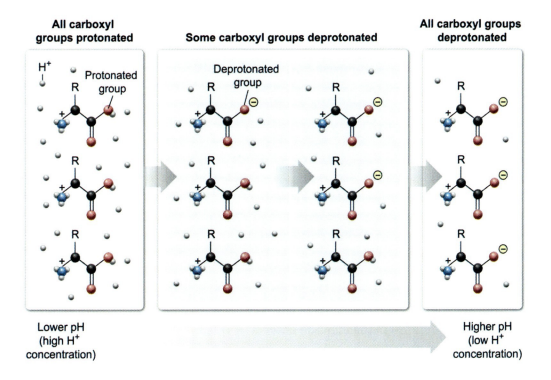

Figure 1.39 Effect of pH on protonation of the carboxyl groups in amino acid backbones.

By definition, the pK_a of an ionizable group is the pH at which 50% of the population is protonated and 50% is deprotonated. When the pH is below the pK_a, the relative abundance of protons in solution makes it more likely for ionizable groups to accept a proton; as a result, decreasing the pH of a solution below the pK_a of a group of interest causes *more* than 50% of its population to be protonated. Similarly, raising the pH above the pK_a causes more than 50% of its population to be *deprotonated* (see Figure 1.40).

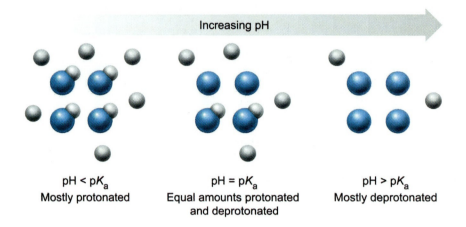

Figure 1.40 Relative amounts of chemical groups that are protonated when pH is less than, equal to, or greater than the pK_a of the chemical group.

The backbone amino and carboxyl groups of each amino acid have slightly different pK_a values depending on the identity of the side chain. For the exam, however, they can be collectively approximated as 9.6 and 2.2, respectively. The pK_a values of these groups along with those of the seven ionizable side chains are summarized in Table 1.1.

Table 1.1 Approximate pK_a values of α-amino, α-carboxy, and ionizable side chain groups of amino acids.

Ionizable group	pK_a
α-Amino group	9.6
α-Carboxyl group	2.2
Arginine side chain	12.5
Lysine side chain	10.5
Tyrosine side chain	10.0
Cysteine side chain	8.0
Histidine side chain	6.0
Glutamate side chain	4.3
Aspartate side chain	3.7

The extent to which an ionizable amino acid group has been protonated or deprotonated can be calculated using the Henderson-Hasselbalch equation (see General Chemistry Lesson 8.4):

$$pH = pK_a + \log\frac{[A^-]}{[HA]}$$

where [A⁻] is the molar concentration of the deprotonated, negative form of the ionizable group and [HA] is the molar concentration of protonated neutral form. Alternatively, if the deprotonated form is neutral (A) and the protonated form is positively charged (HA⁺), this may be expressed as:

$$pH = pK_a + \log\frac{[A]}{[HA^+]}$$

Given the pK_a and pH, the ratio of [A⁻] to [HA] (or [A] to [HA⁺]) can be determined by solving:

$$pH - pK_a = \log\frac{[A^-]}{[HA]} \quad \text{or} \quad pH - pK_a = \log\frac{[A]}{[HA^+]}$$

$$10^{(pH-pK_a)} = \frac{[A^-]}{[HA]} \quad \text{or} \quad 10^{(pH-pK_a)} = \frac{[A]}{[HA^+]}$$

Percent ionization of an amino acid can then be calculated.

$$\%\ \text{ionization} = \frac{[A^-]}{[A^-]+[HA]} \times 100\% \quad \text{or} \quad \%\ \text{ionization} = \frac{[HA^+]}{[A]+[HA^+]} \times 100\%$$

Concept Check 1.2

Calculate the percent ionization of the side chains of tyrosine and histidine, each at pH 7.0.

Solution

Note: The appendix contains the answer.

1.3.04 Isoelectric Points of Amino Acids

Often, amino acids are said to have a particular net charge at a particular pH (eg, lysine has a net charge of +1 at physiological pH). However, based on the Henderson-Hasselbalch equation, at any pH, some amino acids in a population have one net charge, and some have a different net charge. For this reason, it must be recognized that in most cases the amino acids in the sample *predominantly* have a certain net charge, but a smaller subset have a *different* net charge.

For example, histidine is considered neutral at physiological pH because *most* of the histidine molecules in solution are neutral, but a small subset are positively charged. This means that at physiological pH, the histidine molecules collectively have a slight positive charge on average, although most *individual* histidine molecules are neutral.

Given this, any population of amino acids can be made to have a net positive charge by sufficiently *decreasing* the pH, and any population can be made to have a net negative charge by sufficiently *increasing* the pH. Similarly, every amino acid has a specific pH at which the population of amino acids, on average, has a net charge of 0. This pH is called the isoelectric point (pI) of the amino acid of interest.

The pI of any amino acid can be calculated by averaging two pK_a values: the pK_a *below* which the amino acid is predominantly positive and the pK_a *above* which the amino acid is predominantly negative. For example, below a pH of 6.0, histidine is predominantly positive, and above a pH of 9.6, histidine is predominantly negative (see Figure 1.41).

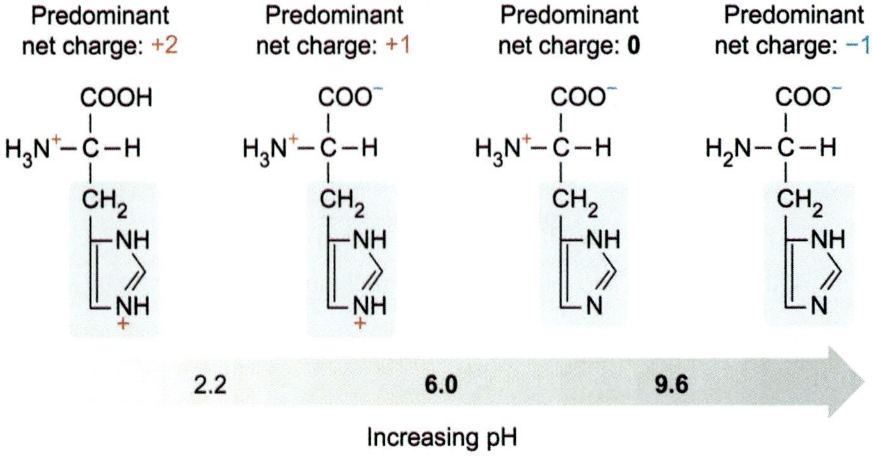

| Predominant net charge: +2 | Predominant net charge: +1 | Predominant net charge: 0 | Predominant net charge: −1 |

(structures of histidine at increasing pH, with transitions at 2.2, 6.0, and 9.6)

Increasing pH →

Figure 1.41 Predominant protonation states and net charges of histidine at different pH levels.

Therefore, these two pK_a values are used in the calculation of pI. The pK_a of the backbone carboxyl group (2.2) is not relevant in this case because it is not directly involved in the transition from charged to neutral. Averaging the side chain and backbone amino group pK_a values gives the pI of histidine as approximately

$$\text{pI} = \frac{6.0 + 9.6}{2} = 7.8$$

Note that the neutral form of histidine is the predominant ionization state throughout the *entire range* of pH values from 6.0 to 9.6, but the charged forms are still present as *minor species* and contribute to average charge. However, at the isoelectric point (ie, 7.8), the levels of the positively charged and negatively charged species are equal and therefore neutralize each other, yielding a population of histidine molecules with an *average* charge of 0.

Figure 1.42 shows the relative amounts of each protonation state of histidine at various pH levels.

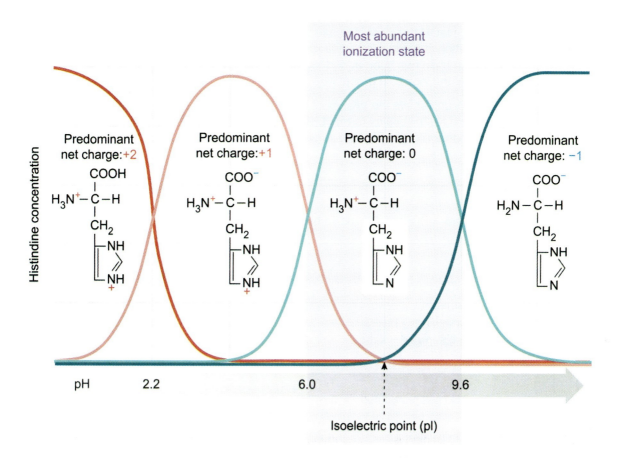

Figure 1.42 At the isoelectric point of histidine, the average charge of all histidine molecules in a sample is 0. Average charge increases as pH decreases and vice versa.

> **Concept Check 1.3**
>
> Calculate the isoelectric points of glutamate, alanine, and lysine.
>
> **Solution**
>
> *Note: The appendix contains the answer.*

1.3.05 Amino Acid Titration Curves

Because amino acids each have at least two ionizable groups (ie, groups that participate in acid-base chemistry), they can be analyzed by titration (see General Chemistry Lesson 8.6). Typically, an amino acid titration is carried out by bringing the solution containing the amino acid to a pH *below* its lowest pK_a (ie, pH <2.2). A strong base such as NaOH is then gradually added, and the pH of the solution is plotted as a function of the amount of NaOH added. The resulting titration curve can help reveal the identity of an unknown amino acid.

When the pH of the solution is *not* near the pK_a of any ionizable group, added NaOH primarily reacts with hydronium ions (H_3O^+), converting hydronium to water and quickly increasing the pH of the solution. However, when the pH of the solution is within 1 pH unit of an ionizable group's pK_a, that ionizable group begins to act as an effective buffer. For this reason, the pH range spanning 1 unit below the pK_a and 1 unit above is called a **buffer region**.

Within a buffer region, NaOH primarily removes protons from the ionizable group instead of from hydronium. Consequently, NaOH does not increase the pH of the solution as quickly in this region, and the buffer region of the titration curve is relatively flat. The midpoint of an ionizable group's buffer region is the point at which enough base has been added to deprotonate exactly half of those groups, and the pH at this point is the pK_a of the ionizable group.

Once the pH is more than 1 unit above the pK_a, most of the ionizable groups have already been deprotonated, and additional NaOH begins to react with protons from hydronium again, which quickly raises the pH outside the buffer region. Figure 1.43 shows an example of a titration curve.

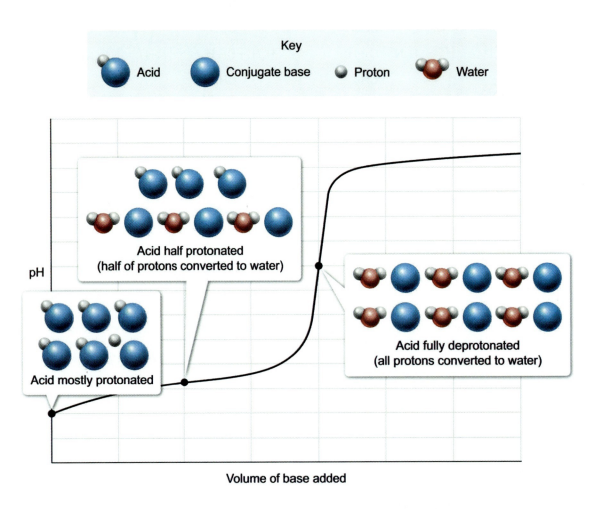

Figure 1.43 Relative amounts of protonated and deprotonated ionizable groups at various points in a titration curve.

The amino acids with nonionizable side chains have two ionizable groups in their backbones: the carboxyl group and the amino group. The carboxyl group has an approximate pK_a of 2.2, so the titration curve is relatively flat from pH ≈ 1.2 to pH ≈ 3.2. The amino group has an approximate pK_a of 9.6, so the curve has another relatively flat region from pH ≈ 8.6 to pH ≈ 10.6.

In addition, as the pH of the amino acid solution approaches the pH of the NaOH solution being used in the titration, the curve flattens again because the pH of the mixture cannot exceed the pH of the NaOH solution being added. Consequently, the amino acids with nonionizable side chains have two buffer regions and one additional flat portion at the end of the curve (see Figure 1.44).

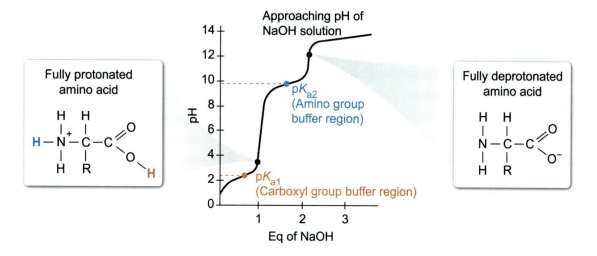

Figure 1.44 Titration curve of an amino acid with a nonionizable side chain.

Each of the ionizable amino acids has the same two buffer regions as any other amino acid. They also have a buffer region spanning 1 pH unit below and 1 above the pK_a of the side chain. Histidine, for example, has an additional buffer region from pH ≈ 5 to pH ≈ 7, as shown in Figure 1.45.

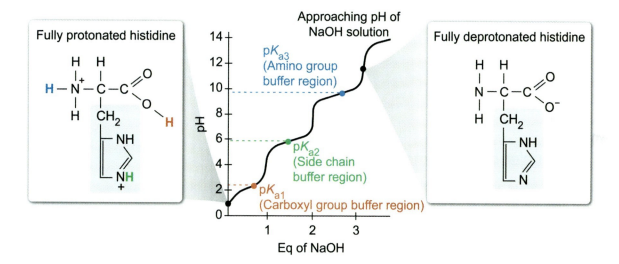

Figure 1.45 Titration curve of histidine.

> ☑ **Concept Check 1.4**
>
> Which amino acids would generate the following titration curves?
>
>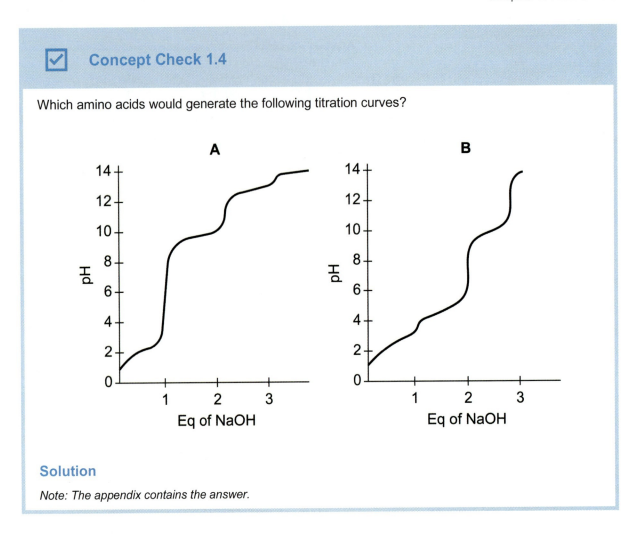
>
> **Solution**
>
> *Note: The appendix contains the answer.*

1.3.06 Nucleophilic Amino Acids

Another important aspect of amino acid chemistry is the ability of a side chain to act as a nucleophile. A nucleophile is a chemical group that is rich in electron density and, as such, is attracted to positive charges (see Organic Chemistry Lesson 5.3). A free electron pair in a nucleophile can "attack" an electrophile (which typically has a partial or full positive charge) to form a covalent bond.

Although nucleophilic attack by an amino acid side chain is not acid-base chemistry in and of itself, it is *related* to the acid-base properties of ionizable groups. Many of the ionizable amino acid side chains can act as nucleophiles; however, their fully protonated forms are *weak* nucleophiles. The nucleophilicity of these functional groups is enhanced by deprotonation, which either converts a positively charged group to neutral, or a neutral group to negatively charged, as shown in Figure 1.46.

Many amino acid side chains can act as nucleophiles. However, the most common are serine, threonine, cysteine, tyrosine, and lysine. Of these amino acids, only cysteine is readily deprotonated at physiological pH. However, serine, threonine, tyrosine, and lysine can be deprotonated in enzyme active sites, enhancing their nucleophilicity. All these amino acids act as nucleophiles in important biological reactions such as phosphorylation (S, T, and Y) or ubiquitination (K) (see Lesson 2.4).

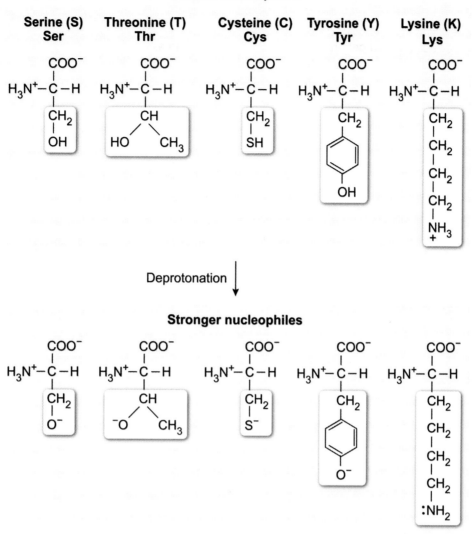

Figure 1.46 Amino acids that commonly act as nucleophiles in biological reactions. Nucleophilicity is improved by deprotonation.

Other amino acids can also act as nucleophiles (eg, asparagine in glycosylation, histidine in certain phosphate transfer reactions). Nucleophilic attack by amino acid side chains will be covered in more detail in lessons that discuss post-translational modifications (Lesson 2.4) and the catalytic mechanisms of enzymes (Lesson 4.3).

Lesson 2.1

Peptides

Introduction

Chapter 1 discussed individual amino acids. Amino acids serve as the building blocks for an important class of molecules called **polypeptides**, which form when the amino group of one amino acid bonds with the carboxyl group of another amino acid. Two amino acids linked together in this manner form a dipeptide. A third amino acid can then bond with the dipeptide to yield a tripeptide, which can link with another amino acid to form a tetrapeptide, and so on. This progression is shown in Figure 2.1.

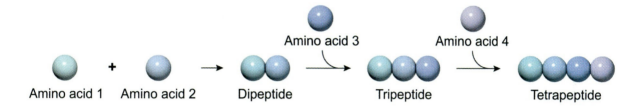

Figure 2.1 Peptides consist of two or more amino acids linked together.

Peptides play several crucial roles in biological systems. This lesson explores the formation, degradation, and characteristics of peptides.

2.1.01 Peptide Bond Formation and Degradation

The amino acids in a peptide are linked to each other through peptide bonds. A peptide bond forms when the carboxyl group of one amino acid backbone reacts with the amino group of another. In this process, the reacting amino acids undergo a net loss of one oxygen atom and two hydrogen atoms, which combine to form water. Therefore, peptide bond formation (Figure 2.2) is an example of a condensation reaction.

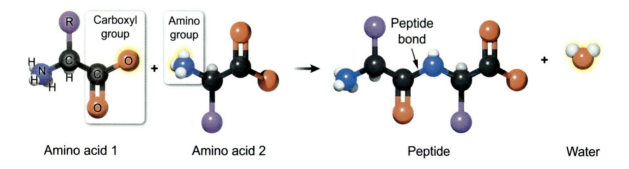

Figure 2.2 Peptide bond formation is a condensation reaction.

Because amino acids lose atoms during bond formation, they are not complete amino acids after the reaction. For this reason, the individual units in a polypeptide are called amino acid residues—they are the *residuals* of amino acids.

Just as peptide bonds form with the accompanying *release* of a water molecule, a peptide bond can be broken when a water molecule is *added* across the bond (ie, water is consumed). This reaction, which is the reverse of condensation, is known as hydrolysis.

Peptides, and therefore peptide bonds, are essential for life. However, under physiological conditions, peptide bond formation is thermodynamically unfavorable and peptide bond hydrolysis is favorable. In other words, without energy input, amino acids tend to remain unlinked. Consequently, *energy input is required* for peptide bond formation to occur in living cells.

Once formed, peptide bonds remain intact for a long time, which allows them to perform their biological functions. This is because, although it is thermodynamically favorable for peptides to hydrolyze, peptide bond hydrolysis has a high activation energy. Therefore, hydrolysis is slow and *peptide bonds are kinetically stable*. For peptide bonds to be hydrolyzed at a significant rate, enzymes such as proteases or the proteasome are required.

☑ Concept Check 2.1

The process of peptide bond formation is *endothermic* overall, yet bond formation itself is always an *exothermic* process. How can this apparent discrepancy be explained?

Solution

Note: The appendix contains the answer.

2.1.02 Peptide Bond Characteristics

Peptide bonds are amide bonds. Therefore, the carbon atom is a carbonyl, is sp^2 hybridized, and can participate in resonance. The nitrogen atom contains a lone pair of electrons that *also* participates in resonance with the adjacent carbonyl.

Consequently, the nitrogen atom is *also* sp^2 hybridized, and the pi electrons in the C=O double bond and the lone electrons on the nitrogen atom are all delocalized across the N, C, and O atoms of the amide group. This electron delocalization allows electrons to be present in a pi bond between the C and N atoms of the peptide bond, as shown in Figure 2.3.

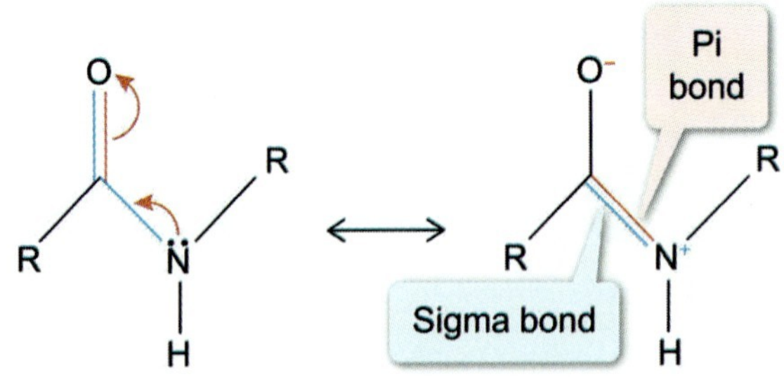

Figure 2.3 Resonance in peptides causes the peptide bond to include pi electrons, giving it partial double bond character.

In other words, although the bond between C and N in a peptide is often drawn as a single bond, in reality, it exhibits significant double bond character. This double bond character has important effects on protein folding. Most importantly, atoms involved in a double bond always exhibit planar geometry (trigonal planar in the case of the C and N atoms in a peptide bond). Together, these factors force *all* atoms linked to the carbon and nitrogen of the peptide bond into the same plane, as shown in Figure 2.4.

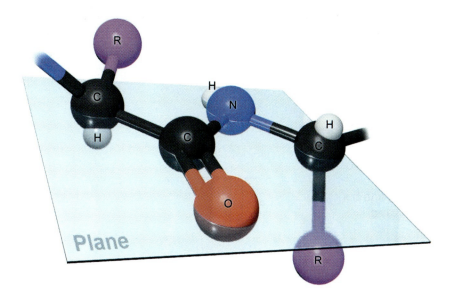

Figure 2.4 The C and N atoms in a peptide bond, and all atoms to which they are directly connected, are coplanar.

In addition to this planarity, the partial double bond character of the peptide bond restricts rotation because double bonds cannot rotate unless the pi bond is temporarily broken. Therefore, the double bond character of peptide bonds limits the conformations that a peptide or protein can adopt. Most peptide bonds adopt the lower-energy *trans* (relative), or *Z* (absolute), configuration, which *minimizes steric overlap* of the groups on either side of the partial double bond (See Figure 2.5).

However, because the side chain of proline connects back to its backbone amine group, both the *cis* (*E*) and *trans* (*Z*) configurations experience similar amounts of steric clashing. Consequently, the peptide bond preceding a proline residue can commonly be found either in the *trans* (*Z*) configuration *or* the *cis* (*E*) configuration.

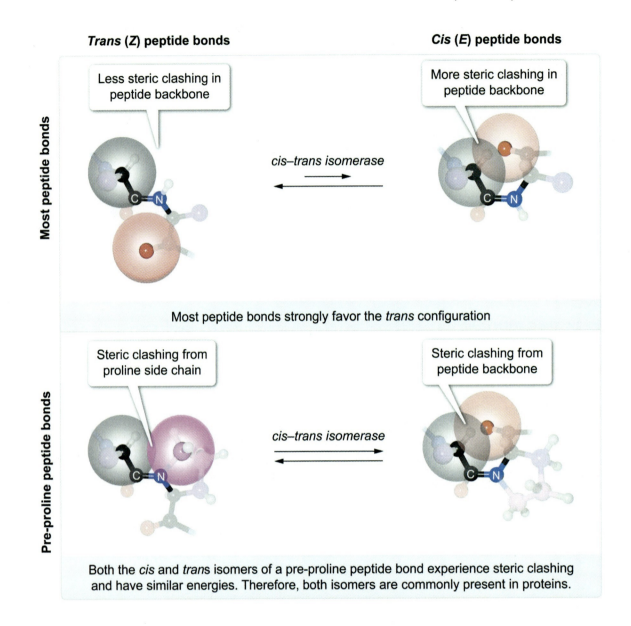

Figure 2.5 The substituents around a peptide bond can be in either a *cis* or *trans* (*E* or *Z*) configuration. Most peptide bonds are *trans* (*Z*), while bonds preceding a proline can be either *cis* or *trans*.

Although peptide bonds cannot rotate, the other bonds in a peptide backbone can. These are the bonds from the nitrogen to the α-carbon and from the α-carbon to the carbonyl.

2.1.03 Peptide Organization

The backbone of each free amino acid contains one amino group and one carboxyl group. When two amino acids form a peptide bond, the carboxyl group of one amino acid and the amino group of the other are consumed and converted into a single amide group. The resulting dipeptide has polarity (ie, the two ends of the dipeptide are distinct). One end contains a free amino group and is known as the amino terminus (N-terminus). The other end contains a free carboxyl group and is called the carboxy terminus (C-terminus).

In biological systems, when another amino acid is added to the peptide, the amino group of the free amino acid reacts with the C-terminus of the growing peptide, consuming one amino group and one carboxyl group to form a new peptide bond. The resulting tripeptide (or tetrapeptide, pentapeptide, etc.) still has one N-terminus and one C-terminus. The N- and C-termini of peptides of different lengths are shown in Figure 2.6.

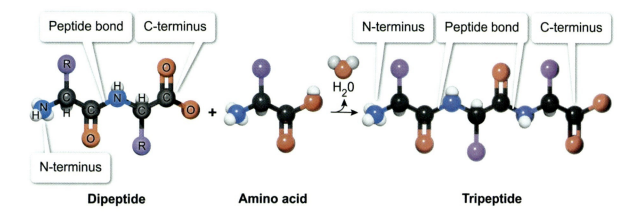

Figure 2.6 A peptide can react with an amino acid to become a longer peptide with more peptide bonds.

Because biological systems synthesize the N-terminus of a peptide first and grow the C-terminus through the addition of amino acids afterward, amino acid sequences are, by convention, written and read *from the N-terminus to the C-terminus*.

For example, the tripeptide ARE contains alanine at the N-terminus, arginine in the middle, and glutamate at the C-terminus. In contrast, the tripeptide ERA contains glutamate at the N-terminus, arginine in the middle, and alanine at the C-terminus. Therefore, although these tripeptides contain the same three amino acids, they are *not* identical and exhibit distinct chemical and biological properties. The structures of these peptides are shown in Figure 2.7.

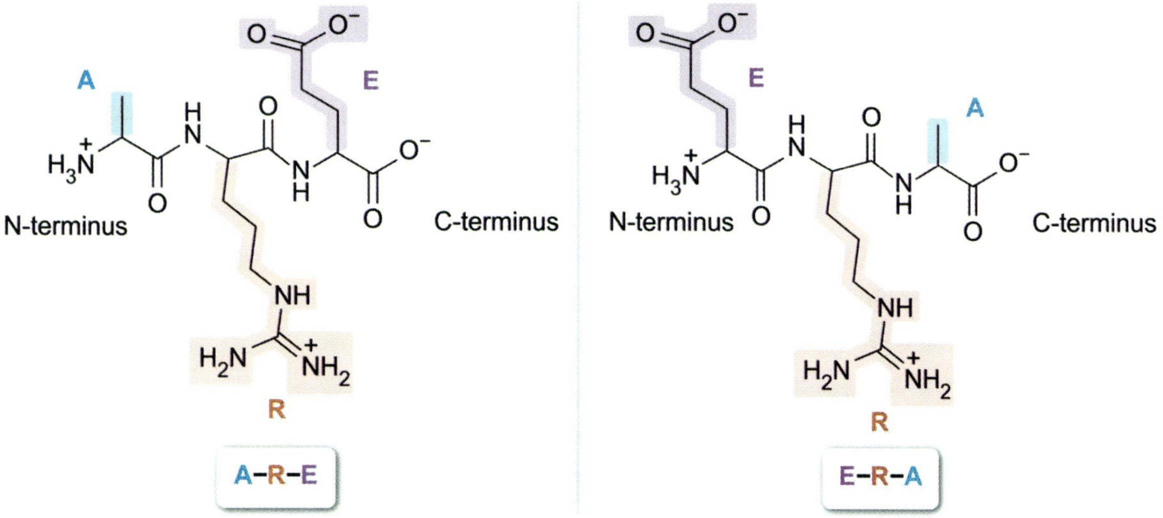

Figure 2.7 The peptides ARE and ERA contain the same three amino acid residues but have distinct structures.

Because the arrangement of amino acid residues in a peptide may change the peptide's properties, even relatively small peptides can exhibit substantial diversity. For example, a tripeptide that contains one A, one R, and one E can be arranged in six different ways (see Figure 2.8), which can be calculated as the factorial of the number of residues being arranged (eg, 3! = 3-factorial = 3 × 2 × 1 = 6).

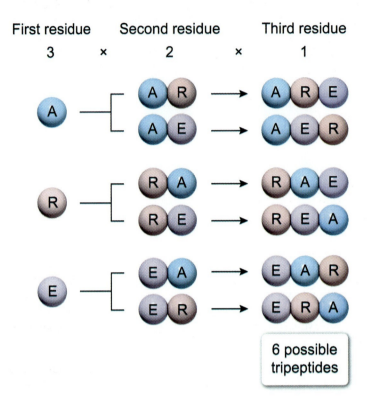

Figure 2.8 The three amino acids in a tripeptide can be arranged in six different ways. Mathematically this is represented as 3! (3-factorial) or 3 × 2 × 1.

This diversity is increased significantly by allowing a wider variety of amino acids. If all 20 proteinogenic amino acids are available to be used in a tripeptide (instead of only A, R, and E) and each can be used multiple times, the tripeptide may be arranged in 20^3 (8,000) different ways. A tetrapeptide (ie, four residues) in which any of the 20 amino acids may be used can be arranged in 20^4 (160,000) different ways.

Therefore, as more amino acids are used, the number of possible arrangements increases exponentially. This helps to explain how peptides and proteins can carry out diverse functions in living cells—the large number of possible arrangements gives rise to many possible sets of biochemical properties.

The amino acid residues in the middle of a polypeptide (ie, those that are not at either end) no longer have backbone amino or carboxyl groups, nor do they have N- or C-termini. However, the peptide bond of an amino acid residue that is **closer to the N-terminus** is called the **N-terminal side** of the residue, and the other peptide bond is called the **C-terminal side** of the residue. This distinction is important in peptide bond hydrolysis because some protease enzymes cleave peptides on the N-terminal side of a specific residue, and some cleave on the C-terminal side.

> **Concept Check 2.2**
>
> Consider the peptide MERGADLFN. The enzyme Arg-C proteinase cleaves peptide bonds on the C-terminal side of arginine, whereas the enzyme Asp-N endopeptidase cleaves peptide bonds on the N-terminal side of aspartate. Two samples of MERGADLFN are prepared. Sample 1 is exposed to Arg-C proteinase, and Sample 2 is exposed to Asp-N endopeptidase. What will be the sequences of the resulting peptides in each sample at the end of the experiment?
>
> **Solution**
>
> *Note: The appendix contains the answer.*

2.1.04 Acid-Base Chemistry of Polypeptides

Because peptides are made from amino acids, and amino acids have ionizable groups (see Lesson 1.3), peptides *also* have ionizable groups. These ionizable groups behave similarly to those of free amino acids, with some exceptions.

pK_a Values of N- and C-Termini

The amino groups of *free* amino acid backbones have pK_a values near 9.6. However, when an amino acid becomes part of a peptide, the pK_a of the amino group at its N-terminus tends to decrease to approximately 8.3. Similarly, the pK_a of the carboxyl group in a free amino acid backbone is typically near 2.2, but the C-terminus of a peptide has an increased pK_a near 3.5 (values specific to a given peptide will usually be given on the exam as needed).

Although the ends of a peptide are predominantly charged, the positive charge of the N-terminus cancels the negative charge of the C-terminus at physiological pH. Therefore, just as free amino acid backbones are zwitterionic, so are peptide backbones.

Net Charge of a Peptide

The predominant net charge of a peptide can be estimated by evaluating the charges of its backbone and its side chains. Because the backbone is zwitterionic at pH levels found in living cells, the charges of its positive N-terminus and negative C-terminus cancel each other. In addition, the only side chains that predominantly contribute positive charge to a peptide at physiological pH are arginine and lysine, and the only side chains that predominantly contribute negative charge are aspartate and glutamate.

As stated in Lesson 1.3, histidine side chains have a pK_a of 6. Therefore, most histidine residues are neutral at physiological pH, and only a minority of them contribute a positive charge. However, if the pH drops below 6 (as in lysosomes), the predominant form of histidine is positively charged. Similarly, cysteine predominantly contributes a negative charge if the pH increases above 8 (as in some peroxisomes), but only if the cysteine side chain is not involved in a disulfide bond. The pK_a of tyrosine is too high to contribute a negative charge to the peptide under conditions that are typically found within cells.

Note that estimations of net charge assume that the side chain pK_a values are not altered by their proximity to other functional groups in the peptide. In large proteins, pK_a values are commonly altered, and the net charge at a given pH must be determined empirically.

> **Concept Check 2.3**
>
> What is the predominant net charge of the peptide ANEQGSKHDIRK at pH 7 and pH 5?
>
> **Solution**
>
> *Note: The appendix contains the answer.*

Isoelectric Point of a Peptide

Just as individual amino acids have isoelectric points (ie, pH value at which the average net charge of a population is 0), so do peptides. The isoelectric point (pI) of a peptide can be estimated in the same way as that of an amino acid—by averaging the pK_a below which the peptide is *positively charged* and the pK_a above which the peptide is *negatively charged*. For short peptides, this can be accomplished relatively easily. Figure 2.9 shows an example estimation of the isoelectric point of the peptide ARE.

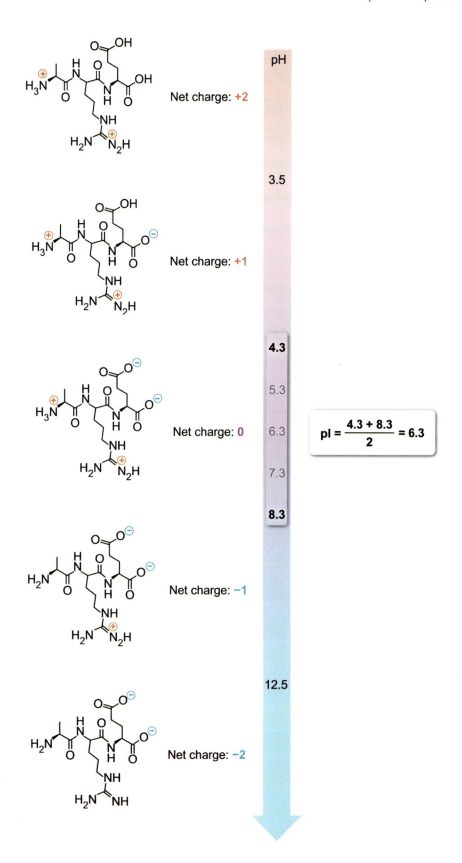

Figure 2.9 Estimation of the isoelectric point of the peptide ARE using pK_a values of the free amino acids.

However, estimation of pI becomes more difficult as the peptide gets longer, both because there are more ionizable groups to consider and because the pK_a values of the groups may change due to interactions with other chemical groups in the peptide.

For large peptides and proteins, the pI can be determined empirically using a technique called isoelectric focusing. This technique uses an electric field to cause the charged peptide or protein of interest to migrate through a gel that contains a pH gradient. Migration continues until the peptide reaches its pI because electrically neutral molecules do *not* migrate in an electric field. For more information on isoelectric focusing, see Lesson 14.1.

Changes to the composition of a peptide alter its pI. In general, addition of an amino acid residue that can carry a positive charge (R, K, and H) *increases* the pI because a higher pH is needed to fully neutralize the added positive charges. Note that arginine, which has the highest pK_a of the three, tends to cause the greatest increase in pI, and histidine causes the smallest increase.

Similarly, residues that can carry a negative charge (D, E, C, and Y) *decrease* the pI because a lower pH is needed to neutralize the negative charges. In this case, aspartate has the lowest pK_a and tends to cause the greatest decrease in pI while tyrosine, with a pK_a of 10.0, rarely has an impact on pI.

Removal of ionizable residues causes the opposite effect; removal of a positively charged residue causes the pI to decrease, whereas removal of a negatively charged residue causes the pI to increase.

Concept Check 2.4

Rank the following tetrapeptides from lowest isoelectric point to highest:

| KALS | DALS | KAES | DAES | KAHS |

Assume the pK_a values of the side chains in each peptide are identical to the pK_a values of the corresponding free amino acids.

Solution

Note: The appendix contains the answer.

Lesson 2.2
Levels of Protein Structure

Introduction

When amino acids join into chains of two or more residues, they become peptides. Chains that contain two residues are dipeptides, those with three residues are tripeptides, and so on. Beyond about 10 residues, these chains are commonly called polypeptides. In general, polypeptides that contain more than 50 residues are called proteins.

As the length of a polypeptide increases, distant residues within the peptide begin to interact with each other noncovalently (eg, through hydrogen bonds or ion-ion interactions). These interactions give rise to several levels of structure: primary, secondary, tertiary, and quaternary structure, depicted in Figure 2.10.

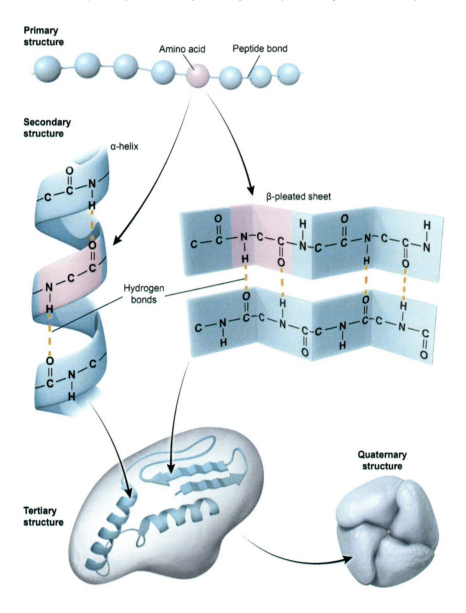

Figure 2.10 Overview of the four levels of protein structure.

Chapter 2: Peptides and Proteins

2.2.01 Primary Protein Structure

Primary protein structure is described by the sequence of amino acid residues within the protein, with each residue linked to its neighbors through peptide bonds (see Figure 2.11).

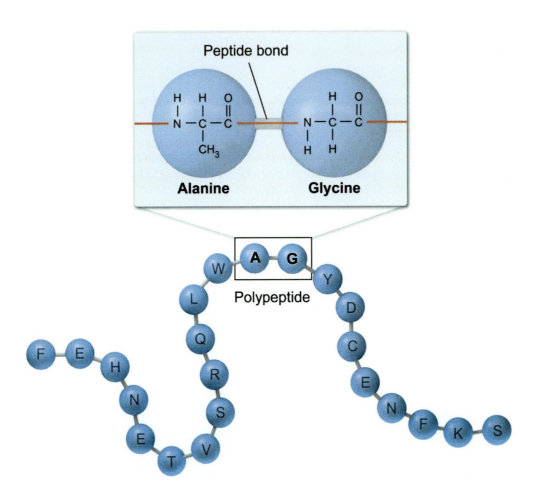

Figure 2.11 The primary structure of a peptide is described by its amino acid sequence and is maintained by peptide bonds between residues.

The amino acid sequence of a protein is typically encoded in the gene that corresponds to the protein (see Biology Chapter 2). Consequently, the primary structure of a protein may be changed by altering the corresponding gene through mutation. Primary structure may also be altered by breaking one or more peptide bonds within a protein or, less commonly, by forming new peptide bonds.

Empirical measurements have shown that some amino acids are more prominent in proteins than others. Based on the observed proportions and the molecular weights of each amino acid residue, the *average* molecular weight of the residues in a protein has been determined to be approximately 110 Daltons (Da) or 0.11 kilodaltons (kDa) per residue (1 Da = 1 atomic mass unit). Using this average, the molecular weight of a protein can be used to estimate the number of amino acid residues in that protein and vice versa.

> ## Concept Check 2.5
>
> A protein of unknown composition is measured to have a molecular mass of 60 kDa. Approximately how many amino acid residues does this protein contain?
>
> ### Solution
>
> *Note: The appendix contains the answer.*

The sequence of amino acid residues in a protein determines the interactions that can form between residues. For instance, a negatively charged residue in one position may interact with a positively charged residue in a distant position, but only if the residues in between adopt a conformation that brings the two charged residues into proximity (see Figure 2.12). In aqueous solution, most proteins quickly adopt a structure that maximizes favorable interactions and minimizes unfavorable interactions between residues.

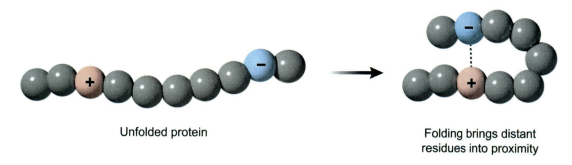

Unfolded protein　　　　　　　　　　Folding brings distant residues into proximity

Figure 2.12 Example of a peptide adopting a structure that facilitates interactions between distant, oppositely charged amino acids.

In other words, the higher levels of protein structure (secondary, tertiary, and quaternary) are fully determined by the primary structure, which dictates which interactions are favorable or unfavorable. Primary structure also determines the degree of sigma bond rotation necessary to allow higher levels of structure. Consequently, in the absence of disruptive forces, a given amino acid sequence always folds into the same shape under physiological conditions. Proteins that have similar primary structures tend to have similar secondary and tertiary structures, and vice versa.

2.2.02 Secondary Protein Structure

Secondary protein structure arises when the functional groups of the peptide *backbone* interact with each other. Specifically, each interaction consists of the N−H group of one backbone amide group forming a hydrogen bond with the C=O group of another backbone amide group.

As discussed in Concept 2.1.02, peptide bonds do not freely rotate, and the atoms linked to the C and N in the peptide bond are all found within a single plane. Only the bonds surrounding each peptide bond can rotate. In the 1950s, Linus Pauling used these facts to predict various conformations that protein backbones might adopt within these constraints. Shortly thereafter, two of these conformations were verified to be present in many proteins. These conformations, shown in Figure 2.13, are known as the α-helix and the β-sheet.

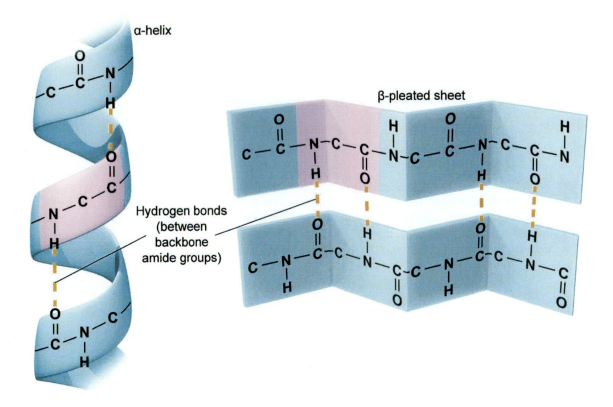

Figure 2.13 The two most common protein secondary structures, α-helices and β-sheets, with hydrogen bonds between backbone groups shown.

α-Helices

α-Helices are right-handed coils with the backbones arranged such that the N–H group of one residue's backbone aligns with the C=O group of a backbone that is four residues away, allowing these groups to hydrogen bond. The side chains project outward from the helix. The features and dimensions of a typical α-helix are shown in Figure 2.14.

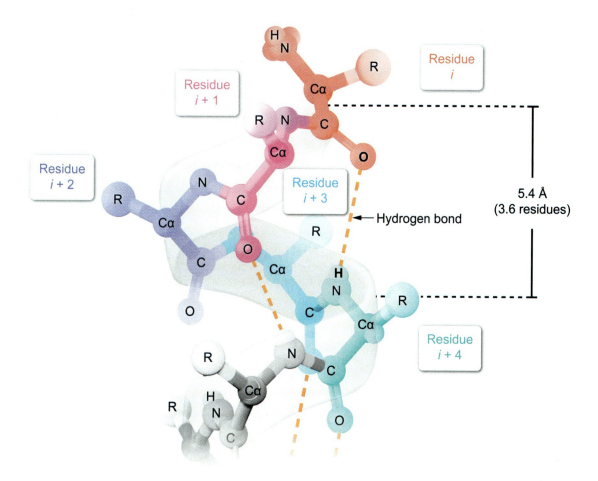

Figure 2.14 Dimensions and backbone interactions of an α-helix.

As with all secondary structures, the hydrogen bonds between backbone functional groups constitute the primary force that holds α-helices together. In addition to these backbone interactions, interactions between side chains may help further stabilize or destabilize α-helices.

Most of the 20 standard amino acids are commonly found within α-helices, with the exceptions of proline and glycine. The small size of glycine's side chain permits greater rotation of the residue due to decreased steric hindrance. In addition, this side chain is too small to have significant stabilizing interactions with other side chains in the helix. Consequently, glycine tends to make protein regions too flexible to maintain the strict features of an α-helix.

Proline causes the opposite problem: because its side chain links to the backbone at two positions instead of one, rotation of the N–$C_α$ bond is restricted and proline is rigid. In addition, proline has no hydrogen atom bonded to its amide nitrogen, so the proline backbone nitrogen cannot hydrogen bond with a carboxyl group on another amino acid. Consequently, proline tends to cause "kinks" that destabilize α-helices.

Chapter 2: Peptides and Proteins

> ☑ **Concept Check 2.6**
>
> Which of the following peptide sequences would be more suitable for formation of an α-helix?
> A) NFEGLQDPSV
> B) DNLSRTIETQ
>
> **Solution**
>
> *Note: The appendix contains the answer.*

β-Sheets

Often called β-pleated sheets, these secondary structural elements consist of multiple amino acid strands (β-strands) that are aligned with each other. The N−H backbone groups in one strand hydrogen bond with the C=O groups of an adjacent strand, and vice versa. Adjacent β-strands may exist in one of two relative configurations: parallel or antiparallel (see Figure 2.15).

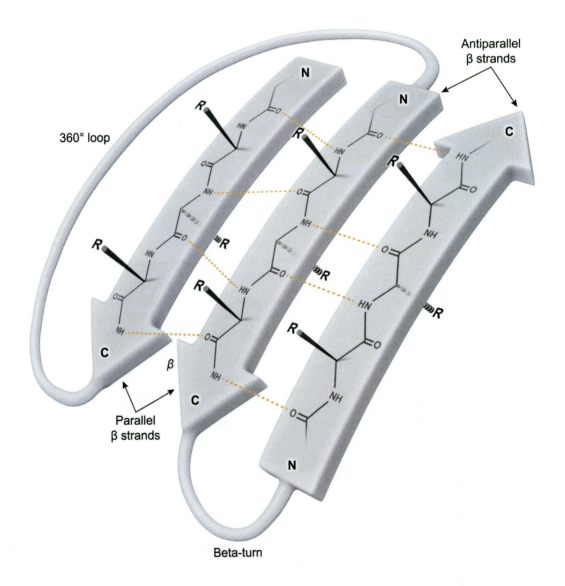

Figure 2.15 Parallel and antiparallel β-strands in a β-sheet.

Antiparallel strands are arranged with the N-terminal end of one strand aligned to the C-terminal end of its neighbor. Adjacent antiparallel strands are often linked to each other by short, highly structured motifs called β-turns (see Figure 2.16). These turns consist of four amino acids arranged in a specific configuration that results in a 180-degree turn. This turn facilitates alignment of the C-terminal end of one strand with the N-terminal end of the next. β-turns commonly contain glycine and/or proline, which facilitate the sharp turn required by these structures due to glycine's flexibility and proline's ability to form *cis* peptide bonds.

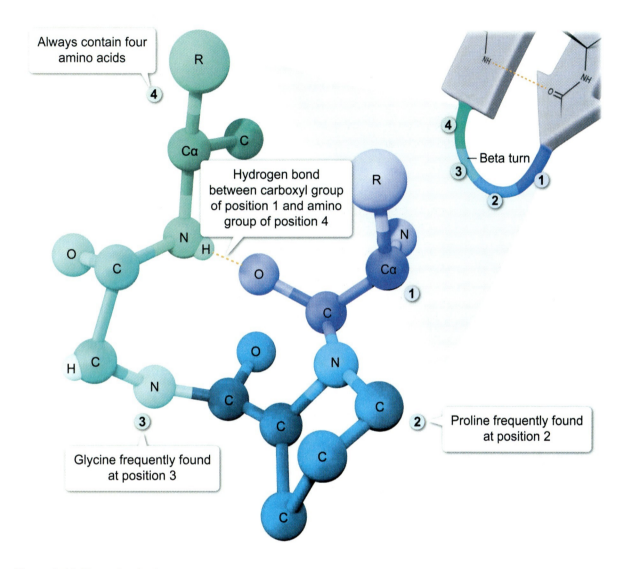

Figure 2.16 Example of a β-turn.

In contrast, parallel strands are oriented such that the C-terminal ends of adjacent strands are aligned with each other, as are the N-terminal ends. Consequently, parallel strands cannot be connected by β-turns and must instead be linked by longer connections.

Importantly, adjacent strands within a β-sheet do not need to consist of amino acids that are near each other in the primary structure. For instance, one strand of the sheet could be followed by flexible, unstructured regions, α-helices, or even other, separate β-sheets that eventually connect to additional strands in the β-sheet of interest. Examples of these structural features are shown in Figure 2.17.

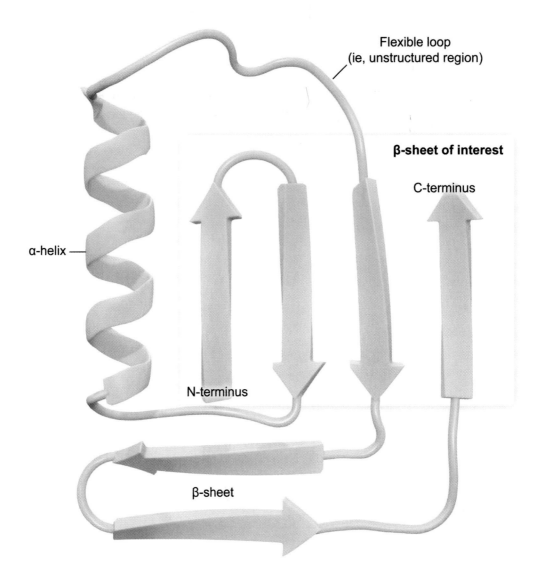

Figure 2.17 Example of a β-sheet in which adjacent strands are *not* near each other in the primary structure but are brought together as the protein folds.

In addition, a single β-sheet may consist of *both* parallel *and* antiparallel strands. As with α-helices, β-sheets may be further stabilized by favorable interactions between side chains, such as ionic interactions or hydrogen bonding.

Concept Check 2.7

In the following image, a β-sheet containing five strands is pictured with the N- and C-terminal ends of each strand marked. The connections between strands are omitted from the image.

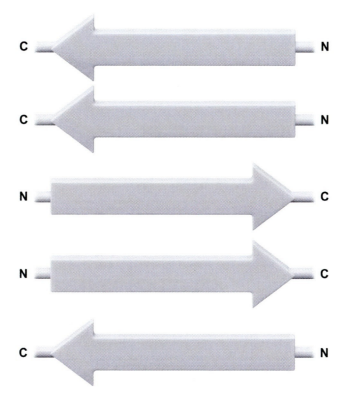

Based on this information, what is the maximum number of β-turns that can be present in this β-sheet without altering the relative orientations of the strands?

Solution
Note: The appendix contains the answer.

The relative amounts of α-helices and β-sheets in a protein, along with the amount of unstructured (ie, flexible loop) regions, can be measured by circular dichroism. For additional information on circular dichroism, see Lesson 14.4.

2.2.03 Tertiary Protein Structure

The tertiary structure of a polypeptide is its three-dimensional folded form, also known as the polypeptide's **native structure**. The biological functions of proteins and polypeptides depend on their ability to interact with molecules in their surroundings. Consequently, a protein must adopt a specific shape to perform its role. In other words, a monomeric (ie, single-polypeptide) protein is functional only when folded into its correct tertiary structure.

Proteins adopt their native forms by arranging themselves into the most energetically favorable configuration possible, often bringing structural elements close together that would otherwise be far apart.

For instance, an α-helix near the N-terminus may be brought into proximity with a β-sheet near the C-terminus when the protein adopts its native form (see Figure 2.18).

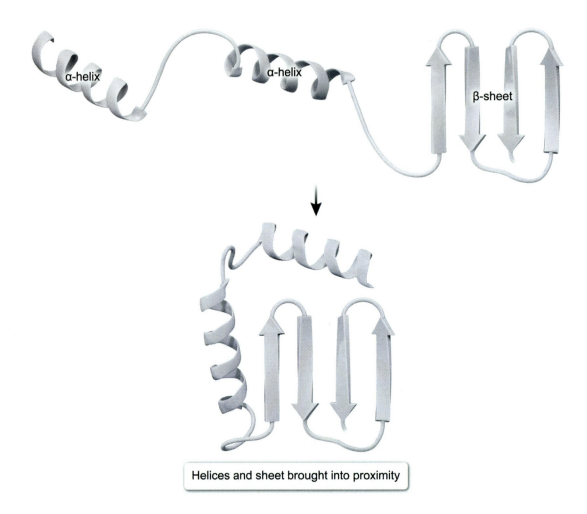

Figure 2.18 Structural elements that are far apart in primary structure may be brought close together when a protein adopts its tertiary structure.

Stabilization of Tertiary Structure

The most energetically favorable form of a protein tends to be a conformation in which hydrophobic side chains are buried in the interior of the protein so that they are not exposed to water. This phenomenon, called the hydrophobic effect, is discussed in greater detail in Lesson 2.3. Due to the hydrophobic effect, the surface of a typical protein consists primarily of hydrophilic residues. These residues may interact with the water molecules in the surrounding environment, or they may interact with each other.

The stability conferred on a protein by burying the hydrophobic residues is often enhanced by multiple types of noncovalent interactions between hydrophilic surface amino acid side chains. These interactions include:

- Hydrogen bonds between polar neutral groups (eg, threonine and glutamine)
- Salt bridges between oppositely charged ionic groups (eg, lysine and aspartate)
- Ion-dipole interactions between one ionic group and one polar neutral group (eg, arginine and serine)

In addition to these noncovalent interactions, some proteins are stabilized by disulfide bonds. Disulfide bonds are unique among the interactions that stabilize tertiary structure because they are covalent bonds.

Disulfide bonds form through oxidation of the sulfur atoms in cysteine side chains and can be broken by reducing agents. Figure 2.19 shows various side chain interactions involved in tertiary structure.

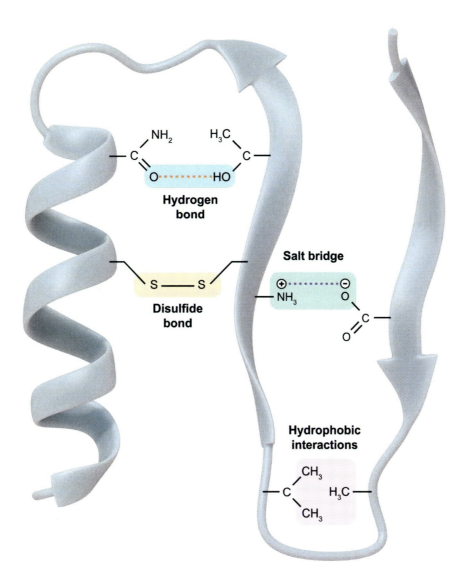

Figure 2.19 Various interactions that stabilize protein tertiary structure.

In most cells, the cytosol is a highly reducing environment (ie, it contains a relatively high concentration of reducing agents) and, consequently, disulfide bonds in general do *not* form in cytosolic proteins. However, the lumen of the endoplasmic reticulum, the Golgi apparatus, and lysosomes, as well as the extracellular space, are all oxidizing environments. Therefore, disulfide bonds *commonly* form in proteins that occupy these environments (eg, secretory proteins).

Protein Domains

The tertiary structure of a protein may include several distinct regions that each fold independently and often carry out different functions. These regions are known as **domains**. Domains are often named by their position within the protein or the characteristics that describe them.

For instance, the domain closest to the N-terminus is often simply called the N-terminal domain. A domain that binds to DNA may simply be called the DNA-binding domain. As with β-sheets in secondary structure, a single domain of a protein may also include stretches of residues that are far apart in the primary structure, which may have intervening sequences between them. An example protein with several domains is shown in Figure 2.20.

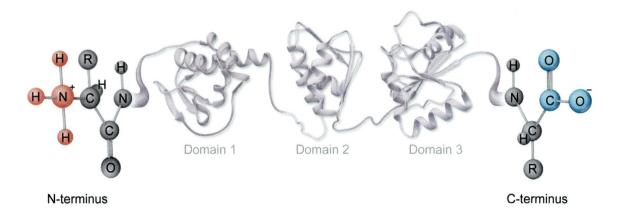

Figure 2.20 Depiction of a protein containing three distinct domains.

One particularly important type of domain is the **transmembrane domain**. Many proteins are integrated into the phospholipid bilayer of a cell or organelle membrane, with some domains of the protein in contact with the cytosol and other domains in contact with the extracellular space or the lumen of the organelle. Transmembrane domains are inserted into the phospholipid bilayer and link the cytosolic domains to the other domains. A transmembrane domain may cross the membrane only once or it may cross multiple times (see Figure 2.21).

Importantly, the amino acids within a transmembrane domain typically do not follow the trend of domains in aqueous environments such as the cytosol. Whereas the surface of a cytosolic domain typically contains hydrophilic residues, the surface of a transmembrane domain typically contains hydrophobic residues. The reason for this is that the surface of a transmembrane domain is *not* in contact with water. Instead, transmembrane domains are in contact with phospholipid tails, which are hydrophobic (see Chapter 9).

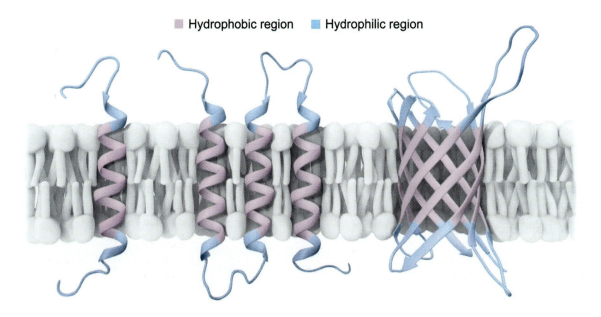

Figure 2.21 Examples of transmembrane domains.

Because separate domains fold and carry out their functions independently, removal of one domain from a protein typically does not affect the folding or function of the other domains. For example, suppose the extracellular domain of a protein binds a molecule and thereby induces the cytosolic domain to catalyze a reaction. If the cytosolic domain is removed, the extracellular domain will likely still be able to bind the molecule; however, the cytosolic reaction will no longer be catalyzed.

> **Concept Check 2.8**
>
> Consider a protein containing a cytosolic domain, a transmembrane region, and an extracellular domain.
>
>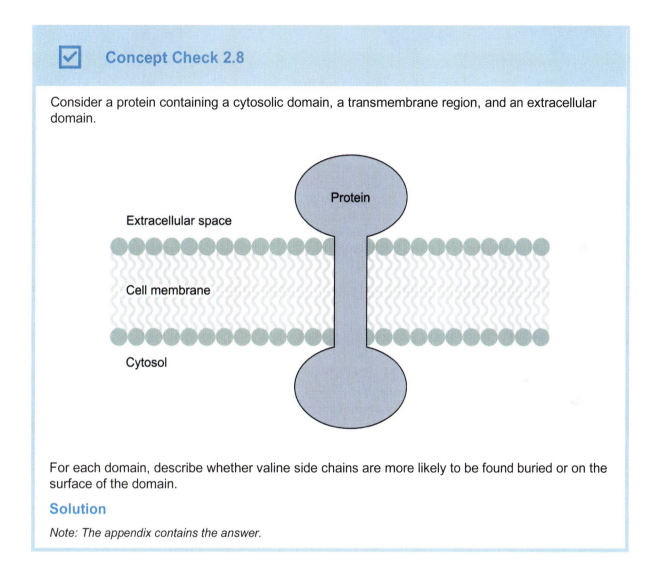
>
> For each domain, describe whether valine side chains are more likely to be found buried or on the surface of the domain.
>
> **Solution**
> *Note: The appendix contains the answer.*

2.2.04 Quaternary Protein Structure

Quaternary structure refers to interactions between several polypeptide units to form a single protein. Not all proteins have quaternary structure; many exist and function as single polypeptides. In proteins that *do* exhibit quaternary structure, each polypeptide is called a subunit. Each subunit is synthesized separately and, as with protein domains, subunits often begin folding into their tertiary structure independently.

After synthesis and initial folding of tertiary structure, the subunits bind to each other, allowing the protein to fold into its final structure (Figure 2.22). Unlike domains, however, individual subunits often cannot function unless they are correctly interacting with the other subunits in the protein.

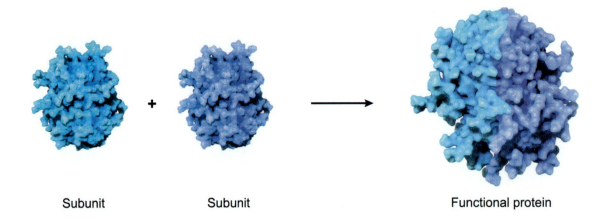

| Subunit | Subunit | | Functional protein |

Figure 2.22 Quaternary structure occurs when multiple polypeptide subunits bind to each other to create one functional protein.

The subunits within a protein are held together by the same forces that stabilize tertiary structure. The interface between two subunits is often lined with hydrophobic residues, which are hidden from water when the subunits interact. In addition, side chains in one subunit may interact with side chains in another through hydrogen bonding, ion-dipole interactions, salt bridges, and (in oxidizing environments) disulfide bonds.

Proteins with quaternary structure are classified by the number and types of subunits they contain (see Figure 2.23). A protein consisting of two subunits is called a dimer, a protein with three subunits is a trimer, a protein with four subunits is a tetramer, and so on. If all subunits in the protein are identical to each other, the prefix *homo–* is added, whereas the prefix *hetero–* is used for proteins in which at least one subunit differs from the others. For instance, a protein composed of two identical subunits is called a homodimer. In contrast, a dimer consisting of two nonidentical subunits is a heterodimer.

Many proteins have more complex structures. For example, hemoglobin consists of four subunits total, with one pair of identical subunits called α-chains and another pair called β-chains. Each α-chain interacts with one β-chain to form a heterodimer, and the two heterodimers interact with each other to form functional hemoglobin. Therefore, hemoglobin is said to be a dimer of dimers.

Some proteins are even more complex, having hundreds of subunits. Rather than denoting the exact number of subunits in these proteins they are typically called oligomers or multimers.

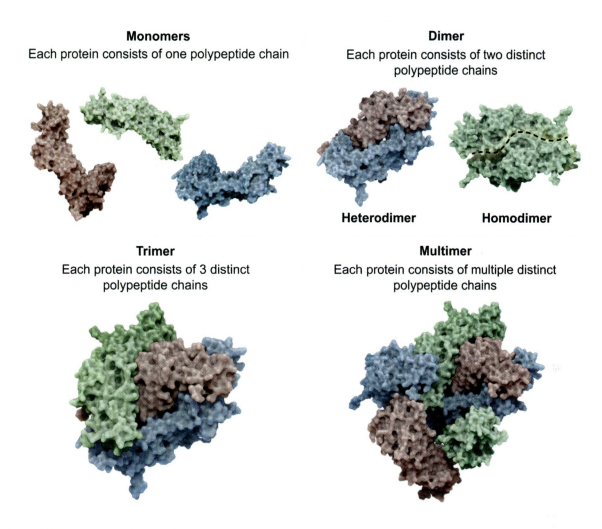

Figure 2.23 Examples showing how monomers may interact to form proteins with quaternary structure.

`Lesson 2.3
Protein Folding and Stability

Introduction

Protein folding is the process by which proteins adopt their functional secondary and tertiary structures. Proteins fold to achieve the lowest possible energy state by maximizing favorable interactions, minimizing unfavorable interactions, and maximizing the total entropy of the system. Under physiological conditions, proteins experience a negative free energy change when they fold, meaning protein folding is spontaneous.

2.3.01 The Hydrophobic Effect

For most proteins, folding involves burying hydrophobic residues in the protein interior (ie, hydrophobic collapse). This occurs due to a phenomenon called the **hydrophobic effect**, which is the dominant driving force in most protein folding. Burying the hydrophobic residues has several energetic effects, some favorable and others unfavorable. These effects include:

- *The entropy of the protein decreases.* By folding into a specific conformation, the protein is unable to adopt as many states as it could before folding. In other words, the protein becomes more ordered. When considering the protein alone this effect is energetically unfavorable, but this effect is more than offset by the much larger entropy *increase* experienced by water.
- *The entropy of the water surrounding the protein increases.* In unfolded proteins, hydrophobic residues are exposed to water. Water surrounds all molecules that are dissolved in it, forming a shell called a solvation layer around them. However, water cannot hydrogen bond with hydrophobic residues. Consequently, water instead forms hydrogen bonds with the other water molecules in the solvation layer, producing a highly ordered (low entropy) structure. The low entropy of water in this scenario is highly unfavorable.

In contrast, water *can* form strong, enthalpy-releasing interactions with *hydrophilic* groups. These bonds include hydrogen bonds and ion-dipole interactions. Consequently, interactions between water and hydrophilic groups are much more favorable than those between water and hydrophobic groups.

When a protein folds and the hydrophobic groups are buried, water no longer forms highly ordered solvation layers around those groups. This allows the entropy of the water in the system to increase, which is highly energetically favorable and is the largest contributor to the energy of protein folding. The change in the entropy of water upon protein folding is depicted in Figure 2.24.

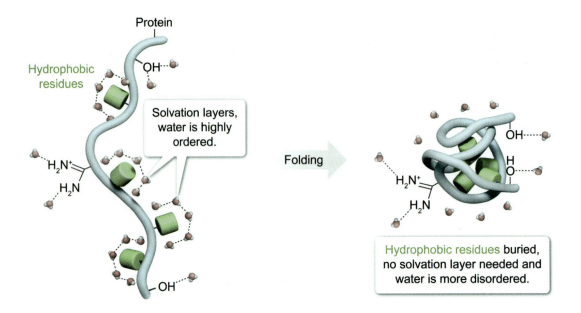

Figure 2.24 The hydrophobic effect hides hydrophobic residues from water, increasing the entropy of the water molecules in the system.

- *Interactions between hydrophobic residues increase.* The hydrophobic residues buried within the protein interact with each other through London dispersion forces. Although these forces are individually weak, the collective strength of all the London dispersion forces together substantially contributes to the energetic favorability of protein folding.

As discussed in Lesson 2.2, interactions between hydrophilic residues also contribute to the stability of a protein's tertiary structure. For example, hydrophilic residues at the surface-interior interface form specific interactions with each other that stabilize the final native conformation (see Figure 2.25).

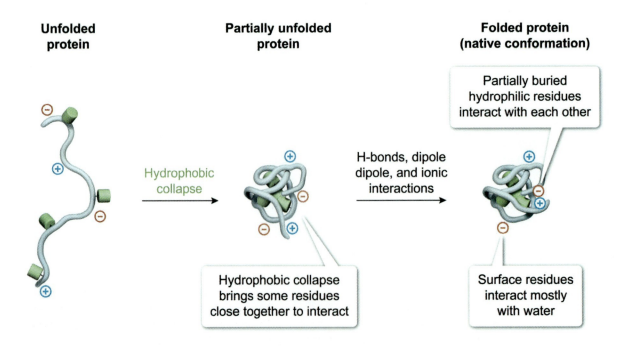

Figure 2.25 Interactions between hydrophilic residues help stabilize native conformation.

Interactions between two hydrophilic side chains have similar favorability to interactions between a side chain and water, and therefore surface side chains often interact with water instead of each other. Consequently, although interactions between surface side chains do make an important contribution to protein structure, they are not the *driving* force in protein folding.

> **Concept Check 2.9**
>
> How would transferring a protein from an aqueous solution to a nonpolar solvent such as hexane most likely affect the folded form of the protein?
>
> **Solution**
>
> *Note: The appendix contains the answer.*

The influence of the hydrophobic effect gives rise to several classes of proteins: fibrous, globular, and intrinsically disordered. Examples of each are shown in Figure 2.26.

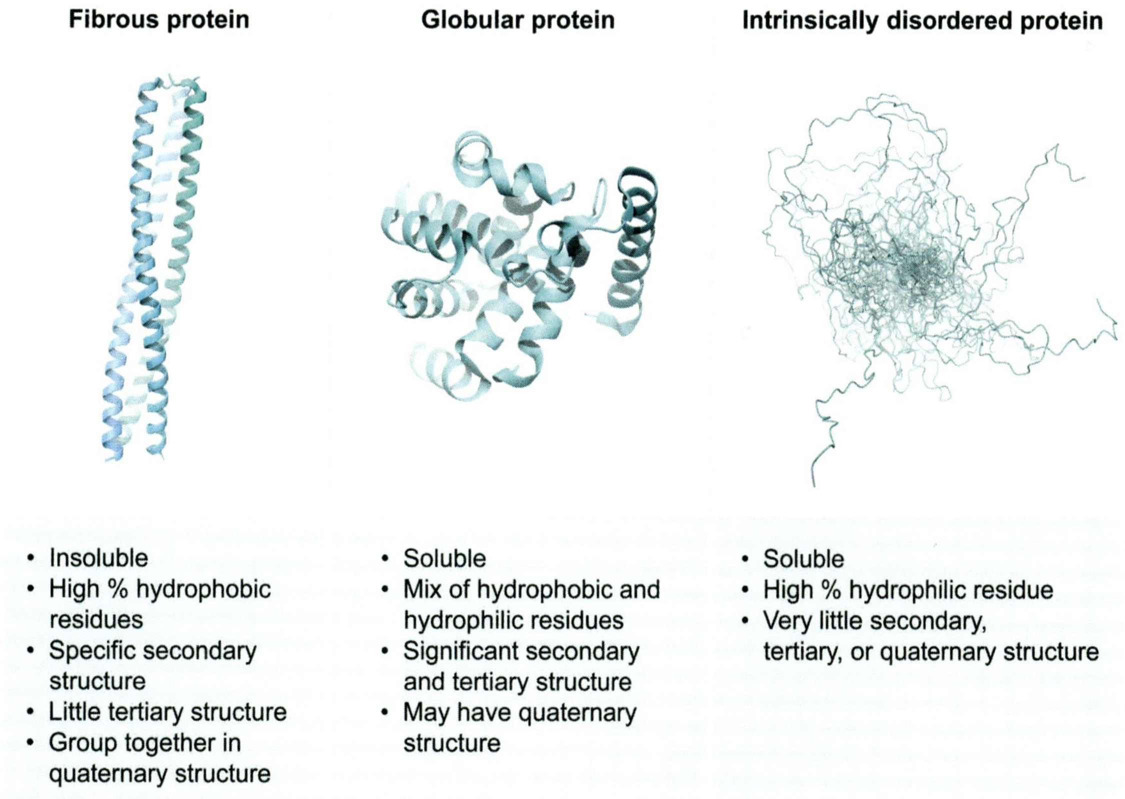

Fibrous protein
- Insoluble
- High % hydrophobic residues
- Specific secondary structure
- Little tertiary structure
- Group together in quaternary structure

Globular protein
- Soluble
- Mix of hydrophobic and hydrophilic residues
- Significant secondary and tertiary structure
- May have quaternary structure

Intrinsically disordered protein
- Soluble
- High % hydrophilic residue
- Very little secondary, tertiary, or quaternary structure

Figure 2.26 Depictions of fibrous, globular, and intrinsically disordered proteins.

Fibrous proteins are insoluble, often due to significant levels of hydrophobic residues on the protein surface. This is frequently the result of a high overall percentage of hydrophobic residues in the primary structure, which prohibits full burial of hydrophobic regions. Fibrous proteins tend to have a single type or very few types of secondary structure and very simple, if any, tertiary structure.

However, fibrous proteins exhibit *significant* quaternary structure as multiple identical polypeptides group together, typically into long, insoluble fibers. Each subunit interacts with hydrophobic residues in adjacent subunits, hiding those residues from water. Examples of fibrous proteins include collagen (a major component of tendons) and myosin fibers involved in muscle contraction. Fibrous proteins typically provide structure to cells and organs.

Globular proteins tend to have a mix of hydrophobic and hydrophilic residues that permits most of the hydrophobic residues to be buried. Consequently, these proteins are typically soluble in water. Globular proteins are perhaps the most common and best-known type of protein and include many enzymes along with proteins that carry out nonenzymatic functions. These proteins may adopt highly complex secondary, tertiary, and quaternary structures.

Intrinsically disordered proteins contain a high percentage of hydrophilic amino acids and may also contain significant amounts of proline, which disrupts α-helices. Because these proteins have relatively few hydrophobic residues, they do not undergo hydrophobic collapse and do not adopt notable tertiary structures. Secondary structure is also often absent from these proteins.

Many proteins contain some regions that are globular and perform specific functions and other regions that are intrinsically disordered (ie, flexible loop regions) that allow for greater motion within the protein.

2.3.02 Energy of Protein Folding

This concept discusses the energy landscape of protein folding. Although the information in this concept is unlikely to be tested *directly* on the exam, comprehension of these principles helps understanding of how protein shape is determined and how shape may change in response to changing conditions, which *are* commonly tested topics.

The free energy associated with protein folding is a state function. Therefore, the native (ie, correctly folded) structure of a protein is the same regardless of the pathway the protein takes to get there. A given protein could follow any of multiple pathways, adopting various intermediate conformations, as it folds.

Interestingly, the number of possible conformations a protein could *theoretically* adopt during folding is so large that, if left to fold by randomly exploring every possibility, a single protein would take longer to fold than the age of the known universe. Yet most proteins fold within a few milliseconds of synthesis.

The only explanation for this observation is that proteins do *not* fold randomly. Instead, local secondary structures tend to form first—amino acid sequences that are conducive to α-helices quickly form α-helices, and those that are conducive to β-strands and turns quickly form β-strands and turns. Note that much of this happens as the protein is being synthesized.

As secondary structures come together to form the tertiary structure, the protein may then explore certain possible conformations, and two identical proteins may take different pathways to reach the same final, folded form. Consequently, the energetics of protein folding are often represented by an energy landscape or funnel as shown in Figure 2.27.

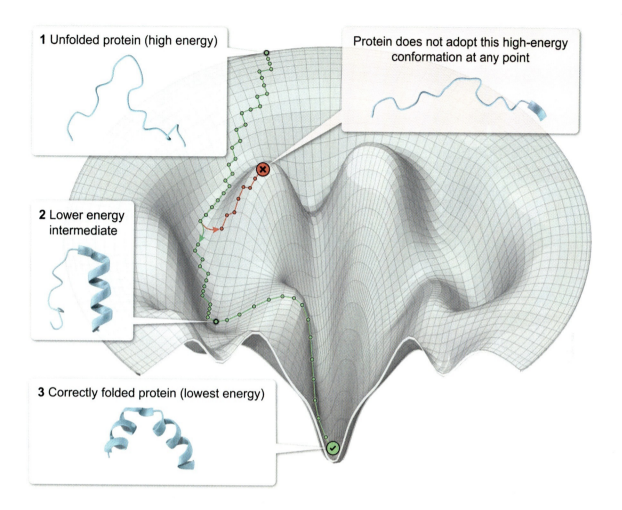

Figure 2.27 Example of a protein folding energy landscape, depicting pathways a protein may take during folding.

However, the conformations that folding proteins can explore are limited, and *most* of the theoretically possible conformations are excluded because adopting them would require too much energy input. Each time the protein goes from a high-energy state to a lower-energy state, it tends *not* to revert to any higher-energy states, allowing the protein to quickly "funnel" to the lowest-energy conformation possible.

2.3.03 Conformational Changes

Although each protein has a lowest-energy conformation, no protein is *entirely* restricted to this single state. Instead, the atoms in proteins and the bonds between them are constantly moving. Therefore, the protein constantly undergoes small changes in shape, called **conformational changes**. In other words, proteins are *dynamic* rather than static.

Many proteins have relatively unstructured regions that regularly undergo substantial shape changes. In addition, although the highly structured regions of a protein are *less* dynamic, they can still undergo noticeable shape changes. Structured regions tend to spend *most* of their time in or near the lowest-energy conformation possible, but at any given moment a small subset of the proteins in solution may briefly undergo a conformational change to a higher-energy state before rapidly returning to the low-energy conformation (see Figure 2.28).

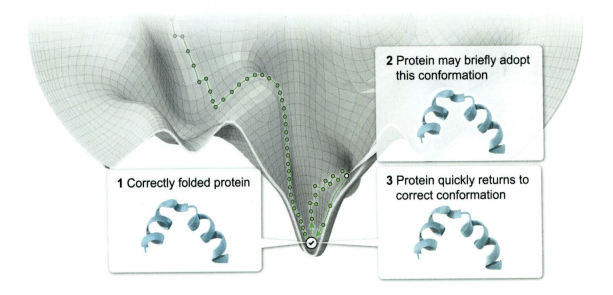

Figure 2.28 Depiction of a protein briefly adopting a higher-energy conformation. If conditions remain constant, the protein will quickly return to its low-energy conformation.

However, changes in the local environment of a protein may alter its energy landscape. In doing so, a higher-energy conformation may be stabilized and become lower in energy. Similarly, the conformation that *was* lowest in energy in the initial conditions may increase in energy. Consequently, changes in a protein's conditions may *induce* conformational changes and cause proteins in solution to adopt a new shape (see Figure 2.29).

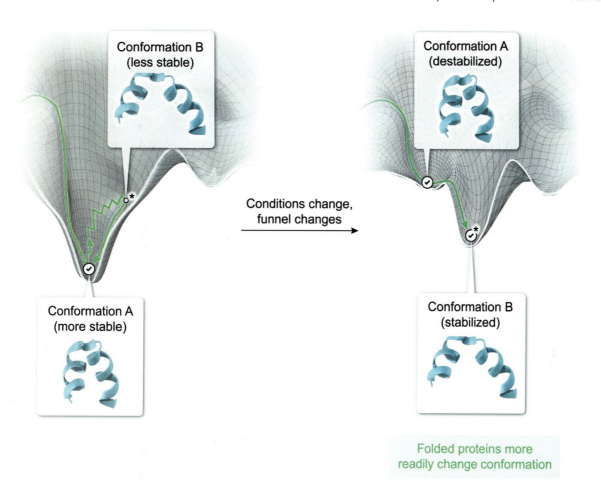

Figure 2.29 Changes in conditions may change the energy landscape, resulting in a new lowest-energy conformation.

Environmental conditions that can alter a protein's most stable conformation (and thereby induce a conformational change) include salt concentration, pH, temperature, and the presence of molecules that bind the protein, called ligands (see Lesson 3.1). Changes in the protein's primary structure (eg, through mutation) and chemical alterations, called post-translational modifications (Lesson 2.4), can also alter the conformation of the protein. The ability of a protein to change shape in response to environmental conditions is critical to its ability to carry out its biological function.

2.3.04 Protein Denaturation and Stability

In some cases, changes in environmental conditions can alter the thermodynamics of protein folding so much that it becomes more favorable for the protein to lose most or all of its tertiary and secondary structure, rendering the protein nonfunctional. In other words, unfolding becomes spontaneous under certain conditions. When this happens, the protein is said to unfold or **denature**. Figure 2.30 shows an example of how an energy landscape might change under denaturing conditions.

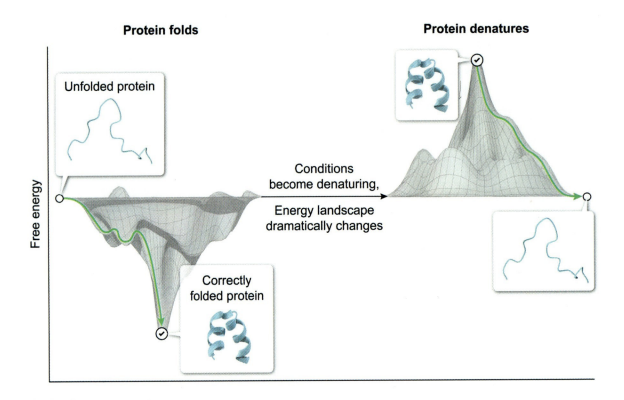

Figure 2.30 Under certain conditions, the energy landscape of folding changes enough to favor spontaneous denaturation.

Proteins can be denatured by several means, including sufficiently large changes in temperature, pH, salt concentration, or the presence of chemicals known as denaturing agents.

Temperature

The temperature of a system measures the average kinetic energy of the molecules within that system; as temperature increases, so does average kinetic energy. In other words, increasing temperature causes the amino acid residues within a protein to move more energetically. The increased motion can cause the intramolecular forces that stabilize protein structure to break (see Figure 2.31), allowing the protein to unfold.

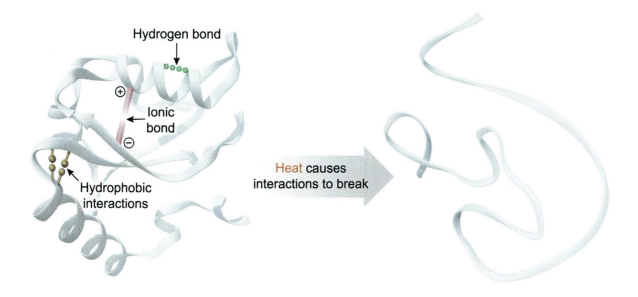

Figure 2.31 Heat can break intermolecular forces, causing protein denaturation.

pH

Although protein tertiary structure is stabilized primarily by the hydrophobic effect, interactions between hydrophilic surface groups can make a significant contribution to structure. Adjusting the pH can alter the protonation state of surface amino acids. For instance, a sufficient increase in pH will deprotonate lysine and arginine residues.

Consider a positively charged lysine residue that interacts with a negatively charged glutamate residue. Deprotonation of lysine at high pH or protonation of glutamate at low pH may disrupt the interaction. Therefore, pH values that are too high or too low can significantly alter the shape of a protein (see Figure 2.32). Although the hydrophobic core is likely to remain mostly intact regardless of pH, the shape of the protein surface may be altered enough to render the protein nonfunctional.

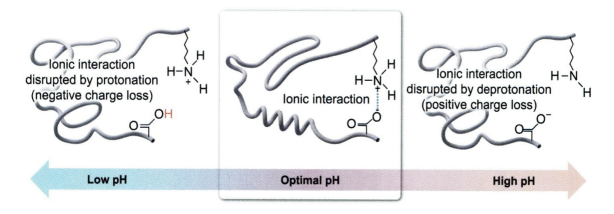

Figure 2.32 Changes in pH interrupt ionic interactions, leading to partial protein denaturation.

Salt Concentration

Like changes in pH, changes in salt concentration may disrupt interactions between charged residues on the protein surface. In this case, the disruption is caused by dissolved salts dissociating into positive cations and negative anions.

A positively charged sodium ion may outcompete arginine or lysine for interactions with glutamate or aspartate, and a negatively charged chloride ion may outcompete glutamate or aspartate for interactions with arginine or lysine. As with pH, altered salt concentration typically has a small or no effect on the hydrophobic core of the protein but disrupts the surface enough to prevent the protein from functioning. Figure 2.33 shows the ions of sodium chloride disrupting an ionic interaction in a protein.

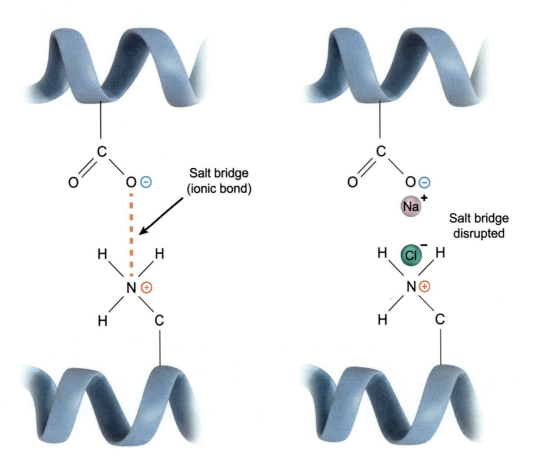

Figure 2.33 Salts can disrupt salt bridges in proteins and thereby cause partial denaturation.

Denaturing Agents

Various other chemicals may be added to a protein-containing solution and cause the proteins to denature. These denaturing agents may work by a variety of mechanisms, but most disrupt the hydrophobic effect (see Figure 2.34).

For instance, sodium dodecyl sulfate (SDS) contains a long hydrophobic tail that can interact with the hydrophobic residues of a protein, disrupting the hydrophobic core that drives protein folding. The hydrophilic end of SDS can simultaneously interact favorably with water. SDS is commonly used as a denaturing agent in protein electrophoresis (see Chapter 14). Other common denaturing agents include urea and guanidinium chloride.

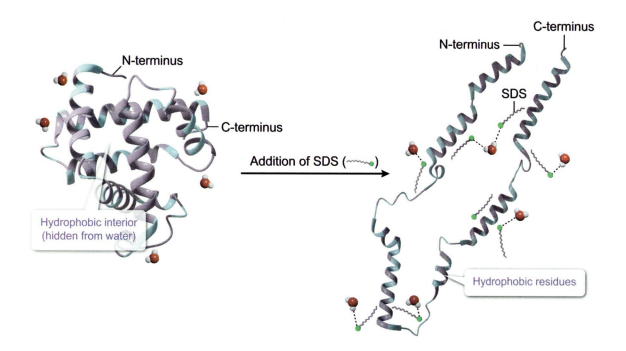

Figure 2.34 Example of a denaturing agent (SDS) interacting with the hydrophobic portions of a protein and causing denaturation.

Conformational Stability

The conformational stability of a protein refers to the free energy change between folded and unfolded forms. A protein that undergoes a large, negative free energy change upon folding is more stable than a protein that undergoes a smaller free energy change. Protein denaturation techniques can be used to measure a protein's conformational stability.

The more stable a protein is in its folded form, the greater the temperature required to denature it. The melting temperature (T_m) of a protein is the temperature at which half of the proteins in solution are denatured and half remain folded. Therefore, a more stable protein has a higher melting temperature.

> **✓ Concept Check 2.10**
>
> A researcher measures the percentage of folded protein in a solution both for the wild-type and for a mutated form of a protein of interest. The following graph shows the melting curves of both protein variants. Based on the graph, does the mutation increase or decrease the protein's conformational stability?
>
>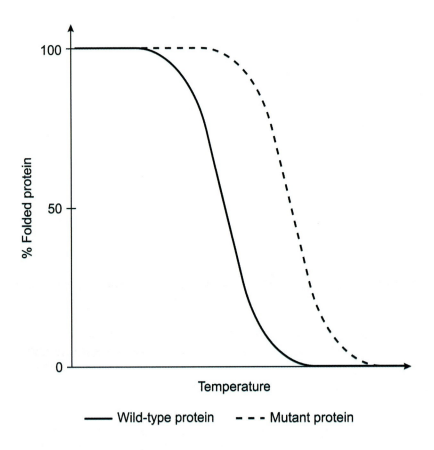
>
> **Solution**
> *Note: The appendix contains the answer.*

Less commonly, the stability of a protein may be measured by examining the concentration of a denaturing agent required to produce denaturation. In general, if a high concentration of denaturing agent is required to denature a protein, the protein is more stable.

2.3.05 Misfolding and Chaperones

Most proteins fold into their native forms quickly. However, some proteins can adopt relatively stable intermediates as they fold. These intermediates correspond to local valleys within the energy landscape. Although such an intermediate is not the *lowest*-energy conformation possible, the protein may become trapped in this nonfunctional conformation because exiting it requires a high activation energy. Proteins that become trapped in stable intermediate conformations are said to be misfolded (Figure 2.35). Most, if not all, cells experience some degree of protein misfolding.

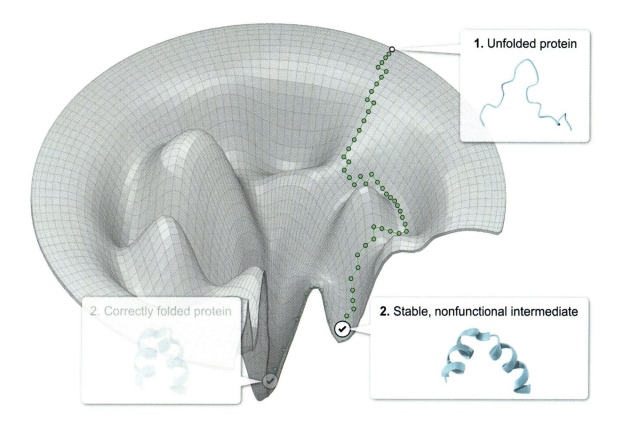

Figure 2.35 Example of a protein adopting a stable but nonfunctional conformation. The protein may become trapped in this state.

Exposure to high temperatures, for example, may cause some proteins to partially or fully denature, as discussed in Concept 2.3.04. Some proteins can refold (ie, renature) once denaturing conditions are removed. However, others cannot renature on their own because they become trapped as relatively stable intermediates before reaching the correctly folded form.

Pathology of Protein Misfolding

Some genetic mutations may result in proteins that essentially *always* misfold. In this case, most of the proteins get degraded and are unable to perform their biological function. This is the basis for a common form of cystic fibrosis.

In other cases, misfolded proteins may aggregate (ie, stick to each other). This occurs primarily because misfolded proteins fail to fully bury their hydrophobic residues. Consequently, the exposed hydrophobic residues in one protein may interact with the hydrophobic residues of another, hiding each set of hydrophobic residues from water (Figure 2.36). Therefore, aggregation shares some features with quaternary structure in that multiple polypeptides interact, often with hydrophobic residues at their interface. However, because the individual polypeptides and their aggregates are not the native, functional form of the protein, aggregation commonly has negative biological consequences.

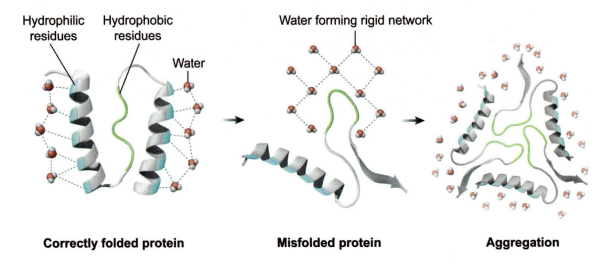

Figure 2.36 Example of protein aggregation.

Protein aggregates may take various forms. In some cases, multiple misfolded proteins assemble into highly ordered fibrous structures known as amyloid fibrils. These fibers are associated with various neurodegenerative disorders, including Alzheimer disease. Other proteins may form much less ordered, amorphous aggregates. These aggregates have also been implicated in certain disease states, including cataract formation.

Certain types of misfolded proteins can act as infectious agents that can be transmitted from one organism to another. These proteins, known as prions, can induce correctly folded proteins to misfold and become prions themselves. This can lead to a catastrophic cascade effect because increasing amounts of misfolded protein induce increasing amounts of correctly folded protein to misfold. Prions are the underlying cause of Creutzfeldt-Jakob disease and its transmissible variant, mad cow disease.

☑ Concept Check 2.11

Identify the labeled groups (ie, 1, 2, 3) in the following diagram as a correctly folded protein, an amyloid fibril, or a prion:

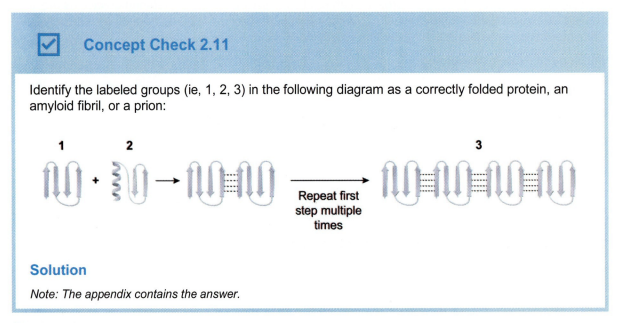

Solution

Note: The appendix contains the answer.

Chaperones

An entire class of proteins, called **chaperones**, exist to aid in the correct folding of other proteins. Chaperones are sometimes also called **heat shock proteins** (HSPs) because they help prevent cellular damage due to high temperatures.

Chaperones provide both newly synthesized and misfolded proteins with an environment that facilitates proper folding. This often involves interactions between hydrophobic residues in the chaperone and hydrophobic residues in the protein to be folded. These interactions protect the hydrophobic residues from water and thereby prevent aggregation while the protein is folding into its proper form.

In addition to helping newly synthesized proteins fold correctly, chaperones can also help misfolded proteins to adopt the correct conformation. Many chaperones help protein aggregates to disaggregate and then help the monomers to fold correctly. This is depicted in Figure 2.37.

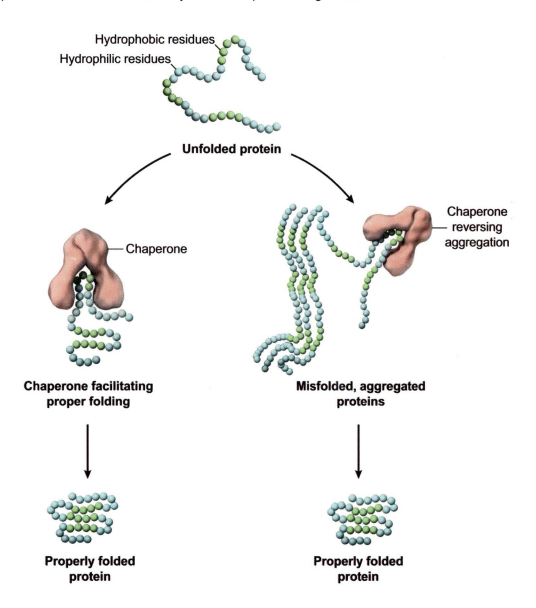

Figure 2.37 Chaperones help unfolded or misfolded proteins to fold correctly.

Lesson 2.4
Protein Modifications

Introduction

All proteins consist of at least one chain of amino acids linked together through peptide bonds. The sequence of amino acids in each chain determines the biological function carried out by that chain when the protein folds correctly. In addition to amino acid sequence, a protein's ability to function often depends on chemical groups other than amino acids. This lesson discusses changes to proteins, both at the amino acid and the non–amino acid levels, that can affect protein function.

2.4.01 Mutations

Genes occasionally undergo random mutations that can change the amino acid sequence of the corresponding protein. Consequently, the amino acid sequence for a given protein can vary from one species to another or between individuals in a species. Within a species, the most common amino acid sequence (ie, the sequence normally found in the wild) is called the wild-type form of that protein. Individuals with a sequence other than the wild-type sequence have a mutant form of the protein.

Protein substitution mutations are typically indicated by the one-letter code of the amino acid from the wild-type protein, followed by its position in the amino acid sequence (in which the N-terminal residue is position 1), and the one-letter code of the new (mutant) amino acid. For example, a Q37L mutation means that in the wild-type protein the 37th amino acid is glutamine (Q), but in the mutant protein it has changed to leucine (L).

> ### ✓ Concept Check 2.12
>
> The following image shows a mutation in a certain peptide sequence. What is the correct designation for this mutation?
>
>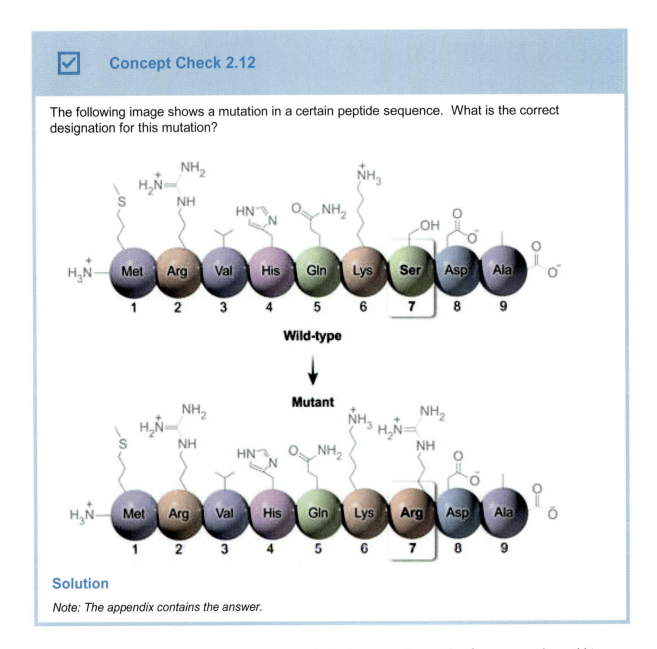
>
> **Solution**
>
> *Note: The appendix contains the answer.*

Different mutations have different effects on a protein's function. A mutation from one amino acid to another with similar chemical properties tends to have a smaller effect than mutations that significantly alter chemical properties. For example, the mutation K45R changes the amino acid at position 45 from one positively charged group (K, lysine) to another (R, arginine).

This mutation is *less* likely to affect protein function than a K45E mutation, which changes a positive charge (K, lysine) to a negative charge (E, glutamate). Mutations that preserve chemical properties are called conservative mutations. Those that do not are called nonconservative.

In addition to the type of mutation (eg, conservative vs. nonconservative), the *location* of a mutation within a protein impacts the effect of that mutation. For example, in flexible, unstructured regions, even nonconservative mutations are less likely to have a significant effect on protein function. In contrast, highly structured regions are more susceptible to disruption by mutation. Furthermore, regions that are critical to protein function may even be affected by conservative mutations. Examples of protein regions with different susceptibilities to mutation are shown in Figure 2.38.

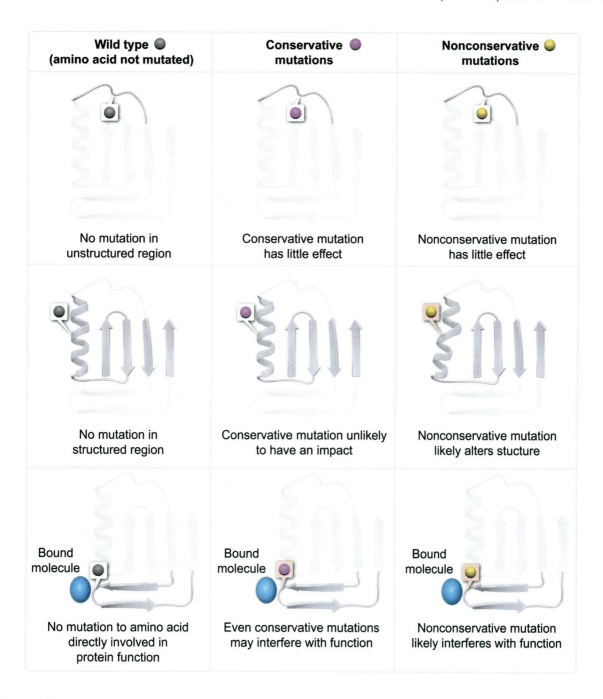

Figure 2.38 Examples of the relation between location and the probable impact of mutations.

Comparing the wild-type sequences for the same protein in different species reveals which amino acids are critical for function and which are not. When sequences of related proteins are properly aligned, the amino acids that are critical to structure and/or function are present in *all* species. These amino acids are said to be **highly conserved**.

In many cases, the chemical properties are important, but any amino acid with those properties can serve (eg, serine and threonine are both acceptable). Amino acids at these positions are also said to be conserved but to a lesser extent than amino acids that must remain unaltered. Amino acids that vary significantly from one species to another are not conserved.

Concept Check 2.13

The wild-type sequences of the protein hexokinase isolated from five different species were aligned. The following image shows a small portion of each sequence.

Species 1 KYRLPDAMRTTQN
Species 2 KFALPENMRTAKS
Species 3 VYDTPENIVHGSG
Species 4 TYCIPAEKMSGSG
Species 5 EVSIPPHLMTGGS

If the amino acid residue at the left of each sequence is position 111, which position is the most critical for hexokinase structure and/or function? (Note: Sequence numbering is based on the sequence of species 1.)

Solution

Note: The appendix contains the answer.

2.4.02 Post-translational Modifications

The process of a ribosome synthesizing a protein is called translation. After a protein is translated, it may be altered by various enzymes acting on it. These enzymes often add chemical groups that are not amino acids to the protein. The resulting alterations are called post-translational modifications. Many types of post-translational modifications exist, including the following:

Phosphorylation

This modification involves the transfer of a phosphate group, usually from ATP, to the side chain of an amino acid residue. Phosphorylation is facilitated by a class of enzymes called kinases (see Concept 4.2.03) and typically occurs on serine, threonine, or tyrosine residues. Each of these amino acids has a hydroxyl group that can act as a nucleophile to attack the γ-phosphate group on ATP, yielding ADP and phosphoserine, phosphothreonine, or phosphotyrosine, respectively (Figure 2.39). Other amino acid side chains can also undergo phosphorylation, but this is much less common.

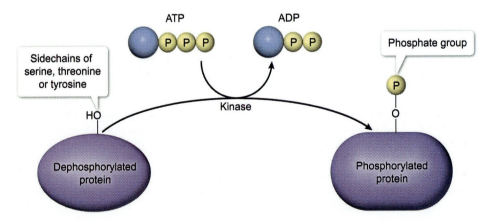

Figure 2.39 General schematic of protein phosphorylation by ATP, facilitated by a kinase enzyme.

Phosphorylation often regulates protein function. The specific effect of phosphorylation depends on the protein being modified; some proteins are activated by phosphorylation, whereas others are inactivated.

Concept Check 2.14

The following graph shows the activity levels of two proteins under different conditions. What is the effect of phosphorylation on each protein?

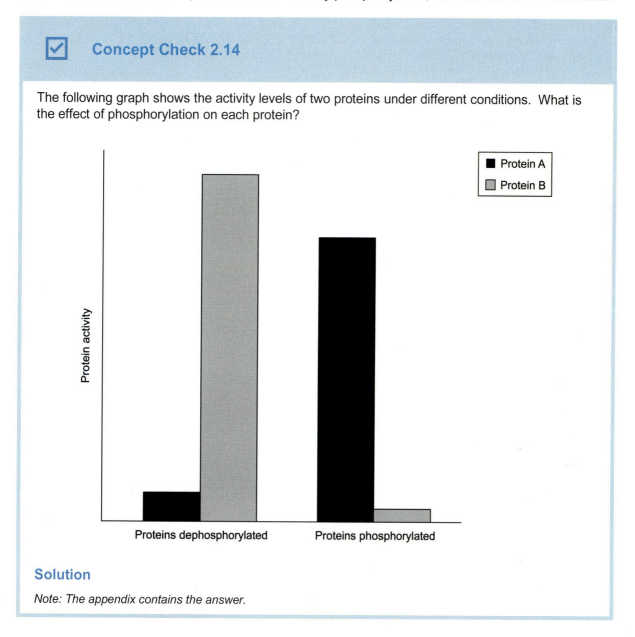

Solution
Note: The appendix contains the answer.

Phosphate groups are negatively charged and therefore typically exert their effect by adding negative charges to a protein. This added negative charge decreases the protein's isoelectric point and may induce conformational changes that alter the protein's ability to interact with its environment and with other molecules.

Interestingly, mutation of a serine or threonine residue to glutamate or aspartate often produces the same effect as phosphorylation. For instance, if a protein is activated by phosphorylation of a specific serine residue, mutating that residue to aspartate can cause the protein to become constitutively active (ie, it cannot be inactivated). The negative charge of the aspartate residue has the same effect as the negative charge of a phosphate group, but aspartate cannot be removed by dephosphorylation. This type of mutation is known as a phosphomimetic mutation because it mimics the effect of a phosphate group.

Glycosylation

Many proteins, particularly those synthesized on the rough endoplasmic reticulum, are modified by glycosylation (ie, the addition of carbohydrate groups). Glycosylation, shown in Figure 2.40, is catalyzed by glycosyltransferase enzymes. The added carbohydrate may consist of a single sugar or many sugars linked together. Additional information on carbohydrates can be found in Chapter 7.

Glycosylation most commonly occurs in the lumen of the endoplasmic reticulum and the Golgi apparatus. The two most common types of glycosylation are *N*-linked and *O*-linked glycosylation, which are distinguished by the amino acid residues on which they take place.

N-linked glycosylation occurs on the –NH$_2$ groups of asparagine side chains. As the glycosylated protein moves from the ER to the Golgi, the complex carbohydrate may be modified by addition or removal of individual sugars. This process helps to direct proteins to the correct destinations (eg, lysosome, extracellular space), and may also facilitate protein folding and activity. *O*-linked glycosylation plays similar roles to *N*-linked glycosylation, but it occurs on the –OH groups of serine or threonine residues.

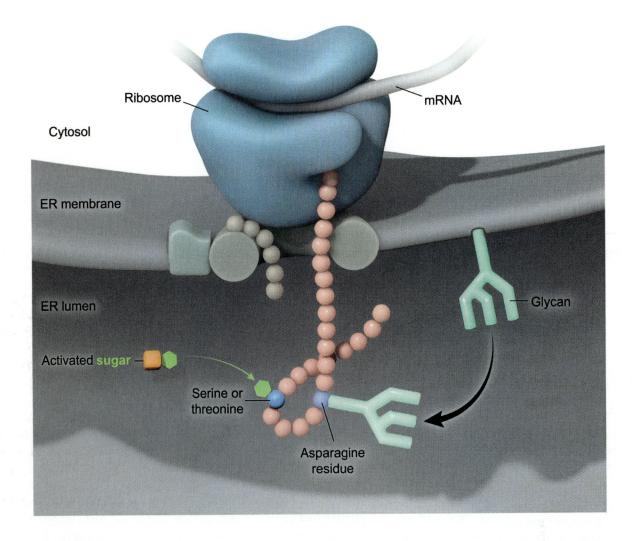

Figure 2.40 Carbohydrates are added to serine, threonine, or asparagine residues in the ER.

Ubiquitination

This modification involves two proteins: the target protein and ubiquitin. Therefore, unlike other post-translational modifications, ubiquitination involves addition of other amino acids (ie, those found in ubiquitin) to an existing protein.

Ubiquitination can occur when the N-terminus or a side chain (often lysine) of the target protein nucleophilically attacks the C-terminus of ubiquitin. Bonds between lysine side chains and ubiquitin C-termini are the most common linkages. This linkage forms an amide bond similar to the peptide bonds between amino acids in a protein backbone. However, because this bond forms between a *side chain* and a backbone group (rather than two backbone groups), it is called an isopeptide bond rather than a peptide bond.

Once one ubiquitin is added to a target protein, another ubiquitin can be added to the first. This process can be repeated multiple times to form a polyubiquitin chain. Ubiquitination serves many biological purposes, but the best characterized is ubiquitin's role in protein degradation. When cytosolic proteins are misfolded or no longer needed, they are commonly modified with a polyubiquitin tag, which then targets them to a protein complex called the proteasome (see Figure 2.41). This complex digests the protein into small peptide and amino acid fragments, which are then recycled to make other proteins.

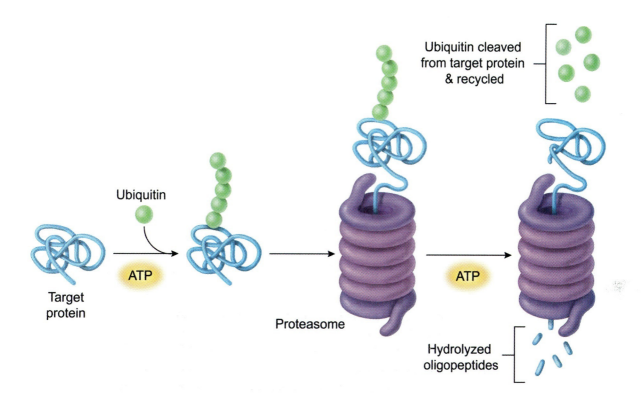

Figure 2.41 Polyubiquitin tags send proteins to the proteasome for degradation.

Acetylation

In addition to modification by ubiquitin, lysine residues in a protein may also be modified by addition of an acetyl group. In these reactions, the side chain nitrogen of lysine acts as a nucleophile to attack the carbonyl carbon of the acetyl group in acetyl coenzyme A, producing acetylated lysine and free coenzyme A. In some cases, the N-terminus of a protein can act as the nucleophile in place of lysine.

Many proteins undergo acetylation. Perhaps the best-known example is histone proteins (Figure 2.42), which bind DNA and are crucial to the structure of chromosomes. When unmodified, the positively charged lysine residues in a histone interact with the negatively charged DNA backbone. However, acetylation neutralizes the positive charge of lysine, releasing the DNA. This process is important in regulating which genes in a cell are transcribed to produce RNA and which are not.

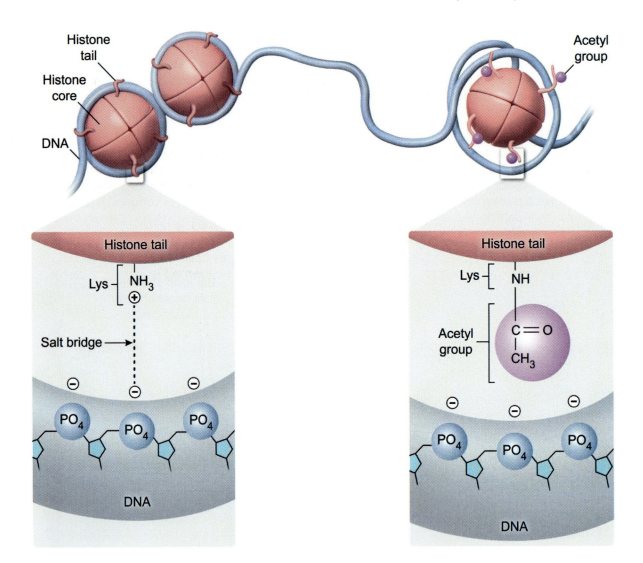

Figure 2.42 Interactions between histones and DNA are disrupted by lysine acetylation.

Lipidation

Lipidation is similar to acetylation in that a nucleophilic amino acid group attacks a carbonyl linked to a hydrocarbon. However, the acetyl group in acetylation contains a single carbon in its hydrocarbon chain, whereas lipidation involves much longer, more complex hydrocarbons. Lipidation occurs on lysine, cysteine, and serine side chains and is involved in many biological processes, including those carried out by certain important metabolic enzymes (see Chapter 12).

Proteolysis

Unlike other post-translational modifications, proteolysis occurs by *removing* groups from proteins. This occurs when a protease enzyme hydrolyzes a peptide bond. Certain proteins must remain inactive until they are transported to the appropriate compartment of a cell or organism or until certain environmental conditions are met. To accomplish this, these proteins are synthesized in an inactive form called a proprotein. Upon reaching the correct environment, the proprotein undergoes proteolysis (also called cleavage) to convert it into its active form.

2.4.03 Cofactors

To function, many proteins require chemical groups other than amino acids. Any non–amino acid group required for protein function is called a cofactor. Cofactors may range in complexity from single metal ions (eg, Ca^{2+}, Mg^{2+}) to complex organic molecules. Organic cofactors are given the additional designation of coenzymes and often assist enzymes in carrying out their reactions. Coenzymes are commonly derived from vitamins.

Some cofactors interact only transiently with their proteins. Other cofactors bind tightly or even covalently to their target proteins. Covalently bound cofactors can be thought of as post-translational modifications. These tightly bound cofactors effectively become part of the enzyme, and, as such, they are called prosthetic groups.

An important coenzyme found in many proteins, often as a prosthetic group, is heme. Heme is derived from the organic molecule porphyrin and contains an iron (Fe) atom at its center, bound to the four pyrrole groups of heme. Heme is involved in several crucial oxidation-reduction reactions and in the transport and intracellular storage of oxygen. It has the characteristic structure shown in Figure 2.43.

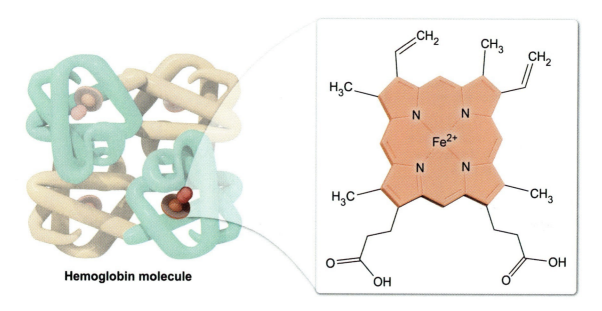

Figure 2.43 Structure of heme, with the porphyrin ring highlighted.

All proteins are initially synthesized using only amino acids and must interact with any cofactors, including prosthetic groups, after translation (and often after folding). Some proteins that require prosthetic groups can be found without these groups in certain circumstances. In this form, these proteins are called **apoproteins**. Upon addition of the correct prosthetic group, an apoprotein is converted into a **holoprotein**. Only the holoprotein form can carry out the correct biological function.

Concept Check 2.15

Based on the given schematic, fill in the blanks in the following sentence using the words "apoprotein," "coenzyme," "holoprotein," and "prosthetic group."

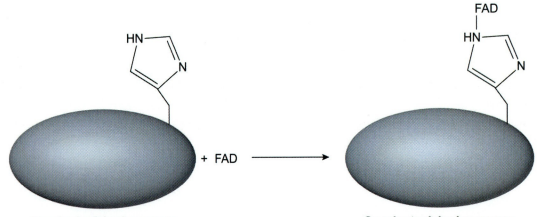

Succinate dehydrogenase (inactive form) → **Succinate dehydrogenase** (active form)

FAD, an organic molecule, is a _____ in succinate dehydrogenase (SDH). Because FAD binds SDH tightly, it acts as a _____. Upon FAD binding, SDH is converted from its _____ form to its _____ form.

Solution

Note: The appendix contains the answer.

Lesson 3.1
Protein-Ligand Interactions

Introduction

Biochemistry places significant emphasis on enzymes, which facilitate the chemical reactions needed to sustain life. This important class of protein is covered extensively in Unit 2. However, many other proteins fulfill important biological roles that do *not* involve chemical reactions. For instance, many proteins help to transport molecules from one location to another within an organism. Others help to store molecules for later use, while still others receive chemical signals from the environment and transmit them to the cell, causing a biological response.

Many nonenzymatic proteins carry out their functions by binding to one or more specific molecules. Any molecule that specifically binds to a protein is called a **ligand** for that protein. This lesson discusses fundamental aspects of protein-ligand interactions.

3.1.01 Ligand Binding

Proteins bind their ligands through noncovalent interactions. These interactions form between the ligand and amino acid residues or prosthetic groups in a region of the protein called the **binding site** or **binding pocket**. The chemical groups in a binding site have a specific orientation and, as a result, often bind with **specificity** (ie, they preferentially bind ligands with specific structural features). Some proteins are *highly* specific and bind only one compound or a few structurally similar compounds. Other proteins are less specific and bind a more diverse set of compounds that share a few key features.

Mutations within the binding pocket of a protein may significantly alter function, particularly for highly specific protein-ligand interactions. In some cases, even conservative mutations may reduce binding. For instance, aspartate and glutamate have nearly identical structures, but the aspartate side chain is shorter than that of glutamate. Therefore, mutating glutamate to aspartate (or vice versa) in a binding pocket may impact binding because the angles required for interactions with the ligand change. Figure 3.1 shows an example of the effects of such a mutation.

Chapter 3: Nonenzymatic Protein Activity

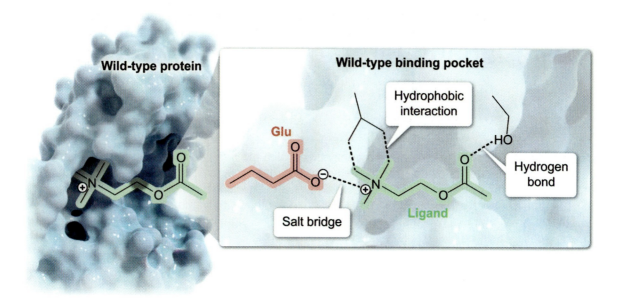

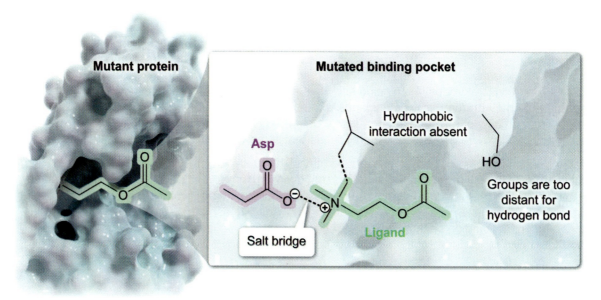

Figure 3.1 Mutations in binding pocket residues can significantly affect binding strength.

Ligands may range in size from small molecules or ions (eg, O_2 binding to hemoglobin) to large molecules (eg, DNA binding to a transcription factor). Frequently, the ligand for one protein is another protein, leading to the formation of a protein-protein complex. This arrangement arises through noncovalent interactions between amino acid residues. In this way, protein complexes are similar to quaternary structure but differ in that each unit in a protein complex can function independently, whereas the individual subunits in a protein with quaternary structure function only when bound together.

Antibodies illustrate several aspects of quaternary structure *and* protein complex formation. An antibody is a protein that consists of four polypeptides—two heavy chains and two light chains—linked together through various noncovalent interactions and disulfide bonds. Without these interactions, an individual chain cannot bind its ligand and the antibody is not fully functional. Because these interactions are required for the structure and function of the antibody, they constitute quaternary structure.

The ligand for an antibody is called an **antigen**. Antigens are often proteins found on the surface of an infectious organism. The specific portion of the antigen to which the antibody binds is called an **epitope**, which may include post-translational modifications such as carbohydrates or phosphate groups. The portion of the antibody that binds the epitope is called the **paratope**. The antibody and the antigen are both functional regardless of whether they are bound to each other, so their binding interactions constitute protein-protein complex formation rather than quaternary structure.

In some cases, a complex may consist of multiple proteins bound to each other. For instance, an antibody may bind its antigen through the paratope (ie, the binding site) while simultaneously binding a cell receptor protein at another site. The cell receptor protein may in turn be linked to many additional proteins. Figure 3.2 depicts an antibody binding both an antigen and a receptor protein.

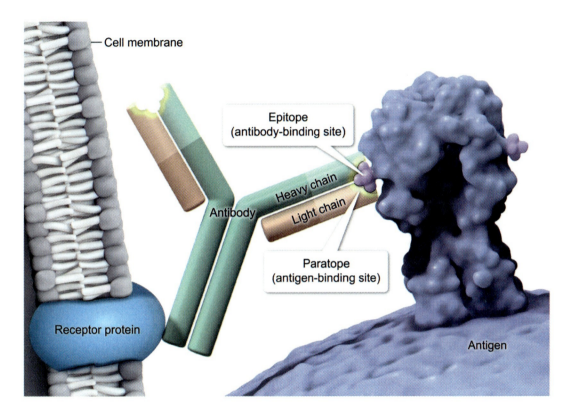

Figure 3.2 Interactions between an antibody and other proteins.

Ligand binding generally causes the protein to undergo a conformational change that results in a biological response. Many drugs are designed to interact with a binding pocket *without* causing the biological response. For instance, cancer proliferation often involves overactivity of the transcription factor Myc, which must bind another protein called Max before binding to DNA to induce transcription. Various drugs that disrupt Myc-Max interactions (Figure 3.3) have been tested for anticancer properties.

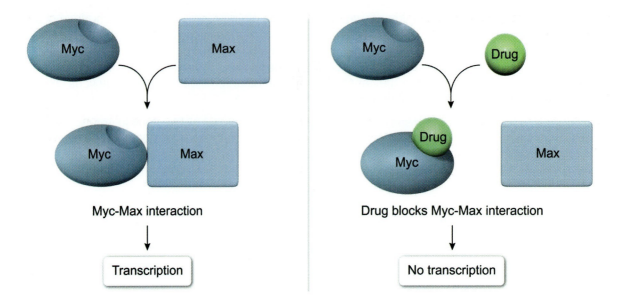

Figure 3.3 A protein-drug interaction disrupting the biological function of a protein.

3.1.02 The Dissociation Constant K_d

Different proteins bind their ligands with different strengths, and a given protein may bind several different ligands, each with a different strength. The overall strength with which a protein binds a ligand is called its **affinity** for that ligand. Strong protein-ligand interactions are high affinity, and weak interactions are low affinity. Typically, the affinity of a protein for its ligand is measured by the dissociation constant K_d. K_d is the equilibrium constant for the process in which the protein-ligand complex (PL) **dissociates** into free protein (P) and free ligand (L).

$$PL \rightleftharpoons P + L$$

As with all equilibrium constants, a large value indicates that products are favored whereas a smaller value indicates greater reactant favorability. Because K_d uses PL as the reactant, a small K_d value indicates that PL formation is favored (ie, dissociation is *not* favored). Therefore, small K_d values correspond to tight (ie, strong) binding and high affinity.

 Concept Check 3.1

The spike protein on the original variant of the SARS-CoV-2 virus contains a glutamate residue at position 484. In later variants, this residue mutated to lysine, reducing the binding strength of antibodies that target an epitope containing this residue. Based on this information, how does this E484K mutation affect the K_d of the interaction between these antibodies and the spike protein?

Solution

Note: The appendix contains the answer.

By definition, an equilibrium constant is the equilibrium concentration ratio of all products to all reactants, with each concentration raised to the power of its stoichiometric coefficient. Therefore, K_d is defined as

$$K_d = \frac{[\text{P}]_{eq}[\text{L}]_{eq}}{[\text{PL}]_{eq}}$$

Each concentration has units of molarity. Because the numerator contains two concentrations multiplied by each other and the denominator contains only one concentration, the units do not fully cancel, and K_d also has units of molarity. However, because K_d is usually small, it is often converted to units of mM, μM, or nM. An example K_d calculation is shown in Figure 3.4.

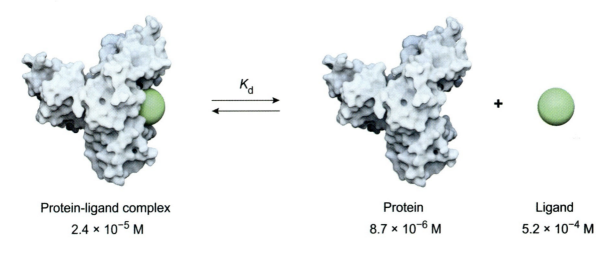

$$K_d = \frac{(8.7 \times 10^{-6} \text{ M})(5.2 \times 10^{-4} \text{ M})}{(2.4 \times 10^{-5} \text{ M})} = 1.9 \times 10^{-4} \text{ M} = 190 \text{ μM}$$

Figure 3.4 Example of a K_d calculation given known concentrations of PL, P, and L at equilibrium.

Protein binding typically behaves as a reversible single-step process—either the protein binds its ligand or the protein-ligand complex dissociates—but no intermediates form in either case. Therefore, the *rate* of complex formation is given as

$$\text{rate} = k_{on}[\text{P}][\text{L}]$$

and the rate of complex dissociation is

$$\text{rate} = k_{off}[\text{PL}]$$

in which k_{on} and k_{off} are the binding and unbinding rate constants, respectively. At equilibrium, binding and unbinding occur at the same rate, and therefore

$$k_{off}[\text{PL}]_{eq} = k_{on}[\text{P}]_{eq}[\text{L}]_{eq}$$

Rearranging this equation yields

$$\frac{k_{off}}{k_{on}} = \frac{[\text{P}]_{eq}[\text{L}]_{eq}}{[\text{PL}]_{eq}} = K_d$$

A small K_d indicates a relatively small (ie, slow) unbinding rate constant or a large (ie, fast) binding rate constant (or both), confirming that a small K_d indicates strong binding affinity. Conversely, a large K_d indicates a relatively large (ie, fast) unbinding rate constant or a small (ie, slow) binding rate constant (or both), confirming that a large K_d indicates weak binding affinity. Note, however, that if only K_d is known, the specific values of the rate constants cannot be determined because various combinations of k_{on} and k_{off} could produce the same K_d (eg, 2/1 = 2, 4/2 = 2).

Concept Check 3.2

A protein binds its ligand with a rate constant $k_{binding}$ of 3.5×10^4 M^{-1}s^{-1} and unbinds with a rate constant $k_{unbinding}$ of 1.4×10^{-1} s^{-1}. What is the dissociation constant K_d for this process?

Solution

Note: The appendix contains the answer.

For any chemical process, the equilibrium constant is related to the standard free energy change $\Delta G°$ of that process. In the case of protein-ligand dissociation, the relationship is given by the equation

$$\Delta G° = -RT \ln(K_d)$$

in which R is the universal gas constant and T is Kelvin temperature of the system. Therefore, if the standard free energy change of dissociation at a given temperature is known, the K_d value can be calculated. More generally, any K_d smaller than 1 M corresponds to a positive (ie, unfavorable) $\Delta G°$ of dissociation. Most biochemically relevant K_d values are well below 1 M, and protein-ligand *dissociation* is usually quite energetically unfavorable. Because the $\Delta G°$ of *binding* has the same magnitude as that of dissociation but the opposite sign, this also means that binding is favorable.

Concept Check 3.3

The following table shows K_d values for a protein-ligand interaction at various temperatures.

Temperature (°C)	K_d (M)
30	3.1×10^{-4}
35	4.8×10^{-4}
40	2.3×10^{-3}

Which temperature corresponds to the most negative value for $\Delta G°$ of binding? (Assume any changes in the $-RT$ portion of the equation have a negligible impact on $\Delta G°$ over the range of temperatures used.)

Solution

Note: The appendix contains the answer.

3.1.03 Binding Curves

The K_d for a protein-ligand interaction can be determined by generating a binding curve. These curves are produced by gradually adding ligand to a known amount of protein and measuring the response. The results are plotted as the fraction of total protein bound (θ) on the y-axis against the concentration of ligand added on the x-axis.

For a single ligand binding to a single pocket, the binding curve fits the equation

$$\theta = \frac{[L]}{K_d + [L]}$$

When plotted, this equation yields a **hyperbolic curve** that rises quickly, then levels off as saturation occurs (ie, when nearly all proteins in solution are bound). When $[L] = K_d$, a substitution can be made so that the equation becomes

$$\theta = \frac{K_d}{K_d + K_d} = \frac{\cancel{K_d}}{2\cancel{K_d}} = \frac{1}{2}$$

In other words, for single protein–single ligand interactions, K_d equals the concentration of free ligand at which 50% of the proteins in solution are bound and 50% are unbound (see Figure 3.5).

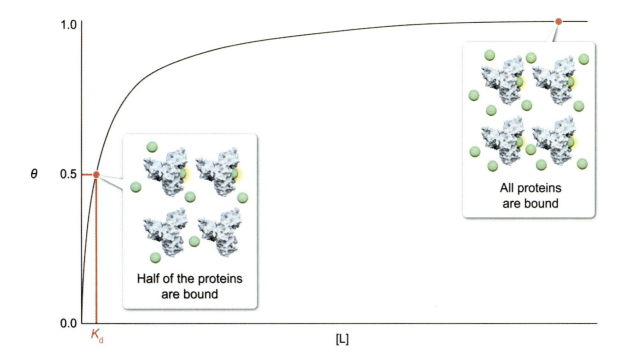

Figure 3.5 Example of a hyperbolic binding curve with K_d marked.

This is consistent with the fact that a lower K_d corresponds to increased binding affinity. A smaller K_d means a smaller amount of ligand is required to bind half the proteins in solution.

For gaseous ligands (eg, O_2), partial pressure is easy to measure and is often used instead of concentration. The partial pressure at which 50% of proteins are bound (ie, the P_{50}) is used instead of K_d in this case, and the binding equation becomes

$$\theta = \frac{P_{gas}}{P_{50} + P_{gas}}$$

The curve generated by this equation looks identical to binding curves that use ligand concentration, except that the x-axis has units of pressure (eg, atm, mm Hg).

Concept Check 3.4

The following graph shows the binding curve for interactions between myoglobin and oxygen under physiological conditions.

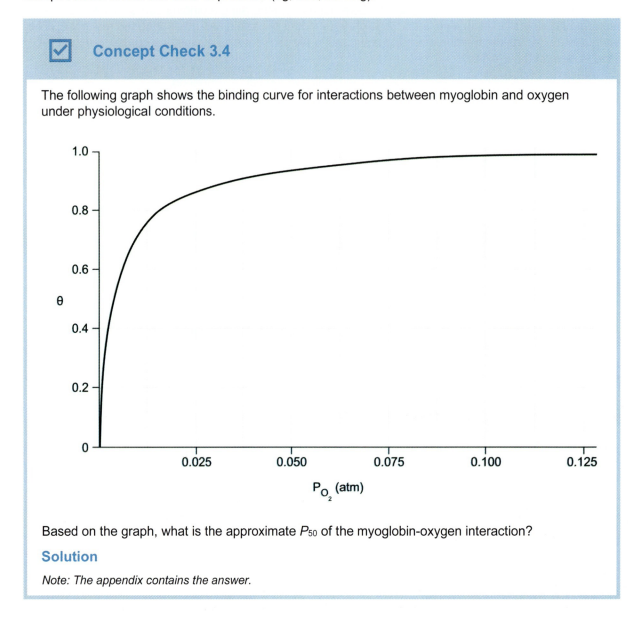

Based on the graph, what is the approximate P_{50} of the myoglobin-oxygen interaction?

Solution

Note: The appendix contains the answer.

Alterations to a protein that change its affinity for its ligand (eg, post-translational modifications) cause changes in the appearance of the binding curve. For example, conditions that increase the affinity cause the K_d of the protein-ligand interaction to *decrease*. On a binding curve, K_d corresponds to a position on the x-axis, and in this case, the position of K_d on the x-axis moves to the left (ie, the curve is left-shifted). In contrast, conditions that *decrease* affinity cause an *increase* in K_d and a right shift on a binding curve. Shifts to the left and right are shown in Figure 3.6.

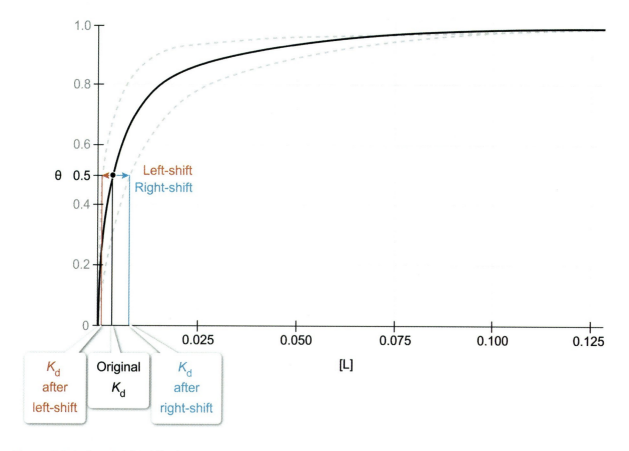

Figure 3.6 Left and right shifts in response to conditions that increase and decrease affinity, respectively.

3.1.04 Allostery and Cooperativity

Binding proteins usually have a primary ligand, which is the ligand that is *required* for the protein to perform its biological function. This ligand binds to the protein at the binding pocket. However, many proteins can bind ligands other than their primary ligand at a site other than the primary binding pocket. Sites on a protein that bind ligands other than the primary ligand are called **allosteric sites** (see Figure 3.7).

These other ligands, called **allosteric effectors**, cause proteins to undergo conformational changes that can affect the whole protein. Even small changes in conformation may realign the functional groups in the binding pocket and alter the affinity for the primary ligand. Some allosteric effectors *increase* a protein's affinity for its ligand while others *decrease* affinity. In this way allosteric effectors can regulate protein-ligand binding by helping proteins to bind and release their primary ligands as needed.

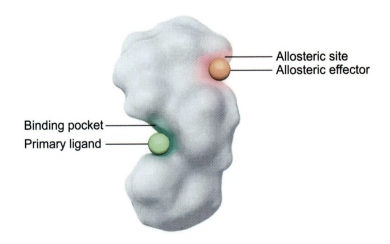

Figure 3.7 An allosteric site and effector compared to the binding pocket and primary ligand.

Concept Check 3.5

A scientist proposes that a specific compound is an allosteric effector for protein P. The scientist further hypothesizes that the compound increases the affinity of P for its ligand L. If the scientist is correct, how will the binding curve of the interaction between P and L change when the compound is added?

Solution

Note: The appendix contains the answer.

One important consequence of allosteric interactions is the phenomenon of binding **cooperativity**. Broadly, cooperativity is any biochemical process that alters the ability of another process to occur. For instance, protein folding and unfolding are cooperative because formation of secondary structures increases the ability of tertiary structure to form. With respect to binding (and unbinding), cooperativity refers to the effect of one binding interaction on other binding interactions by the same protein.

Cooperativity may be positive or negative. Positive cooperativity occurs when binding interactions at one site *increase* the affinity of other sites for their ligands, and negative cooperativity refers to binding interactions that *decrease* the affinity of other binding sites. Thus, allosteric effectors that increase affinity (ie, decrease K_d) induce *positive* cooperativity, and those that decrease affinity (ie, increase K_d) induce *negative* cooperativity.

An especially important form of cooperativity occurs when multiple, identical ligands bind to multiple binding pockets on the same protein. This type of cooperative binding often occurs in proteins with quaternary structure, in which each subunit contains one binding pocket. When one pocket binds its ligand, the other binding pockets may change shape and therefore may change affinity for the ligand. This form of cooperativity is depicted in Figure 3.8.

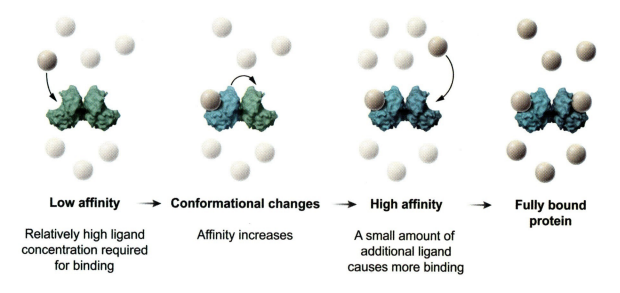

Figure 3.8 Schematic of positive cooperativity in a protein with two pockets that bind the same ligand.

The extent to which cooperativity occurs in proteins with multiple binding sites for the same ligand is measured by the **Hill coefficient** n. A Hill coefficient above 1 indicates positive cooperativity, a Hill coefficient of exactly 1 indicates that no cooperativity occurs, and a Hill coefficient below 1 (but greater than 0) indicates negative cooperativity.

For a protein with x identical binding sites, the Hill coefficient can be no greater than x and no smaller than $1/x$, but it may be anywhere in between. For example, hemoglobin has four oxygen binding sites and a Hill coefficient of 2.8.

The maximum possible Hill coefficient for a protein (eg, 4 in the case of hemoglobin) indicates perfect cooperativity—that is, cooperativity such that either all binding sites bind their ligand simultaneously or none bind at all. In practice, proteins are rarely perfectly cooperative. Table 3.1 summarizes the meanings of various Hill coefficients for a protein with four subunits.

Table 3.1 Interpretation of the Hill coefficient for a protein with four binding sites.

Hill coefficient		Interpretation
$n = 4$		If one site is bound, all four sites must be bound
$1 < n < 4$		Binding at one site **helps** (but does not force) binding at the other sites
$n = 1$		Binding at each site is independent of the others
$\frac{1}{4} < n < 1$		Binding at one site **inhibits** (but does not completely prevent) binding at the other sites
$n = \frac{1}{4}$		Binding is limited to one of the four sites

When a protein has a Hill coefficient n other than 1, the shape of the binding curve changes. In positive cooperativity (ie, $n > 1$), increases in ligand concentration initially cause small changes in binding levels because the affinity is low. This results in a shallow initial slope on the binding curve.

However, as ligand concentration reaches a point where most proteins in solution have one binding site filled, the other binding sites gain higher affinity for the ligand. At this point, small increases in ligand concentration cause a greater increase in the percentage of binding sites that are filled, and the slope of the curve increases dramatically. Once all binding sites are filled, the curve levels off. Consequently, positive cooperativity yields a sigmoidal (ie, S-shaped) curve (see Figure 3.9).

Negative cooperativity ($n < 1$) has the opposite effect. Initially, increases in ligand concentration induce a larger increase in binding than occurs at higher ligand concentrations. As binding continues, other binding sites decrease in affinity for the ligand, so additional ligand causes progressively smaller increases in binding. This yields a curve that looks similar to noncooperative binding, but the curve flattens much more quickly. Because negative cooperativity yields a similar shape to no cooperativity, identification of negative cooperativity from a graph is unlikely on the exam.

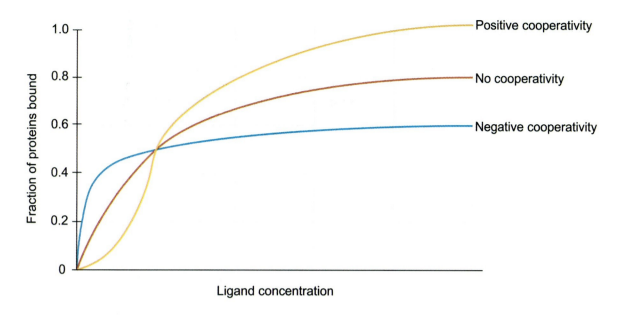

Figure 3.9 Example graphs of protein-ligand binding with positive, negative, and no cooperativity.

Any protein that exhibits binding cooperativity also exhibits unbinding cooperativity. For example, if a protein with positive cooperativity has all sites bound, then removing one ligand makes it easier to remove additional ligands. This can be represented on an unbinding curve as an inverted sigmoid as shown in Figure 3.10.

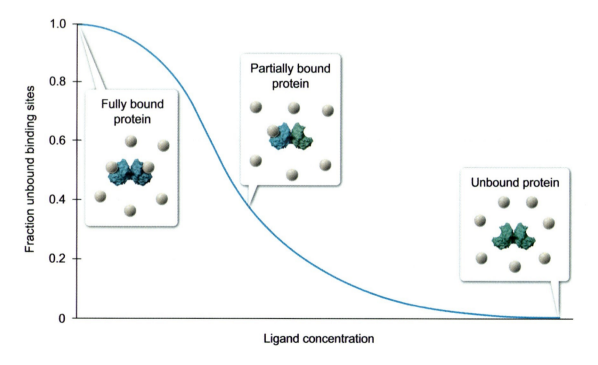

Figure 3.10 Example of a ligand unbinding curve exhibiting positive cooperativity.

Concept Check 3.6

A binding protein is gradually titrated with its ligand, and the resulting binding curve has a sigmoidal shape. Further analysis reveals that the protein has a Hill coefficient of 2.0. What conclusions can be drawn about the number of ligand binding sites in this protein?

Solution

Note: The appendix contains the answer.

Lesson 3.2
Receptor Proteins

Introduction

Cells must constantly monitor and adapt to the conditions of their surroundings. For example, muscle cells typically respond to high levels of extracellular glucose by bringing glucose into the cell and either storing it or breaking it down for energy. In contrast, under low-glucose conditions, cells break down previously stored glucose or begin to use other sources (eg, fat stores) for energy. Many other extracellular conditions (ie, extracellular signals) cause a variety of intracellular responses. Each response is mediated by one or more **receptor proteins**.

A receptor protein is any protein that receives an extracellular signal and, as a result, induces an intracellular response. These extracellular signals may be chemicals that bind the receptor protein (ie, ligands), light that induces a chemical reaction in the receptor, electrical impulses that alter a protein's optimal conformation, or mechanical pressure that distorts the receptor's shape. This lesson focuses primarily on chemical signals, although other signals produce similar responses.

3.2.01 Ligand Binding by Receptor Proteins

Although receptor proteins respond to a variety of signals, the most common signal is the binding of a ligand. Ligands that bind receptor proteins and induce the primary biological response are called **agonists**. In general, receptor proteins are highly specific to their agonist, although similar molecules that bind and induce a response may be synthesized. Interactions between natural or synthetic agonists and a receptor protein are shown in Figure 3.11.

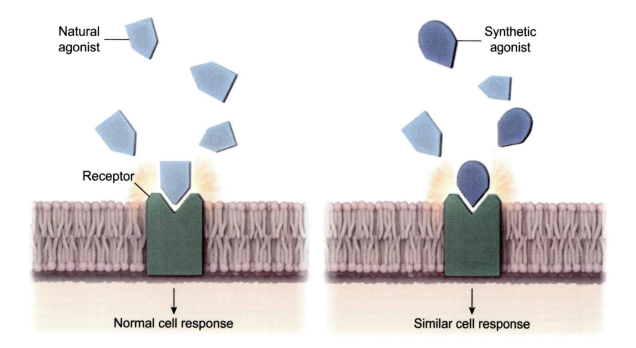

Figure 3.11 Agonist interactions with a receptor induce a cell response.

Like any protein-ligand interaction, the strength with which an agonist binds its receptor can be measured by the dissociation constant K_d. An agonist that binds its receptor with a small K_d value (ie, tight binding) can cause a significant response when only a small amount of agonist is present. Agonists that bind with large K_d values must be present in larger quantities to produce a significant response.

Receptor protein activity can be regulated or altered by another type of ligand called an **antagonist**, which impedes agonist binding (see Figure 3.12). Importantly, binding by an antagonist does *not* cause the same biological response as binding by an agonist.

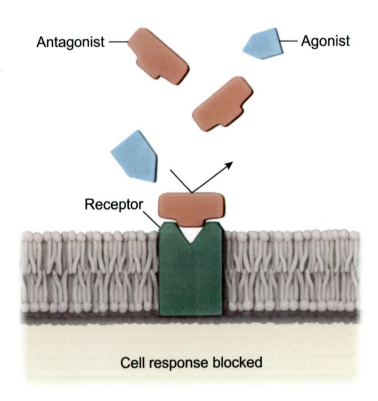

Figure 3.12 Agonists bind to receptors but do not cause the normal biological response.

Antagonists may bind at the same binding site as agonists. In Figure 3.12, the antagonist directly competes with the agonist for binding—when the antagonist is bound, the agonist is physically blocked from binding the receptor, and vice versa. Consequently, adding more agonist can displace the antagonist and reverse its effects.

Alternatively, antagonists may bind receptors at an allosteric site. This may either interfere with agonist binding or with the ability of bound agonist to transmit the signal, or both. Additional agonist does not necessarily displace allosteric antagonists.

Note that allosteric antagonist binding may impede agonist binding without completely abolishing it. In other words, rather than fully blocking agonist binding, an antagonist may merely reduce the affinity of the receptor for its agonist or reduce the intensity of a receptor response.

> [!NOTE] Concept Check 3.7
>
> A scientist measures a certain cellular response in cells cultured with different levels of Compound 1 and Compound 2, each of which binds the same receptor protein. The results are shown in the following graph. Based on the graph, identify Compounds 1 and 2 as an agonist or an antagonist, and determine whether the antagonist most likely binds at the primary binding site or an allosteric site.
>
>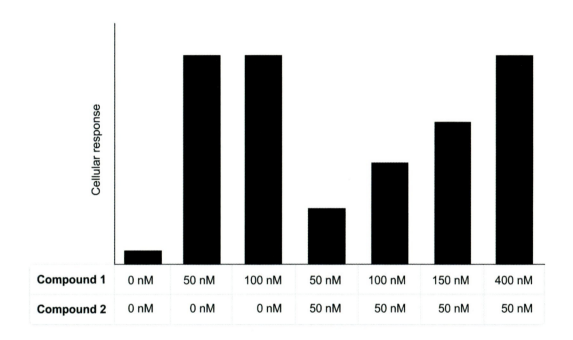
>
> | Compound 1 | 0 nM | 50 nM | 100 nM | 50 nM | 100 nM | 150 nM | 400 nM |
> |---|---|---|---|---|---|---|---|
> | Compound 2 | 0 nM | 0 nM | 0 nM | 50 nM | 50 nM | 50 nM | 50 nM |
>
> **Solution**
> *Note: The appendix contains the answer.*

3.2.02 Transmembrane Receptors

Many extracellular signals are hydrophilic molecules, which cannot easily cross the hydrophobic interior of the membrane phospholipid bilayer. These signals are received by transmembrane receptor proteins (see Figure 3.13). The extracellular domain of the receptor binds the signal (ie, the agonist), which induces a conformational change that extends to the intracellular domain.

The change in shape causes the intracellular domain to interact differently with nearby molecules. For instance, the intracellular domain may bind or release certain ligands or may gain or lose catalytic activity in response to an agonist binding the extracellular domain. Regardless of the details, the intracellular domain eventually modifies the behavior of an intracellular molecule and activates one or more small molecules called **second messengers**. The second messengers then continue the pathway toward the biological response.

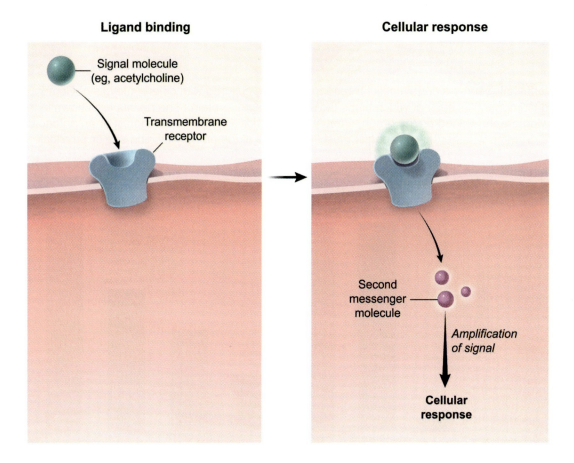

Figure 3.13 Schematic of an agonist binding a receptor and activating a second messenger.

One of the most abundant and best characterized classes of receptor protein is the **G protein–coupled receptor** (GPCR). GPCRs have seven transmembrane (7-TM) regions and interact with a class of proteins called G proteins. In the absence of agonist-GPCR interactions, the G protein binds guanosine diphosphate (GDP) and is inactive. Upon agonist binding, the GPCR intracellular domain undergoes a conformational change that affects the G protein's affinities toward guanosine nucleotides.

As a result of this change, the G protein then releases its GDP molecule and in its place binds guanosine triphosphate (GTP). The G proteins consist of three different subunits (α, β, and γ), making them heterotrimers. When bound to GTP, the α subunit dissociates from the other subunits and gains the ability to interact with an enzyme (ie, the G protein is activated). This interaction alters the activity of the enzyme, ultimately leading to various physiological outcomes. Eventually, the α subunit converts GTP into GDP, reverting itself to its inactive form. A typical GPCR mechanism is shown in Figure 3.14.

Chapter 3: Nonenzymatic Protein Activity

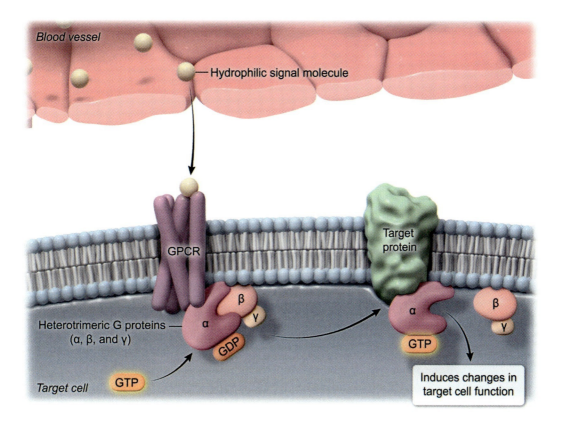

Figure 3.14 General mechanism of GPCR action.

The α subunit may bind different proteins and have different effects, although one of the most common effects is activation of adenylyl cyclase. This enzyme produces cyclic AMP, which acts as a second messenger in many signaling pathways.

GPCRs and other receptors often cause **signal cascades** such as the MAP kinase pathway. In this pathway, agonist binding to a receptor protein leads to phosphorylation of an intracellular protein. This protein then becomes capable of phosphorylating other proteins, which become capable of phosphorylating still more proteins. As a result, a single agonist binding its receptor can cause the phosphorylation of thousands of proteins, yielding a large cellular response. This phenomenon is called **signal amplification** (see Figure 3.15).

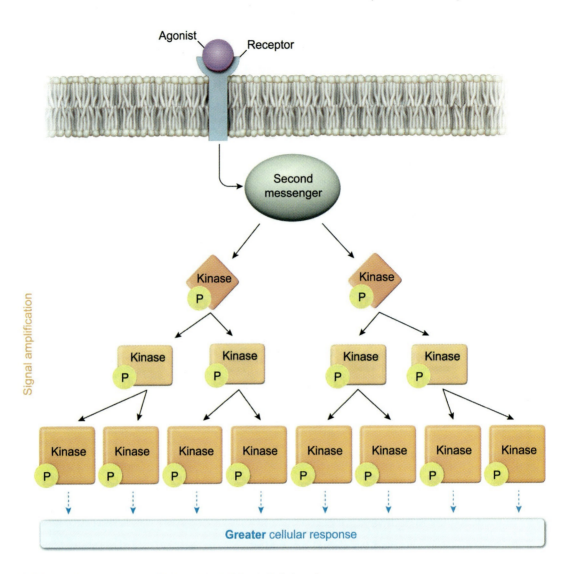

Figure 3.15 Signal cascades amplify the effect of the initial signal.

The result of a signal cascade often involves direct activation or inhibition of existing proteins, or upregulation or downregulation of gene expression. For example, insulin binding to its receptor on a liver cell leads to dephosphorylation of the enzyme phosphofructokinase-2, which activates it and increases glucose consumption.

Simultaneously, insulin binding causes transcription factor modifications that reduce synthesis of the enzyme phosphoenolpyruvate carboxykinase (PEPCK). Decreased PEPCK leads to decreased glucose production.

Concept Check 3.8

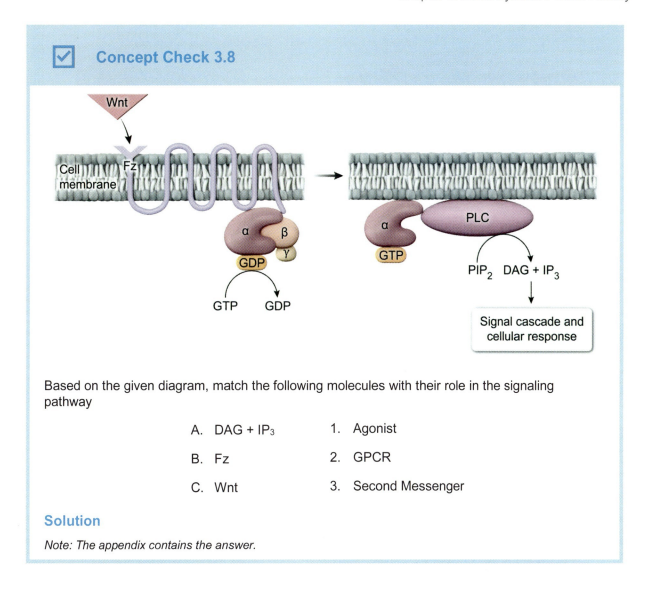

Based on the given diagram, match the following molecules with their role in the signaling pathway

    A. DAG + IP$_3$ 1. Agonist

    B. Fz   2. GPCR

    C. Wnt   3. Second Messenger

Solution
Note: The appendix contains the answer.

3.2.03 Cytosolic Receptors

Some extracellular signals are hydrophobic molecules (eg, steroid hormones, see Lesson 9.3). Because of their hydrophobicity, these molecules dissolve poorly in water. As such, they require carrier proteins to transport them through the bloodstream or other aqueous environments. Once these hydrophobic signals arrive at their target cells, the carrier protein releases them and they readily cross the cell membrane. Therefore, hydrophobic signal molecules do not require transmembrane receptors or second messengers.

Instead, hydrophobic signal molecules commonly interact with receptor proteins in the cytosol. Upon binding, cytosolic receptors are transported into the nucleus, where they act as transcription factors to alter expression of one or more genes (see Figure 3.16).

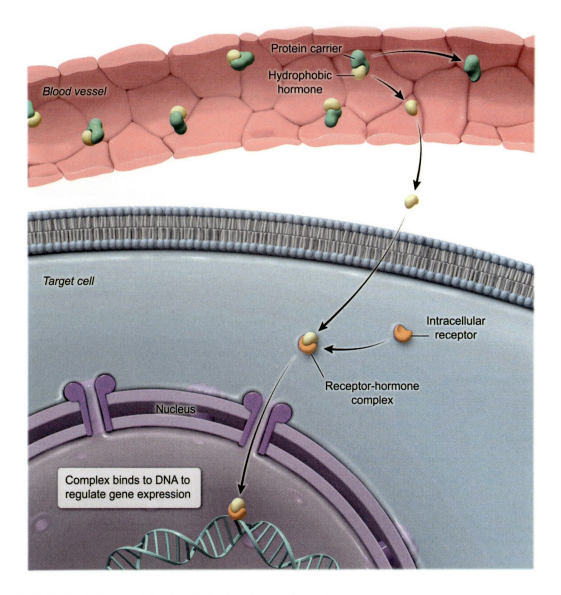

Figure 3.16 Hydrophobic agonists often bind cytosolic receptor proteins.

For instance, cortisol (a steroid hormone) binds to a cytosolic receptor and induces transcription of PEPCK, which stimulates glucose production. Consequently, cortisol in this way has an opposite metabolic effect to that of insulin.

In general, cytosolic receptors act by altering gene expression, which is a relatively slow process. In contrast, signal cascades often lead to much faster responses by activating or inactivating proteins that are *already* present in the cell. Note that although hydrophobic signals do not *require* transmembrane receptors, they *can* interact with transmembrane proteins and may induce signal cascades.

Concept Check 3.9

Which of the following molecules is more likely to interact with a cytosolic receptor in its signaling pathway?

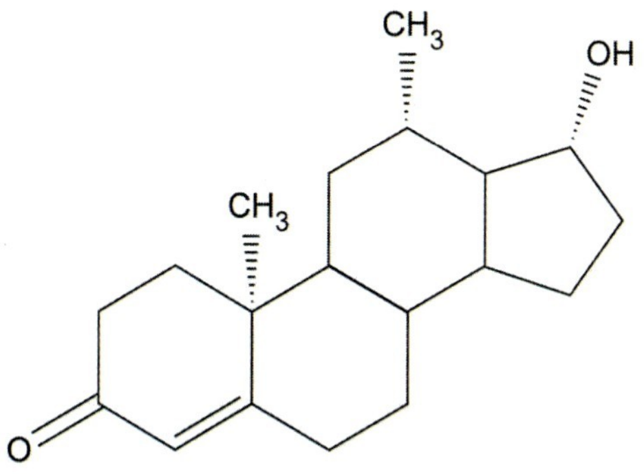

Molecule A

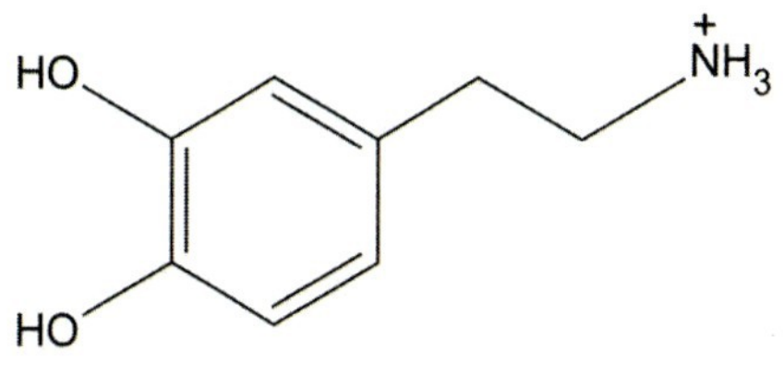

Molecule B

Solution

Note: The appendix contains the answer.

Lesson 3.3
Membrane Transport Proteins

Introduction

A cell membrane separates the interior of the cell from its environment. This allows the cell to control its internal conditions and avoid coming to equilibrium with its surroundings, which would result in cell death. However, to maintain the functions necessary for life a cell must be able to exchange matter with its environment.

Small hydrophobic molecules can cross the cell membrane without aid. However, hydrophilic molecules (eg, ions, glucose) must cross the membrane through transmembrane proteins that provide suitable pathways. These proteins are classified as channels and carriers. Molecules that allow ions to cross a membrane are called **ionophores**. These may act as channels or carriers, and may be proteins, peptides, or even synthetic molecules. This lesson explores the mechanisms and functions of channel and carrier proteins.

3.3.01 Overview of Channels

A channel is a membrane protein with a pore through it. Each channel has a pore of a specific size and composition that allows it to interact with certain molecules (usually ions) but not others. For example, a potassium channel contains a pore that is slightly larger than a potassium ion, and the wall of the pore is lined with polar functional groups and negatively charged amino acid side chains that can interact with the positive charge of potassium specifically.

Like all proteins, ion channels maintain specificity through protein-ligand interactions. For example, Na^+ has the same charge as K^+ and a smaller ionic radius, yet Na^+ ions cannot pass through a potassium channel pore (see Figure 3.17). This is because potassium channels form specific interactions that result in K^+ ion desolvation. In other words, a K^+ ion dissociates from the water molecules surrounding it and instead binds to residues within the pore. In contrast, solvated Na^+ ions are too large to enter the pore, while desolvated Na^+ ions are too small to be stabilized by the protein residues in the pore.

Chapter 3: Nonenzymatic Protein Activity

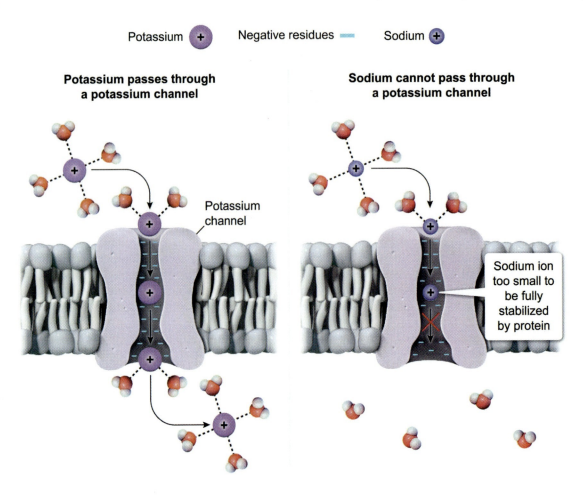

Figure 3.17 A potassium ion desolvates, moves through an ion channel, and resolvates. Other cations cannot form the specific protein-ligand interactions that stabilize potassium.

Solutes can move through channels in either direction. However, if one side of the membrane contains a greater solute concentration than the other, then it is more likely that a solute on the high-concentration side will encounter the pore. Consequently, for *uncharged* solutes, more solutes move through the pore from high concentration to low concentration than the other way around. Accordingly, the *net* flow of uncharged solutes through a pore is *down* the concentration gradient (ie, from high concentration to low concentration). In other words, channels mediate the process of facilitated diffusion.

However, most channels facilitate passage of *ions*, which are charged particles. Consequently, the electric forces exerted by these charges must be considered. For instance, consider a membrane-bound sac (ie, a vesicle) that contains a high concentration of potassium chloride on the inside and a low concentration on the outside. The membrane contains a channel that allows potassium to flow, but not chloride. Under these conditions, potassium will flow from the high concentration inside to the low concentration outside.

Initially, both sides of the membrane have a net charge of 0 as each potassium ion cancels each chloride ion. Once potassium begins to flow out, however, positive charges build up on the outer side of the membrane. The positive charges resist additional flow and drive potassium ions back toward the high-concentration *inner* side of the membrane.

Simultaneously, negative charges build up inside the vesicle due to the anions that remain. These negative charges keep potassium ions away from the channel. Eventually, flow occurs at equal rates in both directions and *net* flow stops, even though the potassium concentrations on each side of the membrane are not equal. Figure 3.18 depicts cation flow and charge buildup across a membrane.

Chapter 3: Nonenzymatic Protein Activity

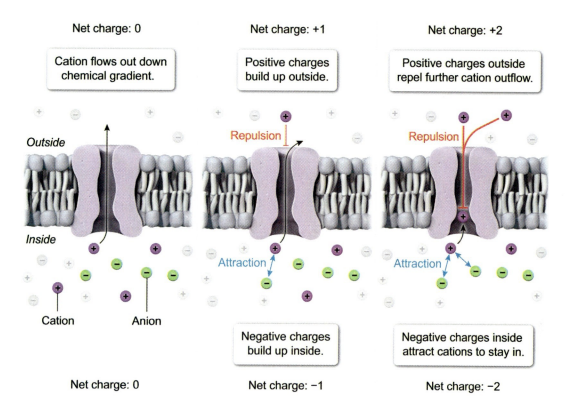

Figure 3.18 As ions flow through a channel, charge separation produces a counteracting force that resists additional flow.

Therefore, rather than flowing down their concentration gradient alone, ions flow down their electrochemical gradient, which involves both concentration *and* charge. The resulting separation of charge across a cell membrane gives rise to a **resting membrane potential** (ie, a voltage across the membrane at electrochemical equilibrium). The voltage required to maintain an uneven distribution of a given ion across a membrane (ie, the equilibrium potential) is given by the Nernst equation:

$$E = \frac{RT}{zF} \ln\left(\frac{[X]_{out}}{[X]_{in}}\right)$$

in which R is the universal gas constant, T is the Kelvin temperature, z is the charge per ion, and F is **Faraday's constant** (~96,500 C/mol). $[X]_{out}$ is the extracellular ion concentration, and $[X]_{in}$ is the intracellular ion concentration. At a physiological temperature of 310 K, the R, T, and F components can be combined, and the ln function can be converted to $\log_{10}$ by dividing by $\log_{10}(e)$. This yields

$$E = \frac{0.0615 \text{ V}}{z} \times \log_{10}\left(\frac{[X]_{out}}{[X]_{in}}\right)$$

Converting this value to millivolts (mV) yields

$$E = \frac{61.5 \text{ mV}}{z} \times \log_{10}\left(\frac{[X]_{out}}{[X]_{in}}\right)$$

Based on this equation, when the intracellular concentration of a positive ion is greater than the extracellular concentration, the equilibrium potential is *negative*. This is because cation outflow causes negative charges to build up on the *inside* until net flow stops (ie, equilibrium is reached). The opposite is true for negative ions: increased intracellular anion concentrations correspond to a *positive* membrane potential due to anion outflow.

Chapter 3: Nonenzymatic Protein Activity

 Concept Check 3.10

A vesicle contains a sodium concentration of 100 mM in its interior. The vesicle is then immersed in a solution containing 10 mM sodium. If a Na⁺ ionophore (ie, a molecule that allows sodium to cross the membrane) is added, what membrane potential is required to prevent sodium ions from diffusing out of the vesicle?

Solution

Note: The appendix contains the answer.

In general, cells exchange more than one type of ion with their environment through various transmembrane proteins. Therefore, the actual membrane potential is a function of *all* the different ions contributing to the charge separation, along with the permeability (P) of each. An ion that cannot cross the membrane has a permeability of 0. Because it cannot cross the membrane, it has no ion flow and therefore no direct effect on membrane potential. As channels permit more flow of a certain ion, the permeability for that ion increases.

Commonly, membrane potential is modeled on the assumption that sodium (Na⁺), potassium (K⁺), and chloride (Cl⁻) are the primary contributing ions. Therefore, the equation to determine the membrane potential of a typical cell is given as

$$E = 61.5 \text{ mV} \times \log_{10}\left(\frac{P_{Na^+}[Na^+]_{out} + P_{K^+}[K^+]_{out} + P_{Cl^-}[Cl^-]_{in}}{P_{Na^+}[Na^+]_{in} + P_{K^+}[K^+]_{in} + P_{Cl^-}[Cl^-]_{out}}\right)$$

In this equation, the chloride ion uses the in:out ratio rather than the out:in ratio because chloride is negatively charged. Typically, cells have a negative resting membrane potential, meaning the interior of the cell is more negatively charged (or less positively charged) than the exterior.

 Concept Check 3.11

A certain cell membrane is permeable to both K⁺ and Cl⁻, each with permeability constants of 0.2. Initially, the membrane is fully impermeable to Na⁺. The intracellular concentration of Na⁺ is less than that of the extracellular concentration. If a stimulus is added that increases the permeability of Na⁺ to 5.0, what effect will this have on the membrane potential?

Solution

Note: The appendix contains the answer.

3.3.02 Types of Channels

To establish a resting membrane potential, some ions must be allowed to diffuse into or out of cells while others are restricted. Therefore, ion flow must be carefully controlled. Consequently, many channels are gated, meaning they are closed under resting conditions and do not allow ion flow. Upon receipt of a particular signal, the channel opens through a conformational change, and ions flow until the gate closes again (eg, due to signal removal). Ion channels are categorized by whether they are gated and what type of signal is required to open the gate. Various channel types are shown in Figure 3.19.

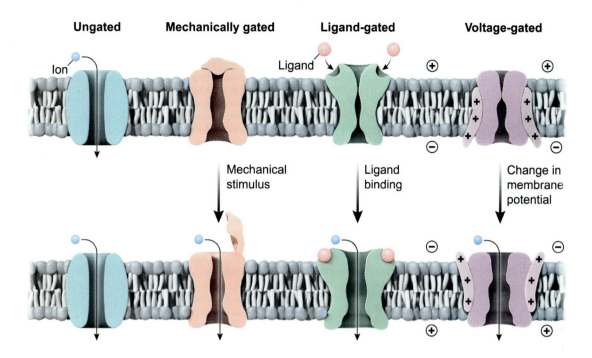

Figure 3.19 Various types of ion channels open in response to different stimuli.

Ungated Ion Channels

Some ion channels are ungated. They are always open and their specific ions constantly flow through them. These channels are also known as leak channels.

Most cells contain ungated potassium channels in their membranes. Typically, the intracellular concentration of potassium is higher than that of extracellular potassium, and ungated potassium channels allow potassium to flow out of the cell. However, because of the resting membrane potential and efforts made by the cell to actively pump potassium *into* the cytosol (see Concept 3.3.03), ungated potassium channels usually do *not* cause potassium to come to equal concentrations on both sides of the membrane.

Other ungated channels are commonly found in the membranes of intracellular organelles. These channels allow ions to flow between the cytosol and the lumen of the organelle.

Ligand-Gated Ion Channels

Ligand-gated ion channels open in response to a ligand binding to the extracellular portion of the protein. In this way, these channels act as **receptor proteins**, with the response being the opening of the gate. Prolonged exposure to the ligand can result in desensitization, which causes the gate to close even in the continued presence of ligand. However, channels that have not been desensitized remain open as long as the ligand is bound. Once the ligand dissociates from the channel, the gate closes and flow stops.

As an example, muscle cells commonly receive signals from neurons in the form of acetylcholine. When acetylcholine binds to channels on the muscle cell membrane, the channels open and allow sodium ions to flow into the cell, which produces a more *positive* membrane potential. Because the resting membrane potential is *negative*, this is called depolarization. A strong enough depolarization can trigger a cascade that results in muscle contraction (see Biology Chapter 17).

Voltage-Gated Ion Channels

Voltage-gated ion channels are sensitive to membrane potential. Each voltage-gated channel has a threshold membrane potential at which it opens. Voltage-gated channels typically open in response to membrane potentials caused by *other channels*. The opening of a voltage-gated channel may either augment or counteract the effect of other channels on membrane potential.

For example, when acetylcholine causes ligand-gated sodium channels to open in muscle cells, the resulting increase in membrane potential (ie, depolarization) causes voltage-gated sodium channels to open, allowing faster entry of sodium into the cell to enhance the depolarization.

Eventually, the voltage-gated sodium channels inactivate and close, and the slower-responding voltage-gated *potassium* channels open. Potassium flows out of the cell, decreasing the membrane potential to resting level (repolarization) or beyond (hyperpolarization).

Mechanically Gated Ion Channels

Mechanically gated ion channels open as a result of physical stimuli such as membrane stretching. These channels are involved in sensations such as touch, in which channels open in response to pressure.

Channels for Uncharged Particles

Most channels facilitate the flow of ions across a membrane, but some channels facilitate the flow of uncharged particles. For example, aquaporins (Figure 3.20) are highly conserved channels that allow the passage of water across a membrane. These channels are critical for osmosis and therefore for regulation of osmotic pressure.

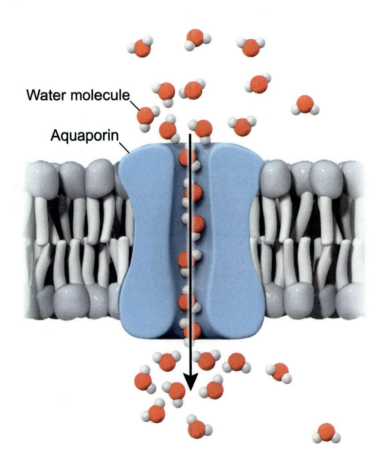

Figure 3.20 Aquaporins are membrane channels that allow water molecules to cross the membrane.

Concept Check 3.12

Two channels, Channel A and Channel B, are tested to determine how they are gated. Cells that express neither channel, one channel, or both channels in their membranes are exposed to Molecule X, an electrically neutral molecule, and the effect on membrane potential was measured. The results are summarized in the following table.

Cell membrane contains	Membrane potential
Neither channel	Does not change
Channel A only	Increases
Channel B only	Does not change
Channels A and B	Increases then decreases

Based on this information, how are Channels A and B most likely gated?

Solution

Note: The appendix contains the answer.

3.3.03 Carrier Proteins

Like channel proteins, carrier proteins are transmembrane proteins that help move solutes from one side of the cell membrane to the other. However, when channels are open, they allow *continuous* solute flow. In contrast, carrier proteins move a *discrete number* of molecules across the membrane each time they operate. In addition, while channels only move solutes *down* their electrochemical gradient, carriers may move solutes down or up their gradients (Figure 3.21).

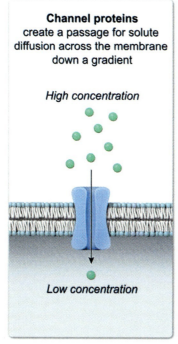
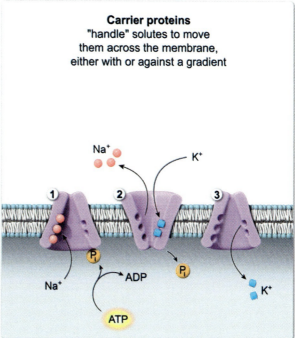

Figure 3.21 Comparison of channel and carrier proteins.

The simplest mechanism for transport by a carrier involves a single solute. Upon binding to the solute (either extracellular or intracellular, depending on the system), the carrier undergoes a conformational change that moves the solute into or out of the cell. Once the solute is released into its new location, the carrier can revert to its resting conformation. Transport of one solute in one direction is called **uniport**.

More complicated mechanisms exist for carriers that transport more than one solute per action. A carrier may bind two solutes, both on the same side of the membrane, and move them to the other side of the membrane. This general mechanism, in which both solutes move in the same direction, is called **symport**. Another mechanism, called **antiport**, involves transport of one solute into the cell while another is transported out of the cell (Figure 3.22).

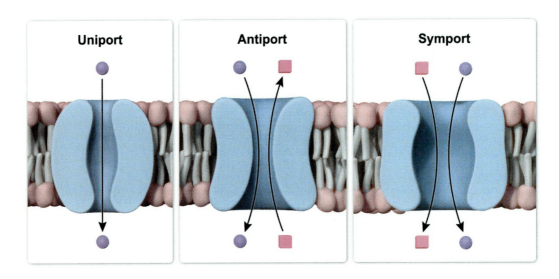

Figure 3.22 Uniport, antiport, and symport.

Carrier proteins may move solutes against their gradient (ie, from low concentration to high concentration) by coupling the movement to a source of energy. This energy coupling often involves hydrolysis of an NTP (eg, ATP, GTP). Carriers that follow this mechanism are classified as translocase enzymes (see Concept 4.2.08), and they facilitate primary active transport.

The sodium-potassium pump is an example of an antiporter that uses primary active transport. This carrier transports three sodium ions out of the cell and two potassium ions into the cell, using ATP hydrolysis to pump both sodium and potassium against their electrochemical gradients.

Other carrier proteins gain the energy needed to move a solute against its gradient by moving *another solute down its gradient*. This mechanism is called secondary active transport. Typically, the solute that moves down its gradient was previously pumped *against* its gradient by a translocase (ie, through primary active transport). Secondary active transport may occur through symport or antiport.

In addition to ions, carrier proteins also transport larger polar molecules across the membrane. GLUT2 and GLUT4, which transport glucose into cells, are carrier proteins. GLUT2 also carries glucose *out* of liver cells during gluconeogenesis (see Chapter 11). Other transporters facilitate movement of amino acids, water-soluble vitamins, and other nutrients.

> ### ✓ Concept Check 3.13
>
> Determine whether a carrier protein, a channel, or either could carry out each of the following functions:
>
> - Bring ions into a cell
> - Release ions from a cell
> - Move a solute across a membrane down its electrochemical gradient
> - Move a solute across a membrane against its electrochemical gradient
> - Allow continual flow of a solute across a membrane
>
> ### Solution
>
> *Note: The appendix contains the answer.*

Chapter 3: Nonenzymatic Protein Activity

END-OF-UNIT MCAT PRACTICE

Congratulations on completing **Unit 1: Amino Acids and Proteins**.

Now you are ready to dive into MCAT-level practice tests. At UWorld, we believe students will be fully prepared to ace the MCAT when they practice with high-quality questions in a realistic testing environment.

The UWorld Qbank will test you on questions that are fully representative of the AAMC MCAT syllabus. In addition, our MCAT-like questions are accompanied by in-depth explanations with exceptional visual aids that will help you better retain difficult MCAT concepts.

TO START YOUR MCAT PRACTICE, PROCEED AS FOLLOWS:

1) Sign up to purchase the UWorld MCAT Qbank
   IMPORTANT: You already have access if you purchased a bundled subscription.
2) Log in to your UWorld MCAT account
3) Access the MCAT Qbank section
4) Select this unit in the Qbank
5) Create a custom practice test

Unit 2 Enzymes

Chapter 4 Enzyme Activity

4.1 Enzymes are Catalysts

- 4.1.01 Enzymes Increase Reaction Rate
- 4.1.02 Enzymes Do Not Affect Thermodynamics or Equilibrium
- 4.1.03 Energetic Coupling of Reactions to Power Endergonic Reactions
- 4.1.04 Enzymes Work in Both Directions
- 4.1.05 Enzymes Are Not Consumed by Their Reactions

4.2 Enzyme Classification

- 4.2.01 Overview of Enzyme Classification
- 4.2.02 Oxidoreductases
- 4.2.03 Transferases
- 4.2.04 Hydrolases
- 4.2.05 Lyases
- 4.2.06 Isomerases
- 4.2.07 Ligases
- 4.2.08 Translocases
- 4.2.09 Receptor Enzymes

4.3 Catalytic Mechanisms of Enzymes

- 4.3.01 Enzyme-Substrate Interactions
- 4.3.02 Lock-and-Key Versus Induced-Fit Theories
- 4.3.03 Implications of the Molecular Recognition Models
- 4.3.04 Enzyme Active Site
- 4.3.05 Enzymes with Multiple Substrates
- 4.3.06 Modes of Catalysis

4.4 Optimization of Enzyme Activity

- 4.4.01 Measuring Enzyme Activity
- 4.4.02 Environmental Conditions for Enzyme Activity
- 4.4.03 Cofactors and Coenzymes

Chapter 5 Enzyme Kinetics

5.1 The Michaelis-Menten Equation

- 5.1.01 Assumptions of the Michaelis-Menten Equation
- 5.1.02 Derivation of the Michaelis-Menten Equation
- 5.1.03 Michaelis-Menten Kinetics in Nonenzyme Systems

5.2 Michaelis-Menten Parameters

- 5.2.01 The Michaelis Constant (K_M)
- 5.2.02 Turnover Number (k_{cat}) and Maximum Reaction Velocity (V_{max})
- 5.2.03 Catalytic Efficiency (k_{cat}/K_M)

5.3 Graphical Representations of Enzyme Kinetics

- 5.3.01 Michaelis-Menten Plots
- 5.3.02 Lineweaver-Burk Plots
- 5.3.03 Cooperative Enzymes

5.4 Enzyme Inhibitors

    5.4.01   Irreversible inhibitors
    5.4.02   Competitive Inhibitors
    5.4.03   Uncompetitive Inhibitors
    5.4.04   Mixed Inhibitors
    5.4.05   Noncompetitive Inhibitors
    5.4.06   Determining Inhibitor Types from Graphical Data

Chapter 6 Enzyme Regulation

6.1 Feedback Regulation

    6.1.01   Overview of Feedback Regulation
    6.1.02   Allosteric Regulation
    6.1.03   Characteristics of Regulated Enzymes
    6.1.04   Positive Feedback

6.2 Regulation by Covalent Modifications

    6.2.01   Zymogens
    6.2.02   Dynamic Covalent Modifications

Lesson 4.1
Enzymes are Catalysts

Introduction

Unit 1 discusses amino acids, the building blocks of proteins; their assembly into polypeptides; and some protein functions, such as ligand binding and ion transport across membranes. One of the most important functions of proteins is to serve as biological catalysts, increasing the rate of the chemical reactions necessary for life. Although other macromolecules, especially RNA (see Chapter 8), can also act as biological catalysts, most biological catalysts fall into a large class of proteins called enzymes. This lesson explores the catalytic properties of enzymes.

4.1.01 Enzymes Increase Reaction Rate

As discussed in General Chemistry Chapter 3, chemical reactions are considered spontaneous if they tend to proceed in the forward direction and equilibrium lies toward the products. However, spontaneous reactions do *not* necessarily proceed at a biologically relevant rate. For example, Concept 2.1.02 describes the hydrolysis of peptide bonds, which is a highly exergonic (ie, spontaneous, $\Delta G \ll 0$) process. Despite this, proteins tend to be stable and do not break apart until prompted by the cell. In other words, peptide bond hydrolysis is favorable (ie, spontaneous), but normally occurs at a *slow rate*.

In general and organic chemistry, slow reactions can often be encouraged to occur faster by subjecting reactants to high temperatures or extreme pH values. However, these extreme conditions would be lethal to most cells. Instead, living organisms employ **biological catalysts** to speed up reaction rates while maintaining the moderate physiological temperature and pH necessary for life. The vast majority of these biological catalysts belong to a class of proteins called **enzymes**.

The rate of a reaction can be calculated as the product of the concentrations of the reactants, each raised to an empirically determined power, and multiplied by a rate constant. The value of the rate constant, in turn, can be calculated via the Arrhenius equation, which describes the inverse relationship between reaction rate and the activation energy.

Activation energy (E_a) is the difference in **free energy** (ΔG) between the reactants and the transition state—large E_a values correspond to a slow reaction, and small E_a values correspond to faster reactions. Enzymes, like other catalysts, increase reaction rates by decreasing the magnitude of the E_a value (Figure 4.1).

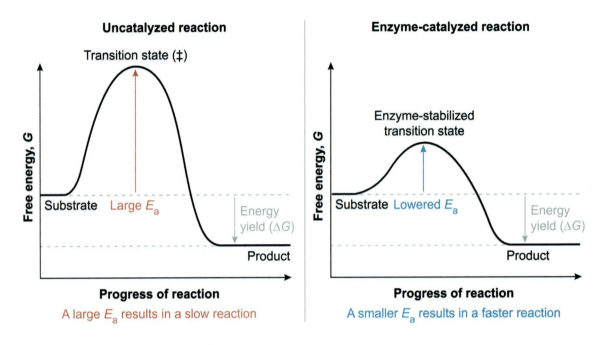

Figure 4.1 The rate of a reaction is affected by its activation energy (E_a). Enzymes and other catalysts speed up the rate of a reaction by lowering the E_a.

Concept Check 4.1

The hydrolysis of a dipeptide is carried out under three conditions. Two conditions contain different enzymes in equal amounts. Both enzymes catalyze the same hydrolysis reaction. The third condition contains no enzyme. Apart from the enzyme (or lack thereof), all three conditions are identical. The change in concentration of the products (ie, free amino acids) is shown in the following graph.

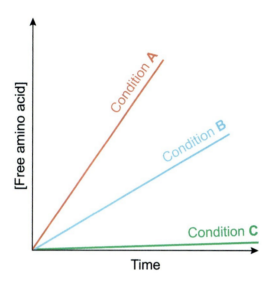

Which two conditions have an enzyme, and which condition does not? Of the two enzyme-containing conditions, what can be stated about the relative free energies of the transition states?

Solution

Note: The appendix contains the answer.

4.1.02 Enzymes Do Not Affect Thermodynamics or Equilibrium

Because enzymes alter the rates of biochemical reactions, they alter reaction **kinetics**. However, enzymes do *not* alter the **thermodynamics** of any biochemical reaction. The thermodynamics of a reaction describe the free energy (G) of its reactants and products and therefore its equilibrium and spontaneity (ie, the sign of ΔG).

None of these factors are affected by the presence of an enzyme or other catalyst. Instead, an enzyme affects reaction rate through its binding interactions with the species involved in the reaction. Free reactant and free product are, by definition, *not* bound to the enzyme; therefore, their free energies are not affected by the presence of an enzyme. Instead, the enzyme interacts only with the reaction *intermediates* and *transition states*, neither of which affects the reaction's overall ΔG.

Consider again the hydrolysis of a peptide bond, which is highly exergonic (ie, it has a large, negative ΔG). This reaction is slow (ie, it has a large activation energy E_a), but it is spontaneous under physiological conditions. The addition of a protease (an enzyme that catalyzes the hydrolysis of a peptide bond) speeds up this reaction by lowering the E_a; however, it does *not* change the reaction spontaneity or equilibrium (Figure 4.2). The hydrolysis of a peptide bond remains exergonic and spontaneous, whereas the reverse reaction, condensation (ie, formation) of a peptide bond, remains endergonic and nonspontaneous.

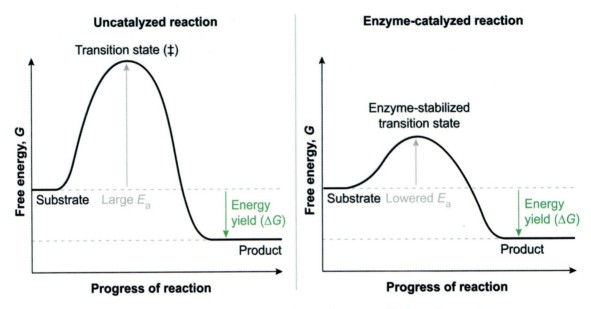

Enzymes do *not* change the thermodynamics or equilibrium of a reaction.

Figure 4.2 Enzymes do not change the free energy (*G*) of the reactants (substrates) or products of a biochemical reaction.

> 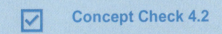 **Concept Check 4.2**
>
> Carbonic anhydrase catalyzes the reaction
>
> $$CO_2(g) + H_2O(l) \rightleftharpoons H_2CO_3(aq)$$
>
> Two aqueous solutions are prepared in an atmosphere containing CO_2. The solutions are identical in all ways except that one contains carbonic anhydrase. Once the solutions have reached chemical equilibrium, what can be said about the concentration of H_2CO_3 in the enzyme-containing vessel, relative to the H_2CO_3 concentration in the vessel without any enzyme?
>
> **Solution**
>
> *Note: The appendix contains the answer.*

4.1.03 Energetic Coupling of Reactions to Power Endergonic Reactions

Given that enzymes do not change the spontaneity of a reaction, how does a cell form peptide bonds given that their condensation is nonspontaneous? To power this endergonic reaction, enzymes can energetically couple it with a separate, exergonic reaction. As with many other endergonic reactions, an enzyme energetically couples peptide bond formation to the exergonic hydrolysis of a nucleoside triphosphate (NTP), allowing the otherwise-endergonic reaction to occur.

Changes in enthalpy (ΔH) of each step of a multistep reaction can be summed, yielding the overall enthalpy change of the net reaction. This is possible because enthalpy is a **state function**—whether the reaction is described over one step or multiple, the overall enthalpy change depends only on the initial and final states of the chemical species.

Similarly, free energy (G) is a state function, and the changes in free energy (ΔG) of multiple reactions can be summed to yield the overall free energy change of the coupled reactions. In the case of the peptide bond, a specialized enzyme called the ribosome couples peptide bond condensation to hydrolysis of guanosine triphosphate (GTP) (see Biology Lesson 2.3). Although peptide bond condensation itself is endergonic, the enzyme-mediated *combination* of condensation and GTP hydrolysis is exergonic overall, allowing peptide bond formation to occur in living organisms (see Figure 4.3).

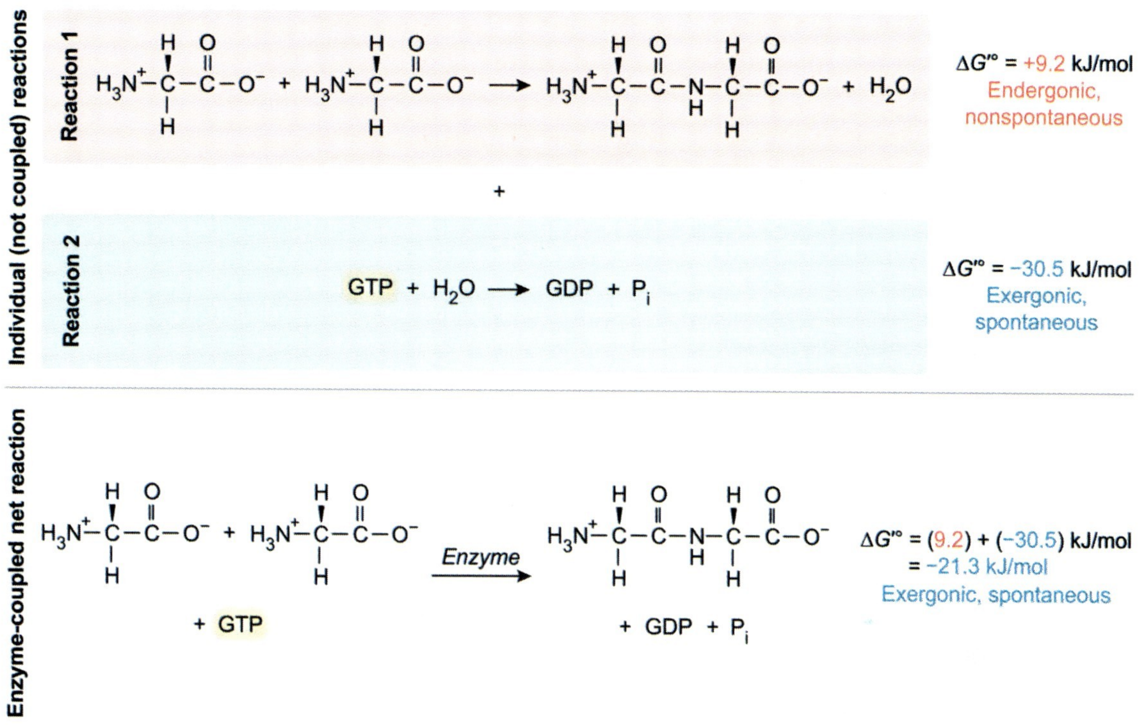

Figure 4.3 Energetic coupling of GTP hydrolysis with peptide bond condensation through an enzyme yields a negative overall free energy change.

Enzyme-coupled peptide bond condensation is *not* simply the reverse reaction of peptide bond hydrolysis. Because of energetic coupling, the formation and hydrolysis of peptide bonds are *separate* reactions catalyzed by *separate* enzymes. Peptide condensation involves both the formation of a peptide bond *and* the conversion of GTP to GDP, whereas peptide hydrolysis uses water to break a peptide bond *without* converting GDP to GTP.

Reactions that reverse one aspect of a process (eg, peptide bond formation and breaking) without reversing other aspects (eg, interconversion of GTP and GDP) are called bypass reactions (see Concept 11.2.01). Because energetic coupling often consumes a cellular energy source (eg, a nucleoside triphosphate such as ATP or GTP), the enzymes in a bypass reaction pair must be regulated to prevent a futile cycle and the wasting of energy.

Concept Check 4.3

The enzyme hexokinase catalyzes the transfer of a phosphate group from ATP to glucose.

Glucose + ATP → Glucose 6-phosphate + ADP

What is the $\Delta G'^\circ$ for this reaction? Relevant $\Delta G'^\circ$ values for two related reactions are listed in the following table.

Reaction	$\Delta G'^\circ$ (kJ/mol)
ATP + H_2O → ADP + P_i	−30.5
Glucose 6-phosphate + H_2O → Glucose + P_i	−13.8

Solution

Note: The appendix contains the answer.

Note that although one of the two processes in a coupled reaction may involve hydrolysis of an NTP and the other involves condensation, it is possible for the overall reaction *not* to involve hydrolysis *or* condensation, even temporarily. For example, phosphorylation of glucose is coupled to ATP hydrolysis, yet no water is used in any step of the mechanism that transfers a phosphate from ATP to glucose. Nevertheless, the $\Delta G'^\circ$ calculated from the sum of ATP hydrolysis and glucose phosphorylation reactions is still an appropriate description of the free energy change of the overall direct transfer reaction.

This is because $\Delta G'^\circ$ is a state function—its value depends only on the initial and final states of the system; it does *not* depend on the path taken to reach those states (Figure 4.4). The initial state (ie, Glucose + ATP) and the final state (Glucose 6-phosphate + ADP) are the same regardless of path, and the reaction intermediates H_2O and P_i are unchanged at the end of the reaction. Therefore, the $\Delta G'^\circ$ value is the same, regardless of whether the mechanism occurs through direct phosphate transfer or through a two-step hydrolysis-condensation mechanism.

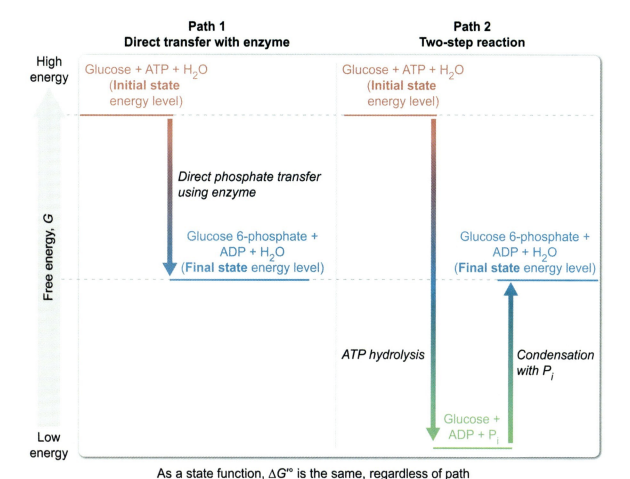

Figure 4.4 Free energy change (ΔG) is a state function whose value depends only on initial and final state.

4.1.04 Enzymes Work in Both Directions

Concept 4.1.01 explains that enzymes catalyze reactions by lowering the free energy of the transition state and therefore the activation energy (E_a) of the *forward* reaction. Concept 4.1.02 explains that enzymes do not change the thermodynamics of a reaction, and therefore the free energies of the reactants and the products are unaffected. Taken together, this means that the E_a of the *reverse* reaction must *also* be lowered by the enzyme, and therefore enzymes catalyze both the forward *and* the reverse reaction (Figure 4.5).

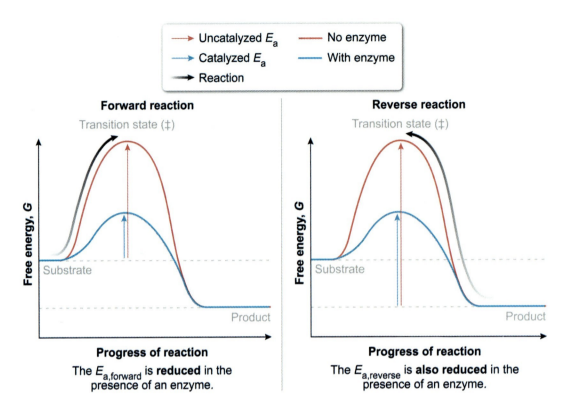

Figure 4.5 Enzymes decrease the activation energies of both the forward and the reverse reactions.

All enzymes can, at least in theory, catalyze reactions in both directions. Most enzyme-catalyzed reactions are reversible and can readily proceed in either direction depending on cellular conditions. Irreversible reactions are technically also catalyzed in both directions (ie, E_a decreases), but the reverse reaction is too energetically unfavorable to proceed. For this reason, irreversible processes require bypass reactions. Metabolic pathways involving both reversible and irreversible reactions are discussed in Unit 4.

The reversibility of a reaction can be determined from its ΔG, the change in free energy under ambient conditions. Reversible reactions have a ΔG near 0 kJ/mol. When ΔG is near 0 kJ/mol, the reaction is near equilibrium, and the forward and reverse reactions both proceed at nearly equal rates (see General Chemistry Chapter 5). The free energy change ΔG of a biochemical reaction is related to the standard free energy change $\Delta G'^\circ$ by the equation

$$\Delta G = \Delta G'^\circ + RT \ln Q'$$

where Q' is the reaction quotient (ie, the starting ratio of products to reactants).

At equilibrium, $\Delta G = 0$ and $Q' = K_{eq}'$, so

$$0 = \Delta G'^\circ + RT \ln K_{eq}'$$
$$\Delta G'^\circ = -RT \ln K_{eq}'$$

For reversible reactions, small changes to the concentrations of either the reactants or products can cause the term ($RT \ln Q'$) to change significantly relative to $\Delta G'^\circ$, which can cause ΔG of the reaction to switch from positive to negative or vice versa. If this occurs, the net direction of the reaction will *reverse* in accordance with Le Châtelier's principle. In contrast, a spontaneous but *irreversible* reaction has a negative ΔG *far* from zero. Even with substantial changes to reactant and product concentrations, the observed ΔG is likely to remain negative, and therefore the net direction of the reaction is unlikely to change (Figure 4.6).

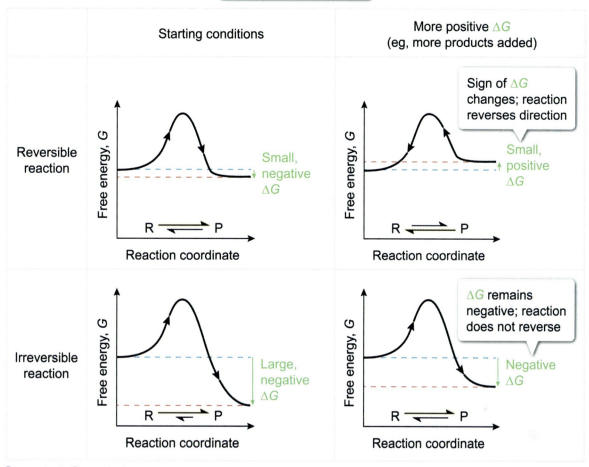

Figure 4.6 Reversible reactions have a ΔG near zero and can reverse directions with small perturbations. Irreversible reactions have a large, negative ΔG and do not easily reverse directions.

Note that the ΔG′° value used in biochemical contexts differs slightly from the standard state free energy change ΔG° used in nonbiological systems. As discussed in General Chemistry Chapter 3, the standard state free energy change (ΔG°) describes the free energy change when all dissolved reactants and products are initially at 1 M concentration. Biochemical standard state, indicated by the prime symbol (′), is similar, except that certain species are held constant at values other than 1 M (eg, water is kept at 55.5 M and protons are kept at 1×10^{-7} M, indicating an aqueous reaction at pH 7.0).

Concept Check 4.4

The enzyme aldolase catalyzes the reaction

$$\text{F16BP} \rightarrow \text{DHAP} + \text{G3P}$$

with a $\Delta G'^\circ$ value of +23.8 kJ/mol. Physiological concentrations of each reactant and product are given in the following table:

Metabolite	Concentration (µM)
F16BP	7.60
DHAP	140
G3P	6.70

Using these values, the reaction quotient Q' is 1.2×10^{-4} and $\ln Q'$ is −9.0. At a temperature of 298 K, RT has a value of 2.48 kJ/mol. At 298 K and physiological concentrations, is the reaction reversible, irreversible, or unlikely to occur? Use ±5 kJ/mol as a cutoff for a ΔG being near zero.

Solution
Note: The appendix contains the answer.

4.1.05 Enzymes Are Not Consumed by Their Reactions

A central property of all catalysts, including enzymes, is that they are *not* consumed by the reactions they catalyze. Enzymes facilitate reactions, but they are not changed by those reactions. This means that an enzyme can catalyze a reaction multiple times without any change in its catalytic ability (Figure 4.7).

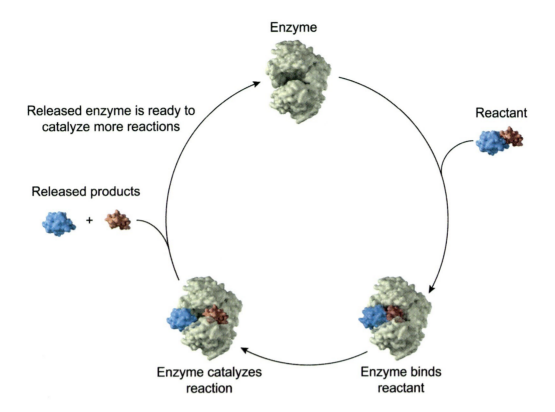

Figure 4.7 Enzymes are not changed by the reactions they catalyze. Upon release of all products, an enzyme is immediately able to catalyze a new reaction.

Although enzymes are unchanged at the *end* of the reaction, they may be *temporarily* changed during the reaction (Figure 4.8). In Lesson 4.3, several mechanisms of enzyme catalysis are discussed, including covalent catalysis, in which a nucleophilic amino acid residue (see Concept 1.3.06) on the enzyme forms a covalent bond to one of the reactants. However, this covalent bond must be broken before the completion of the reaction so that the enzyme can catalyze additional reactions. In other words, any changes to the enzyme during the reaction must be only temporary.

Figure 4.8 Enzymes may form transient bonds within a catalytic mechanism, but these bonds must break by the end of the reaction.

Because the enzyme at the end of a reaction is the same species as it was at the beginning of the reaction, the enzyme can be omitted from the left and right sides of a written net reaction (Figure 4.9). Instead, enzyme names are often written above or below the reaction arrow to indicate their inclusion as a reaction condition.

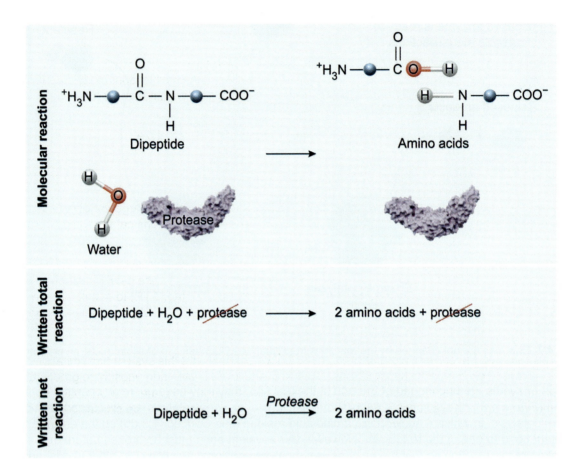

Figure 4.9 In the written net reaction for peptide hydrolysis, "protease" is written above the reaction arrow.

Lesson 4.2
Enzyme Classification

Introduction

Lesson 4.1 discusses many of the catalytic properties of enzymes. The next lessons delve deeper into enzymatic action in living systems and explore enzyme-catalyzed reactions, measurements, and optimization of enzyme activity.

This lesson covers different classes of enzymes and the reactions they catalyze.

4.2.01 Overview of Enzyme Classification

Enzymes catalyze a range of biochemical reactions. These reactions act on a large variety of substrates to produce a similarly large variety of products. However, the *types* of reactions catalyzed are much less varied, and enzymes can be grouped by the type of reaction they catalyze.

Early attempts to classify enzymes involved naming an enzyme based on its reaction and adding the suffix *–ase* (Figure 4.10). For example, enzymes with names that end in *kinase* catalyze reactions in which a phosphate group is transferred from a nucleoside triphosphate (eg, ATP) to a substrate. In contrast, enzymes with names that end in *phosphatase* catalyze reactions in which a phosphate group is *removed* from a substrate by addition of water.

Similar rationale was used to name and group other enzymes; however, this system has limitations that become apparent when classifying larger groups. For example, if kinases add phosphate groups to a substrate while phosphatases break apart an already-phosphorylated substrate, how should phosphorylases—which break apart substrates by adding a phosphate group—be classified?

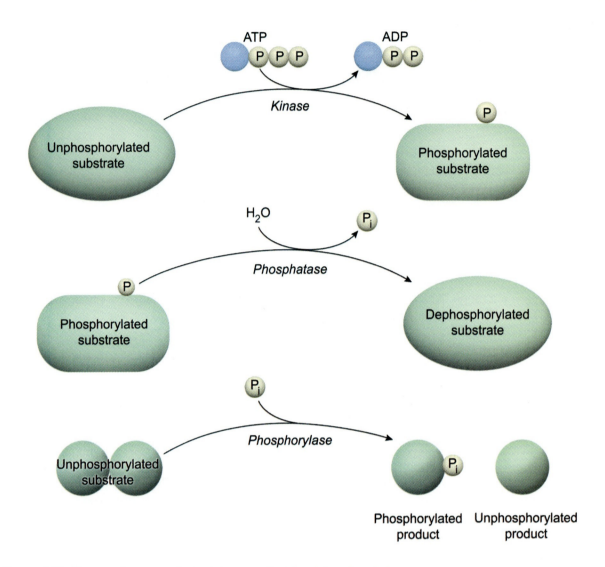

Figure 4.10 Classes of enzymes that catalyze reactions involving phosphate.

In 1961, the Enzyme Commission classified all enzymes into six major classes based on the catalytic mechanism of their major reaction. In 2018, a seventh class was added to this list (Table 4.1).

Table 4.1 The seven major classes of enzymes.

Class	Reaction type	Example mechanism
Oxidoreductases	Oxidation-reduction	$A^- + B \rightleftharpoons A + B^-$
Transferases	Functional group transfer	$A-B + C \rightleftharpoons A + B-C$
Hydrolases	Hydrolytic cleavage of a molecule into two molecules	$A-B + H_2O \rightleftharpoons A-H + B-OH$
Lyases	Group removal or addition with electron rearrangement	$\overset{X}{\underset{\|}{}}\overset{Y}{\underset{\|}{}}$ $A-B \rightleftharpoons A=B + X-Y$
Isomerases	Structural rearrangement of a molecule	$\overset{X\ Y}{\underset{\|\ \|}{A-B}} \rightleftharpoons \overset{Y\ X}{\underset{\|\ \|}{A-B}}$
Ligases	Joining two molecules to form one via bond formation and ATP hydrolysis	$ATP + A + B \rightleftharpoons A-B + ADP + P_i$
Translocases	Primary active transport of a solute across a cell membrane	$ATP + A_{side\ 1} \rightleftharpoons ADP + P_i + A_{side\ 2}$

An enzyme's major classification is determined by the catalytic mechanism of its principal reaction; consequently, many enzymes have been reclassified over time as details about their mechanism have been elucidated through scientific research. Details about the mechanisms that define each major grouping are given in the following concepts of this lesson.

4.2.02 Oxidoreductases

The first major class of enzymes is the **oxidoreductase** class. Members of this class catalyze oxidation-reduction, or redox, reactions, which transfer electrons from one molecule to another (see General Chemistry Lesson 9.1 and Organic Chemistry Lesson 5.7).

Because redox reactions require a substrate to donate electrons *and* a substrate to receive electrons, all oxidoreductases are multisubstrate enzymes. Figure 4.11 shows a typical oxidoreductase-catalyzed reaction, in which a reducing agent and an oxidizing agent (the substrates) react and are converted to oxidized and reduced products, respectively. Several enzyme-catalyzed reactions deviate slightly from this model (eg, by reacting three substrates or releasing three products).

Chapter 4: Enzyme Activity

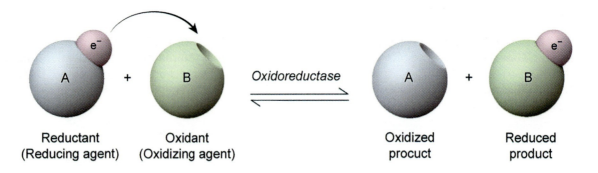

Figure 4.11 A typical reaction catalyzed by an oxidoreductase enzyme. Electrons are transferred from the reducing agent to the oxidizing agent.

In a typical oxidoreductase-catalyzed reaction, one reactant is converted into a product that either *continues down other metabolic pathways* or itself becomes *the final product* of a pathway. In other words, the original substrate is *not* regenerated. This reactant is typically referred to as the true **substrate** of the oxidoreductase enzyme.

The second reactant is changed by the reaction, so it is also technically a substrate. However, this substrate typically cycles between various oxidoreductase enzymes, changing from its oxidized to its reduced state and back, which regenerates it. As such, it is often referred to as a **redox cofactor** (if inorganic) or a **redox coenzyme** (if organic). Figure 4.12 shows how a single redox cofactor can cycle between enzymes to assist various oxidoreductase reactions.

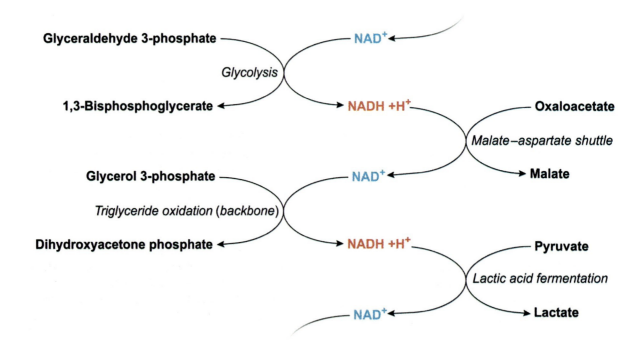

Figure 4.12 Nicotinamide adenine dinucleotide (NAD^+/NADH) cycles between its oxidized and reduced forms within a cell.

Recognizing Oxidoreductase Enzymes and Their Reactions

Enzymes in the oxidoreductase class can be recognized both by the reactions they catalyze and by the name of the enzyme. As discussed previously, oxidoreductase-catalyzed reactions often involve a redox cofactor or coenzyme that changes oxidation state. For example, any reaction that interconverts NAD^+ with NADH, FAD with $FADH_2$, or ubiquinone with ubiquinol can be confidently assigned as a redox reaction catalyzed by an oxidoreductase. Some common redox factors are given in Table 4.2.

Table 4.2 Common redox cofactors and coenzymes used by oxidoreductase enzymes.

Name	Oxidized form (oxidizing agent)		Reduced form (reducing agent)
Nicotinamide adenine dinucleotide (phosphate)	$NAD(P)^+$	$\xrightleftharpoons[-2e^- \text{ (oxidation)}, -2H^+]{+2e^- \text{ (reduction)}, +2H^+}$	$NAD(P)H + H^+$
Flavin adenine dinucleotide	FAD	$\xrightleftharpoons[-2e^- \text{ (oxidation)}, -2H^+]{+2e^- \text{ (reduction)}, +2H^+}$	$FADH_2$
Coenzyme Q	Ubiquinone	$\xrightleftharpoons[-2e^- \text{ (oxidation)}, -2H^+]{+2e^- \text{ (reduction)}, +2H^+}$	Ubiquinol
Transition state metals (eg, iron)	Fe^{3+}	$\xrightleftharpoons[-e^- \text{ (oxidation)}]{+e^- \text{ (reduction)}}$	Fe^{2+}
Disulfides	R–S–S–R'	$\xrightleftharpoons[-2e^- \text{ (oxidation)}, -2H^+]{+2e^- \text{ (reduction)}, +2H^+}$	R–SH HS–R'

Importantly, some reactions involve these molecules but are *not* redox reactions. For example, NAD^+ (nicotinamide adenine dinucleotide) can be used as a source of ADP-ribose to modify a target protein by ADP-ribosylation. In this case, NAD^+ is converted into free nicotinamide and an ADP-ribosyl group, rather than into NADH, and so the enzyme that catalyzes this transfer is not considered an oxidoreductase. Reactions with redox coenzymes are only considered oxidoreductase-catalyzed reactions if the coenzyme changes oxidation state.

Aside from the involvement of a redox cofactor, oxidoreductase-catalyzed reactions can be identified by the oxidation or reduction of the substrate. An alcohol, for example, is oxidized to a carbonyl (ie, a ketone or an aldehyde), and aldehydes are oxidized to carboxylic acids or their derivatives (eg, esters, thioesters). Conversely, carboxylic acids (and their derivatives) are reduced to aldehydes, and carbonyl carbons are reduced to alcohols. Other common substrate oxidations or reactions are given in Table 4.3.

Table 4.3 Common examples of substrate changes catalyzed by oxidoreductase enzymes.

Reduced form		Oxidized form	Notes
—C(OH)(H)—	−2e⁻ (oxidation), −2H⁺ / +2e⁻ (reduction), +2H⁺	C=O	
—C(=O)—H (aldehyde)	+RXH, −2e⁻ (oxidation), −2H⁺ / −RXH, +2e⁻ (reduction), +2H⁺	—C(=O)—X—R	If X = O, product is an ester. If X = S, product is a thioester.
—CH—CH— (with H's)	−2e⁻ (oxidation), −2H⁺ / +2e⁻ (reduction), +2H⁺	C=C	Only redox if *both* carbons gain or lose covalent bonds to hydrogen
R—SH HS—R'	−2e⁻ (oxidation), −2H⁺ / +2e⁻ (reduction), +2H⁺	R—S—S—R'	Disulfides
—CH₂—	+H₂O, −2e⁻ (oxidation), −2H⁺ / −H₂O, +2e⁻ (reduction), +2H⁺	—C(OH)(H)—	Carbon is sp^3 at beginning and end of reaction

The name of an enzyme can also provide an important clue about an enzyme's classification. Any enzyme that includes the word *oxidoreductase* very likely falls into that class, but typically the name will focus on one direction (ie, either oxidation *or* reduction) and include terms such as *oxidase* (eg, cytochrome P450 oxidase), oxygenase (eg, cyclooxygenase), or *reductase* (eg, ribonucleotide reductase).

Additionally, oxidoreductases tested on the exam often contain the term *dehydrogenase*. As shown in Table 4.2 and Table 4.3, the gain or loss of electrons is commonly accompanied by the gain or loss of hydrogen nuclei. Therefore, dehydrogenation reactions (ie, loss of hydrogen nuclei *and* electrons) are oxidation reactions. Conversely, reduction reactions that result in the gain of both electrons *and* hydrogen nuclei are hydrogenation reactions. However, because enzymes catalyze reactions in both directions (see Concept 4.1.04), many enzymes that catalyze reduction of the substrate (ie, hydrogenation) are still referred to as dehydrogenases.

> ☑ **Concept Check 4.5**
>
> Identify the redox coenzyme (if any) in each of the following reactions and whether it becomes oxidized or reduced during the reaction.
>
>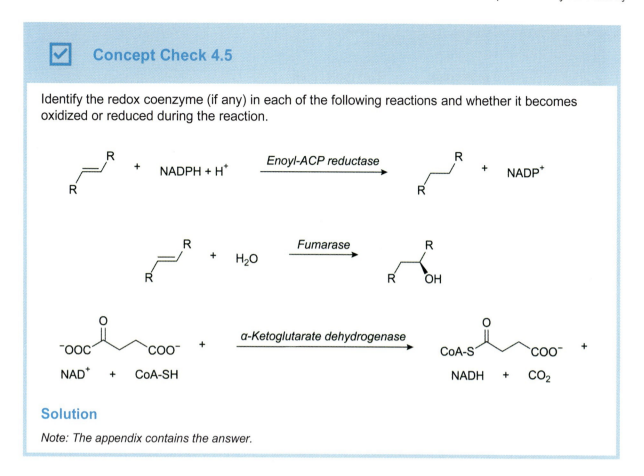
>
> **Solution**
> *Note: The appendix contains the answer.*

4.2.03 Transferases

Enzymes within the transferase class catalyze the transfer of a functional group from one molecule *to another molecule* (Figure 4.13).

Figure 4.13 Schematic of transferase activity.

Transferases require *two substrates* (ie, they are bisubstrate enzymes) and release *two products*. The substrate that loses a functional group during the reaction is often called the functional group *donor*. The substrate that receives the functional group is the *acceptor* (Table 4.4).

Depending on the enzyme, the functional group being transferred can vary greatly in size. Methyltransferases, for example, transfer small $-CH_3$ substituents whereas acyltransferases or aminoacyltransferases transfer much larger fatty acids or amino acids, respectively. Some enzymes can even transfer entire macromolecules such as proteins or polysaccharides.

Table 4.4 Examples of reactions catalyzed by transferases.

Two molecules → Two molecules				Example
Donor	Acceptor			
ATP	+ Substrate	→ ADP	+ Phospho-substrate	Kinases (Phosphoryl group transferred)
Acetyl-CoA	+ Oxaloacetate	→ CoA-SH (Coenzyme A)	+ Citrate	Citrate synthase (Acetyl group transferred)
Glycogen$_n$	+ Inorganic phosphate	→ Glycogen$_{n-1}$	+ Glucose 1-phosphate	Glycogen phosphorylase (Glucose subunit transferred)
NTP	+ RNA$_n$	→ Pyrophosphate	+ RNA$_{n+1}$	RNA polymerase (Nucleotide transferred)

Given the large variation in functional group size, it is not uncommon for the functional group being transferred to be larger than the molecules it is being transferred to or from. This can complicate recognition of a transferase-catalyzed reaction. The following section provides advice on recognizing transferase enzymes and the reactions they catalyze.

Recognizing Transferase Enzymes and Their Reactions

Transferases can be recognized both by the reactions they catalyze and by the name of the enzyme. As previously discussed, transferase reactions typically consume two substrates (one donor and one acceptor) and produce two products. The reaction involves the transfer of a functional group from the donor to the acceptor. One important exception is that water cannot be the acceptor. Transferase-like reactions that would involve water as an acceptor are instead classified as hydrolysis reactions and are catalyzed by hydrolases (Concept 4.2.04).

Many transferases can be identified by their names. For example, acyltransferases are transferases that move fatty acyl groups from one molecule to another. Glycosyltransferases transfer carbohydrates by shifting a glycosidic bond from one molecule to another.

Some transferases do not include the term *transferase* in their name but instead have some other name specifying the subclass of transferase they belong to. Several of these subclasses are given in the next section. Knowing certain subclass names is helpful for identifying transferases. However, the *specific enzymatic reactions described in the following section are only a means of demonstrating functional group transfer and do not need to be memorized.*

Special Classes of Transferases

Kinases are transferases that transfer the γ-phosphoryl group of ATP or another nucleoside triphosphate to a substrate molecule (the acceptor). The result is ADP and a phosphorylated molecule (Figure 4.14). Note that some kinases facilitate the reverse reaction (ie, phosphorylation of ADP using a phosphosubstrate as a donor) under physiological conditions (eg, phosphoglycerate kinase and pyruvate kinase during glycolysis).

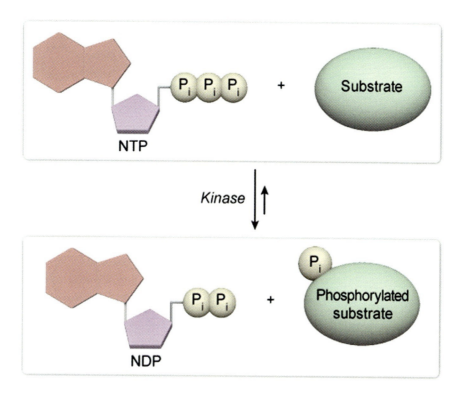

Figure 4.14 Kinases transfer the γ-phosphate group from ATP to a substrate (or vice versa).

Phosphorylases are transferases with an interesting function. They use phosphate groups to break apart another molecule in a process called **phosphorolysis**, similar to hydrolases breaking apart molecules using water (see Concept 4.2.04). A typical phosphorolysis reaction yields two product molecules, one of which is covalently bonded to the phosphate group (Figure 4.15).

Although they are similar to hydrolases, phosphorylases are considered transferases because a group is transferred from the parent substrate to the phosphate group. For example, consider the enzyme glycogen phosphorylase. In the reaction catalyzed by this enzyme, P_i breaks off a glucose subunit from the glycogen polymer (a large molecule composed of multiple glucose units), forming a shortened glycogen and glucose 1-phosphate (two molecules become two different molecules). In this reaction, glycogen acts as a glucose donor, and the glucose is donated to the phosphate, which acts as an acceptor.

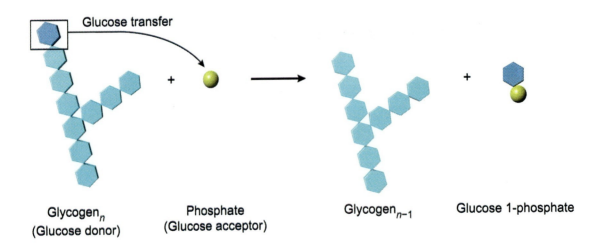

Figure 4.15 Glycogen phosphorylase is an example of a transferase enzyme in which the acceptor is smaller than the transferred group.

Transaminases are enzymes that transfer amine groups. They are also sometimes called aminotransferases. Typically, the transferase moves the α-amine group of an amino acid onto an α-keto acid (Figure 4.16).

Although this process can be considered a redox reaction, the most important physiological result is the amine group transfer. Therefore, transaminases are considered transferases.

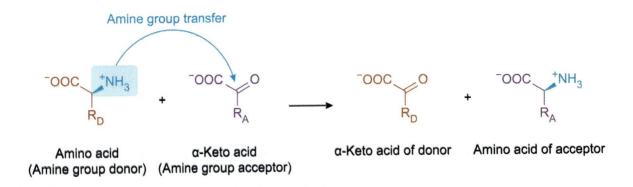

Figure 4.16 Transaminases are examples of transferase enzymes.

Polymerases are another example of transferases. DNA and RNA polymerases transfer a nucleotide from a dNTP or an NTP, respectively, onto a growing nucleic acid chain. The products are a nucleic acid, which has accepted a nucleotide, and a pyrophosphate (PP_i), which has donated a nucleotide (Figure 4.17).

Although polymerases may use NTPs, they are *not* ligases (Concept 4.2.07) because the nucleotide of the NTP is incorporated into the growing molecule. In contrast, the NTPs used by ligases are independent of the two molecules being joined.

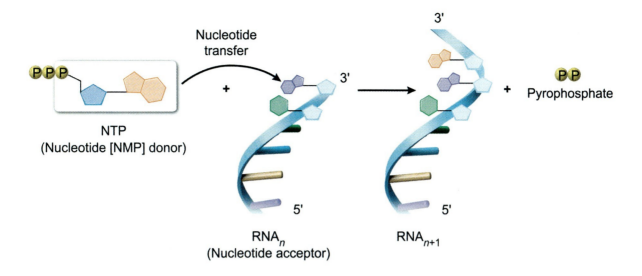

Figure 4.17 Polymerases are examples of transferase enzymes.

4.2.04 Hydrolases

Hydrolases catalyze reactions in which water (H_2O) is used to break apart a sigma bond in a substrate molecule. This type of reaction is known as a hydrolysis reaction (Figure 4.18).

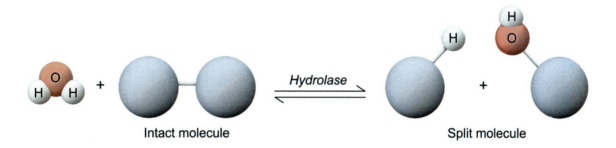

Figure 4.18 Hydrolases use water to break a sigma bond.

Recognizing Hydrolase Enzymes and Their Reactions

Hydrolysis reactions often result in one substrate being broken into two products. However, if a bond within a *cyclic* structure is broken, only one linearized product forms. This occurs with lactonase in the oxidative phase of the pentose phosphate pathway, for example.

Most hydrolases are simply given the name of the substrate or functional group being acted upon with the suffix –ase (eg, esterases hydrolyze esters, glycosidases hydrolyze glycosidic bonds), making it difficult to determine whether an enzyme is a hydrolase based on its name alone. Instead, hydrolases can be recognized by chemical equation of the reaction they catalyze. The following sections list some common hydrolases and potential pitfalls in identifying an enzyme as a hydrolase.

Commonly Encountered Hydrolases

Proteases are hydrolases that typically catalyze the hydrolysis of a peptide bond between two amino acids residues, as shown in Figure 4.19. Proteases are usually sequence-specific and only hydrolyze specific peptide bonds based on the identities or properties of the surrounding amino acids. In contrast to most other enzymes, some proteases have common names that do not end in –ase, so their status as proteases should be remembered. These enzymes include pepsin, trypsin, and chymotrypsin.

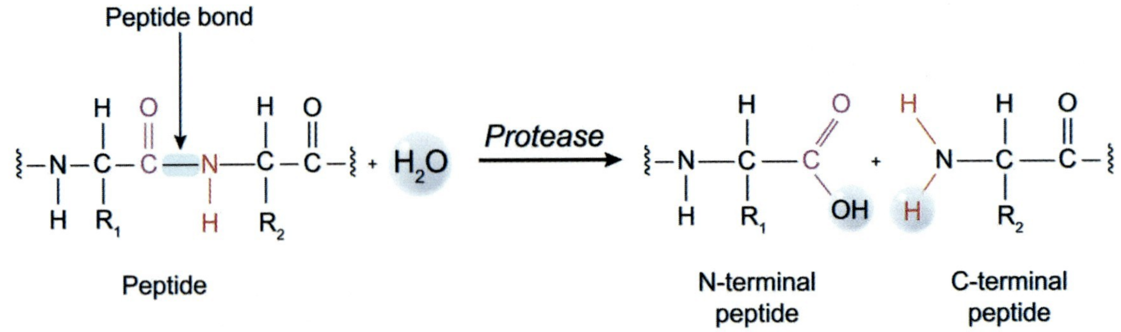

![Peptide bond hydrolysis diagram]

Figure 4.19 Proteases are hydrolases that hydrolyze peptide bonds.

Phosphatases are enzymes that hydrolyze the phosphate group from a phosphorylated substrate. Recall from Concept 4.2.03 that kinases add a phosphate group to substrates through a transfer reaction involving ATP. This involves the removal of a phosphate group from ATP, which is energetically very favorable. The reverse reaction, moving a phosphate back onto ADP to regenerate ATP, is typically energetically *unfavorable*. To circumvent this unfavorable reaction, phosphatases hydrolyze the added phosphate using water; no ATP is involved in a typical phosphatase-catalyzed reaction (Figure 4.20).

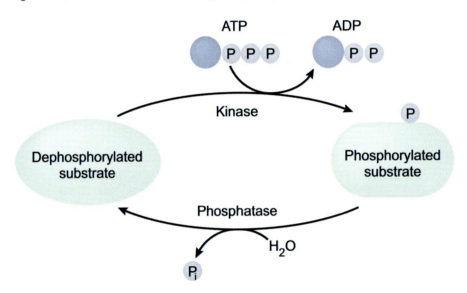

Figure 4.20 Kinases add phosphate groups to a substrate through a transfer reaction involving ATP. Phosphatases remove phosphate groups from substrates by hydrolysis.

Potential Pitfalls in Hydrolase Recognition

Although water is necessary in a hydrolase-catalyzed reaction, water's involvement *alone* is not sufficient to identify a reaction as hydrolysis. For example, some enzymes add water across a double bond. Although the *pi* bond breaks, the atoms involved remain connected afterward through the *sigma* bond, and the sigma bond is *not* broken. These reactions are catalyzed by hydratases (and dehydratases), which are a class of lyase (Concept 4.2.05), *not* hydrolases.

Furthermore, some enzyme-catalyzed reactions *do* involve hydrolysis, but only as a secondary function. These enzymes are therefore grouped in another class. For example, all ligases (Concept 4.2.07) involve hydrolysis of an NTP, but this hydrolysis is coupled to the joining of two substrates together, which is their primary purpose. Many enzymes in other classes similarly hydrolyze an NTP to power an otherwise-endergonic reaction and are classified according to the enzyme's primary function.

Enzymes should only be classified as hydrolases if hydrolysis alone is the main function of the enzyme (eg, phosphatases, proteases, phosphodiesterases, glycosidases). Enzymes that hydrolyze an NTP or another substrate to power a separate reaction are *not* typically classified as hydrolases.

> ☑ **Concept Check 4.6**
>
> The following enzyme-catalyzed reactions occur in glycolysis or gluconeogenesis. Both reactions remove a phosphate group from a substrate, and the relevant phosphate group is shown in red. For each reaction, categorize the enzyme first as a transferase or as a hydrolase, and then further classify it as a kinase, a phosphatase, or a phosphorylase.
>
> A.
>
> B.
>
> **Solution**
>
> *Note: The appendix contains the answer.*

4.2.05 Lyases

Lyases catalyze either the removal or addition of a functional group, depending on the direction of the reaction. Importantly, functional group removal must not be hydrolytic or tied to a redox reaction; otherwise, the enzyme would be classified as a hydrolase (Concept 4.2.04) or an oxidoreductase (Concept 4.2.02), respectively.

Instead, when lyases remove functional groups they leave behind a **double bond** (ie, they catalyze an elimination reaction) or a **cyclic ring**. Similarly, when lyases act in the opposite direction and *add* functional groups, they do so across a double bond (ie, they catalyze an addition reaction) or by breaking open a ring (Figure 4.21).

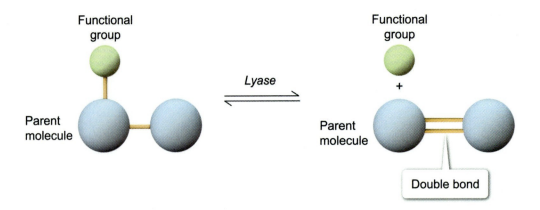

Figure 4.21 Lyases catalyze removal or addition of a chemical group by forming or breaking a double bond or ring.

Similar to transferases (Concept 4.2.03), the functional group involved in a lyase-catalyzed reaction can vary greatly in size. In some cases, the functional group added is as small as a water molecule (H–OH) adding across a double bond of a much larger molecule. In other cases, the group being added (or the group being eliminated) is just as large as or larger than the molecule with the double bond or ring.

Recognizing Lyase Enzymes and Their Reactions

The double bond or the ring is a characteristic feature of a lyase-catalyzed reaction. However, many biomolecules have double bonds or rings that are not involved in every reaction, so classifying an enzyme based only on the *presence* of a double bond or ring in its reaction can lead to errors. Instead, a lyase-catalyzed reaction can be more easily identified by examining the number of substrates and products.

Lyase enzymes catalyze addition and elimination reactions. Addition reactions join two molecules into one, so lyase-catalyzed additions have two substrates and one product. Elimination reactions remove a group from a molecule, so lyase-catalyzed eliminations have one substrate but two products, as shown in Table 4.5.

Table 4.5 Examples of lyase-catalyzed reactions.

	One molecule ⇌ Two molecules	Example
Involves double bonds	[structure: R-CO-CHR-CHOH-R ⇌ R-CO-CH2-R + O=CH-R]	Aldolase
	[structure: R-CHR-CHOH-R ⇌ R-CH=CH-R + H₂O] Note: Although water is involved and breaks a pi bond, all carbon atoms remain covalently bound throughout, so this is not a hydrolase.	Fumarase
Involves ring formation	[structure: ribose-5'-phosphate-PPi → cyclic phosphate + PPi] Note: Although ATP is involved, it is the sole substrate. Phosphate groups are not transferred (transferases), nor is ATP used to join other substrates (ligases) or for active transport (translocases).	Adenylyl cyclase

Examples of Lyases

Aldolase, an enzyme in glycolysis and gluconeogenesis, is an example of a lyase. The aldolase-catalyzed reaction is shown in Figure 4.22. During glycolysis, aldolase acts on fructose 1,6-bisphosphate, eliminating dihydroxyacetone phosphate (DHAP) from glyceraldehyde 3-phosphate (G3P). The double bond formed is a C=O double bond. Note that the "functional group" being eliminated (DHAP) is the exact same size as the molecule left behind (G3P).

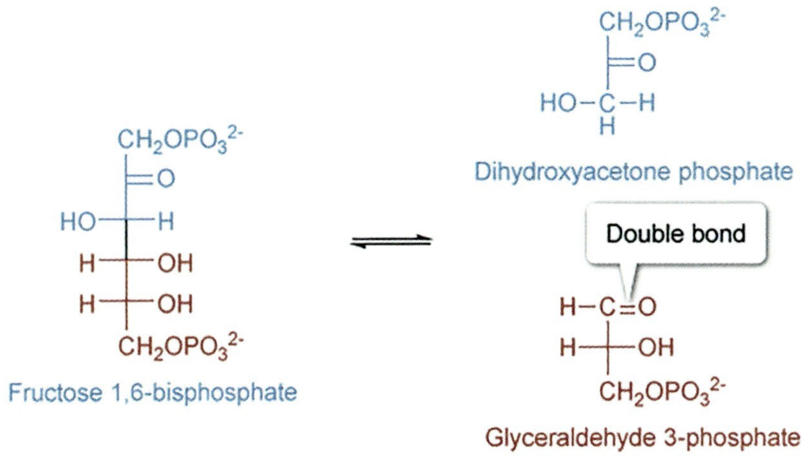

Figure 4.22 The reaction catalyzed by aldolase.

Hydratases and **dehydratases** are enzymes that add water across a double bond or eliminate water to form a double bond, respectively (Figure 4.23). Fumarate hydratase (also known as fumarase) of the citric acid cycle is an example of a hydratase. In fatty acid synthesis, β-hydroxyacyl-ACP dehydratase acts as a dehydratase. Despite the use of water to break a pi bond, the carbon atoms involved in the double bond *remain connected* through their sigma bond, so these reactions are *not* hydrolysis.

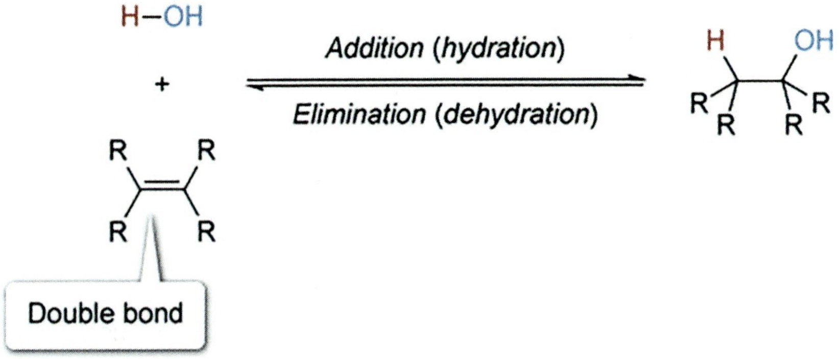

Figure 4.23 Hydration and dehydration reactions are catalyzed by lyases, not by hydrolases.

Nucleotidyl cyclases such as **adenylyl cyclase** are examples of lyases that use elimination reactions to form ring structures instead of double bonds, as shown in Figure 4.24. A single NTP molecule acts as the sole substrate. Pyrophosphate is eliminated, and the remaining nucleoside monophosphate is cyclized into a 3′,5′-cyclic NMP (eg, cAMP).

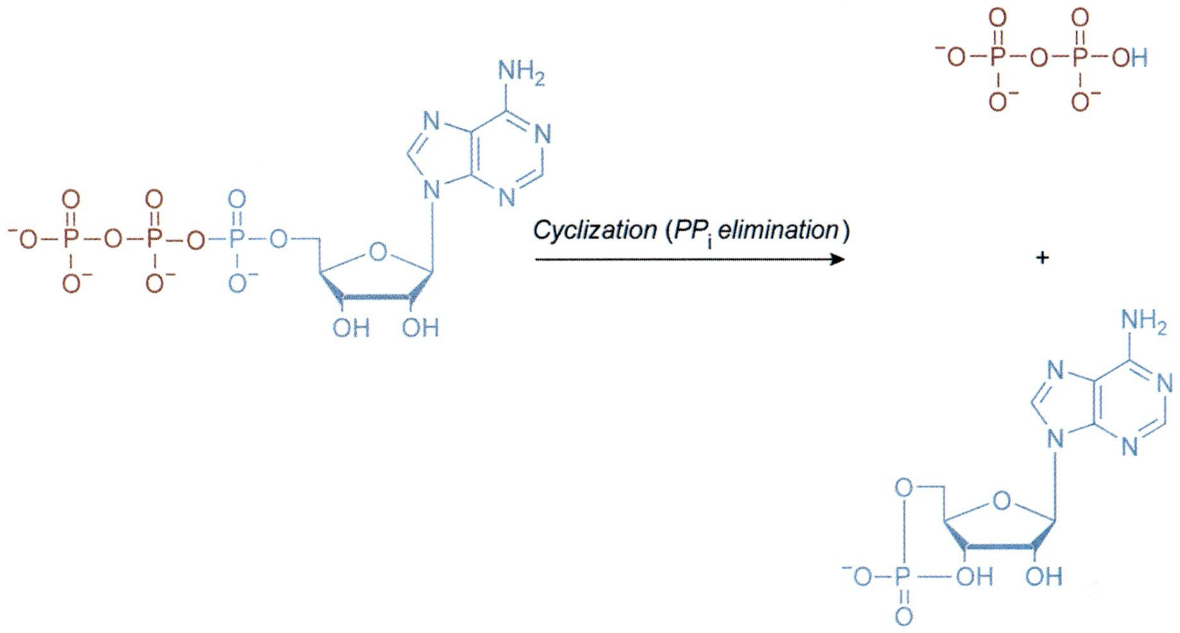

Chapter 4: Enzyme Activity

Figure 4.24 Adenylyl cyclase is an example of a lyase that catalyzes ring formation.

Many **synthases** are also formally classified as lyases. However, "synthase" is an informal classification mostly used to designate enzymes that form larger molecules *without the use of an external energy source such as ATP*. Because lyases can catalyze addition reactions that join two molecules into one product, many lyases that operate primarily in that direction are called synthases.

4.2.06 Isomerases

Isomerases catalyze reactions that change the structure of a single substrate. In other words, isomerases catalyze the changing of a substrate into a different isomer (Figure 4.25). Depending on the enzyme, the catalyzed isomerization reaction can result in a constitutional isomer, a configurational isomer, or a conformational isomer.

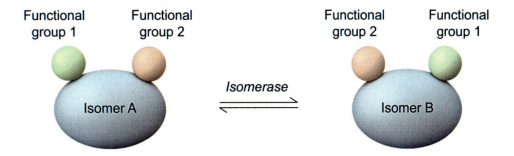

Figure 4.25 Isomerases catalyze the structural rearrangement of their substrate.

Recognizing Isomerase Enzymes and Their Reactions

Isomerases catalyze the structural rearrangement of a molecule with no net functional group addition, elimination, or transfer between molecules. Therefore, most isomerases have only *one substrate* and *one product*. Many isomerases contain the term *isomerase* in their name; however, certain subclasses of isomerases use different terms. These terms will be discussed, along with the type of isomerization reactions they describe, in the following sections.

Chapter 4: Enzyme Activity

Types of Isomerization Reactions: Constitutional Isomerization

Constitutional isomers are isomers that have the same types and numbers of atoms but different covalent bond connectivity between them. The enzyme triose phosphate isomerase, which is involved in glycolysis, is an example of an isomerase that produces constitutional isomers. Specifically, it interconverts an aldose and a ketose by changing the positions of covalent bonds (Figure 4.26).

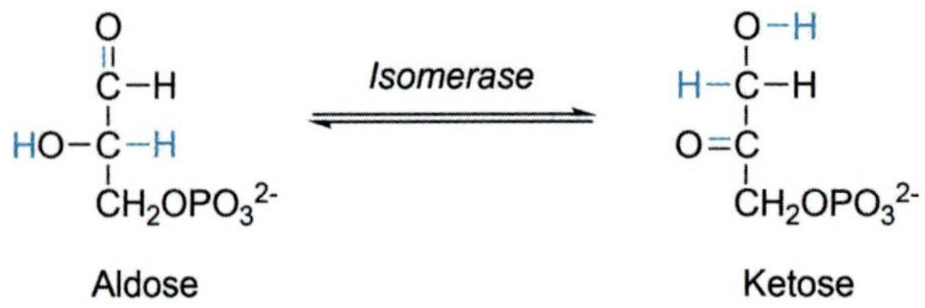

Figure 4.26 The interconversion of aldoses and ketoses is catalyzed by isomerases. The atoms and bonds shown in blue move, producing constitutional isomers.

Mutases are a subclass of isomerase that catalyze *intramolecular* transfer reactions. In other words, they move a functional group substituent from one position to another position *on the same molecule*. An example of a mutase is phosphoglycerate mutase, which plays a key role in glycolysis. Phosphoglycerate mutase transfers a phosphate group from C3 of glyceric acid to C2, or vice versa (Figure 4.27).

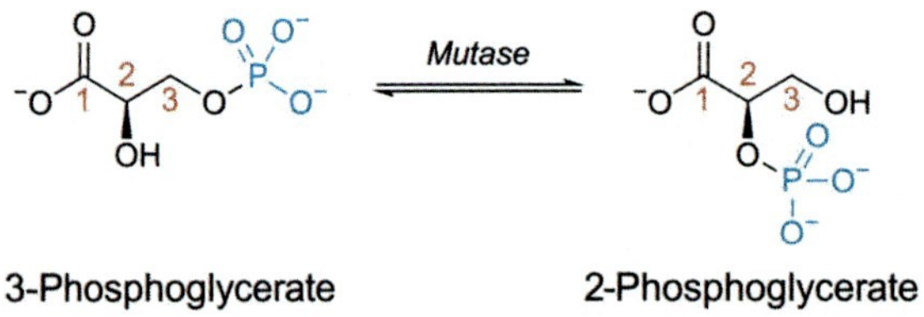

Figure 4.27 Mutases transfer substituents from one position on a molecule to another position on the same molecule.

Protein disulfide isomerases are enzymes that cause a disulfide bond to move from one pair of cysteines to another pair of cysteines from the same molecule, as shown in Figure 4.28. Lesson 2.2 discusses how the formation and breaking of an individual disulfide bond is a redox process. However, because the net oxidation and reduction reactions both occur *on the same molecule* (ie, the protein), enzymes that catalyze these reactiosn are considered *isomerases* rather than oxidoreductases.

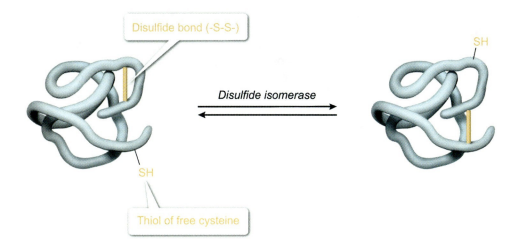

Figure 4.28 Protein disulfide isomerases move disulfide bonds within *the same molecule*.

Types of Isomerization Reactions: Configurational Isomerization

In **configurational isomers**, all the atoms in the molecule remain bonded to the same atoms. However, the *relative placement* of atoms differs, leading to distinct molecules that cannot be superimposed or interconverted through freely rotating bonds. Configurational isomers (see Organic Chemistry Lesson 3.3) include *cis/trans* or *E/Z* isomers around a double bond and *R/S* or L/D isomers around a chiral center.

Peptide bonds have double bond character due to resonance. Therefore, peptide bonds can be described as *cis* or *trans* with respect to the adjacent α-carbons. Lesson 2.1 discusses how most peptide bonds favor the *trans* configuration because it has much less steric clashing than the *cis* configuration. Proline, however, experiences similar clashing—and therefore has similar free energy—in both configurations. Consequently, the peptide bond preceding a proline residue can exist in either the *cis* or the *trans* configuration. The enzyme peptidyl-prolyl isomerase catalyzes the exchange (Figure 4.29).

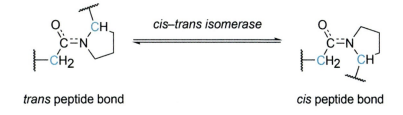

Figure 4.29 A *cis-trans* isomerase interconverts a *trans* peptide bond with a *cis* peptide bond.

Isomerization of a chiral center is another common configurational isomerization, and enzymes that catalyze these reactions often have unique suffix terms to describe them. For substrates with only one chiral center, configurational isomerization produces the enantiomer. If allowed to go to equilibrium, a 1:1 mixture of enantiomers—also known as a racemic mixture—would be produced. Therefore, enzymes that invert the stereochemistry of a substrate's sole chiral center are called **racemases** (Figure 4.30).

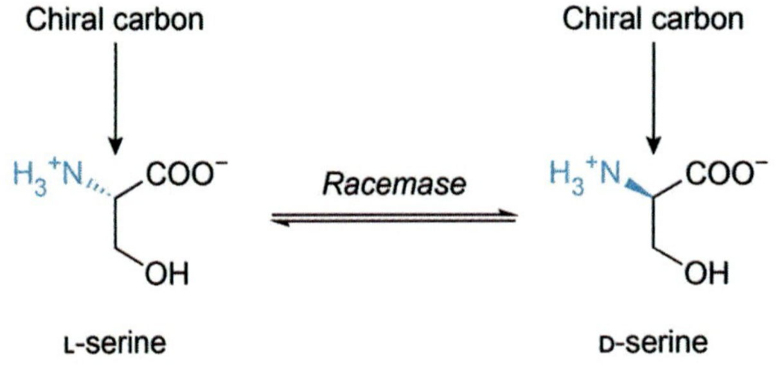

Figure 4.30 A racemase interconverts stereochemistry at a molecule's sole chiral carbon to produce a racemic mixture of enantiomers.

If a substrate has multiple chiral centers, then isomerization of a single center forms a special type of diastereomer known as an epimer (see Chapter 7). Therefore, isomerase enzymes that catalyze these reactions are known as **epimerases** (Figure 4.31).

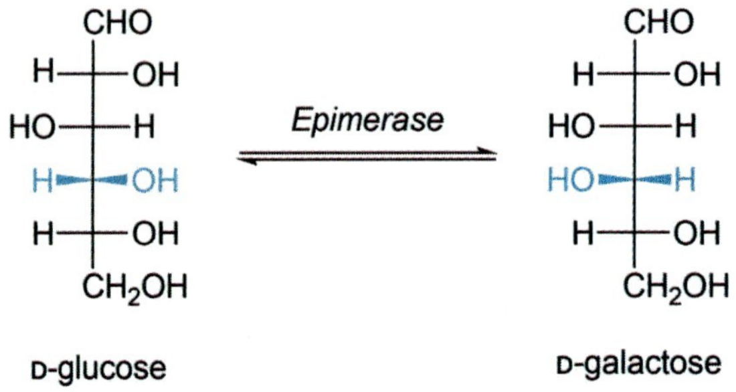

Figure 4.31 In a molecule with multiple chiral centers, epimerases change the stereochemistry at a single chiral carbon.

Monosaccharides, the monomeric unit of carbohydrates (see Chapter 7), gain an additional chiral center when they cyclize. This new chiral center is known as the anomeric carbon, and the two possible epimers are known as the α- and β-anomers. The interconversion of anomers is called mutarotation, and therefore enzymes that catalyze mutarotation are a subclass of epimerases called **mutarotases** (Figure 4.32).

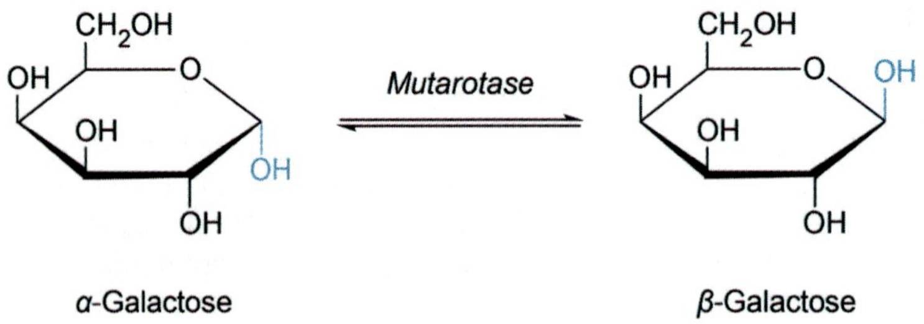

Figure 4.32 A mutarotase catalyzes the stereochemical inversion of a carbohydrate's anomeric carbon.

Concept Check 4.7

Suppose that glucose 6-phosphate can be converted to Product A, Product B, or Product C, and that each conversion occurs through only one enzyme.

Which enzyme is phosphoglucose isomerase, which is phosphoglucomutase, and which is phosphoglucose epimerase?

Solution
Note: The appendix contains the answer.

4.2.07 Ligases

Ligases are the sixth class of enzymes. They catalyze the joining of two molecules using the hydrolysis of ATP (or another nucleoside triphosphate) for the energy needed to do so, as shown in Figure 4.33.

Note that there are two net effects of a ligase-catalyzed reaction: two substrates are joined *and* an NTP molecule is hydrolyzed. Although NTP hydrolysis is an important result of the reaction energetically, it typically is not the main *physiological* result. For this reason, the NTP in a ligase reaction is often called a coenzyme of the ligase (like the redox coenzymes discussed in Concept 4.2.02).

Because NTP hydrolysis is not the primary function of the enzyme, these enzymes are distinct from hydrolases. When classifying enzymes, it is important to consider all processes catalyzed by the enzyme to determine the enzyme's proper category.

Chapter 4: Enzyme Activity

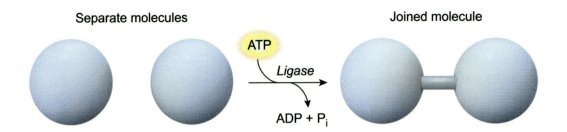

Figure 4.33 Ligases hydrolyze a nucleoside triphosphate (eg, ATP) to covalently join two substrate molecules together.

Recognizing Ligase Enzymes and Their Reactions

Ligase enzymes can be recognized by the reactions they catalyze. Typically, two substrates are joined into one product, and an NTP coenzyme (ie, ATP, GTP, CTP, or UTP) is hydrolyzed to NDP and an inorganic phosphate (P_i). Some ligases instead hydrolyze the NTP into an NMP and pyrophosphate (PP_i). Although NTP hydrolysis is highly exergonic, ligase-catalyzed reactions may also proceed in the reverse direction depending on the free energy change ΔG of the coupled reaction. In this case, ligase reactions break apart a molecule while *forming* NTP.

Ligase enzymes can also be recognized by their names. Many ligase enzymes have the term *ligase* in their name (eg, DNA ligase). Other ligases include the term *synthetase*. As introduced in Concept 4.2.05, the term *synthase* (no "et" in the name) is usually used to describe enzymes that join molecules *without* NTP hydrolysis. In contrast, the term *synthetase* (with the "et") is used to describe enzymes that join molecules *with* NTP hydrolysis. While *synthases* can be found in several enzyme classes, the term *synthetase* is generally synonymous with *ligase*.

One example of a *synthetase* is succinyl-CoA synthetase, an enzyme in the citric acid cycle. The ligation reaction it catalyzes joins succinate and coenzyme A to form succinyl-CoA, while consuming GTP in the process.

$$\text{Succinate} + \text{CoA-SH} + \text{GTP} \rightleftharpoons \text{Succinyl-CoA} + \text{GDP} + P_i$$

Succinyl-CoA synthetase is also an example of a reversible ligase-catalyzed reaction. During the citric acid cycle, succinyl-CoA is broken apart and GTP is formed from GDP and P_i.

$$\text{Succinyl-CoA} + \text{GDP} + P_i \rightleftharpoons \text{Succinate} + \text{CoA-SH} + \text{GTP}$$

> ☑ **Concept Check 4.8**
>
> Consider the following two enzyme-catalyzed reactions.
>
> $$1) \quad X + Y=Z \longrightarrow X-Y-Z$$
>
> $$2) \quad X-Y + Z \xrightarrow[\text{ATP} \quad \text{ADP} + P_i]{H_2O} X-Y-Z$$
>
> Based on these reaction descriptions, fill in the blanks in the following paragraphs:
>
> Reaction 1 synthesizes the molecule XYZ _____ (using / without using) ATP. Therefore, an appropriate name for its enzyme is XYZ _____ (synthase / synthetase). This reaction joins two molecules into one and involves addition across a double bond. Therefore, its enzyme is most likely classified as a _____ (ligase / lyase).
>
> In contrast, Reaction 2 synthesizes the molecule XYZ and _____ (uses / does not use) the energy in ATP to do so. Therefore, an appropriate name for its enzyme is XYZ _____ (synthase / synthetase). ATP is hydrolyzed into ADP + P_i. Despite this, the enzyme is *not* classified as a hydrolase, but rather as a _____ (ligase / lyase), because of its function of joining two molecules together.
>
> **Solution**
>
> *Note: The appendix contains the answer.*

4.2.08 Translocases

The seventh enzyme classification is the translocase class. Translocases are a relatively new classification of enzyme. As such, they may not appear on the exam; however, they still catalyze an important subset of cellular processes and are worth reviewing.

Translocases move a substance across a lipid bilayer membrane (Figure 4.34). Unlike other transporters that mediate passive facilitated diffusion, translocases couple substance movement with a catalyzed chemical reaction. Therefore, translocases are responsible for primary active transport across a membrane.

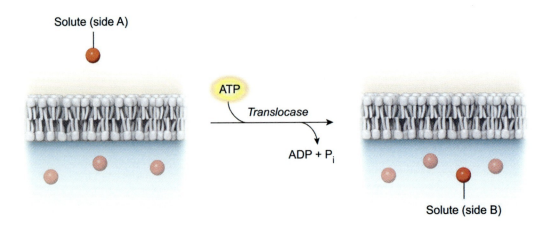

Figure 4.34 Translocases couple the movement of a solute across a membrane with a catalyzed chemical reaction (eg, ATP hydrolysis, NADH oxidation).

Most translocases couple solute movement to either a hydrolysis reaction or to a redox reaction. Consequently, these enzymes were once classified as hydrolases or oxidoreductases, respectively. Complexes I, III, and IV of the electron transport chain are examples of translocases that couple proton movement to the reduction of a redox cofactor.

The sodium-potassium pump (Na^+/K^+ pump) is another example of a translocase that moves multiple solutes. In this case, the Na^+/K^+ pump also moves multiple *types* of solutes to different sides of the membrane. Solute movement is linked to a chemical reaction (ie, ATP hydrolysis), which classifies the Na^+/K^+ pump as a translocase.

Like other enzymes, translocases can also catalyze their reactions in both directions. In other words, instead of hydrolyzing ATP to pump a solute *against* its electrochemical gradient, a translocase can also *form* ATP as a solute moves *down* its gradient. ATP synthase, the final enzyme in the oxidative phosphorylation pathway, is a translocase that operates in this direction.

4.2.09 Receptor Enzymes

Receptor enzymes are not a distinct class of enzymes in that they do not catalyze a distinct type of reaction or reaction-coupled process. Instead, receptor enzymes are each classified under one of the seven classes discussed in the previous concepts of this lesson. However, receptor enzymes also have additional properties that make them worth reviewing.

In general, receptors are the membrane proteins through which cells interact with their external environment. Because the lipid bilayer enclosing the cell is only semipermeable, most external stimuli interact only with the membrane—they do not interact directly with anything inside the cell. Chapter 3 discusses how some membrane proteins bind extracellular agonists (ie, ligands that activate a receptor), which leads to intracellular conformational changes and intracellular effects (eg, ligand-gated ion channels, GPCRs).

Receptor enzymes all act on a similar principle. Each has an extracellular binding domain and an intracellular catalytic (ie, enzyme) domain, as shown in Figure 4.35. The catalytic domain is typically inactive until an agonist binds. Binding leads to a conformational change, which then leads to an increase in catalytic activity. In this way, receptor enzymes exhibit allosteric regulation. Although receptor enzymes can belong to any class of enzyme, many are either kinases (a type of transferase) or guanylyl cyclases (a type of lyase).

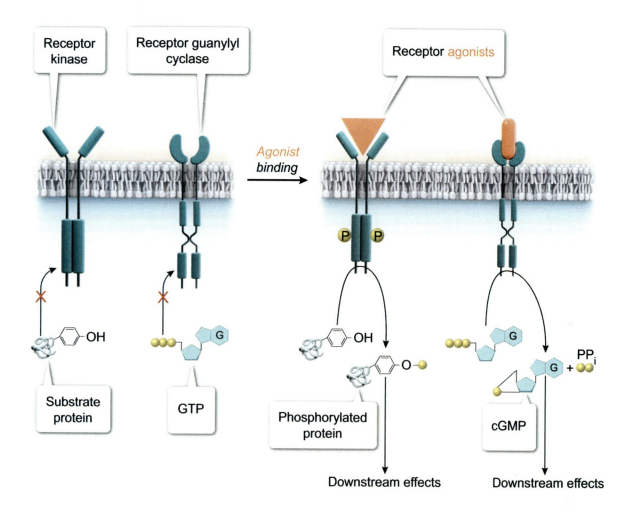

Figure 4.35 Receptor enzymes bind an external signal, which activates their catalytic activity. Activation of receptor enzymes leads to downstream signal propagation.

Of the receptor kinases, many are **receptor tyrosine kinases (RTKs)**. As their name implies, RTKs phosphorylate tyrosine residues when activated. Note that this contrasts with most *soluble* (ie, nonmembrane receptor) protein kinases, which typically phosphorylate serine or threonine residues. Consequently, phosphotyrosine (pTyr) is relatively scarce, which allows it to act as a potent intracellular signal. Downstream proteins in the signaling pathway are "recruited" by pTyr. In other words, they bind to and are stimulated by pTyr on the substrate protein. This leads to propagation of the signal (Figure 4.36).

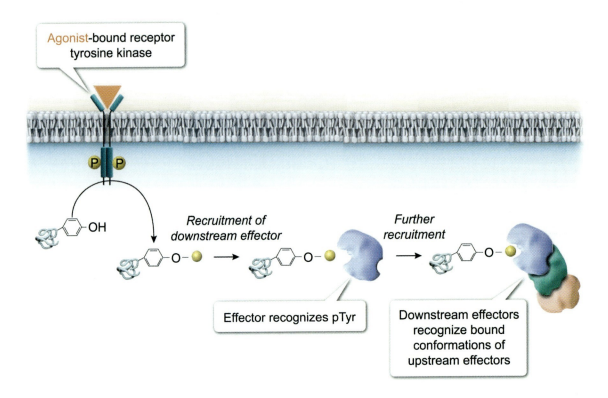

Figure 4.36 Phosphotyrosine "recruits" effector proteins that bind to it. These activated effectors can then recruit downstream effectors, leading to signal transduction.

One important mechanistic feature of RTKs is their **autophosphorylation**. RTKs are typically dimers or become dimers upon agonist binding, bringing the catalytic domains of each monomer close together. Each monomer then phosphorylates a tyrosine residue within the other monomer. This self-phosphorylation further activates the RTK dimer and provides pTyr residues, which recruit the receptor's substrate (Figure 4.37). The insulin receptor, discussed in Unit 4, is an example of an RTK.

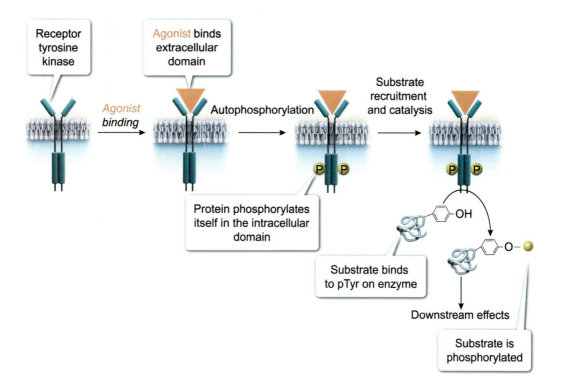

Figure 4.37 The binding of a ligand induces RTKs to undergo autophosphorylation. This helps to recruit substrate and increases the RTK's catalytic activity.

The second major class of receptor enzymes is receptor guanylyl cyclases (Figure 4.38). Receptor guanylyl cyclases are less commonly tested on the exam but play important roles in vasodilation (eg, atrial and brain natriuretic peptide receptors, nitric oxide receptors) (see Biology Lesson 13.1). In addition, receptor guanylyl cyclase action demonstrates the role that receptor enzymes play in cell signaling and signal transduction.

Upon binding their agonist, receptor guanylyl cyclases catalyze a cyclization reaction that converts GTP to 3',5'-cyclic GMP and pyrophosphate (PP$_i$). The resulting cGMP then acts as a second messenger to activate protein kinase G (PKG), which can then phosphorylate various substrates and cause various effects.

One type of guanylyl cyclase enzyme is a soluble (ie, nonmembrane) receptor. This receptor, abbreviated as sGC, resides in the cytosol and binds a ligand that can diffuse across the cell membrane—nitric oxide. Activated sGC then produces cGMP, which activates PKG, leading to downstream effects. The diffusion of an agonist through the membrane is relatively uncommon, but it can also happen with lipid agonists for other receptors (see Lesson 3.2 and Lesson 9.3).

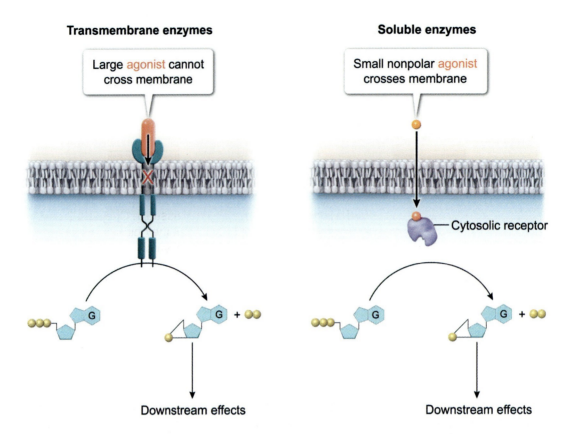

Figure 4.38 Receptor guanylyl cyclases are either transmembrane or soluble enzymes that activate upon agonist binding.

Lesson 4.3
Catalytic Mechanisms of Enzymes

Introduction

As discussed in Lesson 4.1, enzymes are catalysts that affect reaction rate and kinetics without altering thermodynamics and equilibrium. All catalysts share these properties. However, enzymes exhibit many additional properties that differentiate them from chemical catalysts. These properties largely arise because enzymes are proteins and have the properties of proteins discussed in Chapters 2 and 3. This lesson elaborates on some unique properties of enzymes that arise from their biological (rather than chemical) nature and delves deeper into the catalytic mechanisms of enzymes.

4.3.01 Enzyme-Substrate Interactions

As discussed in Lesson 4.1, enzymes catalyze reactions by lowering their activation energy to increase the reaction rate.

$$\text{Reactant} \xrightarrow{Enzyme} \text{Product}$$

Enzymes must physically interact with and bind to reactants before catalysis can occur. In other words, the reactant is a ligand that the enzyme acts upon by catalyzing its chemical change. Because the reactants of enzyme-catalyzed reactions are ligands that are changed through a chemical reaction, enzyme reactant ligands are given the special name of **substrate** (Figure 4.39).

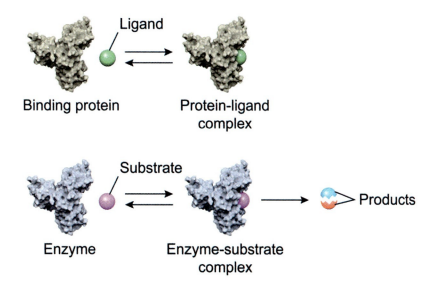

Figure 4.39 Ligands can bind to and unbind from proteins without being chemically changed. Substrates are ligands that bind to enzymes and are subsequently chemically changed.

Ligand binding to proteins is stabilized by a variety of intermolecular forces that depend on the functional groups of the ligand and the protein residues or backbone groups involved. These intermolecular interactions and the formation of noncovalent bonds result in a release of heat and a decrease in enthalpy $(-\Delta H)$. Desolvation and hydrophobic interactions can also stabilize protein-ligand interactions through an increase in entropy of the surrounding water in the system. In this way, the free energy of a system decreases when a protein binds its ligand until equilibrium is reached ($\Delta G_{binding}$).

Recall that enzymes catalyze reactions by decreasing the activation energy (E_a). Because enzymes do not affect equilibrium or thermodynamics, the free energy of the unreacted substrate is unchanged; therefore, the decrease in E_a must come from a decrease in the free energy of the transition state (‡). **Stabilizing protein-ligand interactions** lower the free energy of the enzyme–transition state complex compared to uncomplexed transition state, which leads to an increased reaction rate (Figure 4.40).

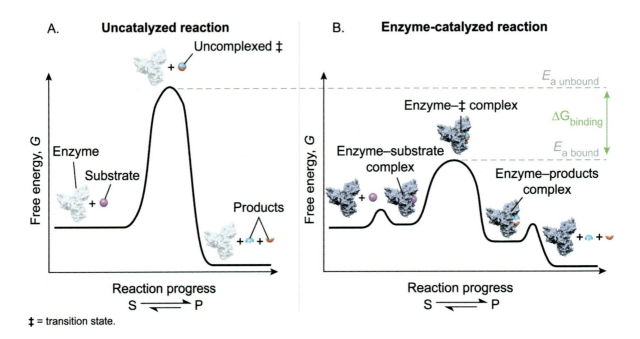

‡ = transition state.

Figure 4.40 Free energy diagrams of an uncatalyzed reaction (A) and an enzyme-catalyzed reaction (B). The activation energy E_a is greatly decreased by binding to the enzyme ($\Delta G_{binding}$).

Note that Figure 4.40B shows the enzyme in three bound states: one bound to the substrate, one bound to the product, and one bound to the transition state ‡. The enzyme-substrate and enzyme-product complexes have similar or higher free energies compared to the substrate and product alone. This relationship is important to ensure that substrates and products *release* from the enzyme (rather than staying bound).

In contrast, the enzyme–transition state complex has a significantly *lower* free energy than the unbound transition state. This shows that the transition state is the species for which the binding interaction energies are strongest (ie, the shift from unbound transition state to bound transition state has the most negative $\Delta G_{binding}$ of the various bound states shown). This implies that the optimal ligand for an enzyme is *neither* its substrate *nor* its product, but rather the transition state (Figure 4.41).

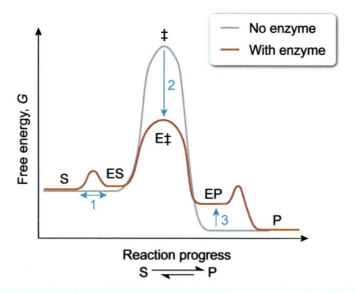

Under the conditions tested:

1 = The free energy of ES is not much different than free E + free S. Substrate binding is near equilibrium.

2 = The enzyme-bound transition state has a much lower free energy than free transition state. The transition state is the optimal ligand.

3 = The free energy of EP is higher than free E + free P. Product binding is not favored.

E = enzyme; P = product; S = substrate; ES = enzyme-substrate complex; E‡ = enzyme–‡ complex; EP = enzyme-product complex.

Figure 4.41 Under physiological conditions, the binding energy $\Delta G_{binding}$ is most negative (ie, most favorable) when the enzyme is bound to the transition state.

✓ Concept Check 4.9

A medicinal chemist is designing a drug against Enzyme X. She plans to design a competitive inhibitor of the enzyme, which means the drug will bind the enzyme in place of the substrate, but it will not react. If the scientist wants to design a drug with the highest binding strength (ie, lowest K_d) possible, should she design the drug to most closely resemble the substrate, the product, or the transition state of the enzyme-catalyzed reaction?

Solution

Note: The appendix contains the answer.

> **Concept Check 4.10**
>
> After designing a candidate drug, structural analysis shows that a negatively charged functional group on the drug interacts strongly with a positively charged residue in the enzyme's substrate-binding site. Given this, which of the following mutant enzyme variants is likely to have the *weakest* interaction with the drug candidate? Assume all mutations are in the substrate-binding site.
>
> A. V56I
> B. K44L
> C. D24E
> D. R39K
>
> **Solution**
>
> *Note: The appendix contains the answer.*

4.3.02 Lock-and-key versus Induced-Fit Theories

Concept 4.3.01 describes how an enzyme's optimal ligand is the transition state of the enzyme-catalyzed reaction, yet to catalyze a reaction, an enzyme must also be able to recognize and bind to its substrates. How does an enzyme recognize the substrates it binds to, and what flexibility does an enzyme have to recognize related molecules? Several models have been proposed to address these questions.

The **lock-and-key** model posits that even when not bound to its substrate, the enzyme displays a protein structure that perfectly complements (ie, matches) its specific substrate. Only the substrate of interest can perfectly fit into the substrate-binding pocket of the enzyme, like a key fitting into a lock, and no conformational change is needed to accommodate the substrate (Figure 4.42).

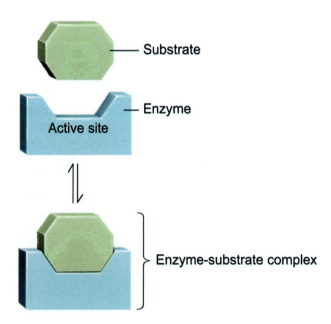

Figure 4.42 The lock-and-key model of an enzyme binding its substrate.

In contrast, the **induced-fit model** proposes that the resting enzyme does *not* perfectly complement the substrate. Instead, the substrate itself causes the enzyme to adopt a structure to accommodate it (Figure 4.43). In other words, the substrate *induces* a conformational change in the enzyme to *fit* around the substrate (see Concept 2.3.03 for more on conformational changes).

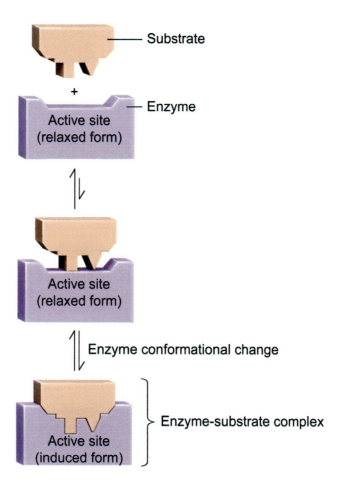

Figure 4.43 The induced-fit model of an enzyme binding its substrate.

A third model, which is related to the induced-fit model, is **conformational selection**. In this model, the unbound enzyme explores a *range* of conformations near the low point of its folding funnel (see Concept 2.3.02). Although the predominant conformation is the low-binding-affinity conformation (ie, a conformation that does *not* fit the substrate), at any given time a small percentage of enzyme is in the high-binding-affinity conformation (ie, the conformation that *does* fit substrate).

Rather than the substrate *inducing* a conformational change, the conformational selection model proposes that the substrate binds to one of the few proteins that has temporarily adopted this high-affinity conformation. Upon binding, the high-affinity conformation is stabilized. As more substrate is added, a higher percentage of the total enzyme population becomes stabilized in this high-affinity conformation (Figure 4.44).

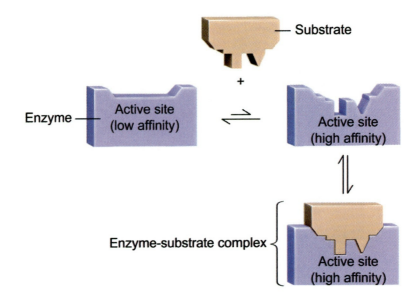

Figure 4.44 The conformational selection model of an enzyme binding its substrate.

Given the three different models of molecular recognition presented, which model is relevant to enzyme-substrate binding? Although some enzymes may exhibit features of one model over the others, all three models contribute useful insight into understanding enzyme-substrate recognition (and protein-ligand recognition in general), and most enzymes can be explained by some mix of all three models.

> **Concept Check 4.11**
>
> For each of the three models of molecular recognition (lock-and-key model, induced-fit model, conformational selection), place the following steps in sequential order, starting from unbound enzyme and ending in high-affinity substrate binding. Not all models will require all steps.
>
> A. Enzyme is in the unbound ground state.
> B. Substrate binds the enzyme with high affinity.
> C. Substrate binds the enzyme with low affinity.
> D. Enzyme undergoes a conformational change.
>
> **Solution**
>
> *Note: The appendix contains the answer.*

4.3.03 Implications of the Molecular Recognition Models

The lock-and-key model was first proposed as a way of explaining the specificity of an enzyme for its substrate. **Specificity** refers to the high selectivity with which an enzyme acts on its own substrate, but *not* on other substrates with related structures. For example, many proteases (enzymes that hydrolyze peptide bonds) act on only specific amino acid sequences. Rather than acting on *all* peptide bonds, proteases accommodate only certain substrates that fit their binding pocket (Figure 4.45).

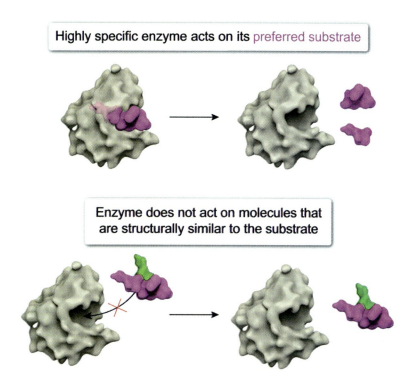

Figure 4.45 Enzymes can be highly specific for their substrates and may fail to act on molecular structures that are very similar to the natural substrate.

Similarly, most amino acids and peptides found in nature display highly specific stereochemistry, as discussed in Unit 1. Most amino acids are chiral, with most physiologically relevant amino acids being the L-enantiomer. Consequently, most proteins (including enzymes) are chiral as well. This leads to one of the important distinctions between chemical catalysts and biological catalysts such as enzymes. Whereas chemical catalysts are mostly *not* stereoselective, enzymes and other biological catalysts usually are *highly* stereoselective. An enzyme that acts on one isomer of a molecule often does *not* act on other isomers of the same molecule (Figure 4.46).

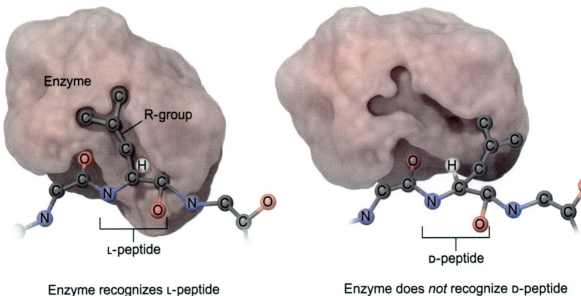

Figure 4.46 Most enzymes are stereoselective so that only the correct stereoisomer (the "key") fits into the substrate-binding pocket (the "lock").

Although the lock-and-key model can explain the high specificity of *some* enzymes, it does not adequately explain why other enzymes are less specific and can recognize a larger variety of substrates. Furthermore, it does not adequately explain an important feature discussed in Concept 4.3.01—that the optimal ligand for an enzyme is not the substrate, but rather the *transition state* of the enzyme-catalyzed reaction.

Both the induced-fit model and the conformational selection model allow for conformational changes in the enzyme, which may permit some enzymes to act on multiple related substrates because they allow for conformational adjustments of the substrate-binding site. Furthermore, these models help explain how a short-lived transition state can be the optimal ligand of the enzyme. Just as the substrate induces a conformational change in the enzyme, the enzyme also induces a conformational change in the substrate toward its transition state (Figure 4.47).

Chapter 4: Enzyme Activity

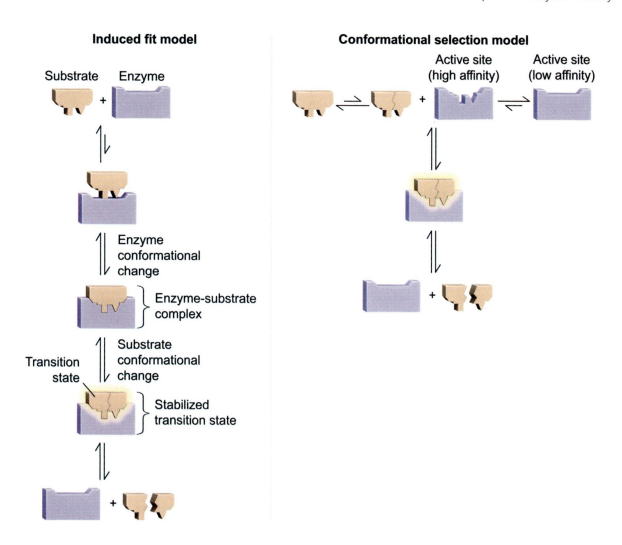

Figure 4.47 Both the induced-fit model (left) and the conformational selection model (right) help explain how enzymes recognize a transition state.

4.3.04 Enzyme Active Site

The previous concepts in this lesson discuss how enzymes recognize and interact with their substrate through the formation of noncovalent bonds. The decrease in enthalpy H associated with the formation of these bonds results in the stabilization of the reaction's transition state, which results in the enzyme's ability to catalyze the reaction. In other words, the binding energy contributes to the activity of the enzyme by helping decrease the activation energy E_a. For this reason, the substrate-binding site of an enzyme contributes a crucial component of an enzyme's **active site** (Figure 4.48).

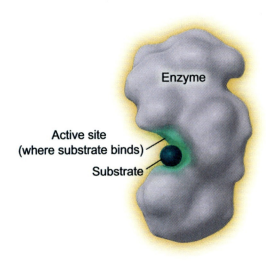

Figure 4.48 The active site of an enzyme is the portion of the enzyme where substrates bind and where catalysis of the reaction occurs.

In addition to binding, some enzymes can employ additional mechanisms to catalyze chemical reactions. These mechanisms, which are covered in more detail in Concept 4.3.06, often require changes in the chemical reactivity of specific enzyme residues. For example, Concept 1.3.06 discusses how the nucleophilicity of an amino acid can be enhanced by deprotonation, yet the pK_a values for free amino acids suggest that the side chains of most nucleophilic amino acids are protonated at physiological pH.

Concept 2.1.04 introduces the idea that the pK_a of an amino acid residue can change due to interactions with nearby functional groups. The tertiary structure of the enzyme creates a **microenvironment** that often brings specific functional groups together, allowing the nucleophilic amino acid to lose a proton at physiological pH values.

As an example, the *catalytic triad* in many serine proteases is a set of three amino acids in the enzyme active site that work together to allow a serine side chain to act as a nucleophile. Although serine's high pK_a (around 13) makes the residue unlikely to lose a proton in aqueous solution, the unique microenvironment of the active site forces a basic histidine residue into close proximity with the serine side chain. This allows serine to act as an acid by giving up its proton to the basic histidine, thereby enhancing serine's nucleophilicity. The protonated histidine, in turn, is stabilized by the third member of the catalytic triad: an acidic residue such as aspartate (Figure 4.49).

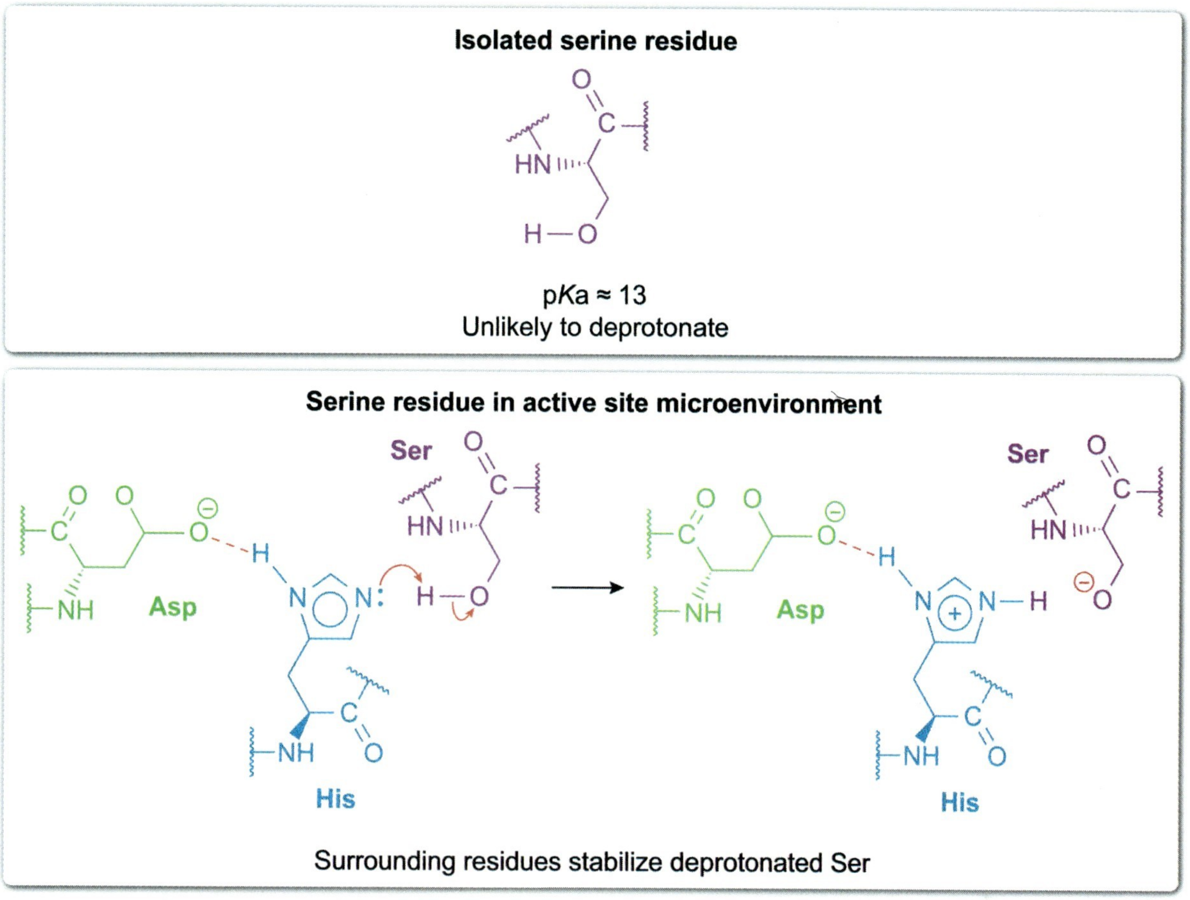

Figure 4.49 The catalytic triad is an example of residues contributing to a unique microenvironment within an enzyme active site.

> ### ✓ Concept Check 4.12
>
> Amino groups (R–NH$_3^+$) can often act as nucleophiles in biochemical systems. For example, the α-amino group of an amino acid acts as a nucleophile during peptide bond formation, and the side-chain ε-amino group of lysine acts as a nucleophile during the isopeptide bond formation reactions of ubiquitination or histone acetylation.
>
> A novel enzyme has been found to catalyze isopeptide bond formation, and the side chain amino group of the substrate lysine is found to interact strongly with specific active site residues. The lysine side chain amino group is most likely to interact with the side chain of which of the following amino acids in the enzyme active site to enhance its nucleophilicity?
>
> A. Leucine
> B. Histidine
> C. Glycine
> D. Methionine
>
> ### Solution
>
> *Note: The appendix contains the answer.*

4.3.05 Enzymes with Multiple Substrates

Importantly, many enzymes catalyze reactions with multiple substrates. Despite this, many enzyme-catalyzed reactions occur at a single active site. In these instances, the active site may have multiple *substrate-binding* subsites that hold the substrates and a singular *catalytic* subsite where covalent bonds are formed and broken. Depending on the specific case, an enzyme may hold multiple substrates at the same time, or it may react with the substrates one by one.

An enzyme that holds two substrates at the same time is said to form a **ternary complex** because it is a complex of *three distinct molecules*: one enzyme and the two substrates together. Some enzymes that form ternary complexes may bind their substrates in an **ordered manner** (ie, substrate A must bind before substrate B). Other enzymes that form ternary complexes bind their substrates in a **random order** (ie, the enzyme can accept either substrate A or substrate B first (see Figure 4.50).

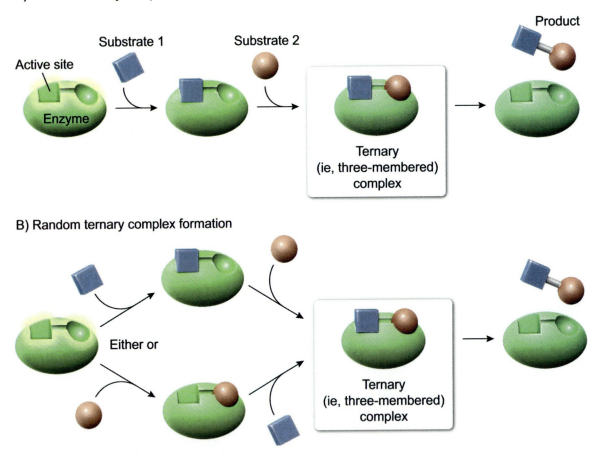

Figure 4.50 A ternary complex is a three-membered complex.

In contrast, other enzymes bind and react with one substrate in one step—forming and releasing one product and yielding a *modified* enzyme as a reaction intermediate—before the modified enzyme binds and reacts with the second substrate. After the second step, the second product is released. As discussed in Concept 4.1.05, enzymes must be unaltered at the end of the reaction, so this second step must also restore the enzyme to its original state. Enzymes that act in this manner are said to follow a **double displacement** or **ping-pong** mechanism (Figure 4.51).

Chapter 4: Enzyme Activity

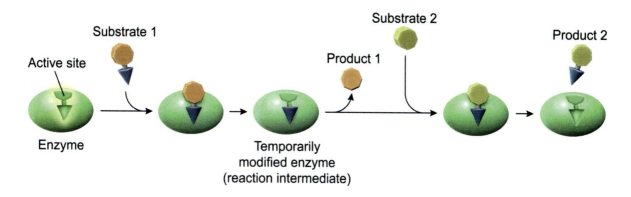

Figure 4.51 A double displacement, also known as ping-pong, enzyme reaction mechanism.

Enzymes may also have multiple active sites that each catalyze separate reactions. These sites are usually found on separate domains or subunits and act independently of each other. For example, the bifunctional enzyme phosphofructokinase-2/fructose-2,6-bisphosphatase (see Chapter 11) has separate active sites on different domains that catalyze separate kinase and phosphatase reactions. Similarly, fatty acid synthase (see Concept 13.1.04) also contains several active sites on different domains to catalyze the various reactions involved in fatty acid synthesis.

Cooperative enzymes also display multiple active sites that each catalyze several instances of the same reaction (see Concept 5.3.05). However, the term "multisubstrate reaction" (eg, ternary complex or double displacement) refers to a reaction involving multiple substrates at a *single* active site, *not* to multiple active sites acting separately.

Chapter 4: Enzyme Activity

> ### ✓ Concept Check 4.13
>
> Transaminases are enzymes that catalyze reactions such as the one shown in the following image: 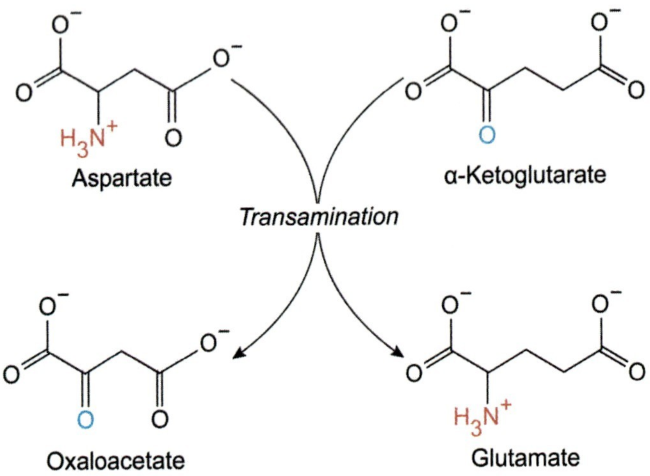
>
> Aspartate + α-Ketoglutarate → Oxaloacetate + Glutamate (Transamination)
>
> A scientist mixes aspartate with the transaminase enzyme. After a brief incubation period, the scientist detects a very small amount of oxaloacetate, despite not having added any α-ketoglutarate. Based on this information, does the transaminase enzyme form a ternary complex, or does it proceed by a double displacement (ping-pong) mechanism? If instead the scientist tested a different two-substrate enzyme but was *not* able to detect any product formation after addition of a single substrate, what could be concluded about the mechanism in this alternate scenario?
>
> #### Solution
>
> *Note: The appendix contains the answer.*

4.3.06 Modes of Catalysis

Effects of Binding, Conformational Changes, and Other Physical Effects

As mentioned in Concept 4.3.01, binding energies between the enzyme and the reaction transition state ‡ lead to a lower free energy G of the enzyme–transition state complex compared to free transition state. This decrease in free energy (and therefore the decrease in activation energy E_a) is a result of negative enthalpy changes ($-\Delta H$) due to formation of noncovalent, stabilizing bonds between the enzyme and the transition state, as well as increases in entropy (ΔS) due to desolvation.

Although the formation of a large enzyme–transition state complex may at first glance seem like an increase in order, which should result in a decrease in entropy, substrate binding forces water to leave the enzyme active site, causing the *water molecules* to become more disordered, which leads to an *increase in entropy* for the whole system (Figure 4.52). This desolvation effect is similar to the hydrophobic effect discussed in Lesson 2.3.

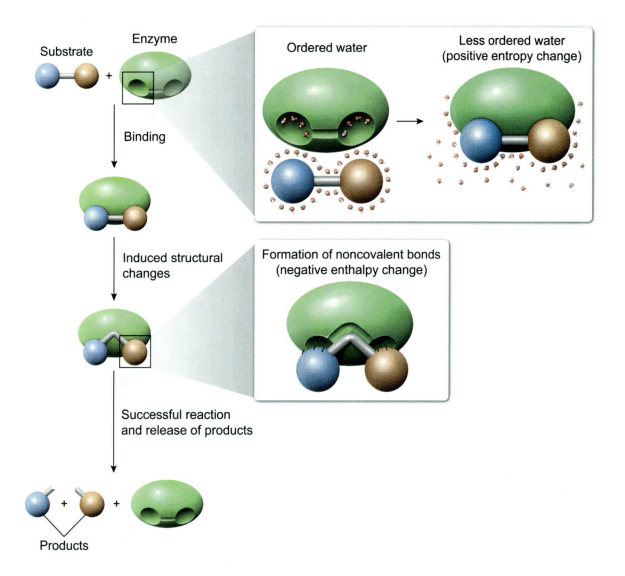

Figure 4.52 Entropy and enthalpy changes contribute to a decreased free energy of the enzyme–transition state complex (decreased E_a).

In addition to the binding energy itself, stabilization of the transition state in the enzyme also results in increased strain of the substrate bonds relative to unreacted substrate. The transition state normally exists only temporarily because this strain gives it a high free energy. Although the enzyme stabilizes the free energy, the strain still primes the substrate for conversion into product (Figure 4.53).

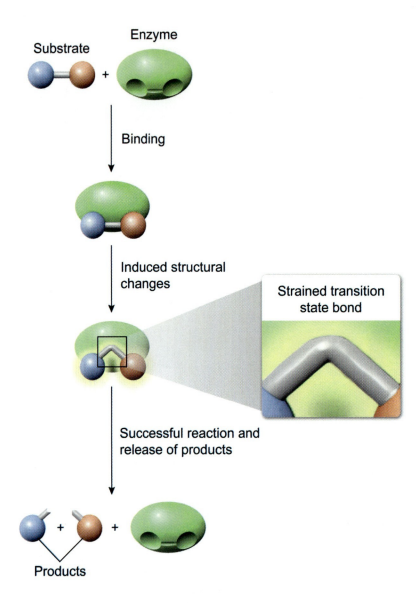

Figure 4.53 The strain in the transition state's bonds primes the substrate for conversion into product.

Enzymes also speed up reactions by physically positioning functional groups in ways that more easily result in productive collisions. Collision theory proposes that chemical reactions only occur if substances collide in the correct orientation with sufficient energy to overcome E_a.

By binding substrates at a single active site, enzymes effectively increase the local concentration of reactants, increasing frequency of collisions. The specific shape of the active site also forces substrates to bind in a specific way, ensuring that the collisions that do occur happen in the correct orientation (Figure 4.54).

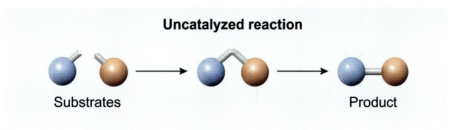

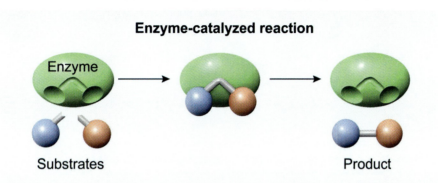

Figure 4.54 Enzymes help catalyze reactions by holding the reacting functional groups in close proximity and in the correct orientation.

Covalent Catalysis

Concept 4.3.04 introduces the catalytic triad of a serine protease as an example of a mechanism in which an enzyme's active site contributes to a microenvironment that changes functional group protonation state and reactivity. In the catalytic triad, a serine residue of the enzyme active site is deprotonated, allowing it to act as a nucleophile and attack the substrate, forming a covalent bond. Because the enzyme forms a temporary covalent bond with a substrate, this is referred to as **covalent catalysis**.

Covalent catalysis can speed up reactions by converting stable functional groups into less stable functional groups. Consider the hydrolysis of a peptide bond by a serine protease. The uncatalyzed hydrolysis reaction involves a water molecule nucleophilically attacking the peptide bond. However, peptide bonds are a subset of amide bonds, which are not very reactive due to resonance stabilization. Therefore, peptide bonds are not easily attacked by a weak nucleophile such as water, even in an enzyme active site.

In contrast, a deprotonated alcohol like serine is a strong nucleophile that can break the amide bond. Doing so replaces the amide with an ester, which *can* be more easily attacked by water (Figure 4.55).

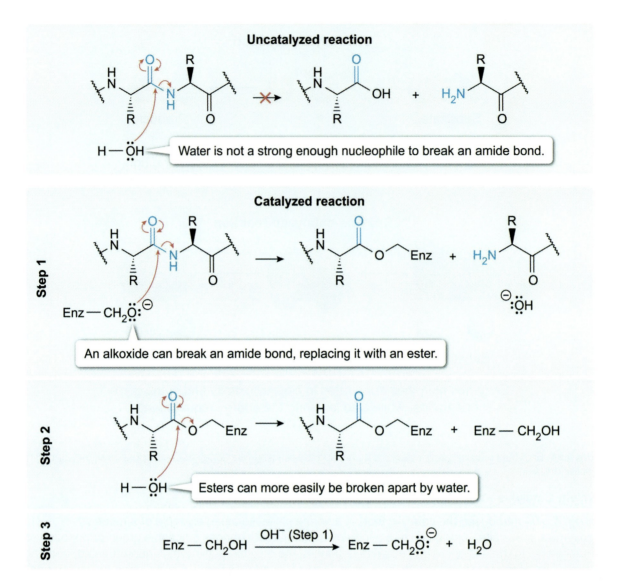

Figure 4.55 Covalent catalysis occurs when the enzyme makes a temporary covalent bond with its substrate.

Note that this covalent catalysis mechanism increases the number of steps needed to complete the reaction. Because free energy is a state function (as described in Lesson 4.1) this does not affect the overall free energy change ΔG of the reaction.

However, it *does* change the reaction from one with a single large E_a representing *one* transition state to a reaction with two smaller E_a values representing *two* transition states. This example also demonstrates that enzymes do not necessarily stabilize the transition state of the uncatalyzed reaction but rather may produce distinct transition state(s) (Figure 4.56).

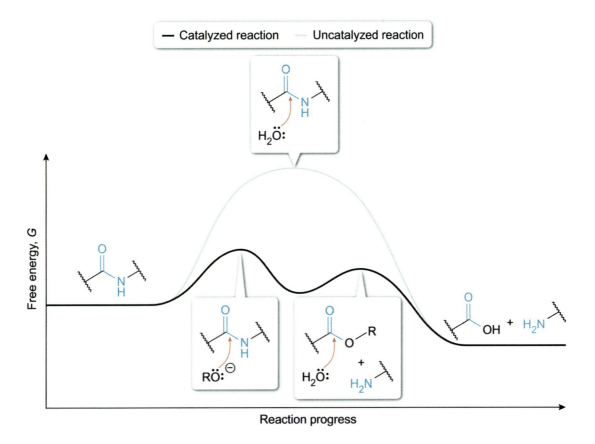

Figure 4.56 Enzyme-catalyzed reactions can proceed through a different reaction pathway than uncatalyzed reactions, which may result in different transition states.

Acid-Base Catalysis

The catalytic triad also demonstrates acid-base catalysis, wherein enzymes catalyze reactions by changing the protonation state, and therefore the reactivity, of various functional groups. In the case of the catalytic triad, the protonation state of the enzyme itself—specifically its active site serine residue—is altered by the active site microenvironment; however, many enzymes can also create an active site microenvironment that affects the *substrate's* protonation state.

If the enzyme itself serves as the Brønsted-Lowry acid (proton donor) or base (proton acceptor), this is known as *general* acid-base catalysis. In contrast, when hydronium or hydroxide ions from solution are used as the acid or base, this is known as *specific* acid-base catalysis. Because histidine's pK_a of 6 is very close to the physiological pH of 7.4, histidine often plays an important role in general acid-base catalysis, although other residues may also contribute depending on the active site microenvironment.

Effects of Cofactors and Coenzymes

Concept 2.4.03 introduced cofactors and coenzymes as important non–amino acid components of proteins. Many enzymes also utilize cofactors or coenzymes in their catalytic mechanism. Metal ions are common inorganic cofactors that assist in catalysis by electrostatically stabilizing the substrate and positioning it in the correct orientation, by serving as a redox center or by acting as a Lewis acid.

Coenzymes are organic cofactors and include molecules such as nicotinamide adenine dinucleotide (NAD^+ and NADH), flavin adenine dinucleotide (FAD and $FADH_2$), coenzyme A (CoA-SH), and coenzyme Q (ubiquinone and ubiquinol). The large variety of coenzymes indicates the large number of roles coenzymes can play. Figure 4.57 depicts the use of several coenzymes in the pyruvate dehydrogenase complex.

For some enzymes, the coenzyme is unchanged by the end of the reaction, and the coenzyme truly serves a catalytic role. Other enzymes modify a coenzyme such that the coenzyme does *not* revert to its original state by the end of the reaction. For this reason, such coenzymes are more accurately described as co-substrates; nevertheless, the name coenzyme persists, likely because the molecules are *eventually* regenerated after several metabolic enzyme-catalyzed reactions in other pathways.

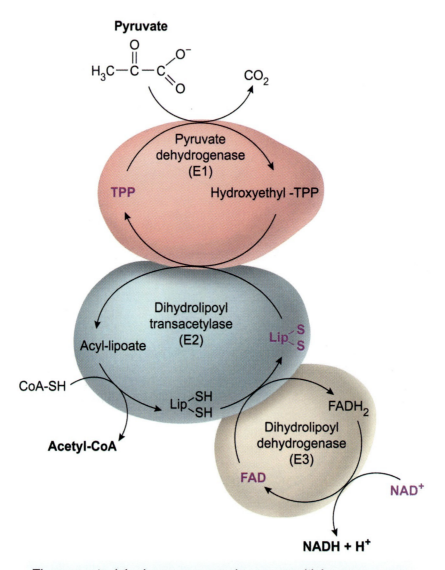

The pyruvate dehydrogenase complex uses multiple coenzymes.

Figure 4.57 The pyruvate dehydrogenase complex is an example of an enzyme complex that uses multiple coenzymes.

Figure 4.57 shows the pyruvate dehydrogenase enzyme complex (PDH), which catalyzes an intermediate step between glycolysis and the citric acid cycle (see Unit 4), as an example of this coenzyme usage. PDH uses the coenzymes thiamine pyrophosphate (TPP), lipoic acid (Lip), flavin adenine dinucleotide (FAD), and nicotinamide adenine dinucleotide (NAD^+). TPP, Lip, and FAD are restored to their original state by the end of the reaction. However, NAD^+ remains converted to NADH at the end of the reaction.

Lesson 4.4
Optimization of Enzyme Activity

Introduction

Enzymes are proteins, and proteins are affected by their environment. This lesson discusses how environmental conditions affect enzyme activity. Although physiological conditions are well controlled, different compartments within a cell or organism can have slightly different environments. These environments have important implications for enzyme activity. Pathological conditions may also alter an enzyme's environment and therefore its activity.

4.4.01 Measuring Enzyme Activity

Enzymes are biological catalysts that accelerate reactions to a rate necessary to sustain life. Therefore, the measurement of an enzyme's activity is, in theory, straightforward. An enzyme's **activity** can be measured as the amount of product produced per unit time. Alternatively, activity can be defined as the amount of substrate *consumed* per unit time. This measurement is often given in units of μmol/min, where 1 μmol/min = 1 enzyme unit (U), but the definition of an enzyme unit may vary depending on the system being studied.

One potential hurdle in the measurement of enzyme activity is that the rate *depends on the amount of enzyme* in the sample. In other words, enzyme activity is an extensive property and varies with enzyme concentration. If not accounted for, this may result in misinterpretation of data.

For example, consider an experiment in which samples of two different isoforms of an enzyme are prepared. The experimenter measures their activity and tries to compare the enzymes' catalytic abilities. Even if equal volumes are tested, a higher concentration of the slower enzyme may cause it to *appear* to have a higher activity than the faster enzyme, leading to an incorrect conclusion (Figure 4.58).

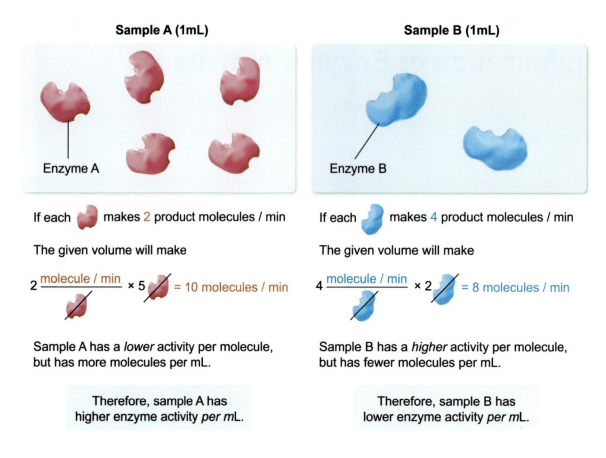

Figure 4.58 Enzyme activity calculations for samples of different enzymes with different concentrations.

Differences in enzyme concentration can come from many sources. These differences may be due to differences in expression level, cell or tissue viability, sample preparation, sample collection, or other factors. To account for enzyme amount, the **specific activity** is typically reported.

Specific activity is calculated by dividing the enzyme activity by the **mass of protein** in the sample, and it is commonly reported in units of µmol·min^{-1}·mg protein^{-1}, or simply U/mg protein. This calculation makes specific activity an intensive property of the protein mixture.

$$\text{Specific activity} = \frac{\text{Enzyme activity}}{\text{Mass of protein}} = \frac{\mu\text{mol/min}}{\text{mg protein}}$$

Note that "mg protein" appears in the denominator of the specific activity equation rather than mol protein. Most protein quantitation assays report *mass* of protein. Therefore, dividing by "mg" simplifies the calculation and avoids any discrepancies between expected and actual molar mass (eg, due to post-translational modifications, alternative splicing, or formation of protein complexes).

In addition, the conversion of mass to moles would only be appropriate if the entire protein mass in the sample were pure enzyme. It is *not* appropriate if there are any proteins other than the enzyme of interest, and samples usually contain other proteins, even after purification. By dividing by total protein mass, specific activity measurements can be calculated independent of enzyme purity. Furthermore, once calculated, specific activity can also be used to track increasing purity during a purification procedure.

Protein purification techniques are covered in detail in Chapter 14. As a brief overview, protein purification begins with the creation of a **crude lysate**, which is a sample of tissue or cells that have been lysed (ie, ruptured). In the crude lysate, all soluble proteins from the cytosol have been released to diffuse freely in the solution. The crude lysate contains not only the protein or enzyme of interest but also many other proteins and enzymes that the cells use to maintain life. The goal of a purification project is to remove unwanted, contaminating proteins while retaining as much of the protein of interest as possible (Figure 4.59).

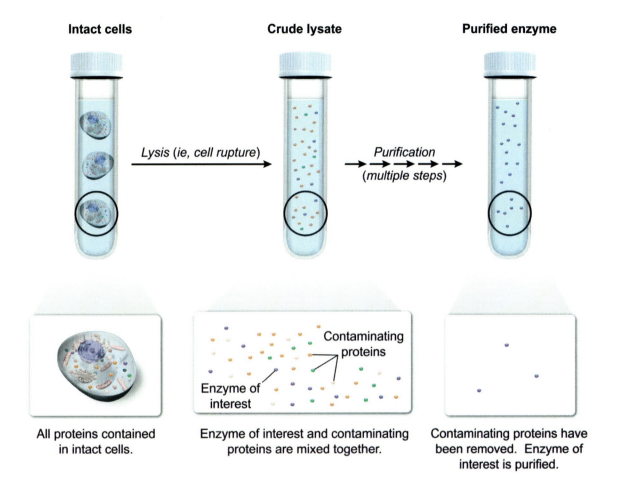

Figure 4.59 Overview of a protein purification process.

Within a purification procedure, the crude lysate contains the maximum amount of the enzyme of interest. Therefore, the crude lysate typically has the highest **total activity** of any purification step. Because some enzyme of interest is inevitably removed by purification steps, the total activity for each subsequent step typically decreases. However, crude lysate also contains the maximum number of *contaminants*. At this stage, the denominator of the activity equation (mg protein) is large, but the contaminating protein does not contribute to enzyme activity. Therefore, dividing by a large total amount of protein yields a low specific activity.

The contaminants are removed in subsequent purification steps while most of the enzyme of interest remains. The result is that the enzyme of interest forms a higher percentage of the total protein mass. Therefore, purification results in an *increased specific activity*.

Table 4.6 shows example measurements from a protein purification procedure. The values in white boxes can be measured directly by the experimenter. From these data, the values in shaded boxes can be *calculated*.

Table 4.6 Purification table showing values collected and calculated for a sample purification.

Purification step	Volume collected (mL)	Protein amount		Activity		Specific activity (U/mg)	Yield (%) (total activity/ total activity of crude lysate)
		Concentration (mg/mL)	Total (mg) (Volume × concentration)	Per mL (U/mL)	Total (U)		
Crude lysate	150	0.34	51	24	3600	71	100
After ion-exchange chromatography	2.2	4.6	10.	1500	3300	330	92
After size-exclusion chromotography	3.5	1.2	4.2	910	3200	760	89

U = enzyme unit = μmol substrate consumed/min.
Shaded boxes represent calculated values.

Total protein mass is the mass of all proteins in the whole volume collected. It is calculated by multiplying the volume by the measured protein concentration in mg/mL. Similarly, **total activity** is calculated by multiplying the volume by the measured activity per mL. Note that the units for volume in the numerator and denominator must cancel to provide a correct value.

$$\text{Total protein mass (mg)} = \text{Volume } (\cancel{mL}) \times \text{Concentration} \left(\frac{mg}{\cancel{mL}}\right)$$

$$\text{Total activity (U)} = \text{Volume } (\cancel{mL}) \times \text{Activity per mL} \left(\frac{U}{\cancel{mL}}\right)$$

The example data for total protein mass in Table 4.6 show how total protein mass *decreases* after each step of the purification.

Specific activity is calculated by dividing enzyme activity by protein mass. From the example data shown in Table 4.6, specific activity can be calculated in two ways. The first way is to divide total activity by the total protein mass.

$$\text{Specific activity} \left(\frac{U}{mg}\right) = \frac{\text{Total activity (U)}}{\text{Total protein mass (mg)}}$$

The second way is to divide the activity per volume by the concentration. Again, the units for volume must cancel out to provide a correct answer.

$$\text{Specific activity} \left(\frac{U}{mg}\right) = \frac{\text{Activity per mL } (U/\cancel{mL})}{\text{Protein concentration } (mg/\cancel{mL})}$$

The example data also demonstrate how specific activity increases with each purification step. To track purity over a purification procedure, the **fold-increase in purity** is calculated by dividing the specific activity of a given step by the specific activity of the crude lysate.

$$\text{Fold-increase in purity} = \frac{\text{Specific activity} \left(\frac{U}{mg}\right)}{\text{Specific activity of crude lysate} \left(\frac{U}{mg}\right)}$$

For example, Table 4.6 shows that size-exclusion chromatography increased the purity 11-fold $\left(\frac{760\ U/mg}{71\ U/mg}\right)$.

The **yield** of a protein purification can be thought of as the amount of enzyme recovered divided by the total amount of enzyme originally present.

$$\% \text{ Yield} = \frac{\text{Total activity}}{\text{Total activity of crude lysate}}$$

Total activity represents the amount of the enzyme of interest. The total activity of the crude lysate is therefore used as the reference for enzyme originally present. Because there is typically at least some loss of correctly folded, functional enzyme at each purification step, each purification step typically *increases* purity and specific activity but *decreases* yield.

✓ Concept Check 4.14

A researcher induced expression of the enzyme glycogen synthase in two separate cultures: one culture of *Escherichia coli* and one culture of *Saccharomyces cerevisiae*. Each culture was lysed, and the lysates were each purified by identical methods. Each lysate yielded 5 mL of purified enzyme, and enzyme activity in each was immediately measured by monitoring UDP-glucose consumption.

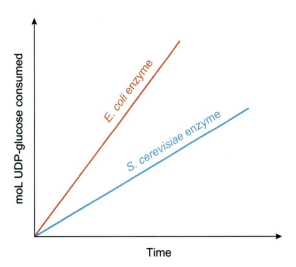

Which sample shows higher activity? What can be said about each enzyme's specific activity?

Solution
Note: The appendix contains the answer.

Concept Check 4.15

Information about the purification of glycogen synthase from *E. coli* is given in the following table.

Purification step	Volume collected (mL)	Protein concentration (mg/mL)	Activity per mL (U/mL)
Crude lysate	500.	1.0	20.
After affinity chromatography	10.	5.0	750

What is the fold-increase in purity of glycogen synthase? What is the percent yield of glycogen synthase?

Solution

Note: The appendix contains the answer.

4.4.02 Environmental Conditions for Enzyme Activity

The activity of an enzyme sample depends not only on the enzyme's purity but also on its environmental conditions. For example, altering the pH, salt concentration, or temperature of an enzyme preparation can have drastic effects on the enzyme's specific activity. Lesson 2.3 discussed how these conditions can induce protein denaturation. Like all proteins, enzymes can denature, and they are also affected by their environment in additional ways. The result is that each enzyme shows peak activity within a narrow range of pH, salt concentration, and temperature. The specific values of this range depend on the identity of each enzyme and constitute that enzyme's optimal conditions.

pH and Salt Concentration

As discussed in Concepts 2.2.03 and 2.2.04, salt bridges play an important role in stabilizing a protein's tertiary and (if present) quaternary structure. Figure 4.60 shows how alteration of pH can affect the protonation state of the residues forming those salt bridges, which disrupts them and can denature the protein. In addition, changes in salt concentration can disrupt salt bridges because the ionized residues may interact with dissolved salt instead of with their appropriate partner residue.

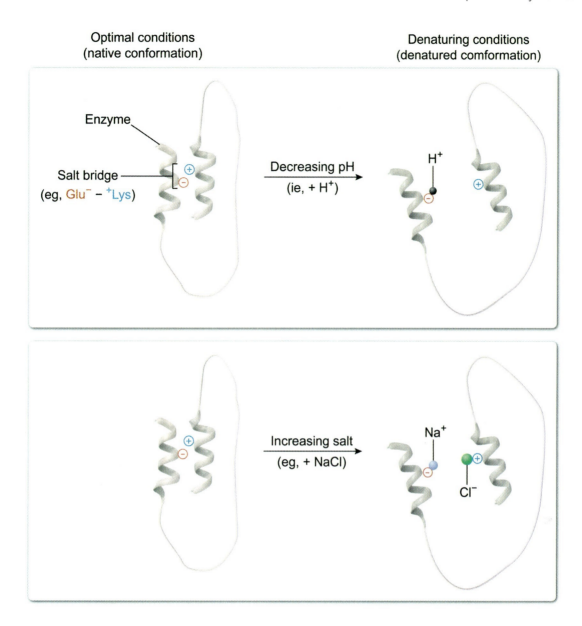

Figure 4.60 pH and salt concentration may lower enzyme activity through disruption of salt bridges, resulting in enzyme denaturation.

In addition to inducing denaturation, altering the pH or salt concentration can also directly affect the enzyme's catalytic mechanism. As discussed in Concept 4.3.05, several modes of catalysis depend on the ionization of active site residues (eg, acid-base catalysis, covalent catalysis, general enzyme-substrate interactions). Although the enzyme active site can alter amino acid pK_a values, the residues are still affected by the surrounding solution's proton and salt concentrations. Changes in these conditions may adversely affect acid-base equilibrium, nucleophilic ability, or substrate binding in the active site.

Because of the effects of pH and salt concentration on enzyme stability, denaturation, and catalytic mechanism, enzymes have evolved to show optimal activity within only a small range. Changes toward either side (too low or too high) can disrupt residue ionization state or correct salt bridge formation, leading to decreased enzyme activity (Figure 4.61).

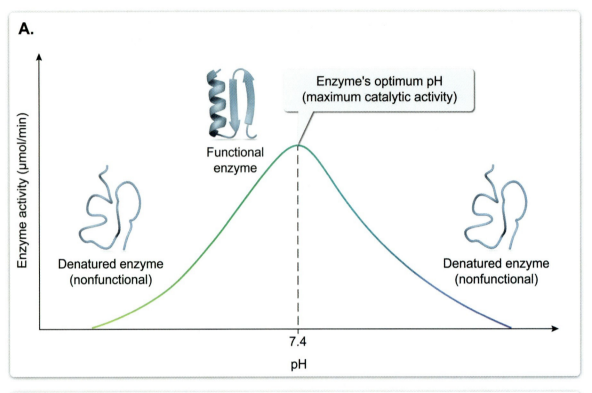

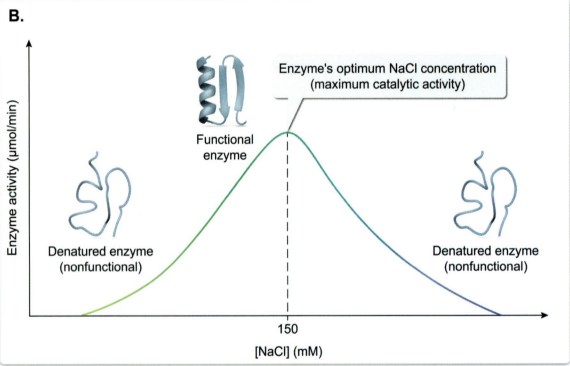

Figure 4.61 The effects of pH (A) and salt level (B) on enzyme activity.

Temperature

Enzymes also have an optimal temperature at which they are most active, as shown in Figure 4.62. Concept 2.3.04 discusses how high temperatures cause proteins to denature and lose their function. Therefore, temperatures above the optimal temperature cause enzyme activity to decrease.

Lower temperatures tend to stabilize proteins and enzymes in their native conformations. However, below-optimal temperatures decrease enzyme activity by decreasing the average energy of the molecules. At low temperatures, fewer molecules can reach the activation energy E_a.

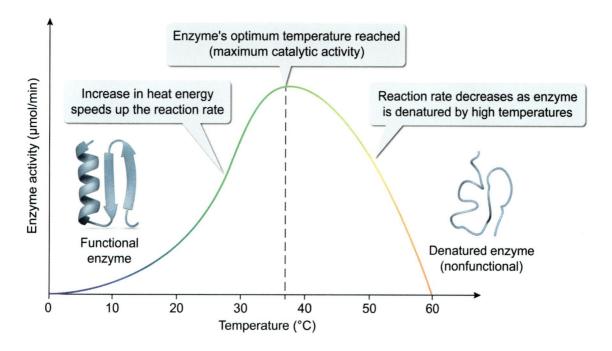

Figure 4.62 The effect of temperature on enzyme activity.

Different Enzymes Have Different Optimal Conditions

All enzymes have *their own* optimal pH, salt concentration, and temperature. The optimal conditions for any given enzyme will likely differ from those for another enzyme, even if they are from the same organism. Although the term "physiological conditions" typically refers to a well-controlled set of standard biochemical conditions (eg, pH 7.4, 37 °C), organisms often have subcompartments in which conditions differ (Figure 4.63).

For example, the cells that line the stomach have a cytosolic pH of 7.4, typical of most mammalian cells. Consequently, their cytosolic proteins likely have an optimal pH at or near 7.4. Conversely, the interior of the stomach is highly acidic. Therefore, the proteins and enzymes that are secreted into the stomach have optimal pH values around 2.0.

Similarly, different intracellular organelles have pH values that differ from the cytosol. For example, lysosomes are acidic whereas the mitochondrial matrix is basic. Consequently, the proteins and enzymes found in those spaces have evolved an optimal pH to match their environment.

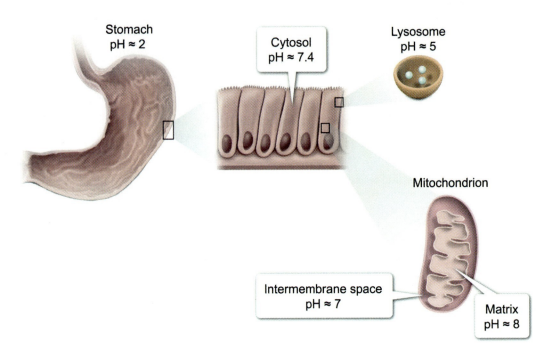

Figure 4.63 Bodies and cells contain subcompartments that maintain differing environmental conditions.

✓ Concept Check 4.16

Thermophiles are organisms that thrive at extreme temperatures (ie, 110–121 °C). In contrast, mesophiles thrive at moderate temperatures (ie, 20–45 °C).

DNA polymerase samples from a thermophile and from a mesophile were purified and tested for activity at various temperatures, yielding the following graph. Which sample was isolated from the mesophile, and which was isolated from the thermophile?

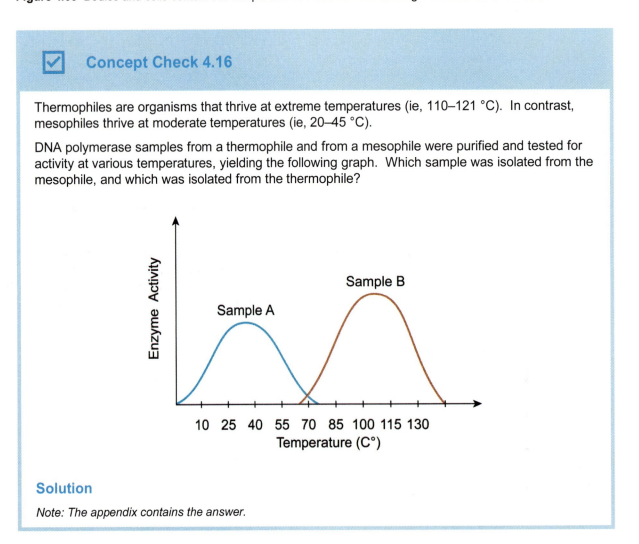

Solution

Note: The appendix contains the answer.

Pathological Conditions Adversely Affect Enzyme Function

Pathological conditions (ie, conditions resulting from an illness or other affliction) also affect conditions within the body and therefore affect enzyme function. For example, alkalosis and acidosis occur when pH values throughout the body become too basic or acidic, respectively. The fever response results in an increased body temperature above baseline (37 °C).

These pathological conditions typically affect enzyme activity and create suboptimal conditions for enzyme function. Sometimes these condition changes may have beneficial side effects; however, large changes are generally harmful. For example, *mild* fevers may stress the germs infecting a host and assist the host's immune response. Prolonged or elevated fevers, however, adversely affect the *host's* enzymes and proteins, resulting in increased risk of damage.

4.4.03 Cofactors and Coenzymes

In addition to being affected by environmental conditions such as pH, salt, and temperature, enzyme activity can also be affected by the presence or absence of specific molecules. The effect of substrate concentration on enzyme activity and enzyme rate is discussed in detail in Chapter 5. This concept focuses on the effects of cofactors and coenzymes.

Concept 2.4.03 introduces cofactors as components that are not amino acids (eg, metal ions) but are required for protein function. Coenzymes are simply organic cofactors. Prosthetic groups are cofactors that are tightly joined to the protein, often by a covalent bond (Figure 4.64).

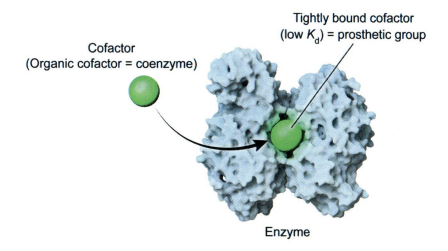

Figure 4.64 Cofactors, coenzymes, and prosthetic groups are components of proteins that are not amino acids but are required for protein function.

Cofactors, coenzymes, and prosthetic groups all play similar roles in enzymes. Cofactors can serve structural purposes and help maintain the enzyme's native conformation, or they can act directly in the catalytic mechanism. For example, metal ions such as Mg^{2+} can help orient substrates in the correct position.

By convention, some molecules are commonly called coenzymes but behave more like cosubstrates. For example, oxidoreductases can convert redox coenzymes such as NAD^+, $NADP^+$, and FAD from their oxidized forms to their reduced forms (Concept 4.2.02). Despite existing in a changed form at the end of the reaction, these molecules are often called coenzymes because they are typically regenerated to their original form by *other* enzymes.

Dietary Vitamins and Minerals Are a Source of Enzyme Cofactors and Coenzymes

The human body cannot synthesize most coenzymes and cofactors from a diet of just carbohydrates, protein, and fat. Therefore, cofactors and coenzyme precursors must be included in the diet as well. Dietary cofactors and coenzyme precursors are known as minerals and vitamins, respectively. An inability to consume enough dietary vitamins and minerals can lead to a loss of enzyme function and adverse health outcomes (Figure 4.65).

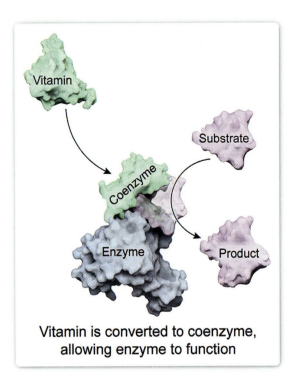 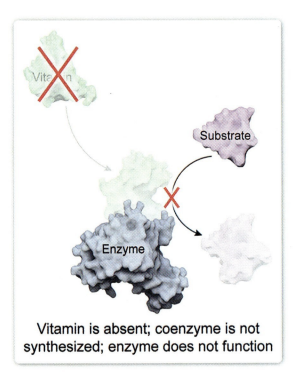

Figure 4.65 Minerals and vitamins are essential to the diet as they are used as cofactors and coenzyme precursors, respectively.

Lesson 5.1
The Michaelis-Menten Equation

Introduction

The study of enzyme kinetics is the study of how quickly an enzyme catalyzes its reaction. Whereas enzyme activity (Lesson 4.4) quantifies the *amount* of substrate consumed (or product produced) per unit time, enzyme kinetics focuses on the *rate* of substrate consumption (or product production) as a function of the amount of substrate and enzyme present.

The kinetics of an enzyme are often described by the Michaelis-Menten equation. This equation relates the reaction velocity V_0 to the maximum possible reaction velocity V_{max}, the Michaelis constant K_M, and the substrate concentration [S].

$$V_0 = \frac{V_{max}[S]}{K_M + [S]}$$

The rate of an enzyme-catalyzed reaction can change based on various factors, including substrate availability (ie, substrate concentration), enzyme concentration, and enzyme identity. Different **isozymes** (ie, different enzymes that catalyze the same reaction) may therefore have different kinetic parameters that allow them to perform different functions in various cells or tissues. A scenario in which two isozymes display different kinetics is graphed in Figure 5.1.

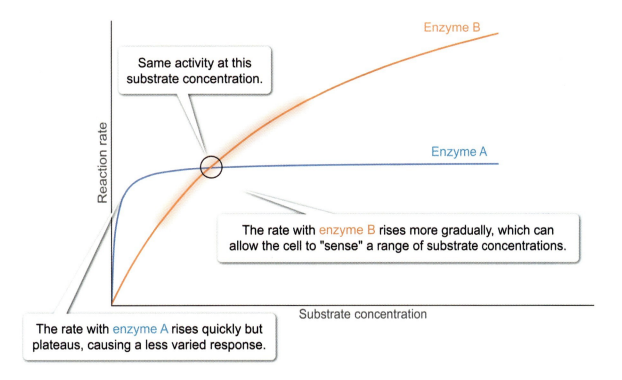

Figure 5.1 Two isozymes may catalyze the same reaction with the *same* activity at one substrate concentration but may have *different* activities if the substrate concentration changes.

The study of enzyme kinetics allows for a succinct description of these differences by quantifying an enzyme's kinetic parameters. Consequently, the exam commonly tests students' understanding of enzyme kinetics. This lesson presents the Michaelis-Menten equation in two sections: a summary of the assumptions that make the Michaelis-Menten equation valid and the derivation of the equation based on these assumptions. A third section discusses Michaelis-Menten kinetics in nonenzyme proteins.

5.1.01 Assumptions of the Michaelis-Menten Equation

The rate of a nonelementary reaction depends on the rate constants for each elementary step as well as the identification of the rate-determining step. Because enzyme-catalyzed reactions involve binding and unbinding in addition to the chemical reaction itself, the reaction scheme for an enzyme-catalyzed reaction can be simplified into three steps: enzyme-substrate binding, catalysis (ie, conversion of substrate to product), and enzyme-product unbinding (Figure 5.2).

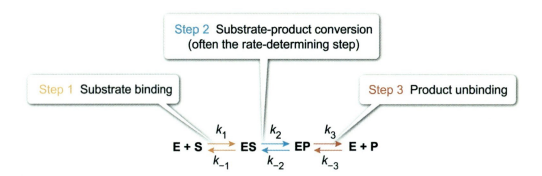

Figure 5.2 Simplified reaction scheme for an enzyme-catalyzed reaction.

In the reaction scheme shown in Figure 5.2, E is the free enzyme, S is the free substrate, ES is the enzyme-substrate complex, EP is the enzyme-product complex, and P is the product. The forward rate constant for Step 1 is k_1, and the reverse rate constant is k_{-1}. Similarly, k_2, k_{-2}, k_3, and k_{-3} are the forward and reverse rate constants for their respective steps. Step 2, in which bound substrate is converted to bound product, is typically assumed to be the slow, rate-determining step, whereas the binding and unbinding rates are assumed to occur much faster.

In reality, enzymes often follow more complex mechanisms than the one presented in Figure 5.2. For example, some enzymes may include multiple catalytic steps or reaction intermediates. Nevertheless, the simplified model is still applicable to many of these enzymes given the appropriate experimental conditions. The more complex models are generally not tested on the exam, and enzymes are therefore assumed to follow Michaelis-Menten kinetics unless otherwise specified.

Importantly, enzymes that exhibit cooperativity do *not* follow Michaelis-Menten kinetics. See Concept 3.1.04 for more information on cooperativity with general protein-ligand interactions and Concept 5.3.03 for more information on the effects of cooperativity on enzyme kinetics specifically.

To derive the Michaelis-Menten equation, several assumptions must be made. These assumptions do not apply to physiological kinetics of an enzyme *in vivo* but apply only to *experimental* measurements of enzyme kinetics *in vitro*. The assumptions and the experimental conditions under which they are valid are irreversibility, the steady state approximation, and the free ligand approximation.

Irreversibility

Michaelis-Menten kinetics assumes that the enzymatic reaction is irreversible—that once product forms (Step 2), it does *not* revert to substrate. Similarly, the subsequent step of product unbinding is irreversible; free product does not bind again to the enzyme. Given these assumptions, the two sequential, irreversible steps can be combined into a single catalytic step, represented by the forward rate constant k_{cat}. No reverse catalytic rate constant applies because this step is assumed to be irreversible.

The combined catalytic step is considered to be the slow, rate-determining step (RDS). Therefore, substrate binding (ie, Step 1), which occurs *prior* to the RDS, *is* considered reversible. This is because the faster rate constants (ie, k_1 and k_{-1}) allow multiple binding and unbinding events to occur prior to catalysis. At small timescales during which the slow catalytic events are unlikely, this binding step is said to be at a **rapid equilibrium**. These simplifications modify the kinetic model to that shown in Figure 5.3.

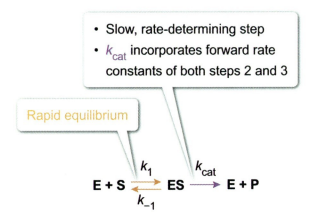

Figure 5.3 Enzyme kinetic mechanism under the irreversibility assumption.

For the irreversibility assumption to be valid during an experiment, *no product is included* in the initial reaction mixture. Without any initial product present, the reverse reaction cannot occur. By measuring only the **initial reaction velocity V_0** (ie, velocity at time $t = 0$), before product has formed, this ensures that no product is present. Consequently, the reaction's velocity is *not* measured near equilibrium (Figure 5.4).

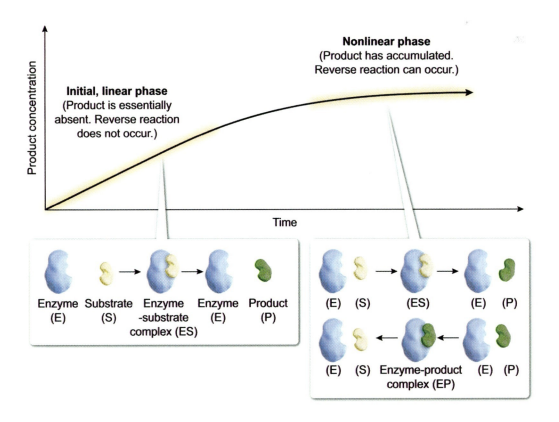

Figure 5.4 The irreversibility assumption for Michaelis-Menten kinetics.

Chapter 5: Enzyme Kinetics

Steady State Approximation

The steady state approximation supposes that concentration of the enzyme-substrate complex [ES] remains constant throughout the experiment (Figure 5.5).

Experimentally, this assumption requires the **initial substrate concentration** to be large enough that it does not appreciably change during the measurement period. If [S] were to appreciably deplete, then the [E] + [S] ⇌ [ES] rapid equilibrium would *also* shift in accordance with Le Châtelier's principle, which would then cause [ES] to change.

In addition, the **initial reaction velocity** V_0 is not measured directly but is instead extrapolated from measurements at later timepoints. Doing this ensures that the data do not include pre–steady state measurements, which would be the reaction rates measured while ES levels were building up but before steady state has been reached.

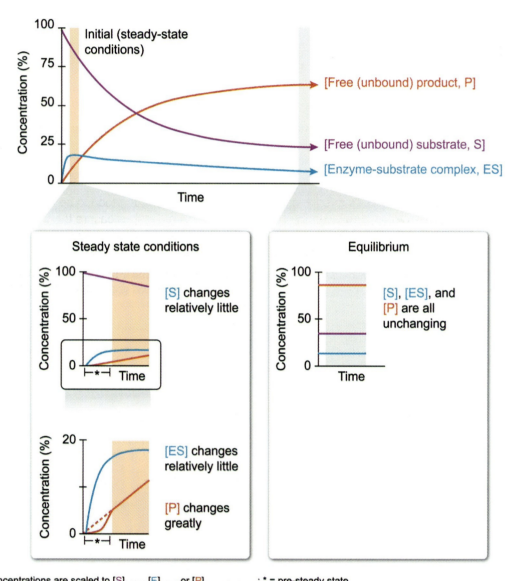

Note Concentrations are scaled to $[S]_{initial}$, $[E]_{total}$, or $[P]_{theoretical\ max}$; * = pre-steady state.

Figure 5.5 The steady state assumption of Michaelis-Menten kinetics.

Note that although steady state and equilibrium are similar concepts, they are *not* the same. For ES to be at equilibrium, *all* reaction steps involved would need to be at equilibrium, including the combined catalytic step—meaning that the forward and reverse reaction steps proceed at equal rates. In such a case, the net reaction rate would be zero.

Instead, the [ES] level is kept steady because any ES consumed by the catalytic step (ie, by conversion of ES to E + P) is quickly regenerated by Step 1. In other words, the rate of ES *destruction* by catalysis is matched by the rate of ES *production* from binding (Figure 5.6).

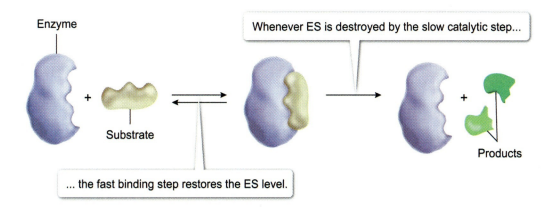

Figure 5.6 According to the steady state approximation, the [ES] level is constant throughout an enzyme kinetics experiment.

Free Ligand Approximation

In the derivation of the Michaelis-Menten equation, the variable [S] is understood to be the concentration of *free* (ie, unbound) substrate. However, in practice experimenters rarely measure the free substrate level. Instead, the free ligand (or for enzymes, free *substrate*) concentration is assumed to be approximately equal to the **initial substrate concentration** used to set up the experiment.

For this assumption to be valid, the substrate concentration must be orders of magnitude greater than the total enzyme concentration (ie, **[S] >> [E$_{tot}$]**). Under these conditions, the substrate concentration [S] does not appreciably decrease even if the enzyme is fully bound (Figure 5.7).

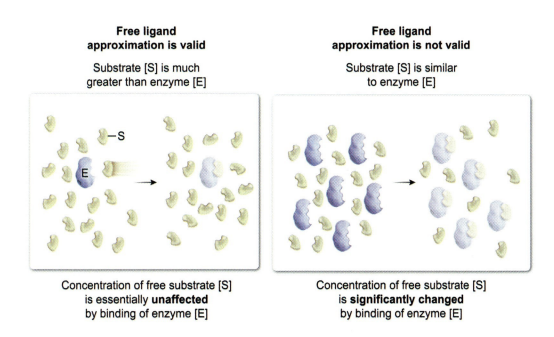

Figure 5.7 The assumption of free ligand approximation for Michaelis-Menten kinetics.

The assumptions necessary to analyze an enzyme with Michaelis-Menten kinetics are summarized in Table 5.1.

Table 5.1 The assumptions of Michaelis-Menten kinetics.

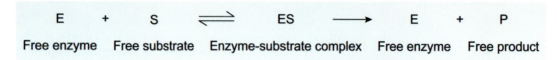

Assumption	Definition	Experimental method to ensure assumption is valid
Irreversibility assumption	Reaction proceeds in only one direction (ie, forward)	• $[P]_0 = 0$: No product in the reaction mixture • Ensures only the forward reaction contributes to net reaction velocity
Steady state assumption	During the experiment, [ES] does not vary, so reaction velocity does not vary	• V_0 = extrapolated initial velocity • Initial velocity ensures measurements occur before equilibrium is reached • Extrapolation ensures pre-steady state kinetics (ie, while ES is accumulating) is avoided
Free ligand approximation	$[S_{free}] \approx [S_{total}]$	• $[E] << [S]$: Total enzyme concentration is much smaller than substrate concentration • Ensures that $[S_{free}]$ is not much different from $[S_{total}]$, even if the enzyme is saturated

✓ Concept Check 5.1

Identify the errors in each of the following statements:

1. Because of the irreversibility assumption, only physiologically irreversible enzymes can be described by Michaelis-Menten kinetics.

2. Under the steady state approximation, [E], [S], [ES], and [P] are all held at steady concentrations throughout the experiment.

3. The free ligand approximation requires [S] to be in vast excess of $[E_{tot}]$ to ensure that the enzyme is saturated throughout the experiment.

Solution

Note: The appendix contains the answer.

5.1.02 Derivation of the Michaelis-Menten Equation

Although the derivation of the Michaelis-Menten equation is *not* directly tested on the exam, knowing its derivation is helpful in gaining a deeper understanding of the equation. Given the rate-determining step of the simplified kinetic model presented in Figure 5.3, the rate law can be used to express the reaction rate V_0 as a function of [ES]:

$$ES \xrightarrow{k_{cat}} E + P \quad \Rightarrow \quad V_0 = k_{cat}[ES]$$

The concentration of ES is difficult to measure or directly control and is consequently inconvenient to include in the final equation. However, the *steady state approximation* can be used to solve for ES in terms of more convenient factors. This substitution starts by recognizing that the rate of ES formation is equal to the rate of ES destruction:

$$\text{rate}_{formation} = \text{rate}_{destruction}$$

The formation of ES is represented only by the forward binding step. ES destruction, on the other hand, can occur through two methods: catalysis and unbinding. The combined rate of ES destruction is simply the sum of the rates of these two processes. Therefore, under steady-state conditions,

$$k_1[E][S] = k_{cat}[ES] + k_{-1}[ES]$$

This equation can be further rearranged to yield the following relationship:

$$\frac{[E][S]}{[ES]} = \frac{k_{cat} + k_{-1}}{k_1}$$

Being a function of only constants, the right side of this equality is itself a new constant. This combined constant, called the Michaelis constant, is denoted as K_M (sometimes stylized as K_m):

$$K_M = \frac{[E][S]}{[ES]}$$

To remove the fraction, this can be rearranged as:

$$K_M[ES] = [E][S]$$

The concentration of free (ie, unbound) enzyme [E] is also difficult to measure or control. However, [E] equals the total enzyme concentration [E_{tot}] minus the concentration of enzyme that is bound by substrate [ES]:

$$[E] = [E_{tot}] - [ES]$$

Therefore, the right side of this equation can be substituted for [E] in the previous equation, such that:

$$K_M[ES] = ([E_{tot}] - [ES])[S]$$

And simplifying,

$$K_M[ES] = [E_{tot}][S] - [ES][S]$$

The [ES] terms can all be collected on the same side of the equation by adding [ES][S] to both sides:

$$K_M[ES] + [ES][S] = [E_{tot}][S]$$

Factoring out [ES] yields

$$[ES](K_M + [S]) = [E_{tot}][S]$$

and solving for [ES] yields

$$[ES] = \frac{[E_{tot}][S]}{K_M + [S]}$$

This can then be substituted into the rate law ($V_0 = k_{cat}$[ES]) to give:

$$V_0 = k_{cat} \cdot \frac{[E_{tot}][S]}{K_M + [S]}$$

Both the total enzyme concentration [E_{tot}] and the total substrate concentration can be controlled by the experimenter. Based on the free ligand approximation, the free substrate concentration [S] is approximately equal to total substrate, so [S] can also be effectively controlled by the experimenter. Therefore, the rate of an enzyme-catalyzed reaction can be expressed as a function of variables that can be controlled, allowing for models to be developed and predictions to be made.

Given that $V_0 = k_{cat}$[ES], the maximal possible [ES] value—and therefore the maximal possible reaction velocity V_{max}—occurs if *all* enzymes present are bound to substrate (ie, if [ES] = [E_{tot}]). Therefore,

$$V_{max} = k_{cat}[E_{tot}]$$

This value can be substituted into the previous equation to give the most common form of the Michaelis-Menten equation:

$$V_0 = \frac{V_{max}[S]}{K_M + [S]}$$

Alternatively, because reaction velocity depends on the amount of enzyme bound to substrate, the Michaelis-Menten equation can also be derived in a manner similar to that by which the binding equation was derived in Lesson 3.1. This alternate derivation also highlights the similarities between the two relations, including the similarities in application between K_d and K_M the shapes of their described curves and plots, and their response to cooperativity (Lesson 5.3).

 Concept Check 5.2

A given enzyme has a K_M of 5 mM. When prepared with 15 mM substrate, the initial reaction velocity (V_0) is measured to be 75 μM/min. What is the V_{max} for this enzyme preparation? If the total enzyme concentration were to double, what would be the new V_{max} value and the new V_0 value when measured against 15 mM substrate?

Solution

Note: The appendix contains the answer.

5.1.03 Michaelis-Menten Kinetics in Nonenzyme Systems

Although the Michaelis-Menten model was derived specifically for enzymes, many other biochemical systems can be described as having similar response profiles. As previously discussed, the equation describing the fraction of protein bound (θ) is similar to the Michaelis-Menten equation. Therefore, responses that depend on protein-ligand interactions have a similar profile to Michaelis-Menten kinetics. This similarity also extends the effects of reversible inhibitors on protein responses (see Lesson 5.4).

Membrane transporters (see Concept 3.3.03) have similar response profiles to Michaelis-Menten kinetics. For transporters, their relevant equation is:

$$J_0 = \frac{J_{\max}[C]}{K_t + [C]}$$

where J is the flux (ie, rate of transport) of the solute, $[C]$ is the concentration of the solute being transported on a specified side of the membrane, and K_t is the transport constant. J, $[C]$, and K_t are analogous to V, $[S]$, and K_M in the Michaelis-Menten equation, respectively.

Chapter 5: Enzyme Kinetics

Lesson 5.2

Michaelis-Menten Parameters

Introduction

In addition to its dependent and independent variables (ie, V_0 and [S], respectively), the Michaelis-Menten equation involves two parameters that are constant within an experiment: V_{max} and K_M. As described in Concept 5.1.02, V_{max} is related to the catalytic rate constant k_{cat}, and k_{cat} can be further combined with K_M to form a new constant that describes an enzyme's intrinsic catalytic efficiency.

This lesson explains the Michaelis-Menten parameters, their derived constants, and what these values reveal about an enzyme's properties.

5.2.01 The Michaelis Constant (K_M)

K_d and K_M are both constants that have units of concentration (eg, millimolar [mM]). Although the Michaelis constant K_M is not a true binding constant, it can be used to describe the **affinity** of an enzyme for its substrate. Just as K_d indicates the ligand concentration at which half the proteins in a solution are bound, analysis of the Michaelis-Menten equation demonstrates that if the substrate concentration [S] equals the K_M value, then the reaction proceeds at half-maximal velocity (ie, ½V_{max}):

$$V_0 = \frac{V_{max}[S]}{K_M + [S]}$$

Given: [S] = K_M

$$V_{0,[S]=K_M} = \frac{V_{max} \cdot K_M}{K_M + K_M}$$

$$V_{0,[S]=K_M} = \frac{V_{max} \cdot K_M}{2\,K_M}$$

$$V_{0,[S]=K_M} = \frac{V_{max}}{2}$$

$$V_{0,[S]=K_M} = \frac{1}{2}V_{max}$$

An extension of this relationship can estimate V_0 as a percentage of V_{max} by expressing [S] as a multiple (n) of K_M:

Given: [S] = $n \cdot K_M$

$$V_{0,[S]=nK_M} = \frac{V_{max} \cdot nK_M}{K_M + nK_M}$$

$$V_{0,[S]=nK_M} = \frac{V_{max} \cdot nK_M}{(1+n)\,K_M}$$

$$V_{0,[S]=nK_M} = \frac{V_{max} \cdot n}{1+n}$$

$$V_{0,[S]=nK_M} = \frac{n}{n+1} \cdot V_{max}$$

Such that when

$$[S] = 1\,K_M, \qquad V_0 = \frac{1}{2}V_{max}$$

$$[S] = 2\,K_M, \qquad V_0 = \frac{2}{3}V_{max}$$

$$[S] = 3\,K_M, \qquad V_0 = \frac{3}{4}V_{max}$$

Because reaction velocity is proportional to the amount of enzyme bound to substrate, this also means that half the enzymes in solution are bound to substrate under experimental conditions when $[S] = K_M$, and two thirds of the enzymes in solution are bound when $[S] = 2\,K_M$. As $[S]$ approaches infinity, the enzymes become fully bound (ie, saturated), and V_0 approaches V_{max}.

The relationship between the K_M value and affinity is *inverse*. A small K_M value means that only a small concentration of substrate is needed to reach half-maximal velocity, and therefore the enzyme has a *high* affinity for its substrate. In contrast, a large K_M value means a large concentration of substrate is needed to reach half-maximal velocity, and therefore the enzyme has a *low* affinity for its substrate (Figure 5.8).

This definition of K_M—that it equals the substrate concentration that corresponds to half-maximal velocity—is used even for enzymes that do *not* follow conventional Michaelis-Menten kinetics (eg, cooperative enzymes; see Concept 5.3.03).

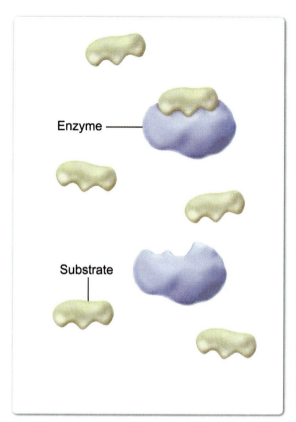 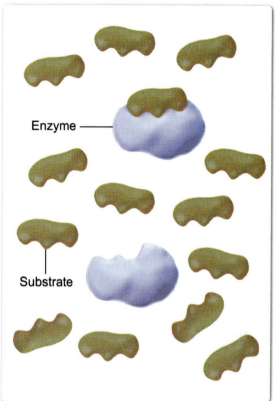

Low K_M
- Few substrate molecules are needed to bind enzyme
- High affinity

High K_M
- Many substrate molecules are needed to bind enzyme
- Low affinity

Figure 5.8 K_M is a measure of an enzyme's affinity for its substrate.

5.2.02 Turnover Number (k_{cat}) and Maximum Reaction Velocity (V_{max})

Maximal Reaction Velocity (V_{max})

V_{max} is the maximal reaction velocity that a given amount of enzyme can achieve within a set of experimental conditions. It is measured in units of concentration of product produced per unit time (eg, micromolar per minute [μM/min]). This velocity is only achieved if the enzyme is *saturated* with substrate. In other words, if an enzyme is operating at its V_{max}, then all enzyme active sites are bound to (and acting on) substrate (Figure 5.9).

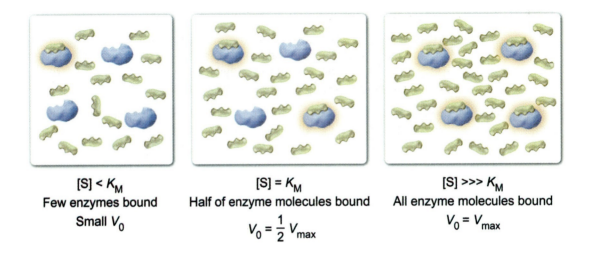

Figure 5.9 V_{max} is achieved when all enzyme molecules are bound to substrate (ie, enzyme is saturated).

Experiments can be performed to determine V_{max}. However, V_{max} is also an extensive property, which means that it *depends on the amount* of enzyme used in an experiment. An increased enzyme concentration corresponds to an increased V_{max}. Despite this, scientists often want to describe enzyme properties in ways that are intrinsic to the enzyme itself and *not* dependent on the amount used in an experiment. Therefore, the measured V_{max} is often converted into the intensive variable k_{cat}.

Turnover Number (k_{cat})

As described in Lesson 5.1, the reaction velocity (ie, reaction rate) can be described as a rate law that multiplies the concentration of the enzyme-substrate complex by the catalytic rate constant:

$$V_0 = k_{cat}[ES]$$

Because maximal reaction velocity is reached when all enzyme molecules (ie, E_{tot}) are saturated with substrate (ie, [ES] = [E_{tot}]), k_{cat} can be related to the measured V_{max} parameter through the equation:

$$V_{max} = k_{cat}[E_{tot}]$$

Rearranging this equation to solve for k_{cat} yields:

$$k_{cat} = \frac{V_{max}}{[E_{tot}]}$$

Like other first-order rate constants, k_{cat} has units of reciprocal time (eg, min^{-1}, s^{-1}).

The catalytic rate constant k_{cat} describes how many substrate molecules an enzyme can convert to product (or "turn over") per unit time, assuming the enzyme is in *saturating substrate conditions*. For this reason, k_{cat} is also known as the enzyme's **turnover number** (Figure 5.10).

Chapter 5: Enzyme Kinetics

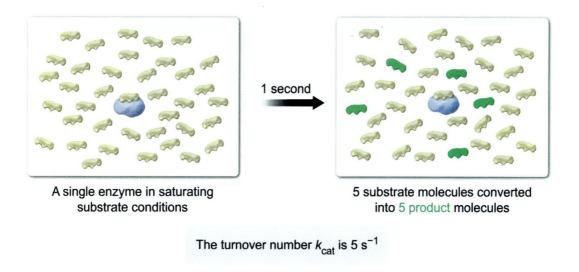

Figure 5.10 The turnover number (k_{cat}) is the number of reactions catalyzed per enzyme per unit time.

5.2.03 Catalytic Efficiency (k_{cat}/K_M)

Catalytic Efficiency

The Michaelis constant (K_M) and the turnover number (k_{cat}) provide two separate intensive parameters to describe an enzyme's ability to catalyze chemical reactions. To facilitate the comparison of different isozymes (ie, different enzymes that catalyze the same reaction), these parameters can be combined into a single constant called the enzyme's **catalytic efficiency**, also sometimes known as the *specificity constant*.

The catalytic efficiency of an enzyme is calculated by dividing the enzyme's k_{cat} value by its K_M value. As k_{cat} increases (meaning an enzyme converts more substrates per unit time when saturated), the catalytic efficiency also increases. As K_M *decreases* (meaning less substrate is needed to reach half-maximal velocity), catalytic efficiency also *increases* (Figure 5.11).

Because of these characteristics, catalytic efficiency is a succinct way to describe the combined effects of both kinetic parameters with one value.

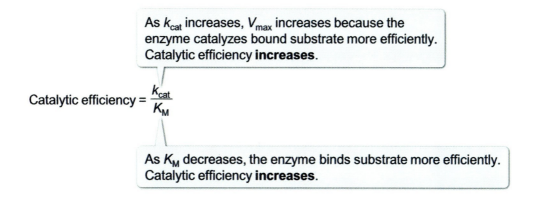

Figure 5.11 The catalytic efficiency of an enzyme.

The catalytic efficiency can also be thought of as being similar to a rate constant when substrate concentrations are well below the K_M value. If [S] << K_M, then the denominator of the Michaelis-Menten equation (K_M + [S]) can be approximated as K_M. Figure 5.12 shows how this approximation is derived.

$$V_0 = \frac{V_{max}[S]}{K_M + [S]}$$

If $[S] <<< K_M$, $K_M + [S] \approx K_M$

Therefore, $$V_0 \approx \frac{V_{max}[S]}{K_M}$$

$V_{max} = k_{cat}[E_{tot}]$

Therefore, $$V_0 \approx \frac{k_{cat}[E_{tot}][S]}{K_M}$$

Catalytic efficiency $= \frac{k_{cat}}{K_M}$

Rearranging the equation, $$V_0 \approx \frac{k_{cat}}{K_M}[E_{tot}][S]$$

When [S] is low, V_0 is proportional to catalytic efficiency

Figure 5.12 The approximation of the Michaelis-Menten equation, given low substrate concentration [S].

A higher catalytic efficiency value implies a faster rate. At extremely high values of catalytic efficiency (ie, greater than 10^8 M^{-1}s^{-1}), enzymes are said to have reached "catalytic perfection." This condition can occur if the k_{cat} is so large that the enzyme catalyzes the reaction almost as soon as enzyme and substrate meet, or if the K_M is so small that the substrate is unable to unbind before being catalyzed, or some combination of both. For catalytically perfect enzymes, *diffusion*—not binding nor catalysis—is the rate-limiting step.

Despite the connotations of the term "perfection," it is important to note that a high catalytic efficiency does *not* necessarily make an enzyme better. An organism may evolve an enzyme isoform to have a higher K_M (ie, lower affinity) if certain substrates need to be shared across several biochemical pathways. Alternatively, a lower k_{cat} may be desirable to minimize errors (eg, during DNA replication) or to minimize the production of toxic reaction intermediates.

Concept Check 5.3

Suppose that four different 5-mL samples of *Escherichia coli* cell culture are expressed (A through D), and a specific enzyme is purified from each sample without any further concentration or dilution steps. Each sample has a different enzyme mutation. Match the mutation description with the expected effect on the kinetic parameter.

- A. A mutation that increases activation energy (E_a) but does not affect substrate affinity or expression level.
- B. A mutation that increases expression level but does not affect affinity or E_a.
- C. A mutation that decreases substrate affinity but does not affect E_a or expression level.
- D. A mutation that increases substrate affinity but does not affect E_a or expression level.

- I. Increased catalytic efficiency
- II. Increased V_{max}
- III. Decreased k_{cat}
- IV. Increased K_M

Solution

Note: The appendix contains the answer.

Lesson 5.3

Graphical Representations of Enzyme Kinetics

Introduction

The Michaelis-Menten equation can be graphed directly by plotting reaction velocity (V_0) as a function of substrate concentration [S]. The Michaelis-Menten parameters may be difficult to discern from this graph, and it is often convenient to use a Lineweaver-Burk plot instead. This is a double reciprocal plot in which $1/V_0$ is plotted as a function of $1/[S]$ to yield a straight line.

Some enzymes do not follow Michaelis-Menten kinetics. Many enzymes exhibit cooperativity and follow Hill kinetics, which yield a sigmoidal (S-shaped) curve rather than a hyperbolic curve.

This lesson describes several graphs associated with enzyme kinetics and how these plots can be used to determine important enzyme parameters.

5.3.01 Michaelis-Menten Plots

To interpret experimental kinetics data, scientists often plot the data on a graph. The **Michaelis-Menten plot**, which compares reaction velocity V_0 on the *y*-axis against substrate concentration [S] on the *x*-axis, shares many features with the binding curves introduced in Concept 3.1.03 (Figure 5.13). Both plots are hyperbolic, and both plots feature a rapid rise followed by a plateau as the protein sample becomes saturated.

This concept discusses the generation of Michaelis-Menten plots and how they can be used to find the kinetic parameters defined in Lesson 5.2.

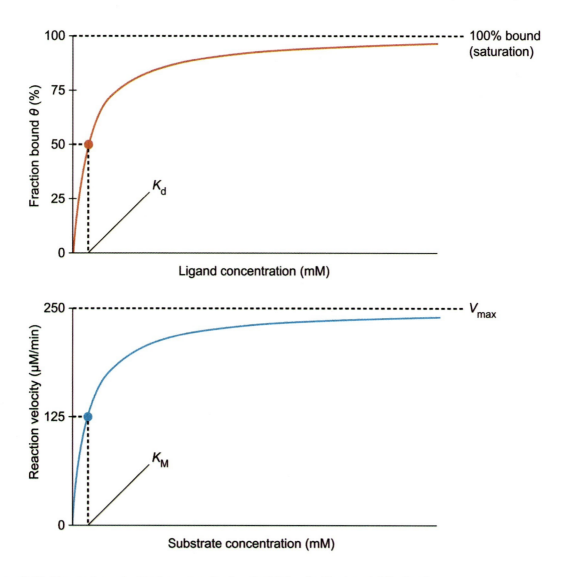

Figure 5.13 Comparison of a binding curve (top) and a Michaelis-Menten plot (bottom).

To construct a Michaelis-Menten plot, the reaction velocities must be measured at different substrate concentrations in experiments that satisfy the assumptions in Concept 5.1.01 (ie, [S] >> [E_{tot}], [P]$_0$ = 0, velocity measured at or near time t = 0). In a graph that plots product concentration versus time, the reaction velocity (ie, reaction rate) can be calculated as the *slope* of the activity curve.

Velocity measurements are taken for multiple samples, each with different initial substrate concentrations but the same total enzyme concentration. All other conditions are held constant. Each velocity is plotted against the substrate concentration that produced it, and the data are fit to a curve based on the Michaelis-Menten equation (Figure 5.14).

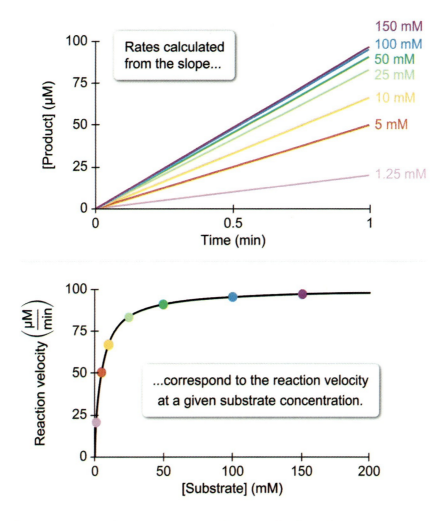

Figure 5.14 Steady-state reaction velocities are measured in several independent experiments (top), then plotted against substrate concentration to generate the Michaelis-Menten plot (bottom).

Once the plot is generated, the Michaelis-Menten parameters can be determined by analyzing the plot. The V_{max} value is the horizontal asymptote—in other words, it is the value the curve approaches as substrate concentration increases. The K_M value can be determined by dividing the V_{max} value in half and determining the substrate concentration that results in a point on the curve with that value (Figure 5.15). Note that K_M is *not* equal to ½V_{max}, but rather it is equal to the *substrate concentration* at which ½V_{max} occurs.

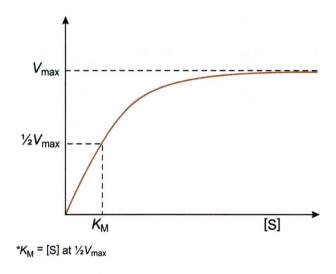

*K_M = [S] at ½V_{max}

Figure 5.15 V_{max} and K_M can be determined from a Michaelis-Menten plot.

The turnover number k_{cat} cannot be determined directly from a Michaelis-Menten plot. However, k_{cat} can be calculated if the enzyme concentration used to generate the plot is known by dividing V_{max} by the total enzyme concentration.

Similarly, the catalytic efficiency cannot be determined directly but must be calculated from k_{cat} and K_M. However, Figure 5.16 shows that the initial, linear phase of the Michaelis-Menten plot has a slope that is *proportional* to the catalytic efficiency. Specifically, the slope is approximately equal to the ratio of V_{max} to K_M (ie, $\frac{V_{max}}{K_M}$). This relationship becomes more important when comparing two or more Michaelis-Menten plots.

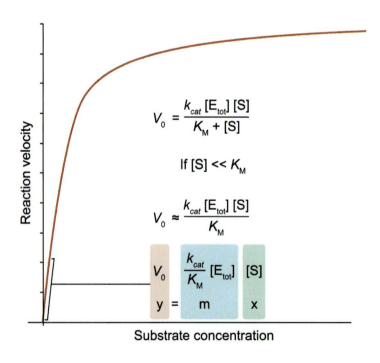

Figure 5.16 The catalytic efficiency is proportional to the initial slope of a Michaelis-Menten plot.

Comparing Multiple Michaelis-Menten Plots

Michaelis-Menten plots can be compared to examine enzyme kinetics from separate samples. This includes comparisons of different isozymes or the performance of a given enzyme under different conditions.

Like the binding curves discussed in Concept 3.1.03, left or right shifts of the Michaelis-Menten curve imply that K_M has decreased or increased, respectively. Therefore, a left shift (lower K_M) implies an *increased* affinity, and a right shift (higher K_M) implies *decreased* affinity (Figure 5.17).

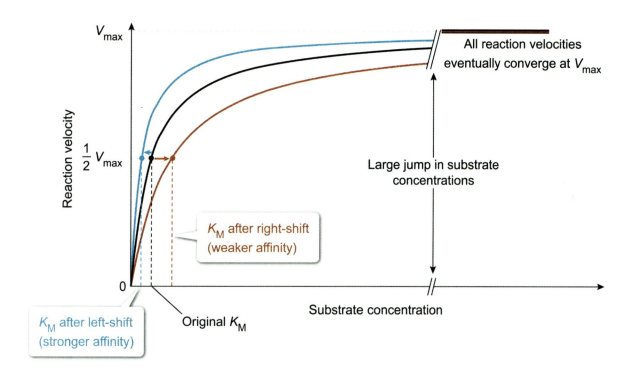

Figure 5.17 Differences in K_M are reflected as left or right shifts on a Michaelis-Menten plot.

In a similar manner, vertical shifts are indicative of a change in V_{max}. An increase in V_{max} is represented as a shift upward, and a decrease in V_{max} is represented as a shift downward (Figure 5.18).

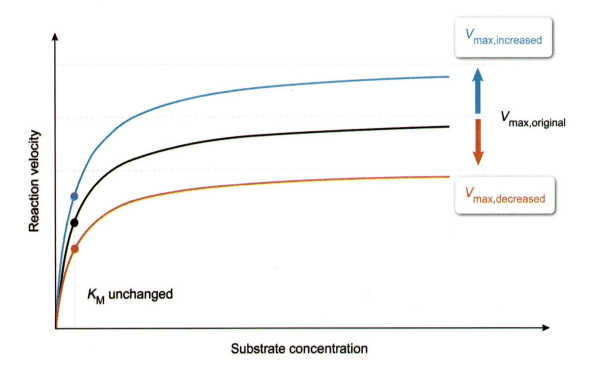

Figure 5.18 Differences in V_{max} are reflected as vertical shifts on a Michaelis-Menten plot.

Assuming that the total enzyme concentration is held constant, a change in V_{max} applies to k_{cat} as well. In other words, an increase in k_{cat} is reflected by a vertical shift upward, and a decrease in k_{cat} is reflected by a vertical shift downward. However, if a constant total enzyme concentration *cannot* be assumed, then no conclusions can be drawn about k_{cat} based on vertical shifts of the curve.

As discussed in the previous section, the V_{max}/K_M ratio is approximately equal to the initial slope of the Michaelis-Menten plot. If a constant total enzyme concentration can be assumed, then the slope is proportional to catalytic efficiency. Therefore, a decreased catalytic efficiency is represented by a decreased slope, which appears as a clockwise shift of the linear phase (Figure 5.19). Conversely, an increased catalytic efficiency is represented by an increased slope and a counterclockwise shift of the linear phase.

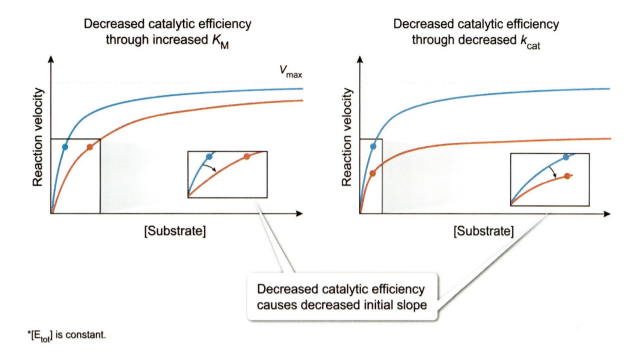

Figure 5.19 Changes in the initial slope (ie, when [S] < K_M) of a Michaelis-Menten graph represent changes in catalytic efficiency.

Physiological Relevance of Enzymes with Different Kinetics

Concept 5.2.03 introduced the idea that enzymes with higher catalytic efficiencies are not necessarily better because enzymes with lower efficiencies may simply play a different role. Consider the two enzymes that produce the Michaelis-Menten plots shown in Figure 5.20. This pair of plots represents two different isozymes that catalyze the same reaction but with different kinetics.

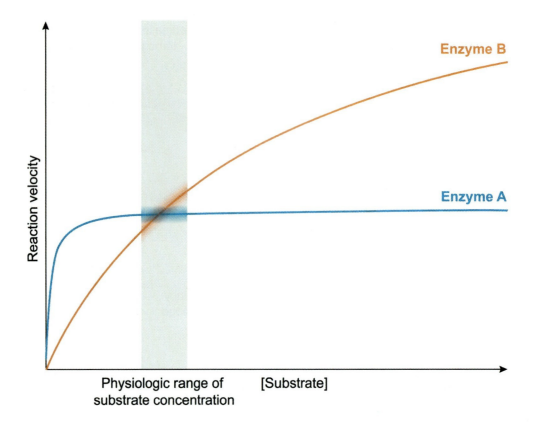

Figure 5.20 The Michaelis-Menten plots of two isozymes.

It is evident that Enzyme A reaches saturation more quickly than Enzyme B. Within the physiologic range of values for substrate concentration, Enzyme A always operates near its V_{max}. Enzyme B, on the other hand, has a much larger K_M, and therefore is within its *linear phase* throughout the physiologic substrate range. Which enzyme a cell uses is dictated by whether the cell needs the enzyme to be sensitive or insensitive to the level of substrate.

For example, red blood cells need quick access to glucose and therefore use the enzyme hexokinase, which is insensitive to glucose levels (like Enzyme A) to trap glucose as soon as it enters the cell. Conversely, the pancreas regulates its hormone production based on glucose level. It uses the related enzyme glucokinase, which shows a wider variation in velocity in response to changes in glucose concentration (like Enzyme B).

Transporters have kinetics similar to Michaelis-Menten kinetics and can operate on a similar principle. For example, red blood cells use the glucose transporter GLUT1, which is insensitive to glucose levels, whereas the pancreas uses GLUT2, which is more sensitive to glucose levels.

> **Concept Check 5.4**
>
> Hexokinase-2 is an isoform of hexokinase with an expression level that is upregulated by insulin signaling. Upon upregulation of hexokinase expression, how does its Michaelis-Menten plot change (ie, left, right, upward, or downward shift, increased or decreased initial slope)? How do the parameters (ie, K_M, V_{max}, k_{cat}, catalytic efficiency) change?
>
> **Solution**
>
> *Note: The appendix contains the answer.*

5.3.02 Lineweaver-Burk Plots

Before computers, the fitting of enzyme kinetics data to a Michaelis-Menten curve was difficult. Even *with* computers, a Michaelis-Menten curve can be difficult to interpret visually because the velocity only ever *approaches* (but never *reaches*) V_{max}. Several transformations of the Michaelis-Menten equation have been proposed that linearize the data, making the data both easier to fit as well as easier to interpret and analyze (Figure 5.21).

The Lineweaver-Burk equation is one of the most common linearizations of the Michaelis-Menten equation. Its corresponding plot results in recognizable data points (ie, the *y*- and *x*-intercepts) that relate to the values of V_{max} and K_M, and its slope is related to the V_{max}/K_M ratio. Consequently, many questions on the exam that involve enzyme kinetics data present Lineweaver-Burk plots as opposed to traditional Michaelis-Menten plots.

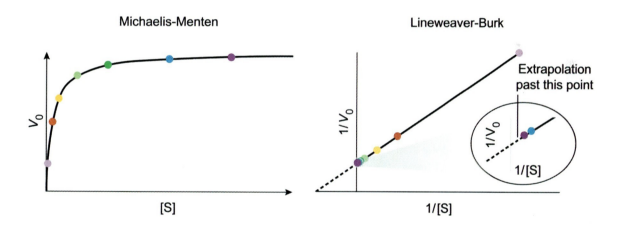

Figure 5.21 The Lineweaver-Burk plot is a linearization of Michaelis-Menten data.

The Lineweaver-Burk plot is a double reciprocal plot of its Michaelis-Menten counterpart. Instead of V_0 being plotted on the *y*-axis and [S] on the *x*-axis, the reciprocals are plotted: $1/V_0$ on the *y*-axis and $1/[S]$ on the *x*-axis. Mathematically, the Lineweaver-Burk equation can be derived by taking the reciprocal of both sides of the Michaelis-Menten equation:

$$V_0 = \frac{V_{max}[S]}{K_M + [S]}$$

$$(V_0)^{-1} = \left(\frac{V_{max}[S]}{K_M + [S]}\right)^{-1}$$

$$\frac{1}{V_0} = \frac{K_M + [S]}{V_{max}[S]}$$

$$\frac{1}{V_0} = \frac{K_M}{V_{max}[S]} + \frac{[S]}{V_{max}[S]}$$

$$\frac{1}{V_0} = \frac{K_M}{V_{max}} \cdot \frac{1}{[S]} + \frac{1}{V_{max}}$$

The resulting equation is in slope-intercept form (ie, $y = mx + b$). The dependent variable y is $1/V_0$ and the independent variable x is $1/[S]$. The slope (m) is K_M/V_{max} and the y-intercept (b) is $1/V_{max}$. Further analysis reveals that the x-intercept is equal to $-1/K_M$.

Although the Michaelis-Menten plot never reaches V_{max}, which can make visual determination of both V_{max} and K_M difficult, the line of best fit for a Lineweaver-Burk plot provides definite reference points (ie, the intercepts) to easily identify these kinetic parameters (Figure 5.22).

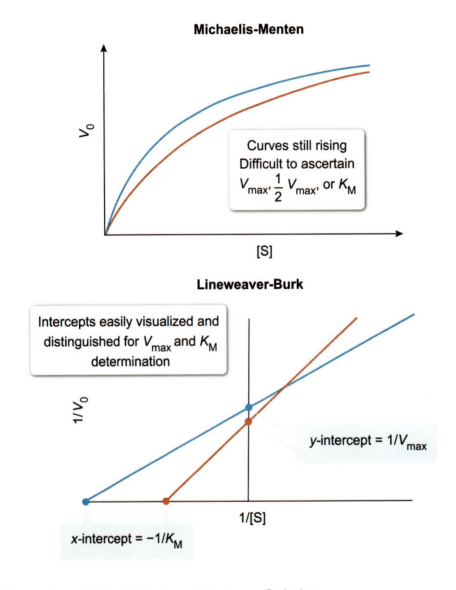

Figure 5.22 Comparison of Michaelis-Menten and Lineweaver-Burk plots.

Comparing Multiple Lineweaver-Burk Plots

Because they provide definite reference points for determining V_{max} and K_M, Lineweaver-Burk plots are more useful than Michaelis-Menten plots for comparing enzyme kinetics from separate samples. Changes in V_{max} appear as vertical shifts of the y-intercept, changes in K_M appear as horizontal shifts of the x-intercept, and changes in the V_{max}/K_M ratio (and therefore catalytic efficiency, assuming total enzyme concentration is constant) appear as changes in the slope.

However, because the Lineweaver-Burk plot is a double reciprocal plot, the direction of these shifts is **inversely related** to the direction of the change, as shown in the figures that follow. An *increase* in V_{max} results in a *decrease* in its reciprocal $1/V_{max}$, and therefore the y-intercept shifts *downward*. Conversely, a *decrease* in V_{max} results in an *increase* in its reciprocal and an *upward* shift of the y-intercept (Figure 5.23).

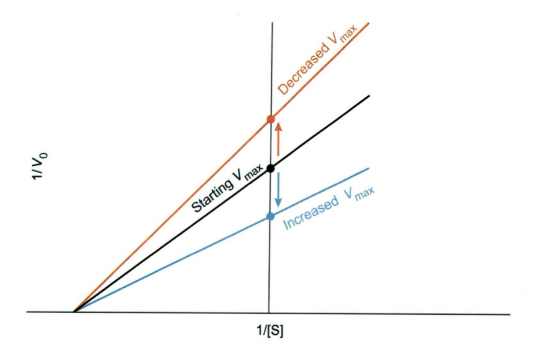

Figure 5.23 The y-intercept changes in the opposite direction of the change in V_{max}.

An *increase* in K_M results in a *decrease* in its reciprocal, and therefore the x-intercept ($-1/K_M$) shifts to the *right*, back *toward* the origin. Conversely, a *decrease* in K_M results in an *increase* in its reciprocal and a shift to the *left*, *away* from the origin (Figure 5.24).

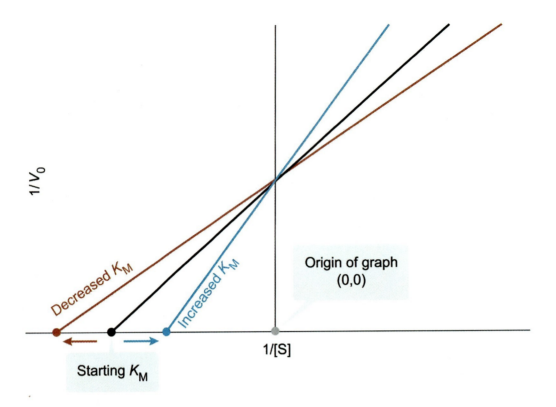

Figure 5.24 With respect to the origin, the x-intercept changes in the direction opposite the change in K_M.

The slope of a Lineweaver-Burk plot is $\frac{K_M}{V_{max}}$, or equivalently, $\frac{K_M}{k_{cat}[E_{tot}]}$. Therefore, the slope is inversely proportional to catalytic efficiency. Assuming that [E_{tot}] is kept constant, an *increase* in catalytic efficiency results in a *decreased* slope (ie, a shallower rise). Conversely, a *decrease* in catalytic efficiency results in an *increased* slope (ie, a steeper rise) (Figure 5.25).

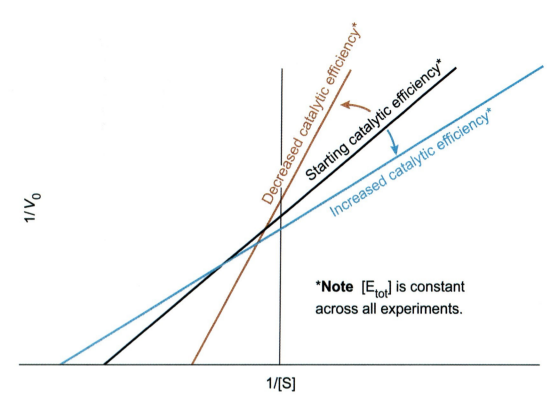

Figure 5.25 Assuming [E$_{tot}$] is constant, the slope changes in the direction opposite the change in catalytic efficiency.

> ☑ **Concept Check 5.5**
>
> Transaminase's kinetic data are measured with respect to glutamate (Glu) under two conditions: with 10 mM and with 100 mM oxaloacetate (OAA). Other than the concentration of OAA, both sets of experiments were performed under the same conditions, including the same [E$_{tot}$]. How does the increased concentration of OAA affect transaminase's measured V_{max}, k_{cat}, K_M, and catalytic efficiency values?
>
>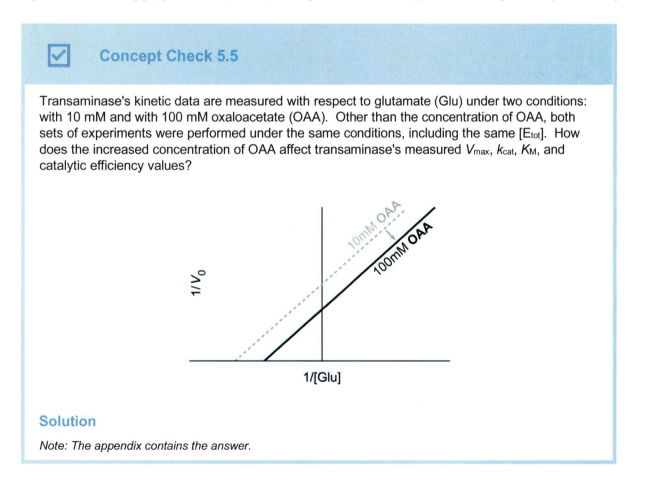
>
> **Solution**
> *Note: The appendix contains the answer.*

5.3.03 Cooperative Enzymes

As discussed in Concept 3.1.04, cooperativity in proteins occurs if the binding of one ligand causes a conformational change that influences the protein's ability to bind subsequent ligands. Enzymes can show cooperativity if the binding of one substrate affects that enzyme's ability to bind a second (or third, or fourth) substrate molecule.

Enzymes typically show cooperativity either by binding multiple copies of the same substrate or by binding multiple different substrates at the same time (eg, a bisubstrate enzyme that forms a ternary complex; see Concept 4.3.05). The exam focuses primarily on binding to multiple copies of the same substrate, so that behavior is the focus of this lesson.

Enzymes that bind multiple copies of the same substrate usually have multiple catalytic domains. Typically, this occurs in multimeric complexes (eg, dimers, trimers). The Michaelis-Menten plots for cooperative enzymes change in ways like those described in Concept 3.1.04.

In positive cooperativity (ie, binding of the first substrate facilitates binding subsequent substrates, Hill coefficient $n > 1$), the Michaelis-Menten curve becomes sigmoidal (ie, S-shaped). In negative cooperativity (ie, binding the first substrate hinders binding to subsequent substrates, $0 < n < 1$), the plot shows a sharp initial rise followed by a more gradual increase toward V_{max} (Figure 5.26). The Lineweaver-Burk plot also changes (from linear to nonlinear), although interpretation of cooperativity in Lineweaver-Burk plots is rarely required by the exam.

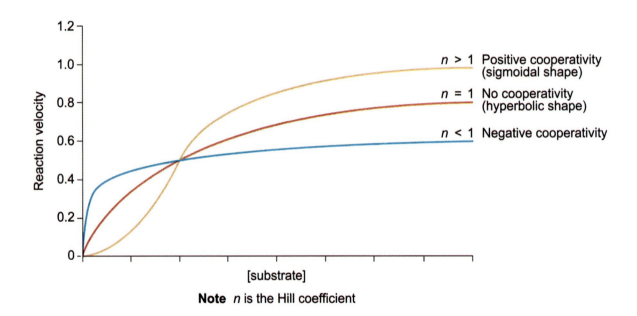

Note n is the Hill coefficient

Figure 5.26 Michaelis-Menten plots for an enzyme showing cooperativity with regard to the same substrate molecule.

Concept 5.3.01 introduces the idea of certain enzymes acting as "sensors" of substrate levels because a small change in substrate concentration can produce a large change in reaction velocity. Positive cooperativity can enhance this property by causing the curve to be steeper near the enzyme's K_M (Figure 5.27). This sharp response to substrate levels is similar to the benefits of cooperativity seen with hemoglobin for oxygen binding and delivery (see Biology Lesson 13.1).

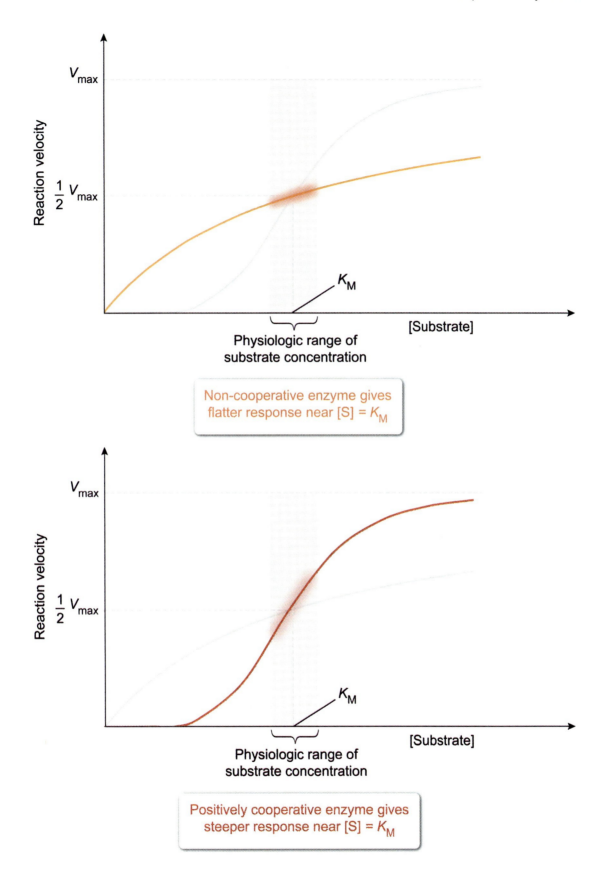

Figure 5.27 The sigmoidal curve of a positively cooperative enzyme allows for a steeper change in reaction velocity with regard to changes in substrate concentration.

Lesson 5.4
Enzyme Inhibitors

Introduction

Enzymes can be inhibited by small molecules that bind either the active site or an allosteric site and thereby alter the reaction kinetics. Enzyme inhibition can act in a reversible (or dynamic) manner or in an irreversible manner. As the name implies, dynamic, reversible inhibition can be undone once the molecule that caused it is removed.

Reversible inhibitors are classified as competitive, uncompetitive, or mixed (with noncompetitive being a special case of mixed inhibition) based on their mechanism of action. In contrast, irreversible inhibition *cannot* be undone.

This lesson describes the mechanisms of irreversible and reversible inhibition, and the graphs they produce are examined.

5.4.01 Irreversible Inhibitors

Irreversible inhibition results from a *permanent* alteration to an enzyme that *decreases* that enzyme's activity. Typically, these alterations are the result of **irreversible covalent modifications** to the enzyme or from very tight (ie, low K_d) binding (Table 5.2).

Because the modification is permanent, irreversible inhibition can only be overcome by synthesizing more unaltered enzyme; therefore, irreversible inhibition is energetically costly. Aside from protein degradation, irreversible inhibition is relatively uncommon as an *endogenous* (ie, native to the organism) means of regulation. Instead, irreversible inhibitors typically come from *exogenous* sources such as synthetic drugs or natural toxins.

Table 5.2 Examples of reversible and irreversible inhibition.

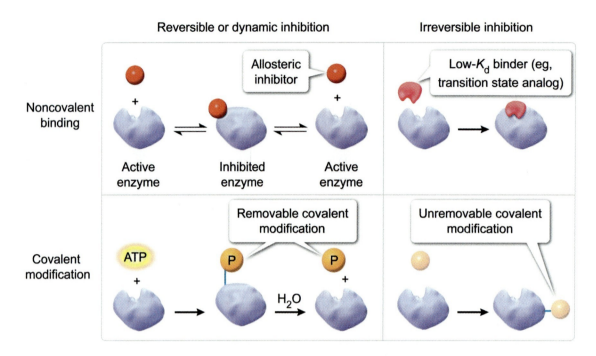

The energetic cost associated with irreversible inhibitors ensures that they have powerful effects *in vivo*. Irreversible inhibitors can also be very useful in the study of enzymes *in vitro*. Many irreversible inhibitors specifically target their enzymes by binding to the active site. As discussed in Concept 4.3.01, the optimal ligand for an enzyme is the reaction's transition state. Therefore, a noncovalent inhibitor that binds with a K_d value low enough to be considered irreversible is likely a transition state analog, which can provide important insight into a reaction mechanism.

Other irreversible inhibitors can participate in the first part of a covalent catalysis mechanism (see Concept 4.3.06) but are *unable to complete* the process. In other words, the irreversible inhibitor is subject to nucleophilic attack by an active site residue, forming the *modified enzyme*, but the reaction *does not regenerate* the unmodified enzyme at the end of the reaction (Figure 5.28).

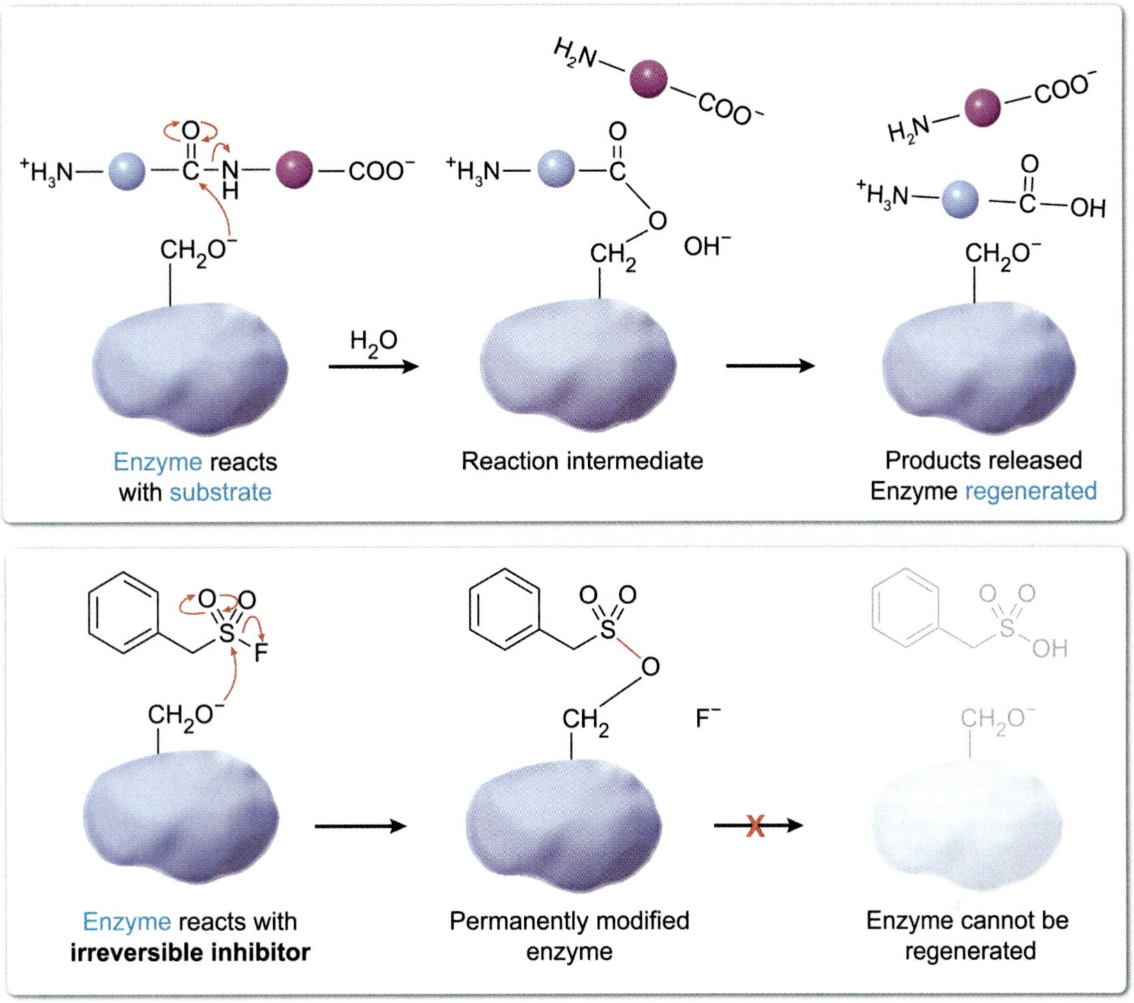

Figure 5.28 Irreversible inhibitors do not regenerate the unmodified enzyme (compare to Figure 4.8).

Because inhibitors that act in this manner participate partly in the catalytic mechanism, they are sometimes called **mechanism-based inhibitors**. Another name for irreversible inhibitors is **"suicide inhibitors"** because they are unable to inhibit additional enzymes after they have reacted once. Importantly, because irreversible inhibitors do not regenerate the unmodified enzyme, they are *not true substrates* of the enzyme.

Kinetic Effects of Irreversible Inhibitors

Because many irreversible inhibitors require a chemical reaction to occur, irreversible inhibitors generally require *more time* to take full effect than reversible inhibitors. Consequently, the inhibitory effect of irreversible inhibitors can be enhanced by an **incubation period** (ie, a period *before activity measurements are taken* during which the enzyme and the inhibitor have time to react) (Figure 5.29).

Because irreversible inhibitors effectively remove functional enzyme from the system, an irreversible inhibitor decreases the V_{max} of the enzyme. If an inhibitor competes with the substrate for the active site, there may be a brief period during which the apparent K_M seems to increase; however, once all the inhibitor has reacted with enzyme, any remaining enzyme will have an unaltered K_M.

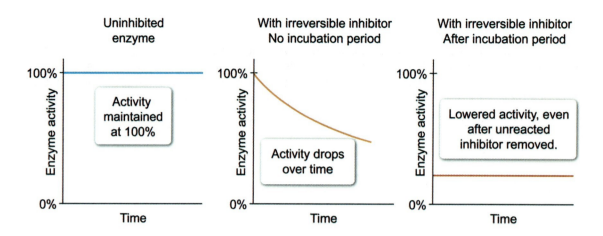

Figure 5.29 The effect of an irreversible inhibitor increases over time.

Concept Check 5.6

A sample of a specific enzyme is separated into three aliquots, and each aliquot is treated with a different molecule for 10 minutes. After treatment, the enzyme is purified to remove unreacted treatment molecules, and the enzyme activity is measured again. The results are shown in the table provided.

	Molecule 1	Molecule 2	Molecule 3
Specific activity before treatment	100 nmol/min/mg enzyme	100 nmol/min/mg enzyme	100 nmol/min/mg enzyme
Observations after 10-minute treatment	Specific activity drops to 75 nmol/min/mg enzyme	Specific activity drops to 90 nmol/min/mg enzyme and a new molecule (molecule 4) is detected.	Specific activity drops to 50 nmol/min/mg enzyme
Specific activity after purification post-treatment	75 nmol/min/mg enzyme	100 nmol/min/mg enzyme	100 nmol/min/mg enzyme

Which of the molecules is most likely an irreversible inhibitor?

Solution

Note: The appendix contains the answer.

5.4.02 Competitive Inhibitors

Unlike irreversible inhibition, reversible inhibition does *not* cause a permanent change to the affected enzyme. Although some covalent regulatory mechanisms are dynamic and can be reversed (see Concept 6.2.02), this concept focuses on reversible regulators that inhibit their target enzyme through noncovalent means (ie, through binding-induced conformational changes).

Broadly speaking, reversible inhibitors bind their target enzyme at one of two places: at the enzyme active site or at a different (ie, allosteric) site. The measured kinetics of the enzyme can differ based on where the inhibitor binds and whether the inhibitor binds free enzyme, the enzyme-substrate complex, or both. Consequently, reversible enzyme inhibitors are typically classified into one of several classes: competitive inhibitors, uncompetitive inhibitors, mixed inhibitors, and noncompetitive inhibitors.

As the name implies, competitive inhibitors "compete" with the substrate for the free enzyme. Most commonly, this occurs if the competitive inhibitor is a **substrate analog** (ie, a molecule that structurally looks like, but is *not*, the substrate). Substrate analogs bind the enzyme active site but do *not* react. Importantly, an enzyme can bind only its substrate *or* a competitive inhibitor but *cannot bind both at the same time*, as shown in Figure 5.30.

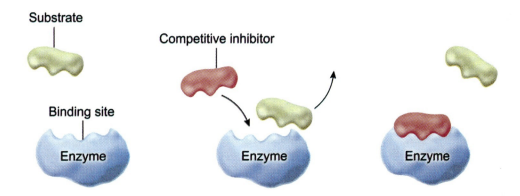

Figure 5.30 Most competitive inhibitors resemble and bind to the same site as the substrate. The enzyme can bind one or the other but cannot bind both at the same time.

Put another way, the competitive inhibitor binds *the free enzyme* but *not* the enzyme-substrate complex. This relationship is shown schematically in Figure 5.31. Note that some competitive inhibitors can bind at an allosteric site so long as they induce a conformational change that completely prevents substrate binding. In other words, any reversible inhibitor that binds free enzyme only (ie, does *not* bind to enzyme-substrate complex) is a competitive inhibitor, whether it binds at the active site or an allosteric site. However, any reversible inhibitor that binds the active site *must* be competitive, and this form of competitive inhibition is the most likely to appear on the exam.

Chapter 5: Enzyme Kinetics

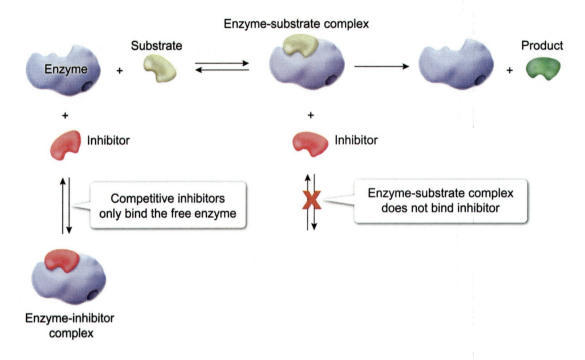

Figure 5.31 Competitive inhibitors bind only free enzymes.

Because competitive inhibitors bind free enzymes and block substrate binding, they not only prevent the enzyme from interacting with its substrate but also effectively decrease the concentration of free enzyme [E] by converting it to the enzyme-inhibitor complex (EI). In doing so, competitive inhibitors shift the substrate-binding equilibrium to the left (ie, toward unbinding of substrate) to increase free enzyme levels in accordance with Le Châtelier's principle. This shift effectively decreases the apparent affinity of the enzyme for its substrate, which in turn *increases* the apparent value of the Michaelis constant K_M (Figure 5.32).

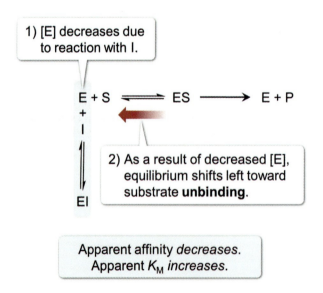

Figure 5.32 Competitive inhibitors cause an increase in apparent K_M.

In contrast, V_{max} is *not changed* by a competitive inhibitor. The maximal enzyme reaction velocity V_{max} is achieved when the enzyme is fully saturated with substrate molecules. Although the binding of a competitive inhibitor blocks the simultaneous binding of substrate, competitive inhibitor binding is *reversible*, which means they can unbind to *be replaced by substrate*. At high enough substrate concentrations, the substrate outcompetes the inhibitor, resulting in a fully active, substrate-saturated enzyme population. More substrate is needed to reach saturation (ie, higher K_M), but the *velocity at saturation is unchanged* (Table 5.3).

Table 5.3 High substrate concentrations can outcompete a competitive inhibitor.

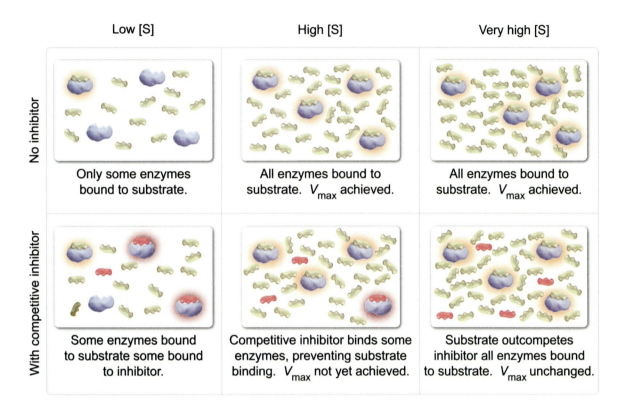

5.4.03 Uncompetitive Inhibitors

Unlike competitive inhibitors, which only bind free enzymes, *uncompetitive* inhibitors *only bind the* **enzyme-substrate complex** and thereby prevent the inhibitor-bound enzyme-substrate complex from completing its reaction. Inhibitor binding does *not* compete with substrate binding but neither is it completely independent of substrate binding; instead, <u>un</u>competitive inhibitor binding requires binding to the <u>un</u>ion of the enzyme and substrate together (Figure 5.33).

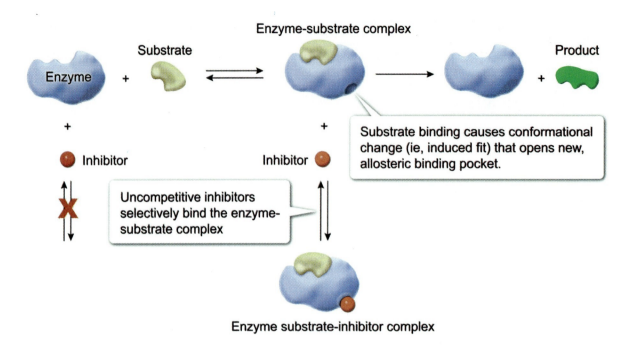

Figure 5.33 Uncompetitive inhibitors bind only the enzyme-substrate complex.

By binding to the enzyme-substrate complex (ES), uncompetitive inhibitors effectively lower the ES concentration ([ES]), which affects both the binding step and the catalytic step. According to Le Châtelier's principle, removal of ES causes the rapid equilibrium of the binding step to shift to the right, toward production of more ES and therefore *toward substrate binding*. This result is measured as an *increase* in apparent affinity, which corresponds to a *decrease* in the apparent K_M value (see Figure 5.34).

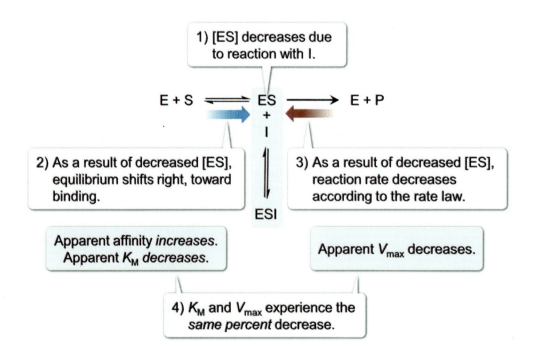

Figure 5.34 Uncompetitive inhibitors decrease both K_M and V_{max} by the same factor.

In addition, the catalytic step is also affected by the decrease in [ES]. Although the combined catalytic step is not at equilibrium under experimental conditions, the decrease of reactant concentration leads to a *decrease in rate* (V_0), as expected according to the rate law. Unlike competitive inhibitors, this decrease *cannot* be countered by increasing the substrate concentration. Even at saturating substrate concentrations, uncompetitive inhibitors can bind the enzyme-substrate complex to inhibit the reaction, which leads to a *decrease in V_{max}* (Table 5.4).

Table 5.4 Uncompetitive inhibitors decrease reaction rate, even at saturating substrate concentrations.

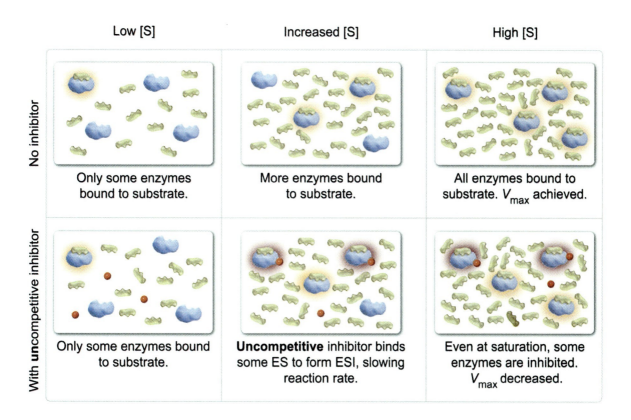

Importantly, uncompetitive inhibitors decrease both the apparent K_M and V_{max} values by *exactly the same factor*, as noted in Figure 5.34. In other words, an uncompetitive inhibitor that causes a 50% decrease in K_M will also cause a 50% decrease in apparent V_{max}. Another uncompetitive inhibitor that causes a 10% decrease in V_{max} will also cause a 10% decrease in K_M.

This effect means that uncompetitive inhibitors result in no change to the **V_{max}/K_M ratio**. Because [E_{tot}] should be unchanged across experiments, this also means that although k_{cat} has decreased, the apparent catalytic efficiency (ie, k_{cat}/K_M) is unchanged by an uncompetitive inhibitor.

5.4.04 Mixed Inhibitors

Whereas competitive inhibitors bind only free enzyme and uncompetitive inhibitors bind only the enzyme-substrate complex, mixed inhibitors can bind *either* **free enzyme** *or* the **enzyme-substrate complex** to prevent the reaction (Figure 5.35).

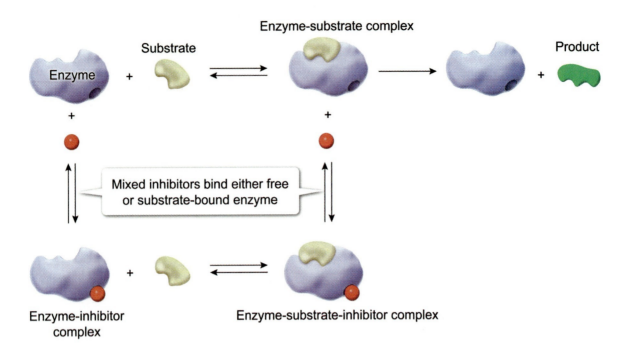

Figure 5.35 Mixed inhibitors bind either the free enzyme or the enzyme-substrate complex.

As with uncompetitive inhibitors, mixed inhibitors can bind the enzyme-substrate complex. Therefore, mixed inhibitors still inhibit the reaction at saturating substrate concentrations, meaning that mixed inhibitors always cause a *decrease in* V_{max}.

However, the effects of a mixed inhibitor on the apparent K_M value are more variable and depend on the **relative affinities** of the inhibitor for free enzyme and enzyme-substrate complex. As discussed in Concept 3.1.02, affinities can be described by an equilibrium dissociation constant (eg, K_d). For enzyme inhibitors, this dissociation constant is sometimes denoted as K_i. For mixed inhibitors, K_i is often used to describe the affinity of the inhibitor for the free enzyme, whereas K_i' is used to describe the affinity for the enzyme-substrate complex.

Recall that competitive inhibitors bind only the free enzyme. Therefore, if a mixed inhibitor has a stronger affinity for the free enzyme (ie, $K_i < K_i'$), then the inhibitor is more "competitive-like." Consequently, this type of mixed inhibitor produces an *increase* in the apparent K_M value. In contrast, if the mixed inhibitor has a stronger affinity for the enzyme-substrate complex (ie, $K_i' < K_i$), then the inhibitor is more "uncompetitive-like" and produces a *decrease* in the apparent K_M.

Note that K_M cannot decrease by a greater factor than V_{max} does because the lower limit of affinity—the point at which inhibitor binding to free enzyme is so weak that it does not bind at all—is represented by *true* uncompetitive inhibition, which sets the lower bound for the K_M decrease. Consequently, mixed inhibitors that prefer the enzyme-substrate complex cause K_M to decrease by a *smaller* factor than the decrease in V_{max}. For example, if V_{max} decreases by 50%, K_M must decrease by *less* than 50%.

The effects of mixed inhibitors on apparent K_M are summarized in Figure 5.36.

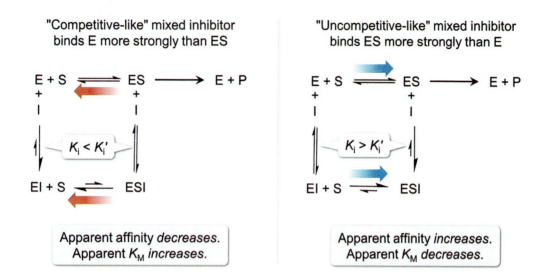

Figure 5.36 A mixed inhibitor lowers V_{max} and can either increase (if it binds free enzyme more strongly) or decrease (if it binds the enzyme-substrate complex more strongly) the apparent K_M.

5.4.05 Noncompetitive Inhibitors

A special subclass of mixed inhibitors binds the free enzyme and the enzyme-substrate complex with *equal affinities* (ie, $K_i = K_i'$). As with other mixed inhibitors, V_0 is decreased at all substrate concentrations, including saturating substrate concentrations (ie, V_{max} is decreased). However, because these inhibitors *do not prefer* either the free enzyme or the enzyme-substrate complex, the apparent K_M value is neither increased nor decreased but remains *unchanged* by the inhibitor.

Another way to think about this subclass of inhibitors is that substrate binding does not affect how well the inhibitor binds, and similarly inhibitor binding *does not affect* how well the substrate binds. The two events (ie, substrate binding and inhibitor binding) are *completely independent*. To denote this special property, a member of this subclass of mixed inhibitors is often called a purely **noncompetitive inhibitor** (Figure 5.37).

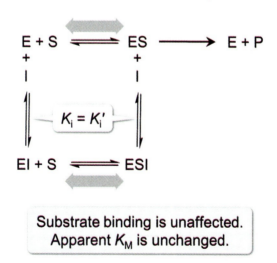

Figure 5.37 Noncompetitive inhibitors bind the free enzyme and the enzyme-substrate complex with equal affinities and leave K_M unchanged.

Summary

Competitive inhibitors are *usually* substrate analogs that bind at the enzyme's primary binding site (although some bind at allosteric sites). All other inhibitor classes *must* bind at an allosteric site. Whereas competitive inhibitors can be outcompeted (leaving V_{max} unchanged), the ability of the other classes to bind the enzyme-substrate complex allows them to inhibit the enzyme even when the enzyme is saturated with substrate, which causes V_{max} to decrease. The effect on apparent K_M for each class depends on the relative affinity of each inhibitor for the free enzyme and the enzyme-substrate complex.

The features of each inhibitor class are summarized in Table 5.5.

Table 5.5 Reversible enzyme inhibitors.

Type of inhibition		Inhibitor binds	Effect on K_M	Effect on V_{max}
	Competitive	E only	Increases	No change
Mixed (3 versions)	Favors free enzyme	E more than ES	Increases	Decreases
	Noncompetitive (special case of mixed)	E and ES equally	No change	Decreases
	Favors ES complex	ES more than E	Decreases	Decreases
	Uncompetitive	ES only	Decreases	Decreases by same factor as K_M

*E = free enzyme; ES = enzyme-substrate complex.

 Concept Check 5.7

The effects of a certain reversible inhibitor on an enzyme are tested. The measured V_{max} and K_M values are shown in the table provided.

	No inhibitor	With inhibitor
V_{max}	100 µM/min	50 µM/min
K_M	1.0 mM	0.75 mM

How should the reversible inhibitor be classified? What are the relative affinities (ie, higher affinity, lower affinity, no affinity at all) of the inhibitor toward the enzyme versus the enzyme-substrate complex?

Solution

Note: The appendix contains the answer.

Chapter 5: Enzyme Kinetics

5.4.06 Determining Inhibitor Types from Graphical Data

The Michaelis-Menten plot was introduced in Concept 5.3.01 as a method of representing kinetic data by plotting V_0 on the y-axis against [S] on the x-axis. The Lineweaver-Burk plot introduced in Concept 5.3.02 is a linearization of the Michaelis-Menten plot that is particularly useful in interpreting *changes* in kinetic data, such as those caused by reversible inhibition.

This concept explores the graphical changes in both Michaelis-Menten and Lineweaver-Burk plots brought about by inhibitors.

Competitive Inhibitors

As discussed in Concept 5.4.02, competitive inhibitors *increase apparent K_M and leave V_{max} unchanged*. On a Michaelis-Menten plot, this is represented as a right shift of K_M with *no change* in the horizontal asymptote (ie, V_{max}). Because V_0 approaches (but never reaches) the V_{max} value, an unchanged V_{max} may be difficult to identify unless sufficiently high [S] values are tested, as shown in Figure 5.38 (left).

Because the Lineweaver-Burk plot extrapolates the line of best fit beyond the collected data points, changes in the K_M and V_{max} values can be clearly identified by examining the x- and y-intercepts, respectively. The increase in K_M changes the x-intercept (ie, $-1/K_M$) by shifting it rightward, *toward the origin*. Because V_{max} is unchanged, the y-intercept (ie, $1/V_{max}$) is *unchanged*. This results in an inhibited data curve that intersects the uninhibited data curve at the y-intercept. These changes are reflected in Figure 5.38 (right).

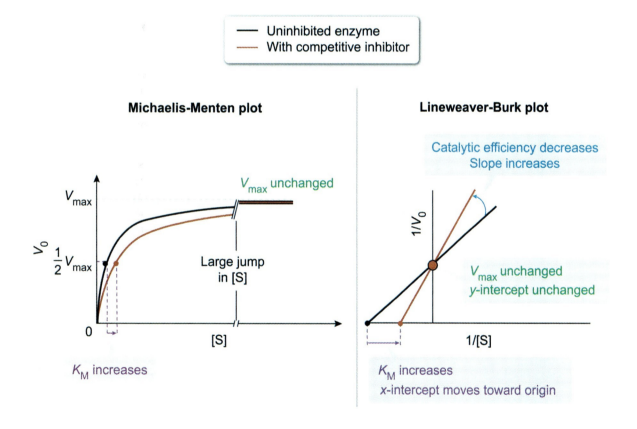

Figure 5.38 Changes to the Michaelis-Menten plot (left) and Lineweaver-Burk plot (right) caused by the addition of a competitive inhibitor.

Figure 5.38 also showcases some additional features of competitive inhibition on the kinetic plots. The slope of the Lineweaver-Burk plot (ie, K_M/V_{max}) increases, indicative of the *decrease* in the V_{max}/K_M ratio and in catalytic efficiency (ie, k_{cat}/K_M). This change is also noticeable in the Michaelis-Menten plot as a decrease in the initial slope (ie, when [S] << K_M).

Uncompetitive Inhibitors

Uncompetitive inhibitors bind the unified enzyme-substrate complex and *decrease both K_M and V_{max} by exactly the same factor*. On a Michaelis-Menten plot, the horizontal asymptote representing the V_{max} value is lowered, and K_M is shifted to the left. An example of uncompetitive inhibition on a Michaelis-Menten plot is shown in Figure 5.39 (left). However, with simple visual inspection of a Michaelis-Menten plot, it can be difficult to discern between an uncompetitive inhibitor and an "uncompetitive-like" mixed inhibitor (which *also* decreases both K_M and V_{max} but by *different* factors).

On a Lineweaver-Burk plot, decreases in the K_M and V_{max} values cause both the *x*- and *y*-intercepts, respectively, to shift *away from the origin*—more specifically, the *x*-intercept (ie, $-1/K_M$) shifts left and the *y*-intercept (ie, $1/V_{max}$) shifts upward. Because both parameters shift by the same factor, the slope of the Lineweaver-Burk plot (ie, K_M/V_{max}) is *unchanged*, resulting in the inhibitor data producing a line that is *parallel* to the uninhibited data. An example of uncompetitive inhibition on a Lineweaver-Burk plot is shown in Figure 5.39 (right).

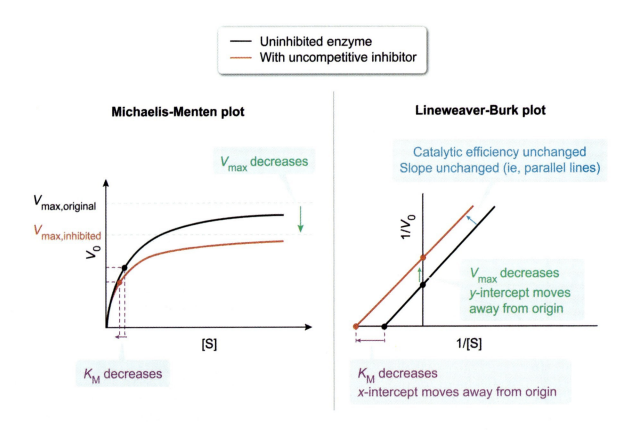

Figure 5.39 Changes to the Michaelis-Menten plot (left) and Lineweaver-Burk plot (right) caused by the addition of an uncompetitive inhibitor.

Noncompetitive Inhibitors

Noncompetitive inhibitors are a subclass of mixed inhibitors that cause a *decrease in V_{max}* with *no change to K_M*. On a Michaelis-Menten plot, the horizontal asymptote representing the V_{max} value is lowered and the K_M is unchanged. An example of noncompetitive inhibition on a Michaelis-Menten plot is given in Figure 5.40 (left). However, because the reference V_{max} value changes between the uninhibited and inhibited curves, pinpointing and comparing the K_M values may be difficult unless they are explicitly given.

On a Lineweaver-Burk plot, the decrease in V_{max} causes the y-intercept (ie, $1/V_{max}$) to shift *away from the origin* (more specifically, it shifts upward). In contrast, the x-intercept (ie, $-1/K_M$) is unchanged because K_M is unchanged, resulting in an inhibited data curve that intersects the uninhibited curve at the x-intercept. A Lineweaver-Burk plot showing noncompetitive inhibition is shown in Figure 5.40 (right).

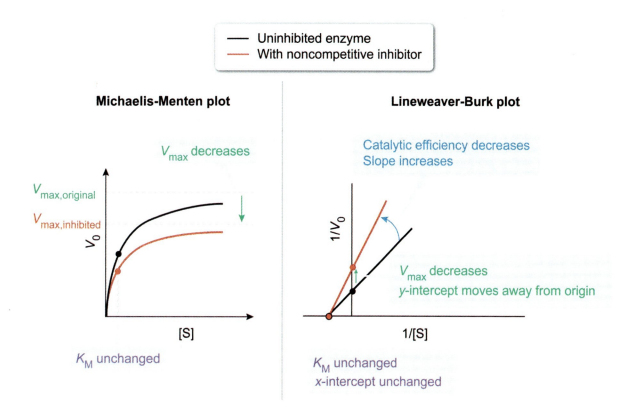

Figure 5.40 Changes to the Michaelis-Menten plot (left) and Lineweaver-Burk plot (right) caused by the addition of a noncompetitive inhibitor.

As with competitive inhibitors, the Lineweaver-Burk plot of an enzyme with a noncompetitive inhibitor shows an *increased* slope (ie, increased K_M/V_{max}), which corresponds to its decreased V_{max}/K_M ratio and decreased catalytic efficiency.

Mixed Inhibitors

Mixed inhibitors that are *not* purely noncompetitive come in two major subclasses: "competitive-like" mixed inhibitors, which *increase* the apparent K_M, and "uncompetitive-like" mixed inhibitors, which *decrease* the apparent K_M. In both subclasses, the V_{max} decreases, so the horizontal asymptote that represents V_{max} on a Michaelis-Menten plot *also* decreases. K_M shifts to the right for "competitive-like" mixed inhibitors and to the left for "uncompetitive-like" mixed inhibitors. Example Michaelis-Menten plots are given in Table 5.6 (left).

On a Lineweaver-Burk plot, the decrease in V_{max} causes the y-intercepts (ie, $1/V_{max}$) to shift *away from the origin* for both subclasses of mixed inhibitor—more specifically, they shift upward.

For "competitive-like" mixed inhibitors, which increase the apparent K_M, the x-intercept shifts *toward the origin* (ie, to the right). For "uncompetitive-like" inhibitors, which *decrease* the apparent K_M, the x-intercept shifts *away from the origin* (ie, to the left). Note that the x-intercept cannot shift beyond the point where the inhibited and uninhibited curves are parallel. If the curves did become parallel, this would indicate true uncompetitive inhibition. Example Lineweaver-Burk plots depicting mixed inhibition are shown in Table 5.6 (right).

Table 5.6 Changes to the Michaelis-Menten plot (left) and Lineweaver-Burk plot (right) caused by the addition of a mixed inhibitor.

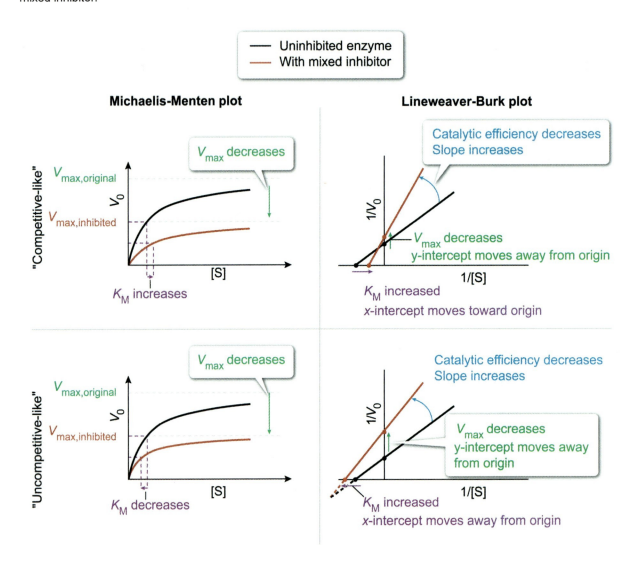

Note that both the "competitive-like" and the "uncompetitive-like" mixed inhibitors cause an increase in the slopes of Lineweaver-Burk plots, indicating a reduction in apparent catalytic efficiency and resulting in nonparallel lines that intersect at points that are not on either axis. "Competitive-like" mixed inhibitors intersect at some point above the *x*-axis, whereas "uncompetitive-like" mixed inhibitors intersect below the *x*-axis.

Concept Check 5.8

An enzyme sample is divided into three aliquots, and two aliquots are treated with different molecules. One aliquot is left untreated. The Lineweaver-Burk plots of treated and untreated enzyme aliquots are shown in the figure provided.

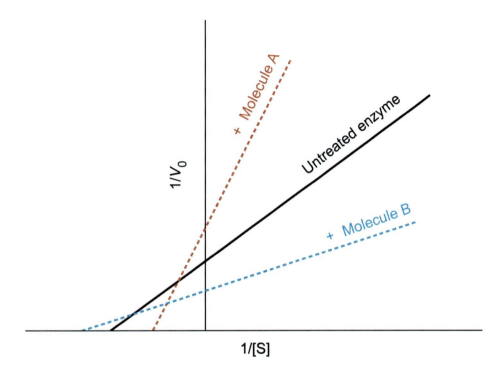

Which of the molecules is a reversible inhibitor? What subclass of reversible inhibitor does that molecule belong to?

Solution

Note: The appendix contains the answer.

Lesson 6.1
Feedback Regulation

Introduction

As discussed in Lesson 5.1, enzyme reaction rate (V_0) and maximum reaction rate (V_{max}) are affected by the total concentration of enzyme present [E_{tot}]. In living cells, [E_{tot}] depends on expression level. By altering an enzyme's rate of synthesis or degradation, a cell can alter enzyme amount and therefore control (or **regulate**) enzyme activity.

However, regulation of activity through enzyme synthesis and degradation is slow and energetically costly. To conserve resources, many enzymes are regulated by small molecules or by post-translational modifications, which allow for quick changes in enzyme activity. This chapter discusses the various ways enzyme activity is regulated outside of enzyme expression. The first lesson of this chapter details the phenomenon of feedback regulation. For a deeper exploration of regulation through expression level, see Biology Chapter 2.

6.1.01 Overview of Feedback Regulation

One of the fundamental ways an organism regulates an enzyme's *net* reaction rate is through its substrates or products. As the products of a reaction accumulate, the need for the reaction to occur generally decreases, so the net reaction rate tends to decrease. Conversely, if product levels decrease, the net reaction rate typically increases. This type of self-regulation is an application of the law of mass action and Le Châtelier's principle.

Self-regulation of the *net* reaction rate, however, applies only to *reversible* enzymes, which operate near equilibrium. The rate and direction of **irreversible enzymes**, on the other hand, are less affected by relative substrate and product concentrations. Because these enzymes act *far from equilibrium*, the reverse reaction is unlikely to occur, so a buildup of products will *not* cause the net rate to change significantly. To avoid buildup of potentially toxic intermediates or the wasting of resources, irreversible enzymes require separate mechanisms to quickly slow their activity once sufficient product has been made (Figure 6.1).

Chapter 6: Enzyme Regulation

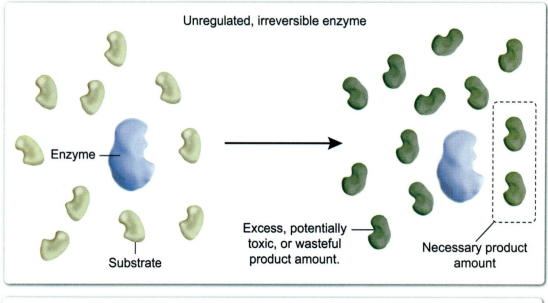

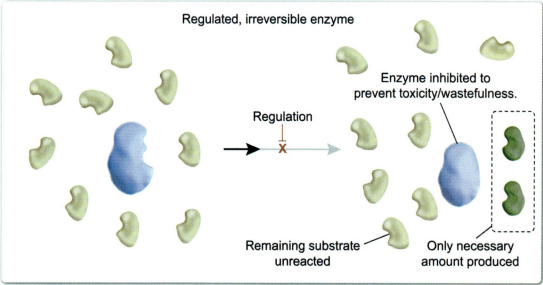

Figure 6.1 Regulation can be used to stop or slow enzymes that catalyze irreversible reactions before equilibrium is reached.

Enzyme regulation often occurs when a product molecule directly alters the activity of the enzyme that produces it or when the product molecule regulates an enzyme that catalyzes an *earlier* step in the metabolic pathway. These products "feed backward" to regulate enzyme activity; therefore, this process is known as **feedback regulation**.

Most commonly the end product of an enzyme or a pathway has a negative (ie, inhibitory) effect on the enzyme it regulates. This type of regulation is known by various names, including **negative feedback**, **feedback inhibition**, or **end-product inhibition**. Figure 6.2 shows examples of feedback inhibition in the pentose phosphate pathway (left panel) and in glycolysis (right panel). The details of the metabolic pathways used as examples in this lesson are discussed further in Unit 4.

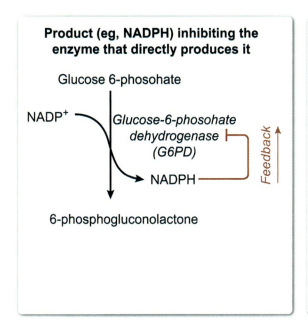
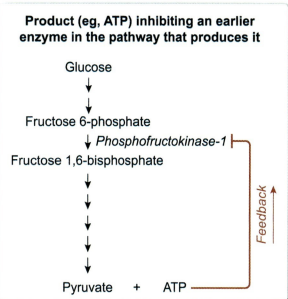

Figure 6.2 Feedback inhibition involves a product molecule inhibiting an enzyme at the same or earlier step in the pathway that produced it.

Although the forward reactions of both reversible and irreversible enzymes may slow in response to interaction with a product, different mechanisms cause these changes. For reversible enzymes, net activity decreases not due to *inhibition* of the enzyme's catalytic ability (ie, k_{cat}), but because product molecules bind the active site and are *converted back into substrates*. In other words, the net reaction rate decreases because the reverse reaction rate increases, *not* because of a change in the enzyme's ability to function.

In contrast, the reverse reaction does *not* occur in feedback inhibition of irreversible enzymes. Therefore, the product does not bind the *active* site but instead binds a separate, **allosteric site** (Figure 6.3). This allosteric feedback inhibition *does* decrease the enzyme's catalytic ability.

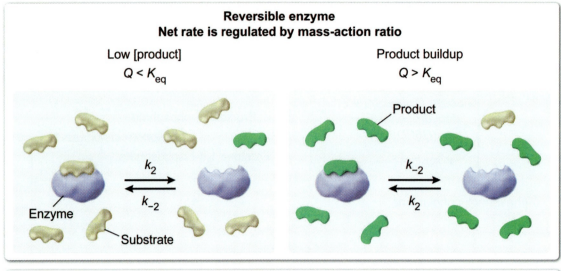

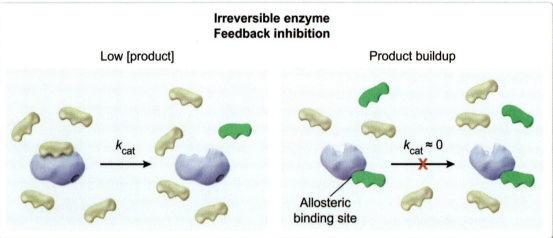

Figure 6.3 Feedback inhibition occurs through an allosteric mechanism to decrease catalytic activity.

6.1.02 Allosteric Regulation

Allosteric effectors bind enzymes at a site separate from the active site. The feedback inhibitors described in Concept 6.1.01 and most of the reversible inhibitors described in Lesson 5.4 are examples of small-molecule allosteric *inhibitors*. However, allosteric effectors can also *stimulate* or *activate* enzymes, and the effectors can be small molecules, large proteins, or any other molecule that binds to the target enzyme (Figure 6.4).

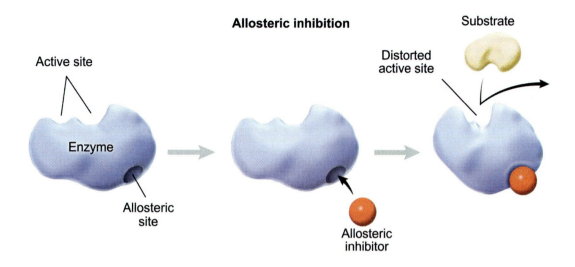

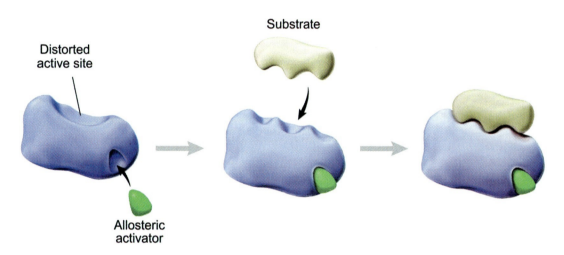

Figure 6.4 Allosteric effectors bind enzymes at a site separate from the active site to inhibit or activate enzyme activity.

Allosteric effectors (also called allosteric modulators) affect enzyme activity by stabilizing a conformational change through induced fit or conformational selection (Concept 4.3.02). This conformational change typically affects the enzyme's affinity for its substrate (ie, K_M), the turnover number (ie, k_{cat}), or both.

A single enzyme can be regulated by multiple allosteric modulators. For example, phosphofructokinase-1 (PFK1) is the most heavily regulated enzyme of glycolysis because it is the first irreversible enzyme that catalyzes a committed step of the pathway. Consequently, PFK1 is allosterically feedback inhibited by ATP, one of the main products of glycolysis, as well as by citrate, which is an early metabolite in a pathway further downstream (ie, the citric acid cycle).

PFK1 is also allosterically *activated* by AMP and ADP (which would accumulate if ATP stores were low), as well as by another molecule called fructose 2,6-bisphosphate (F2,6BP). PFK1 and glycolysis regulation are discussed further in Chapter 11. An overview of PFK1 regulation is shown in Figure 6.5.

Chapter 6: Enzyme Regulation

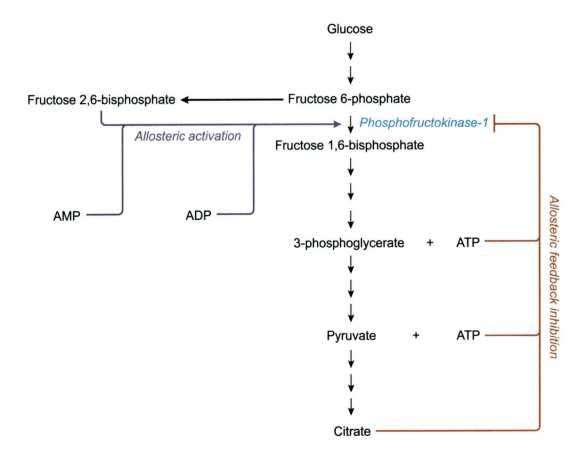

Figure 6.5 In glycolysis, the enzyme phosphofructokinase-1 is affected by many allosteric modulators.

Some allosteric modulators can override others. For example, the molecule F2,6BP allosterically *stimulates* PFK1, whereas ATP allosterically *inhibits* it. F2,6BP is more potent, with a smaller K_d for its allosteric site. However, sufficiently high concentrations of ATP can still inhibit PFK1 and force dissociation of F2,6BP. In this way, allosteric modulators can act as competitive or "competitive-like" mixed inhibitors of each other (Figure 6.6).

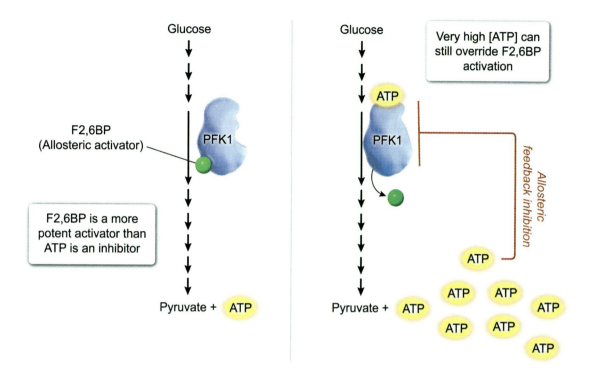

Figure 6.6 Enzymes can be affected by multiple allosteric regulators. The enzyme integrates the signals from these modulators to yield a fine-tuned response.

Concept Check 6.1

An enzyme is treated with an allosteric modulator. The kinetics of treated and untreated enzyme are shown in the following double reciprocal plot. Is the modulator an activator or an inhibitor? How does the modulator affect the enzyme's K_M and V_{max} values?

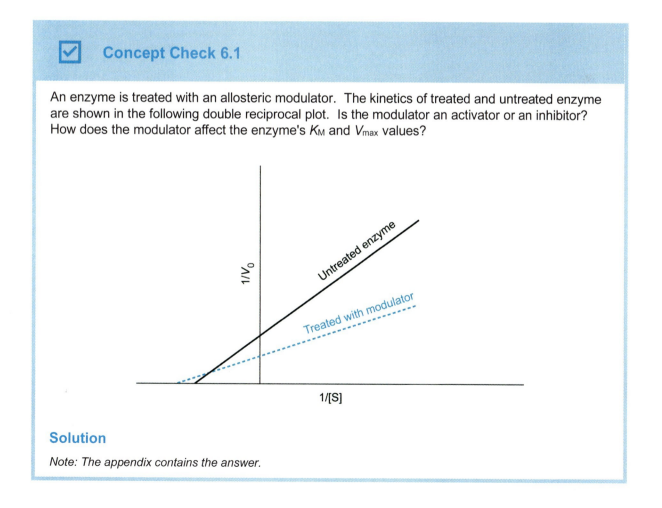

Solution

Note: The appendix contains the answer.

6.1.03 Characteristics of Regulated Enzymes

Many metabolic pathways contain multiple enzymes that catalyze irreversible reactions. Although all these enzymes typically show at least some level of regulation, certain irreversible enzymes are subject to stricter regulation than other irreversible enzymes. Regulation of these enzymes can come from feedback inhibition, as well as other types of regulation discussed later in this chapter, such as post-translational modifications (Lesson 6.2).

The enzymes that are most regulated tend to act *early* in a metabolic pathway and commit the metabolites to the pathway. For example, glycolysis contains three enzymes that catalyze irreversible reactions: hexokinase, phosphofructokinase-1, and pyruvate kinase. Of these, phosphofructokinase-1 (PFK1) is the most tightly regulated.

PFK1 is the third step (ie, third enzyme) of the 10-step glycolytic pathway and is therefore an early step of the process. If pyruvate kinase (step 10) were feedback inhibited but PFK1 were not, then the metabolites between them would build up. This would result in wasted energy as those metabolites have been *committed* to glycolysis (ie, the metabolites are unlikely to proceed down any other useful pathway except for glycolysis), but they *cannot finish* the pathway.

Although PFK1 is an *early* step of glycolysis, it is not the *first* irreversible step. Hexokinase catalyzes the first step of the glycolytic pathway, but its product (glucose 6-phosphate) can proceed down *many* pathways (eg, the pentose phosphate pathway, glycogen synthesis [see Unit 4]). Consequently, although hexokinase catalyzes the first step in glycolysis, the subsequent metabolites are *not* committed to glycolysis, and hexokinase is a less useful enzyme to target to regulate glycolysis (Figure 6.7).

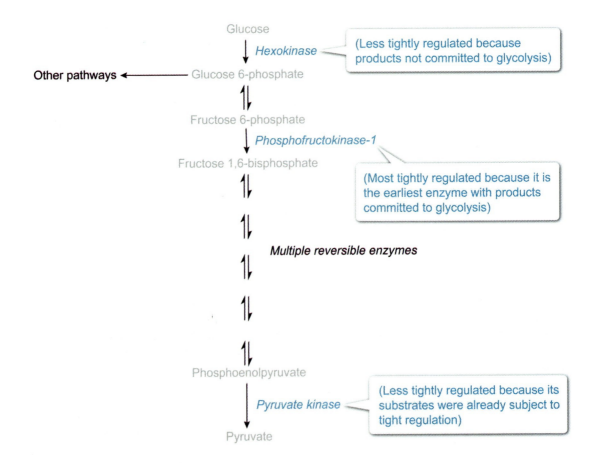

Figure 6.7 The most tightly regulated enzyme in a pathway is typically the earliest irreversible enzyme that catalyzes a committed step.

Concept Check 6.2

The following figure is a hypothetical biochemical metabolic map.

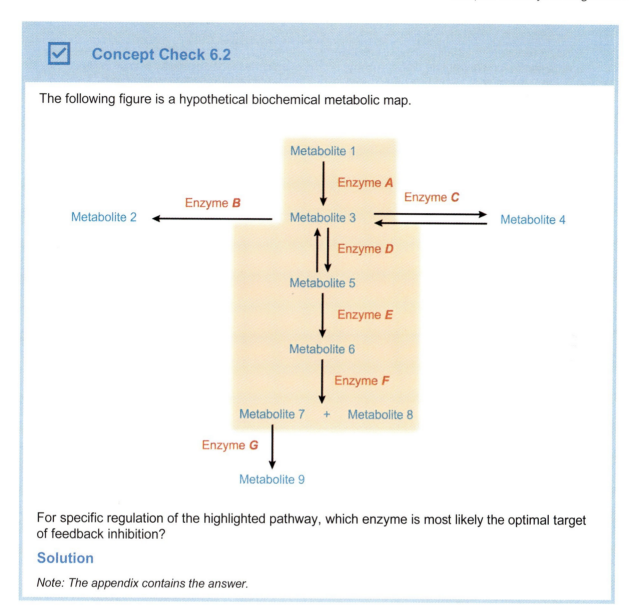

For specific regulation of the highlighted pathway, which enzyme is most likely the optimal target of feedback inhibition?

Solution

Note: The appendix contains the answer.

6.1.04 Positive Feedback

In addition to feedback inhibition (also known as negative feedback), another type of feedback regulation, called **positive feedback**, also exists. Like feedback inhibition, positive feedback occurs when a product molecule feeds backward to regulate an *earlier* enzyme. However, for *positive* feedback regulation, the product molecule *stimulates* or *activates* the enzyme. Positive feedback regulation is also known as **feedback activation**. Feedback activation is used when the presence of a product molecule indicates that *more*, not less, product is required, as in activation of digestive enzymes such as pepsin or in the clotting cascade (see Figure 6.8).

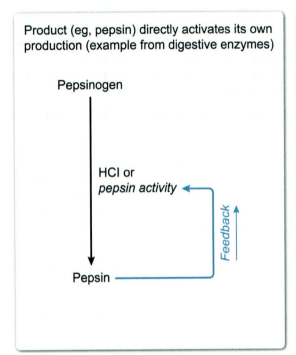
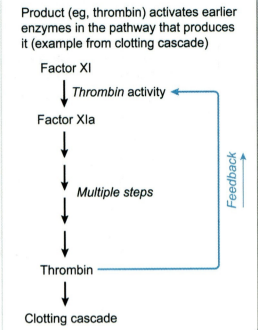

Figure 6.8 Positive feedback occurs when a product molecule activates an enzyme at an earlier step in the pathway that produced it.

Enzymes are also sometimes controlled by feedforward regulation. In contrast to feedback regulation, feedforward regulation occurs when a product feeds *forward* and regulates a *later* step. In metabolic pathways, this type of regulation is typically stimulatory and is known as feedforward *activation*. Feedforward activation can help minimize buildup of metabolites that have already been committed to a certain pathway.

For example, although PFK1 is the most tightly regulated enzyme of glycolysis because it is the earliest committed enzyme, pyruvate kinase is also an irreversible enzyme that catalyzes a committed step of glycolysis. To prevent a wasteful buildup of the metabolites between PFK1 and pyruvate kinase, the immediate product of PFK1 (ie, fructose 1,6-bisphosphate) can feed forward to activate pyruvate kinase. This process of feedforward activation helps to ensure that pyruvate kinase activity is sufficiently high to keep up with PFK1 activity (see Figure 6.9).

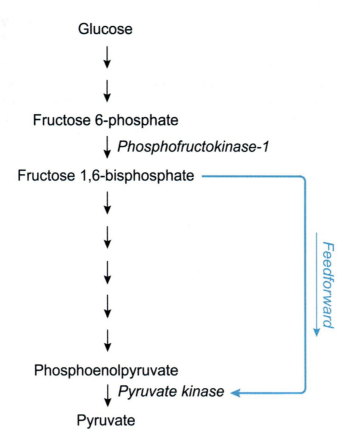

Figure 6.9 In feedforward activation, a product molecule activates a later enzyme in the metabolic pathway.

Lesson 6.2
Regulation by Covalent Modifications

Introduction

In addition to allosteric regulation (eg, feedback inhibition), enzymes are commonly regulated by covalent modifications. Lesson 2.4 discussed many post-translational modifications that a protein may undergo. This lesson discusses the effects of those modifications on enzyme activity specifically.

6.2.01 Zymogens

Proproteins are proteins that are synthesized in an inactive form and then converted to an active form by proteolytic cleavage. For example, proinsulin is cleaved to form the active hormone insulin, and fibrinogen is cleaved to form the active clotting factor fibrin (see Biology Concept 13.1.05). Zymogens are proproteins that generate an enzyme upon cleavage. Zymogens are also known as proenzymes.

Pepsinogen is an example of a subclass of a zymogen. Specifically, pepsinogen can proteolytically cleave itself under certain conditions, and this cleavage irreversibly converts it into its active form, pepsin (Figure 6.10).

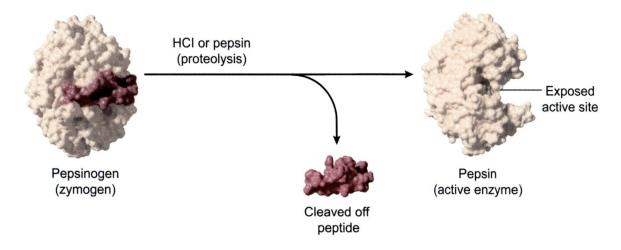

Figure 6.10 The inactive protein pepsinogen is proteolytically cleaved to form the active enzyme pepsin.

Pepsin can then cleave additional pepsinogen and is an example of an enzyme that activates its own zymogen precursor. Although pepsin acts in this way, not all enzyme-zymogen pairs follow this pattern. Many zymogens are *activated* by separate enzymes; for example, trypsinogen is cleaved into its active form, trypsin, by the enzyme enteropeptidase.

Many enzymes that would be harmful if they became active too early are synthesized as zymogens to ensure that they stay inactive until needed. It is for this reason that zymogens are commonly found in digestion pathways (eg, pepsinogen, trypsinogen, chymotrypsinogen), clotting pathways (eg, prothrombin, factor X), and apoptotic pathways (eg, procaspase). Aberrant early activation of zymogens can have pathological consequences.

The synthesis of enzymes as inactive zymogen precursors allows for their rapid and energy-efficient activation when needed. Unlike allosteric activation, a separate regulatory molecule is *not* needed for every single zymogen molecule to be activated, and, unlike some other post-translational covalent modifications such as phosphorylation, proteolytic hydrolysis is an exergonic process that does *not* require an external energy source such as ATP.

6.2.02 Dynamic Covalent Modifications

Irreversible inhibitors can stop an enzyme's catalytic activity by covalently modifying an individual molecule, and most of these inhibitors are exogenous to the organism (eg, drugs, toxins). Concept 6.2.01 introduces endogenous covalent modification of zymogens through proteolysis, which irreversibly activates the molecule to form an active enzyme.

Both irreversible inhibition and zymogen activation are examples of **post-translational modifications**—covalent modifications that occur on a peptide after the sequence has been accurately translated by a ribosome (Figure 6.11). Although many post-translational modifications (PTMs) are irreversible, this concept focuses instead on covalent PTMs that are **dynamic**—these modifications can be added, but also removed, to dynamically regulate a target enzyme's activity.

Note that sometimes dynamic covalent modification is referred to as reversible covalent modification. However, this does *not* refer to the *thermodynamic* reversibility of a single enzymatic reaction. Instead, it refers to the ability to undo a covalent modification through the action of a *separate* enzymatic reaction.

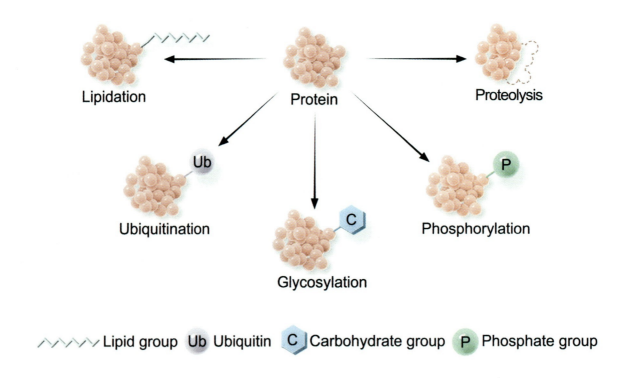

Figure 6.11 Examples of protein post-translational modifications.

Phosphorylation is one of the most common regulatory dynamic PTMs. As discussed in Concept 2.4.02, phosphorylation is the addition of a phosphate group to a protein, typically through the side chain hydroxyl of a serine, threonine, or sometimes tyrosine residue. The addition of the phosphate group is most commonly mediated by a kinase enzyme through transfer of the γ-phosphate of ATP (see Concept 4.2.03). The phosphate group is most commonly removed by hydrolysis using a phosphatase enzyme (see Concept 4.2.04).

Although both kinase and phosphatase reactions are considered thermodynamically irreversible when considered individually, they can work together to control the reversible phosphorylation status of a target protein, as shown in Figure 6.12. Phosphorylation status can therefore be used to regulate the activity of the target protein.

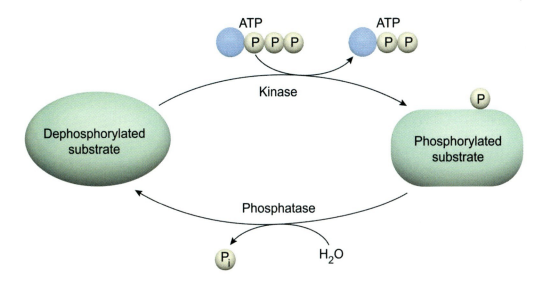

Figure 6.12 Kinases phosphorylate proteins and phosphatases dephosphorylate proteins to regulate their activity.

Because enzymes are proteins, the phosphorylation status of an enzyme can also control its level of enzyme activity. The specific effect of a phosphorylation event depends on the enzyme being modified (Figure 6.13).

Some enzymes are "switched on" when phosphorylated; others are "switched off" by phosphorylation. Some enzymes stay "on" regardless of phosphorylation status but are simply switched into a *more* active or *less* active state. Many enzymes have multiple phosphorylation sites, and phosphorylation at one site may have a different or opposing effect to phosphorylation at a different site.

Mechanistically, phosphorylation affects an enzyme's activity because the modification induces a conformational change in the enzyme. This conformational change may affect the enzyme's turnover number (k_{cat}), its ability to bind its substrate (ie, its K_M value), or both. An enzyme is turned "on" if its catalytic efficiency (ie, k_{cat}/K_M) rises, and an enzyme is turned "off" if its catalytic efficiency drops to nearly zero.

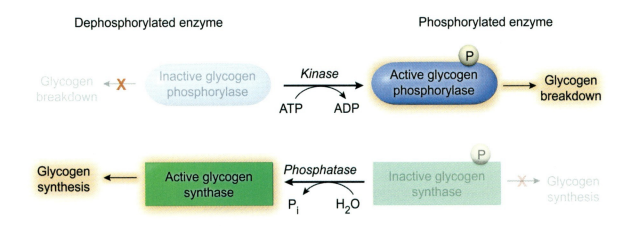

Figure 6.13 Some enzymes are activated by phosphorylation (eg, glycogen phosphorylase), and some are inactivated by phosphorylation (eg, glycogen synthase).

Covalent regulation of enzymes offers an alternate means of regulating enzymes that supplements feedback mechanisms (Lesson 6.1) and other allosteric mechanisms. The unique features of covalent modification allow this type of regulation to either enhance or override allosteric signaling to offer greater control over the metabolic activity of a cell or organism.

Concept 3.2.02 explains that **signal cascades** can propagate a cellular response. Changes in phosphorylation status often play a role in the ability of these cascades to amplify the original signal. This is because each kinase or phosphatase involved can act on multiple target substrates.

For example, suppose one molecule of Kinase A phosphorylates and activates 10 molecules of Phosphatase B. Each Phosphatase B can then dephosphorylate and activate 10 molecules of Kinase C, resulting in 100 molecules of active Kinase C from a single activated Kinase A. In contrast, a single feedback or allosteric modulator can affect only a single enzyme and only when the modulator is bound to the enzyme (Figure 6.14).

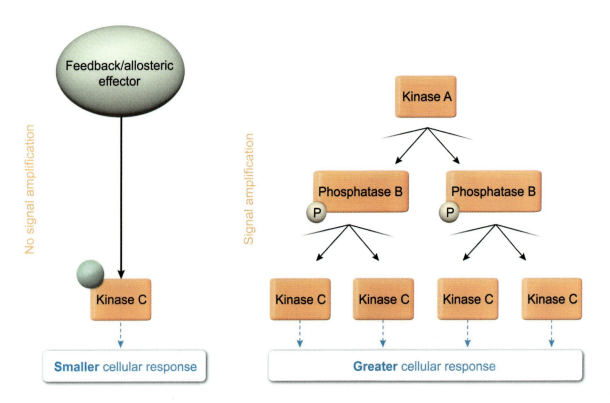

Figure 6.14 Kinases and phosphatases, which covalently modify and regulate their targets, can amplify a cellular signal.

In addition to acting on multiple units of the same target substrate, many of the kinases and phosphatases involved in covalent regulation can also affect multiple *distinct* targets. In other words, many kinase and phosphatase enzymes can act on several types of substrate and can therefore affect multiple pathways. For example, protein kinase A (PKA) can phosphorylate and regulate enzymes involved in glycolysis, glycogen synthesis, glycogen breakdown, fatty acid metabolism, and other pathways (Unit 4).

As explained in Concept 6.1.01, glycolysis is partially regulated through allosteric feedback mechanisms (discussed further in Chapter 11). Regulation through dynamic covalent modifications, such as phosphorylation, can be used to override feedback control in some cases.

For example, in the fed state (ie, after an influx of dietary glucose), a hepatocyte (ie, a liver cell) will likely have sufficient or excess ATP stores, which would normally cause feedback *inhibition* of glycolysis. However, because high blood glucose is damaging to the circulatory system, phosphatases are activated to covalently regulate glycolysis enzymes (eg, pyruvate kinase) and other related enzymes (eg, phosphofructokinase-2) such that glycolysis is *stimulated* (Figure 6.15).

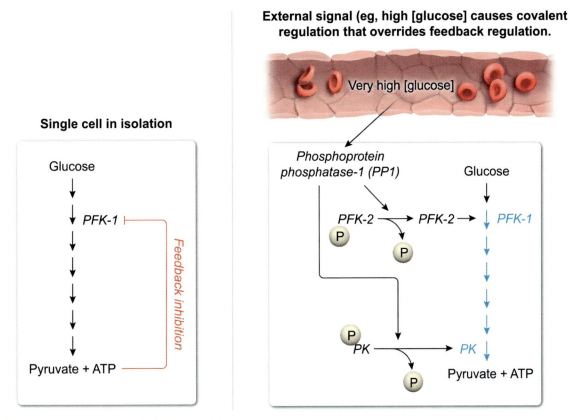

PFK-1 = phosphofructokinase-1; PK = pyruvate kinase.

Figure 6.15 Covalent regulation of enzymes can be used to override feedback inhibition of enzymes.

Because dynamic covalent regulation of enzymes can be used to amplify signals, affect multiple pathways, and override feedback control, effectors of covalent regulation (eg, kinases, phosphatases) are often central players in hormonal regulation of metabolic processes.

Although this chapter has focused primarily on phosphorylation, many other types of dynamic covalent regulation of enzymes exist as discussed previously (Figure 6.11). These modifications work in a similar way as phosphorylation—they cause conformational changes that affect an enzyme's catalytic efficiency to turn an enzyme "on" or "off" or to fine-tune their level of activity.

Chapter 6: Enzyme Regulation

END-OF-UNIT MCAT PRACTICE

Congratulations on completing **Unit 2: Enzymes**.

Now you are ready to dive into MCAT-level practice tests. At UWorld, we believe students will be fully prepared to ace the MCAT when they practice with high-quality questions in a realistic testing environment.

The UWorld Qbank will test you on questions that are fully representative of the AAMC MCAT syllabus. In addition, our MCAT-like questions are accompanied by in-depth explanations with exceptional visual aids that will help you better retain difficult MCAT concepts.

TO START YOUR MCAT PRACTICE, PROCEED AS FOLLOWS:

1) Sign up to purchase the UWorld MCAT Qbank
   IMPORTANT: You already have access if you purchased a bundled subscription.
2) Log in to your UWorld MCAT account
3) Access the MCAT Qbank section
4) Select this unit in the Qbank
5) Create a custom practice test

Unit 3 Carbohydrates, Nucleotides, and Lipids

Chapter 7 Carbohydrates

7.1 Monosaccharides

    7.1.01   Biologically Relevant Monosaccharides
    7.1.02   Monosaccharide Classification
    7.1.03   Monosaccharide Stereochemistry
    7.1.04   Monosaccharide Derivatives

7.2 Complex Carbohydrates

    7.2.01   Glycosidic Bonds
    7.2.02   Disaccharides
    7.2.03   Polysaccharides

Chapter 8 Nucleotides and Nucleic Acids

8.1 Nucleotides

    8.1.01   Nitrogenous Bases
    8.1.02   Chemical Modifications of Nitrogenous Bases
    8.1.03   Nucleosides and Nucleotides
    8.1.04   Overview of Nucleotide Function

8.2 Nucleic Acids

    8.2.01   Nucleic Acid Formation
    8.2.02   Base Pairing and Secondary Structure in Nucleic Acids
    8.2.03   Chargaff's Rules
    8.2.04   Double Helix Stability

Chapter 9 Lipids

9.1 Energy Storage Lipids

    9.1.01   General Properties of Lipids
    9.1.02   Fatty Acids
    9.1.03   Triacylglycerides

9.2 Structural Lipids

    9.2.01   Glycerophospholipids
    9.2.02   Sphingolipids
    9.2.03   Cholesterol
    9.2.04   Waxes
    9.2.05   Organization of Structural Lipids
    9.2.06   Properties of Lipid Bilayers
    9.2.07   The Fluid Mosaic Model
    9.2.08   Determinants of Fluidity

9.3 Signaling Lipids

    9.3.01   Signaling Lipids Derived from Hydrolyzable Membrane Lipids
    9.3.02   Terpenoids as Signaling Molecules

Lesson 7.1

Monosaccharides

Introduction

Carbohydrates are a class of molecules consisting of hydrated carbons (ie, carbons that have combined with water in a 1:1 ratio) that plays many important roles in biology. For example, carbohydrates are a source of energy, help provide cellular structure, assist in protein folding, and are involved in cell-cell recognition and immune responses. Errors in carbohydrate processing can lead to a variety of disease states including degenerative disorders (eg, muscular dystrophy) and connective tissue disorders (eg, Ehlers-Danlos syndrome).

The simplest carbohydrates are called monosaccharides. These consist of an uninterrupted carbon chain in which one carbon is a carbonyl and the other carbons are alcohols (Figure 7.1).

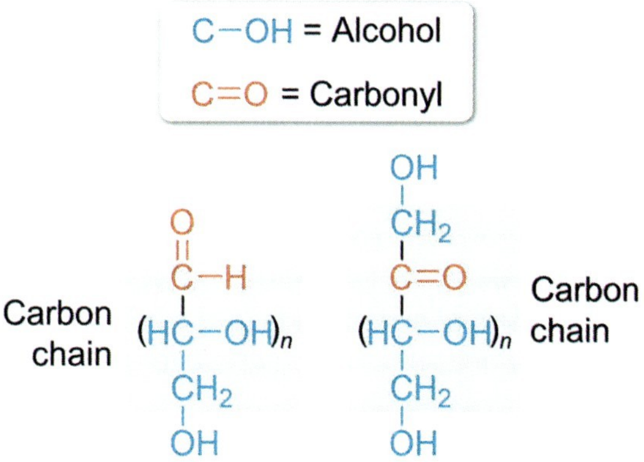

Figure 7.1 Structural features of monosaccharides.

This lesson examines the structural features of monosaccharides and gives an overview of their biological functions.

7.1.01 Biologically Relevant Monosaccharides

A simple monosaccharide is a molecule that meets the following criteria:

- It contains at least three carbon atoms.
- It has the molecular formula $(CH_2O)_n$.
- It can exist as a linear carbon chain.
- One carbon in the linear chain is either an aldehyde or a ketone.
- The other carbons in the linear chain are primary or secondary alcohols.

Many molecules meet these criteria and are classified as monosaccharides; however, relatively few monosaccharides play significant biological roles. As with the amino acids, memorization of certain biologically relevant monosaccharides is important for the exam. This concept provides an overview of several biologically important carbohydrates and the roles they play.

The most abundant monosaccharide in nature is glucose, which has the formula $C_6H_{12}O_6$. Like most monosaccharides, glucose is given a D or L designation to indicate its stereochemistry (see Concept

7.1.03). Most biologically important carbohydrates can be biochemically synthesized using glucose as the starting material.

Several other biologically important monosaccharides have the same molecular formula as D-glucose ($C_6H_{12}O_6$) and differ from D-glucose only in the arrangement of the atoms, as detailed later in this chapter. In other words, these molecules are isomers of D-glucose. The main biologically important isomers of D-glucose include D-fructose, D-mannose, and D-galactose. Figure 7.2 shows each of these sugars in a Fischer projection (see Organic Chemistry Lesson 3.3 for more information on Fischer projections).

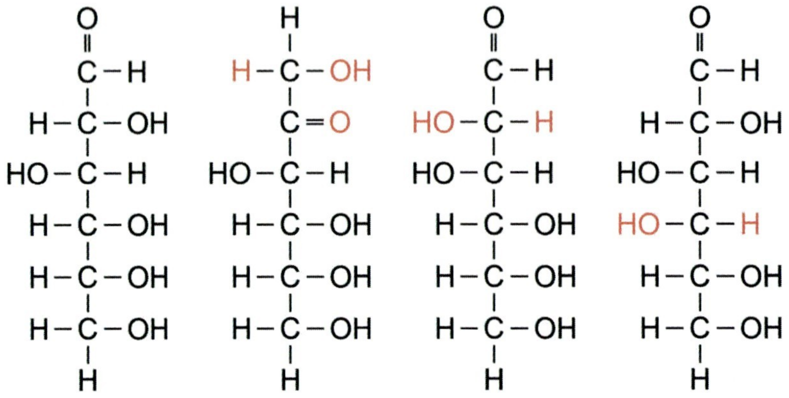

D-Glucose D-Fructose D-Mannose D-Galactose

Figure 7.2 Structures of several monosaccharides with the formula $C_6H_{12}O_6$. The positions at which each sugar differs from glucose are marked in red.

D-Ribose ($C_5H_{10}O_5$) is another important monosaccharide; it is a component of nucleotides, which can combine to form DNA and RNA (see Chapter 8). Other carbohydrates with the same molecular formula as ribose ($C_5H_{10}O_5$) include D-ribulose and D-xylulose, which both play roles in the pentose phosphate pathway (see Lesson 11.3). Figure 7.3 shows the Fischer projections of D-ribose, D-ribulose, and D-xylulose.

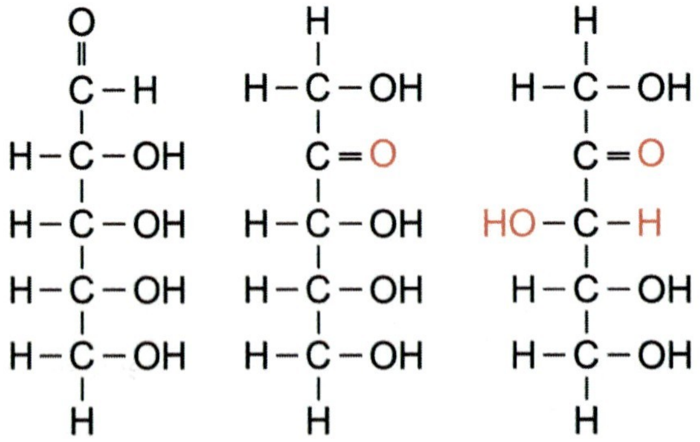

D-Ribose D-Ribulose D-Xylulose

Figure 7.3 Monosaccharides with the formula $C_5H_{10}O_5$. The positions at which each sugar differs from D-ribose are marked in red.

Two smaller carbohydrates, D-glyceraldehyde and dihydroxyacetone, also play important roles in metabolism. These molecules both have the formula $C_3H_6O_3$, and Fischer projections of each are shown in Figure 7.4. Note that dihydroxyacetone has no chiral centers, so it is not given a D or L designation.

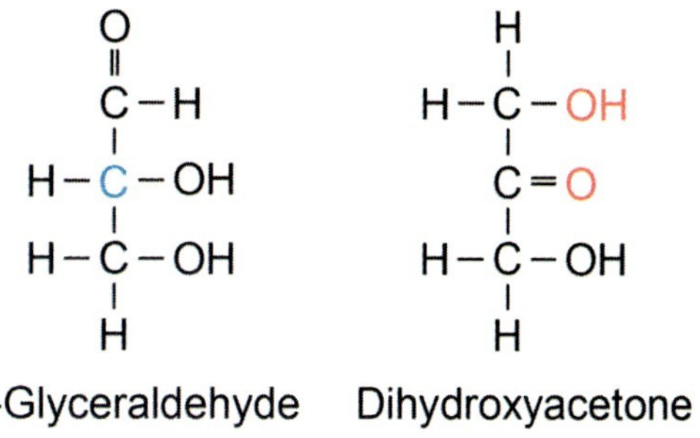

```
        O                    H
        ‖                    |
        C—H            H—C—OH
        |                    |
    H—C—OH             C=O
        |                    |
    H—C—OH          H—C—OH
        |                    |
        H                    H
   D-Glyceraldehyde    Dihydroxyacetone
```

Figure 7.4 Monosaccharides with the formula $C_3H_6O_3$. The positions at which dihydroxyacetone differs from D-glyceraldehyde are marked in red, and the chiral carbon of D-glyceraldehyde is marked in blue.

Phosphorylated derivatives of both D-glyceraldehyde and dihydroxyacetone are found in glycolysis (Chapter 11).

7.1.02 Monosaccharide Classification

Monosaccharides are classified according to the number of carbons they contain, the functional groups within them, and whether they are cyclized. This concept explores each type of classification.

Classification by Number of Carbons

The empirical formula for a monosaccharide, $(CH_2O)_n$, can give rise to many molecular formulas depending on the value of n. Monosaccharides are classified according to their molecular formula. This general classification uses the Greek prefix for the number of carbons in the molecule followed by the suffix –*ose*. For example, any monosaccharide with the molecular formula $C_3H_6O_3$ (ie, three carbons) is a triose. The classifications for monosaccharides of various lengths are summarized in Table 7.1.

Table 7.1 Monosaccharide classification by number of carbons.

Number of carbons (*n*)	Classification
3	Triose
4	Tetrose
5	Pentose
6	Hexose
7	Heptose
8	Octose
9	Nonose
10	Decose

Based on this information, glucose, fructose, mannose, and galactose are hexoses; ribose, ribulose, and xylulose are pentoses; and glyceraldehyde and dihydroxyacetone are trioses. Many enzymes and metabolic pathways are named for the type of carbohydrate on which they act. For example, hexokinase can act on multiple hexoses (eg, glucose, fructose, mannose). Similarly, the pentose phosphate pathway produces and manipulates pentoses with phosphates attached.

Classification by Anomeric Carbon Position

Another important monosaccharide classification relies on the position of the **anomeric carbon** (ie, the carbon with two bonds to oxygen). In linear form, the anomeric carbon is either an aldehyde or a ketone (Figure 7.5). Monosaccharides that contain an aldehyde are called **aldoses**, and those that contain a ketone are called **ketoses**. In aldoses, the anomeric carbon is designated carbon 1. In biologically relevant ketoses, the anomeric carbon is found at carbon 2.

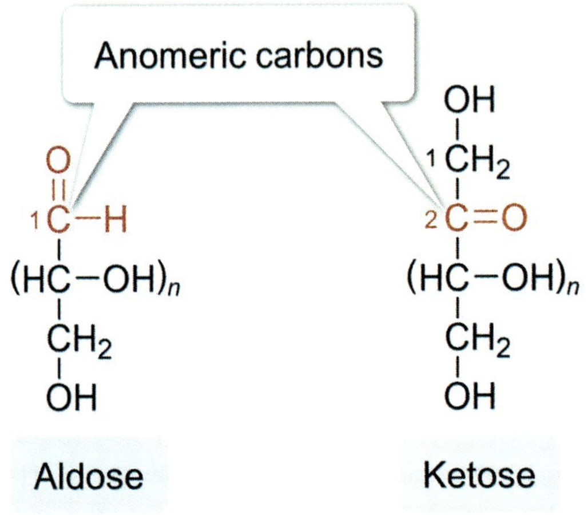

$$\text{Aldose:} \quad {}_1\text{C}(=\text{O})\text{H} - (\text{HC-OH})_n - \text{CH}_2\text{OH}$$

$$\text{Ketose:} \quad \text{CH}_2\text{OH} - {}_2\text{C}=\text{O} - (\text{HC-OH})_n - \text{CH}_2\text{OH}$$

Figure 7.5 General structures of aldose and ketose sugars.

In their linear form, glucose, galactose, mannose, ribose, and glyceraldehyde all contain aldehydes, making them aldoses. In contrast, fructose, ribulose, xylulose, and dihydroxyacetone each contain a ketone at carbon 2 and therefore are ketoses. Note that ketoses often contain the suffix –*ulose*. For example, ribulose is the ketose form of ribose, and xylulose is the ketose form of xylose. This naming convention can be helpful in identifying ketoses (although exceptions such as fructose exist).

Carbohydrates are often classified by both the number of carbons they contain *and* the position of the anomeric carbon. For instance, a monosaccharide with six carbons and an aldehyde (eg, glucose) is an **aldohexose**. A monosaccharide with five carbons and a ketone (eg, ribulose) is a **ketopentose**. Table 7.2 shows the combined classifications of several biologically important monosaccharides.

Table 7.2 Select biologically relevant monosaccharides characterized by number of carbons and position of the anomeric carbon.

	Trioses (C$_3$H$_6$O$_3$)	Pentoses (C$_5$H$_{10}$O$_5$)	Hexoses (C$_6$H$_{12}$O$_6$)
Aldoses	D-Glyceraldehyde (aldotriose)	D-Ribose (aldopentose)	D-Glucose (aldohexose)
Ketoses	Dihydroxyacetone (ketotriose)	D-Ribulose (ketopentose)	D-Fructose (ketohexose)

Chapter 7: Carbohydrates

☑ Concept Check 7.1

The diagram provided shows two monosaccharides.

```
                           H
                           |
                       H—C—OH
                           |
                          C=O
                           |
                       HO—C—H
         O                 |
         ||            H—C—OH
         C—H               |
         |             H—C—OH
       H—C—OH              |
         |             H—C—OH
       H—C—OH              |
         |                 H
       H—C—OH
         |
         H

      D-Erythrose      D-Sedoheptulose
```

Characterize each by the number of carbons and the position of the anomeric carbon.

Solution

Note: The appendix contains the answer.

Classification by Type of Cyclization

Many carbohydrates are classified based on whether and how they are cyclized. Aldoses that contain at least five carbons can readily cyclize into five- or six-membered rings.

Upon cyclization, the carbonyl carbon is converted into either a hemiacetal (aldoses) or a hemiketal (ketoses). Note that although a cyclized aldose no longer has an aldehyde group, it is still classified as an aldose. Similarly, cyclized ketoses are still classified as ketoses.

The five-membered rings formed by carbohydrates resemble the molecule furan; therefore, when carbohydrates adopt this conformation, they are said to be in their **furanose form**. Similarly, the six-membered rings formed by carbohydrates resemble pyran, and molecules in this conformation are in their **pyranose form**.

These terms can be combined with other classification terms. For instance, an aldose in its five-membered ring form is an aldofuranose, and a ketose in its six-membered ring form is a ketopyranose. Different types of cyclization in aldohexoses and ketohexoses are shown in Figure 7.6.

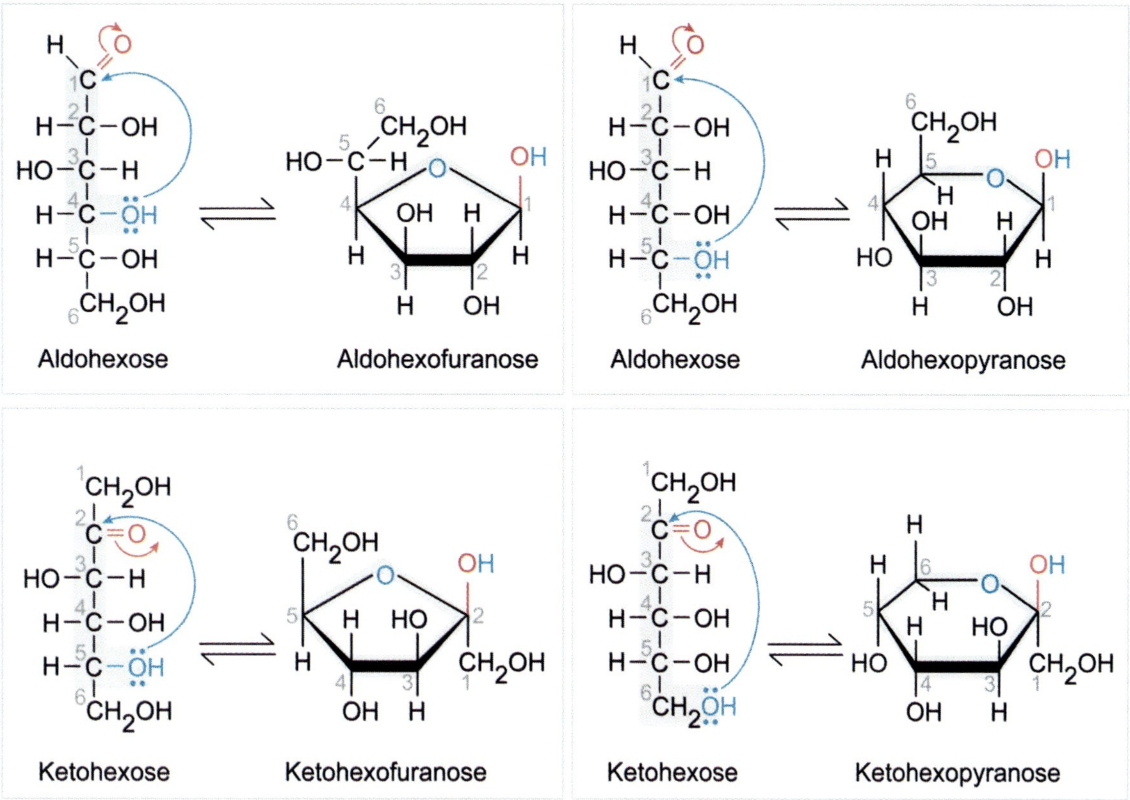

Figure 7.6 Cyclization in aldohexose and ketohexose sugars.

Note that in the cyclic form, the anomeric carbon still has two bonds to oxygen. However, instead of a *double bond* to *one* oxygen atom, the anomeric carbon now has *two single bonds to two different* oxygen atoms (see Figure 7.7)

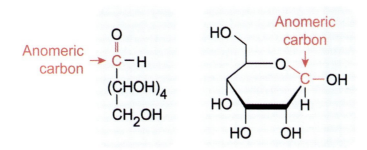

Figure 7.7 The anomeric carbon has two bonds to oxygen whether the monosaccharide is in linear or cyclic form.

Certain monosaccharides have a greater tendency to adopt a furanose or pyranose form. For instance, glucose, galactose, and mannose tend to adopt pyranose conformations. In contrast, fructose and ribose often adopt furanose forms, especially when incorporated into other molecules. However, each of these molecules *can* adopt either form. For instance, cyclized glucose may be classified as **glucofuranose** *or* **glucopyranose**, depending on the cyclic form it adopts. Linear (ie, noncyclic) glucose is neither a furanose nor a pyranose.

It is important not to confuse the number of members in a ring with the designation of the sugar as a pentose or a hexose. The number of members in the ring and the number of carbons in the linear chain can differ for two reasons. First, one of the members of the ring is always an oxygen atom. Second, not all carbon atoms are necessarily part of the ring. The designation of a monosaccharide as a pentose or hexose depends *only* on the number of carbons in the chain and *not* on how they are arranged. In contrast, designation as a furanose or pyranose depends on *only* the type of ring formed.

For instance, although fructose often forms a five-membered ring, it is a hexose (six carbons), not a pentose (five carbons). Fructose in the five-membered ring form is a hexofuranose. Figure 7.8 illustrates this principle.

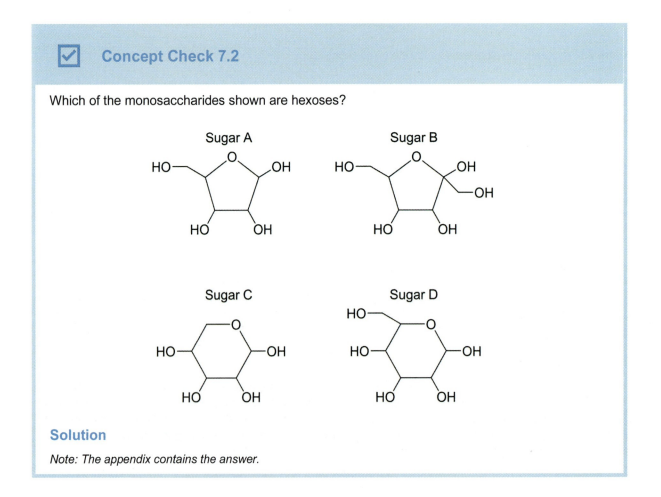

D-Fructofuranose has a five-membered ring... ...but contains six carbons.
Therefore, it is a hexofuranose.

Figure 7.8 The number of members in a ring determines whether a sugar is a furanose or a pyranose, not whether it is a pentose or a hexose.

☑ **Concept Check 7.2**

Which of the monosaccharides shown are hexoses?

Sugar A

Sugar B

Sugar C

Sugar D

Solution

Note: The appendix contains the answer.

7.1.03 Monosaccharide Stereochemistry

Some monosaccharides have different molecular formulas (eg, glucose versus ribose). Others have the *same* molecular formula but different functional groups (eg, glucose and fructose), and are constitutional isomers. However, many carbohydrates have the same molecular formula *and* the same functional groups at each position. These molecules differ from each other only in their stereochemistry (ie, they are stereoisomers). This concept covers various stereochemical aspects by which monosaccharides can differ.

L-Sugars and D-Sugars

Monosaccharides are classified based on the similarity of their stereochemical configurations to those of L- and D-glyceraldehyde, which are themselves monosaccharides. These designations are assigned based on the configuration of the chiral carbon that is farthest from the anomeric carbon (ie, the *second-to-last* carbon in the linear chain), as shown in Figure 7.9. Note that the *final* carbon in any monosaccharide chain is a primary alcohol bound to *two H atoms*, so it is achiral and does *not* have stereochemistry.

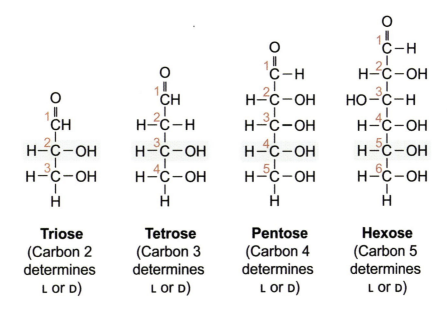

Figure 7.9 The assignment of L or D stereochemistry is determined by the configuration of the chiral carbon with the highest number.

For monosaccharides, when the last chiral center has an *R* absolute configuration, it is a D-sugar, and when it has an *S* configuration, it is an L-sugar (see Organic Chemistry Lesson 3.3 for an explanation of *R* and *S* stereochemistry). Note that this fact applies *only* to unmodified monosaccharides and should *not* be used to determine whether *other* molecules are in the L or D form since modifications may alter the Cahn-Ingold-Prelog priority of each substituent. In the Fischer projection of a monosaccharide in linear form, the L isomer shows the hydroxyl group of the last chiral center on the left, whereas in the D form the hydroxyl group is on the right.

Importantly, the L and D forms of a given monosaccharide are enantiomers. For instance, in L-glucose *every* chiral center differs in configuration from that of D-glucose. Flipping only the *last* chiral center of D-glucose does *not* convert the sugar to L-glucose but instead forms a monosaccharide called L-idose. To convert D-glucose to L-glucose (or vice versa), *every* chiral center must be flipped. Figure 7.10 shows several biologically important monosaccharides and their enantiomers.

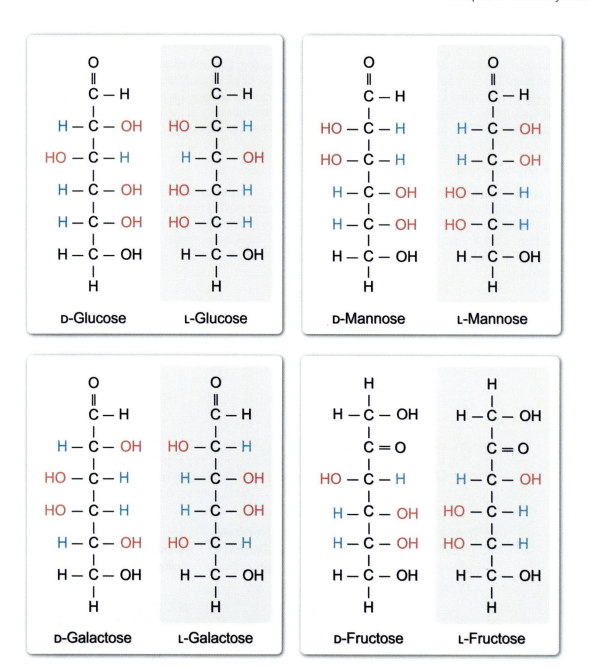

Figure 7.10 The D and L forms of several monosaccharides.

In contrast to the amino acids (which are predominantly found in the L form), naturally occurring monosaccharides are almost exclusively D isomers.

Concept Check 7.3

Arabinose is a monosaccharide that, unlike most sugars, is more abundant in nature in its L form than its D form. The diagram provided shows the structure of D-arabinose. Based on this diagram, draw the structure of L-arabinose.

$$\begin{array}{c} \text{O} \\ \parallel \\ \text{C}-\text{H} \\ | \\ \text{HO}-\text{C}-\text{H} \\ | \\ \text{H}-\text{C}-\text{OH} \\ | \\ \text{H}-\text{C}-\text{OH} \\ | \\ \text{H}-\text{C}-\text{OH} \\ | \\ \text{H} \end{array}$$

Solution
Note: The appendix contains the answer.

Epimers

D-Glucose, D-mannose, and D-galactose are aldohexoses. These each contain six carbons, each have the anomeric carbon at position 1, and the carbons at positions 2 through 6 each have one alcohol group. These molecules differ from each other by their configurations at *some* but *not all* stereocenters, making them diastereomers (see Organic Chemistry Lesson 3.3).

Specifically, D-mannose and D-galactose each differ from D-glucose at only one stereocenter. Diastereomers that differ at *only one position* are called epimers. An understanding of the structure of each sugar is facilitated by memorizing the structure of D-glucose and applying the relationship of the other molecules to it (see Figure 7.11).

The Fischer projection of D-glucose shows the hydroxyl groups of carbons 2, 3, and 4 pointing to the right, left, and right, respectively. D-Glucose and D-mannose differ only in the orientation of the hydroxyl and hydrogen groups at carbon 2 (ie, the hydroxyl group of D-mannose points to the left). Because this difference occurs at carbon 2, D-glucose and D-mannose are called C2 epimers.

D-Galactose and D-glucose also differ from each other at a single chiral center, but the difference occurs at carbon 4 instead of carbon 2. Therefore, D-galactose and D-glucose are called C4 epimers (Figure 7.11). Note that although D-galactose and D-mannose are both epimers of D-glucose, they differ *from each other* at *both* carbon 2 *and* carbon 4. Therefore D-galactose and D-mannose are *not* epimers of each other but instead can only be classified as diastereomers.

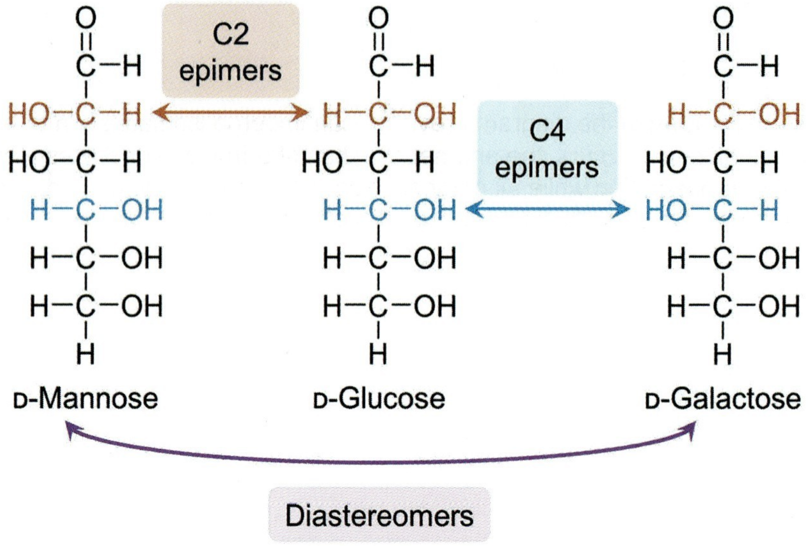

Chapter 7: Carbohydrates

```
        O                              O                              O
        ‖         ┌──────────┐         ‖         ┌──────────┐         ‖
        C—H       │   C2     │         C—H       │          │         C—H
        │         │ epimers  │         │         │   C4     │         │
    HO—C—H  ←─────┴──────────┴────→  H—C—OH      │ epimers  │      H—C—OH
        │                              │         │          │         │
    HO—C—H                          HO—C—H       └──────────┘      HO—C—H
        │                              │                              │
     H—C—OH                         H—C—OH  ←──────────────────→  HO—C—H
        │                              │                              │
     H—C—OH                         H—C—OH                         H—C—OH
        │                              │                              │
     H—C—OH                         H—C—OH                         H—C—OH
        │                              │                              │
        H                              H                              H
    D-Mannose                      D-Glucose                      D-Galactose
```

Diastereomers

Figure 7.11 Stereochemical relationships between D-glucose, D-mannose, and D-galactose.

D-Ribose has epimers that play biologically important roles, but these epimers are unlikely to be tested on the exam. The Fischer projection of D-ribose can be memorized by recognizing the similarity of the word *ribose* to "*right*-bose." In D-ribose, the hydroxyl group on every chiral center points to the right in a Fischer projection (see Figure 7.12).

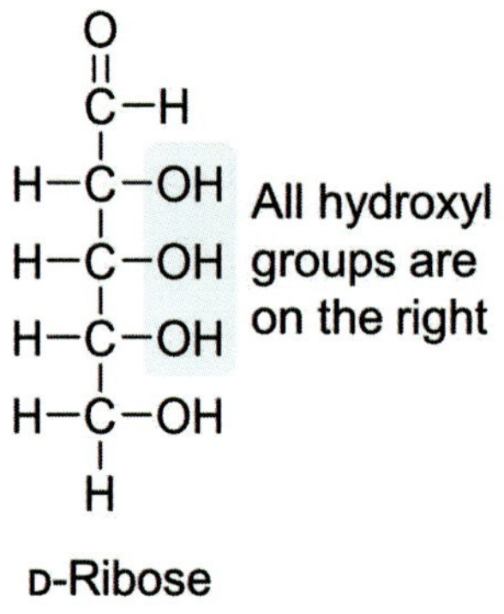

```
        O
        ‖
        C—H
        │
     H—C—OH    All hydroxyl
        │
     H—C—OH    groups are
        │
     H—C—OH    on the right
        │
     H—C—OH
        │
        H
     D-Ribose
```

Figure 7.12 The Fischer projection of D-ribose shows all hydroxyl groups on the right.

D-Glyceraldehyde also shows the hydroxyl group of its only chiral carbon on the right (refer to Figure 7.4).

Concept Check 7.4

D-Fructose, a ketose shown in the diagram provided, can undergo isomerization to become an aldose. When isomerization occurs, the anomeric carbon of D-fructose becomes a chiral center and may adopt either orientation while all other stereocenters remain unchanged. Based on this, which two epimers are produced by the isomerization of D-fructose?

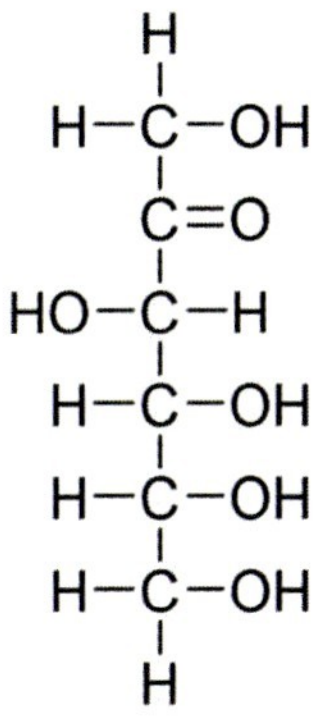

$$\begin{array}{c} H \\ | \\ H-C-OH \\ | \\ C=O \\ | \\ HO-C-H \\ | \\ H-C-OH \\ | \\ H-C-OH \\ | \\ H-C-OH \\ | \\ H \end{array}$$

D-Fructose

Solution
Note: The appendix contains the answer.

Stereochemistry of the Anomeric Carbon

The cyclic form of a monosaccharide can be shown as a Fischer projection by connecting the attacking hydroxyl group to the anomeric carbon (Figure 7.13).

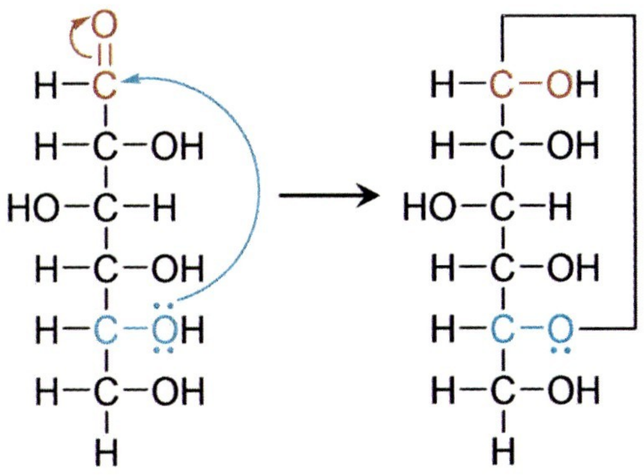

Linear D-glucose → Cyclized D-glucose

Figure 7.13 Fischer projections of D-glucose in linear and cyclic forms.

However, this projection requires displaying a bond with several bends in it, and it is often more convenient to show a cyclic carbohydrate as a **Haworth projection**. These projections can be formed after rotating the Fischer projection 90 degrees clockwise. Furanose rings are typically drawn as pentagons with the oxygen atom at the top, and pyranose rings are typically shown as hexagons with the oxygen in the upper right corner.

Any hydroxyl groups that were on the left in the Fischer projection point upward in the Haworth projection, and hydroxyl groups that pointed to the right in the Fischer projection point downward in the Haworth projection. For instance, D-glucose has hydroxyl groups on carbons 2, 3, and 4 pointing down, up, and down, respectively. In D-fructose, the hydroxyl groups of carbons 3 and 4 point up and down, respectively (Figure 7.14). Note that this form does not apply to the carbon that contributes its oxygen atom to the ring, or to any carbons with a higher number than that carbon.

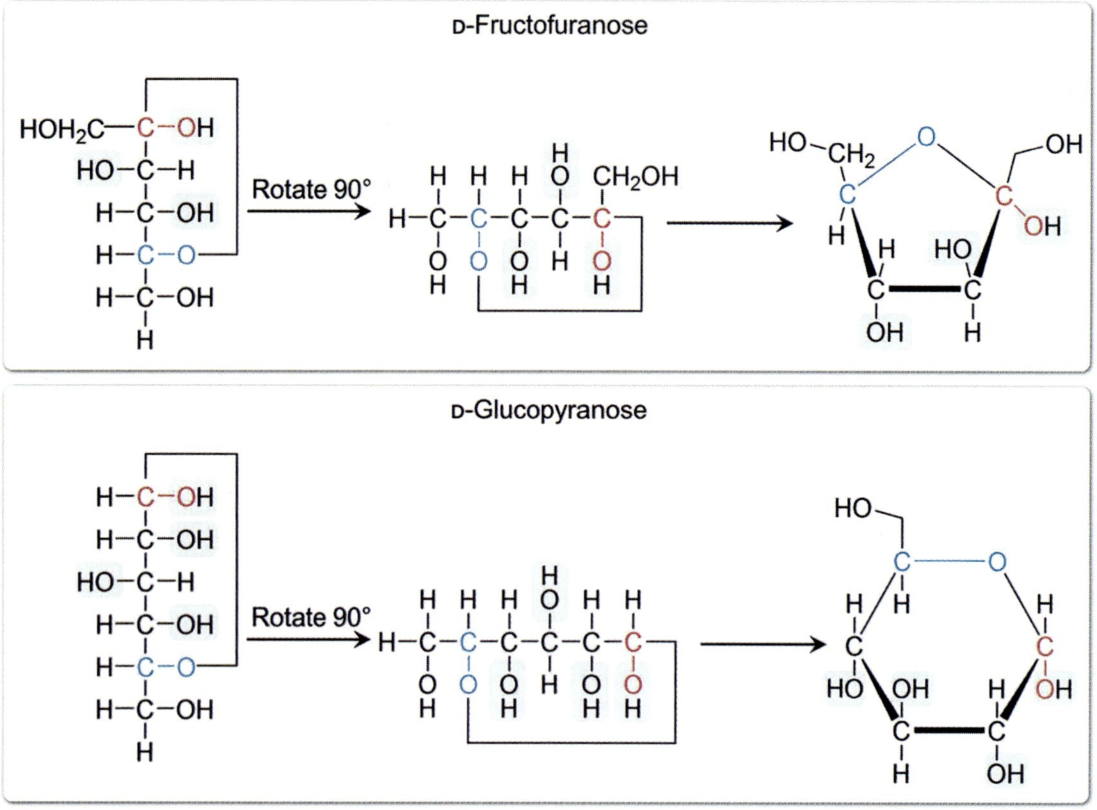

Figure 7.14 Conversion of furanose and pyranose sugars from Fischer to Haworth projections.

Monosaccharides may also be shown in **wedge-dash form**, in which any hydroxyl group that is down in a Haworth projection is a dash, and any hydroxyl group that is up in a Haworth projection is a wedge. In D-glucose, carbons 2, 3, and 4 have hydroxyl groups on a dash, a wedge, and a dash, respectively (Figure 7.15).

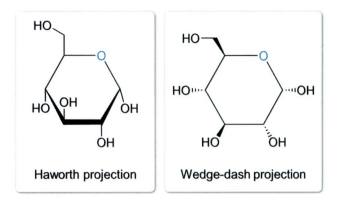

Figure 7.15 Examples of Haworth and wedge-dash projections of cyclic D-glucose.

In linear monosaccharides, the anomeric carbon is achiral. Upon cyclization, it can be attacked from either side, after which it adopts a tetrahedral arrangement with four distinct substituents, making it chiral. Therefore, a given cyclic sugar can take on one of two forms, and may interconvert between forms through the process of mutarotation.

The two cyclic forms are special types of epimers called **anomers** and are designated as the α and β anomers. For the exam, these designations can be made based on the relative positions of the hydroxyl on the anomeric carbon and the carbon with the highest number. If these two functional groups are *trans* to each other (ie, the groups are on opposite sides of the ring), the carbohydrate is in its α form. If they are *cis* (ie, the groups are on the same side of the ring), it is in its β form. Therefore, when shown in the typical Haworth projection the anomeric carbon of an α-D-sugar is shown with the hydroxyl group pointing down, and a β-D-sugar is shown with the hydroxyl group pointing up (Figure 7.16).

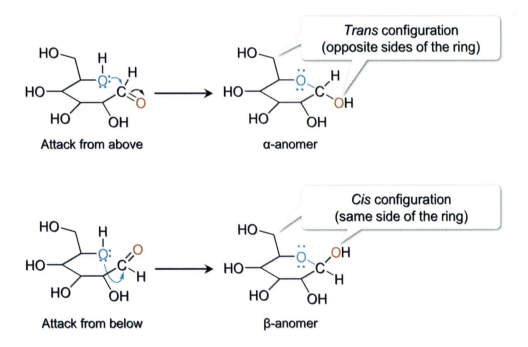

Figure 7.16 Formation of α and β anomers by cyclization.

Chapter 7: Carbohydrates

> ☑ **Concept Check 7.5**
>
> The image provided shows a cyclic form of D-ribose as a Fischer projection. 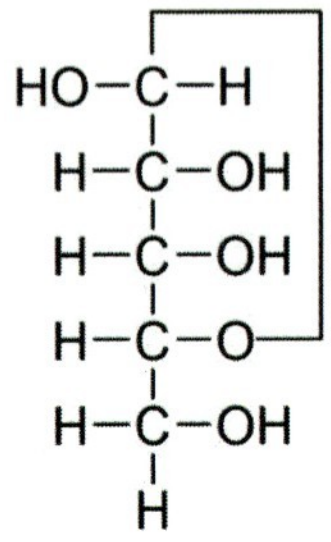
>
> $$\begin{array}{c} \text{HO}-\text{C}-\text{H} \\ \text{H}-\text{C}-\text{OH} \\ \text{H}-\text{C}-\text{OH} \\ \text{H}-\text{C}-\text{O} \\ \text{H}-\text{C}-\text{OH} \\ \text{H} \end{array}$$
>
> Convert the diagram to a Haworth projection and determine whether it is the α or β anomer.
>
> **Solution**
>
> *Note: The appendix contains the answer.*

Because switching from the D form to the L form requires *all* chiral centers to flip, Haworth projections of α-L-sugars show the hydroxyl group of the anomeric carbon pointing up, and projections of β-L-sugars show it pointing down. Figure 7.17 compares α-D-glucose to α-L-glucose.

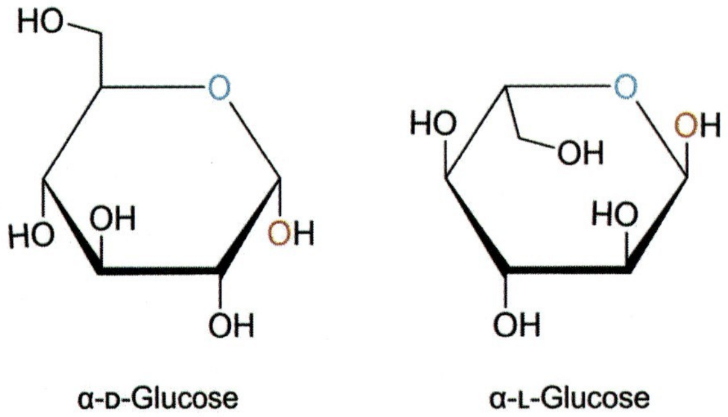

Figure 7.17 Comparison of α-D-glucose and α-L-glucose. All chiral centers are flipped.

Because of mutarotation, the α and β anomers of a free monosaccharide can interconvert and exist in equilibrium with each other. For instances in which the stereochemistry of the anomeric carbon is not important, the anomeric hydroxyl can be represented as extending horizontally (ie, neither pointing upward nor downward) in a Haworth projection, as being connected by a simple line in wedge-dash representation (ie, neither wedge nor dash), or as being connected by a wavy bond to indicate the lack of stereospecificity (Figure 7.18).

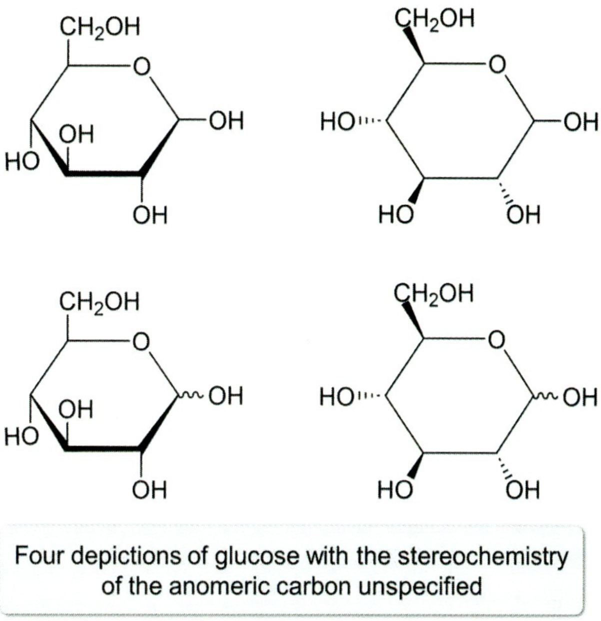

Four depictions of glucose with the stereochemistry of the anomeric carbon unspecified

Figure 7.18 Different representations of D-glucose. In these representations, the stereochemistry of the anomeric carbon is left unspecified.

Each of the stereochemical designations is important in characterizing polysaccharides (discussed in Lesson 7.2). Slight changes to any of these configurations can significantly alter the biological function of a complex carbohydrate and may result in disease states.

7.1.04 Monosaccharide Derivatives

Some monosaccharides are abundant in their unmodified forms, that is, forms with the formula $(CH_2O)_n$. However, many exist much more abundantly in chemically modified forms called **monosaccharide derivatives**. Several biologically important monosaccharide modifications are explored in this concept.

Although monosaccharides have various possible stereochemical designations (eg, D, L, α, β), they are often referred to without these designations. Unless otherwise specified, each is assumed to be in the D form, and the orientation of the anomeric carbon is only specified as needed. Most of the structures in this concept do not need to be memorized, but they may be helpful in understanding descriptions of various sugar derivatives.

Phosphosugars

One of the most common monosaccharide modifications is **phosphorylation**, in which at least one hydroxyl group on the monosaccharide is replaced by a phosphate. Most commonly, phosphorylation occurs when a primary alcohol acts as a nucleophile to attack an incoming phosphate group. Any other hydroxyl group (ie, the secondary alcohols in the sugar) may also attack a phosphate group, but this is less common.

All the biologically important monosaccharides listed in Concept 7.1.01 can exist in one or more phosphorylated forms in living cells. These are named according to the carbon to which the phosphate group becomes bonded. For example, glucose is commonly phosphorylated on the hydroxyl group of carbon 6, and this form is called glucose 6-phosphate. Certain reactions transfer the phosphate group to the hydroxyl group of carbon 1 to form glucose 1-phosphate. A few phosphorylated forms of glucose are shown in Figure 7.19.

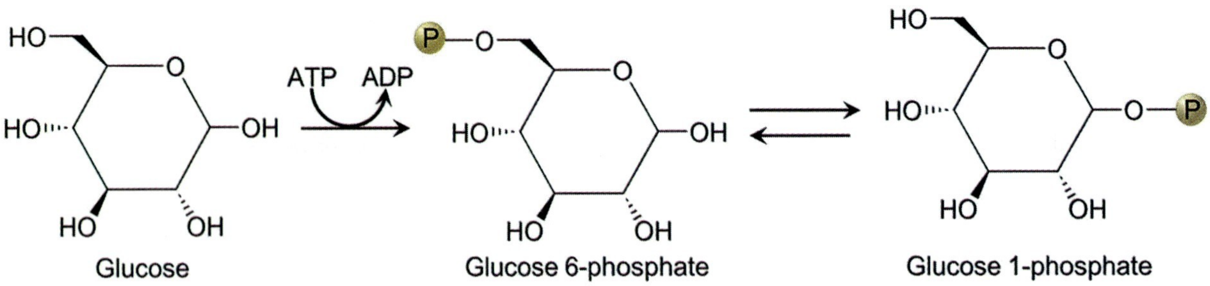

Figure 7.19 Phosphorylation of glucose to form glucose 6-phosphate and isomerization to form glucose 1-phosphate.

Monosaccharides that are phosphorylated at more than one position are denoted with a prefix. For example, fructose is commonly phosphorylated on both carbon 1 and carbon 6, and in this form, it is called fructose 1,6-bisphosphate. The prefix *bis–* indicates that two *different* positions in fructose each contain a separate phosphate group. This nomenclature contrasts with the term *diphosphate*, which indicates that *one* position on the sugar is linked to a functional group consisting of *two* phosphate groups connected to each other.

For example, in uridine diphosphate glucose, glucose is linked to uridine through a bridge composed of two phosphate groups. Figure 7.20 illustrates the difference between bisphosphate and diphosphate sugars. Similarly, the term *trisphosphate* indicates three phosphate groups in separate positions, whereas *triphosphate* indicates three phosphates connected to each other.

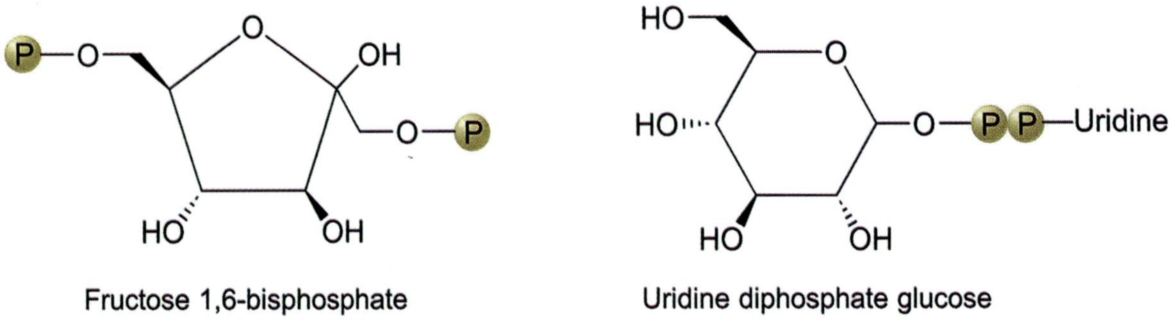

Figure 7.20 Bisphosphate and diphosphate designations indicate the relative positions of phosphate groups.

Chapter 7: Carbohydrates

> ☑ **Concept Check 7.6**
>
> Identify the monosaccharide derivatives shown and draw them as Fischer projections.
>
>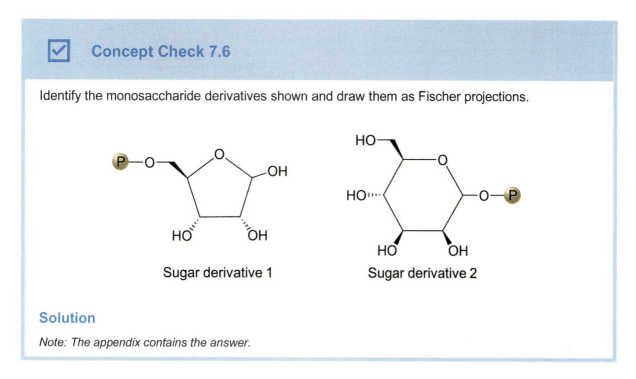
>
> **Solution**
>
> Note: The appendix contains the answer.

Deoxy Sugars

A deoxy sugar is a sugar in which one hydroxyl group is replaced by a hydrogen atom through a redox reaction. The most important example of this modification is 2-deoxyribose (often simply called deoxyribose), in which carbon 2 has been reduced and its hydroxyl group has been replaced (see Figure 7.21).

The backbone of DNA contains a repeating pattern of deoxyribose and phosphate groups (see Chapter 8). Other deoxy sugars play roles in intercellular communication and immune responses. Figure 7.21 shows the difference between ribose and deoxyribose.

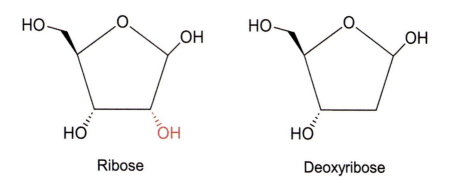

Figure 7.21 Ribose and deoxyribose.

A few dideoxysugars, from which *two* hydroxyl groups have been removed, are involved in some biological and biotechnological processes. One of the most important dideoxysugars in laboratory techniques is 2,3-dideoxyribose, a modified form of deoxyribose in which the hydroxyl groups of both carbon 2 *and* carbon 3 have been replaced. The dideoxysugar 2,3-dideoxyribose forms the sugar backbone of the 2′,3′-dideoxynucleotides used in DNA sequencing (see Biology Lesson 4.1).

Chapter 7: Carbohydrates

Oxidized and Reduced Sugars

The carbons in monosaccharides can undergo either oxidation or reduction to modify the sugar. Oxidation of an aldehyde yields a carboxylic acid, and aldoses that have undergone oxidation of the anomeric carbon have become aldonic acids. These sugar derivatives are commonly named by removing the *–ose* suffix from the name of the monosaccharide and replacing it with *–onic acid*. When deprotonated, the suffix becomes *–onate* or *–erate*.

Biologically important aldonic acids include 3-phosphoglycerate (glycolysis) and 6-phosphogluconate (pentose phosphate pathway). Figure 7.22 shows these molecules.

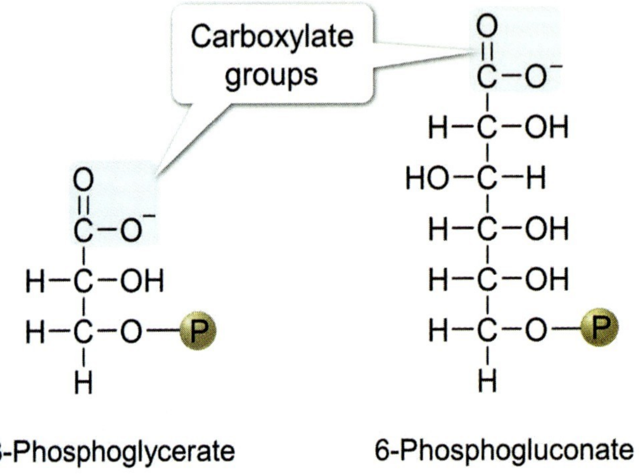

Figure 7.22 Examples of phosphorylated aldonic acids.

Other carbons in a monosaccharide can also undergo oxidation. Most commonly, the final carbon (ie, the primary alcohol) is oxidized to a carboxylic acid. Sugars that undergo this change are named similarly to aldonic acids but instead of changing the suffix to *–onic acid*, it becomes *–uronic acid*. For instance, glucose becomes glucuronic acid (Figure 7.23), which is found in some glycoproteins.

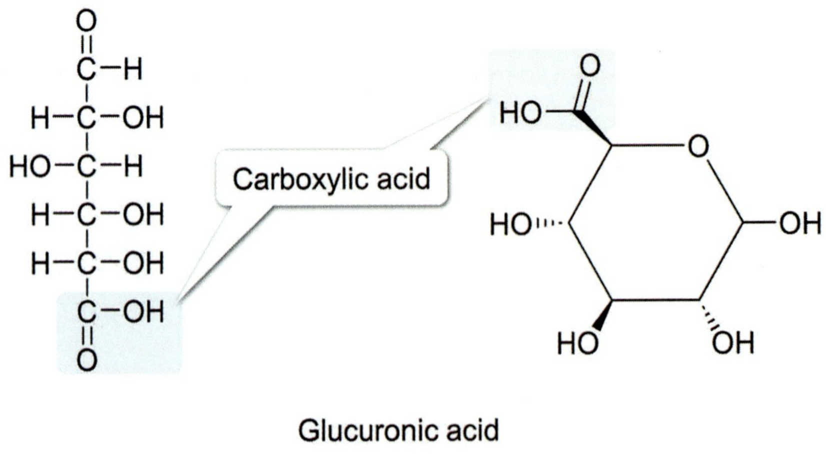

Figure 7.23 Glucuronic acid, a form of glucose in which carbon 6 is oxidized to a carboxylic acid, shown in both linear and cyclic forms.

Reduction of an aldehyde yields a primary alcohol. When an aldose is reduced in this way, it produces a sugar alcohol. Glycerol, an important component of phospholipid and triacylglyceride synthesis, is an example of a sugar alcohol (reduced glyceraldehyde). Ribitol (reduced ribose) is a component of the redox cofactors FMN and FAD. Figure 7.24 shows the structures of these sugar alcohols.

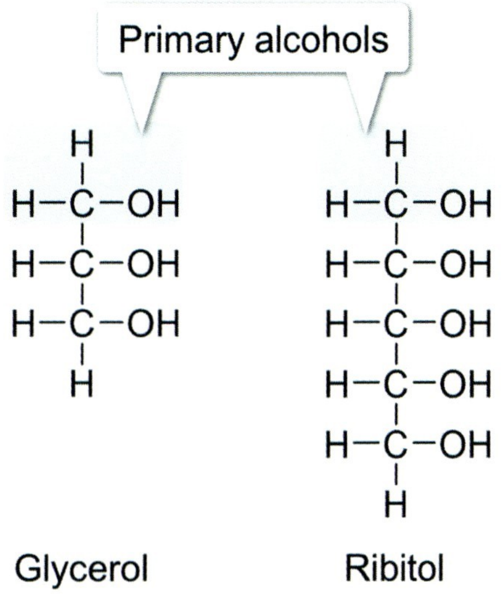

Figure 7.24 Examples of sugar alcohols.

Amino Sugars

Monosaccharides can also be modified by the replacement of a hydroxyl group with an amino group. In living cells, amines tend to become protonated and positively charged. To remove this charge, amino sugars are commonly acetylated, which converts the highly basic amine to a less basic amide. Typically, amination occurs on carbon 2 of an aldose or, occasionally, carbon 3 of a ketose.

Monosaccharides that undergo amination are given the suffix –amine (eg, glucose becomes glucosamine). Addition of an acetyl group to the N atom of the amine yields N-acetylglucosamine, sometimes abbreviated as GlcNAc (Figure 7.25). N-acetylgalactosamine (GalNAc) and N-acetylmannosamine (ManNAc) are commonly found in polysaccharides along with GlcNAc.

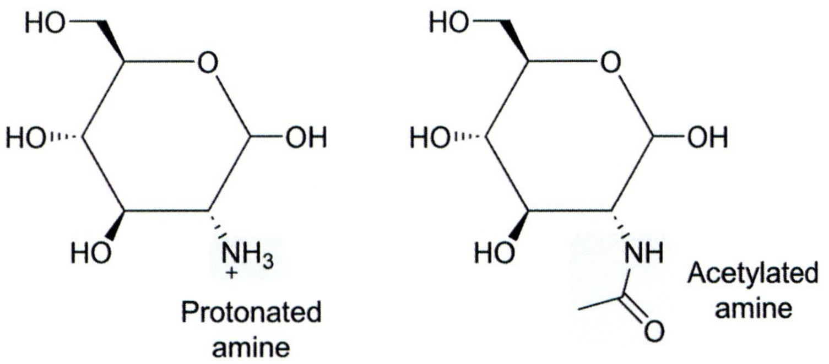

Figure 7.25 Examples of amino and N-acetylamino sugars.

Lesson 7.2

Complex Carbohydrates

Introduction

Lesson 7.1 discussed various biologically important monosaccharides and the systems used to classify them. Just as amino acids combine to form polypeptides and proteins, monosaccharides can combine to form more complex carbohydrates.

The linkage between two monosaccharides is called a **glycosidic bond** (or a glycosidic linkage). Monosaccharides can combine through glycosidic bonds to form disaccharides, trisaccharides, and longer chains, including highly complex polysaccharides with multiple branch points. Polysaccharides are also commonly known as **glycan chains**. In addition to forming glycosidic bonds with other carbohydrates, a carbohydrate may form a glycosidic bond with another biomolecule such as a protein, a nitrogenous base, or a lipid. In this way, many biomolecules may be modified by linkages to complex glycan chains.

This lesson explores the ways in which monosaccharides can form glycosidic bonds and the important structures formed by these linkages.

7.2.01 Glycosidic Bonds

A glycosidic bond is a chemical linkage between the anomeric carbon of a carbohydrate and a nucleophilic atom of another molecule. Often the other molecule is another carbohydrate but it may also be a protein, a lipid, or the nitrogenous base of a nucleoside. This concept covers several types of glycosidic bonds and their classification.

Glycosidic Bonds Between Two Carbohydrates

Two monosaccharides form a glycosidic bond and become a disaccharide when the hemiacetal or hemiketal carbon on one monosaccharide condenses with a hydroxyl group on the other, as shown in Figure 7.26. Formation of the glycosidic bond releases water and therefore constitutes a condensation reaction, which can be reversed by hydrolysis.

Just as amino acids become amino acid residues upon peptide bond formation (see Lesson 2.1), monosaccharides become monosaccharide residues (also called monosaccharide units) when glycosidic bonds form.

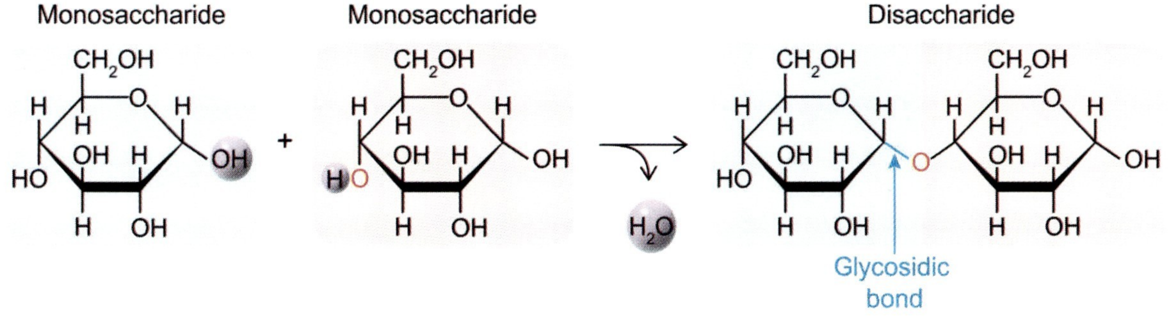

Figure 7.26 An example of glycosidic bond formation between two monosaccharides to form a disaccharide.

Chapter 7: Carbohydrates

A disaccharide can become a trisaccharide by combining with a third monosaccharide. This process may be repeated to form carbohydrates consisting of many monosaccharide residues. These complex carbohydrates are known as **oligosaccharides** (meaning "a few sugars"), **polysaccharides** (meaning "many sugars"), or simply as **glycans**.

Due to the loss of water, carbohydrates consisting of more than one monosaccharide residue do *not* have the general formula $(CH_2O)_n$. Instead, disaccharides have the formula $C_n(H_2O)_{n-1}$, trisaccharides have the formula $C_n(H_2O)_{n-2}$, and so on. Figure 7.27 shows structures of carbohydrates that follow these molecular formulas.

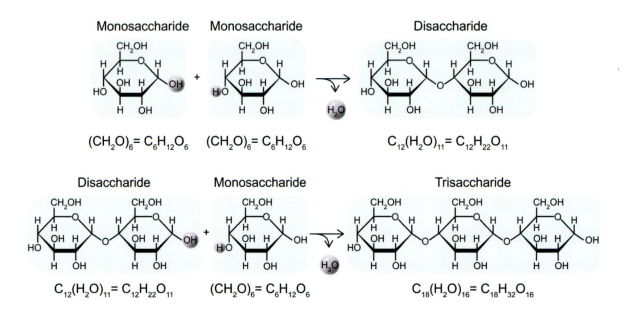

Figure 7.27 Examples of changes in the general molecular formula when glycosidic bonds form.

✓ **Concept Check 7.7**

What is the molecular formula of a tetrasaccharide formed from one glucose molecule ($C_6H_{12}O_6$), one ribose molecule ($C_5H_{10}O_5$), one erythrose molecule ($C_4H_8O_4$), and one fructose molecule ($C_6H_{12}O_6$)?

Solution

Note: The appendix contains the answer.

Glycosidic bond formation converts the hemiacetal (of a cyclic aldose) or the hemiketal (of a cyclic ketose) into an acetal or a ketal, respectively. The acetal or ketal can no longer open to linearize or allow for mutarotation. Therefore, once a monosaccharide unit has contributed its anomeric carbon to a glycosidic bond, it is locked in cyclic form and has a fixed anomeric configuration (see Figure 7.28).

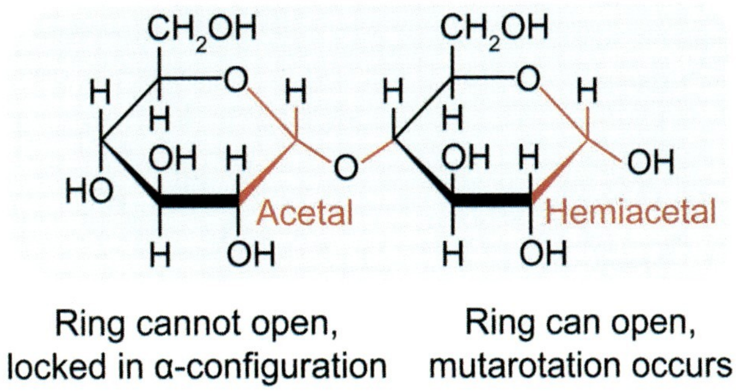

Chapter 7: Carbohydrates

Ring cannot open, locked in α-configuration   Ring can open, mutarotation occurs

Figure 7.28 Monosaccharide units that contain an acetal (or ketal) group because of a glycosidic bond are locked in a cyclic form.

Unlike amino acids, which generally form linear polymers, glycans often have branches. For instance, a single aldohexose unit may be linked to one carbohydrate through its anomeric carbon (carbon 1), to another carbohydrate through carbon 4, and to yet another through carbon 6. Each of these monosaccharide units may *also* have branches, resulting in highly complex structures. Figure 7.29 shows an example.

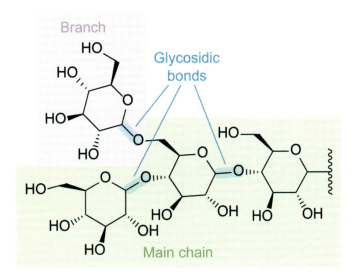

Figure 7.29 A single monosaccharide unit may participate in multiple glycosidic bonds, forming complex branched structures.

To distinguish glycosidic bonds at different positions within a monosaccharide unit, these are classified based on the positions of each carbon and the stereochemistry (α or β) of any anomeric carbons involved. For example, if the anomeric carbon of an aldose (ie, carbon 1) is in the α configuration and linked to carbon 4 of another monosaccharide unit, the linkage is classified as an **α-1,4-glycosidic** bond. Similarly, a ketose linked through a β-anomeric carbon (carbon 2) to carbon 6 of another monosaccharide has a **β-2,6-glycosidic linkage**. Figure 7.30 shows examples of different glycosidic linkages.

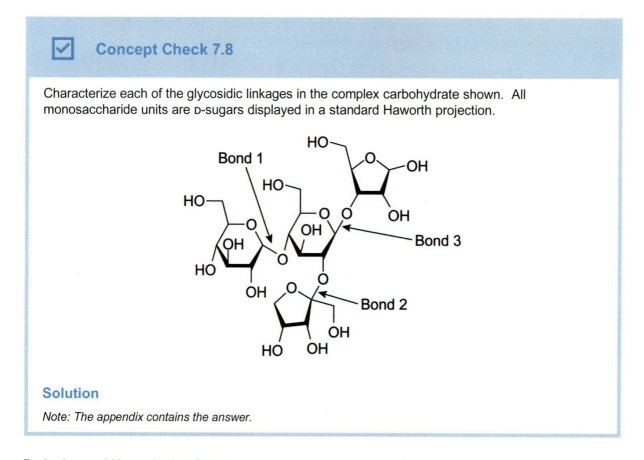

Figure 7.30 Examples of α-1,4-glycosidic and β-2,6-glycosidic linkages.

☑ Concept Check 7.8

Characterize each of the glycosidic linkages in the complex carbohydrate shown. All monosaccharide units are D-sugars displayed in a standard Haworth projection.

Solution
Note: The appendix contains the answer.

Reducing and Nonreducing Sugars

To form a glycosidic bond, one of the carbohydrates involved must have a free anomeric carbon (ie, an anomeric carbon *not* involved in a glycosidic bond). Historically, carbohydrates with free anomeric carbons have been classified as **reducing sugars** because they can reduce silver ions (Ag^+) to silver metal. The reaction that produces this effect is called the Tollens test.

The phrase "reducing sugar" is a misnomer in some ways because *any* primary or secondary alcohol group can potentially act as a reducing agent. However, when the phrase was first introduced, it referred only to the ability of carbohydrates to reduce Ag^+, which carbohydrates can accomplish *only* through an aldehyde group.

All free monosaccharides are reducing sugars. For aldoses, reduction occurs when the aldehyde group of a linear monosaccharide donates electrons to the silver ions, and the aldehyde becomes a carboxylic acid. The ketone group in a ketose cannot be oxidized any further, but *ketoses can act as reducing sugars* because under the Tollens test reaction conditions they can tautomerize, which converts them to aldoses.

Carbohydrates with no free anomeric carbon are classified as **nonreducing sugars**. Trehalose (shown in Figure 7.31) is an example of a nonreducing sugar because both anomeric carbons are involved in the glycosidic bond, so neither is free. Note that to align the anomeric carbons, the Haworth projection of one of the glucose units is rotated 180 degrees.

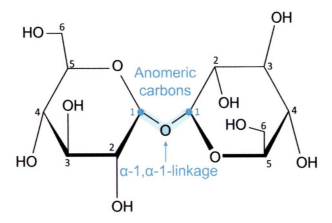

Figure 7.31 Structure of trehalose, a nonreducing sugar.

Because nonreducing sugars have a glycosidic bond involving *two* anomeric carbons, characterization of the glycosidic bond requires indicating the configurations of both. For instance, in the case of trehalose, the glycosidic bond is an α-1,α-1 linkage.

Reducing and nonreducing terminology can also be used to describe the polarity of complex sugars. Just as peptides have N-terminal and C-terminal ends, complex carbohydrates have **reducing and nonreducing ends**. The reducing end of a complex carbohydrate is the end that terminates with an anomeric carbon. The other end, with no free anomeric carbons, is the nonreducing end.

Figure 7.32 shows the reducing and nonreducing ends of carbohydrates. Each monomer within a complex carbohydrate also has reducing and nonreducing ends.

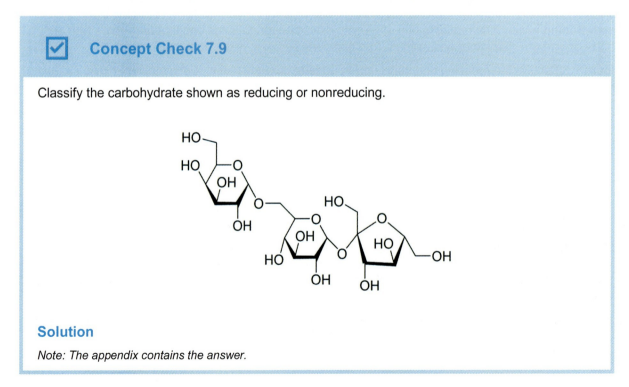

Figure 7.32 Nonreducing and reducing ends of complex carbohydrates. (A) The reducing end of the chain can open and reduce Ag^+. (B) Reducing and nonreducing ends of a monosaccharide unit.

For two carbohydrates to link through a glycosidic bond, at least one of them must have a free reducing end. However, note that glycosidic bond formation is a condensation reaction and *not* a redox reaction. Consequently, although ketoses must tautomerize to aldose form to give a positive Tollens test (which involves a redox reaction), they do *not* need to tautomerize to form glycosidic bonds.

☑ Concept Check 7.9

Classify the carbohydrate shown as reducing or nonreducing.

Solution

Note: The appendix contains the answer.

Chapter 7: Carbohydrates

Glycosidic Bonds Between a Carbohydrate and a Noncarbohydrate

In addition to glycosidic linkages between two carbohydrates, a glycosidic bond can also form between the reducing end of one carbohydrate and the nucleophile in another molecule, creating a molecule known as a glycoside. Some important examples include the glycosylation of proteins discussed in Lesson 2.4, the formation of a bond between ribose and a nitrogenous base (see Chapter 8), and the formation of a bond between a carbohydrate and a lipid (see Chapter 9). In some cases, the anomeric carbon in these glycosides is bound to one oxygen and one nitrogen instead of two oxygens. Examples are shown in Figure 7.33.

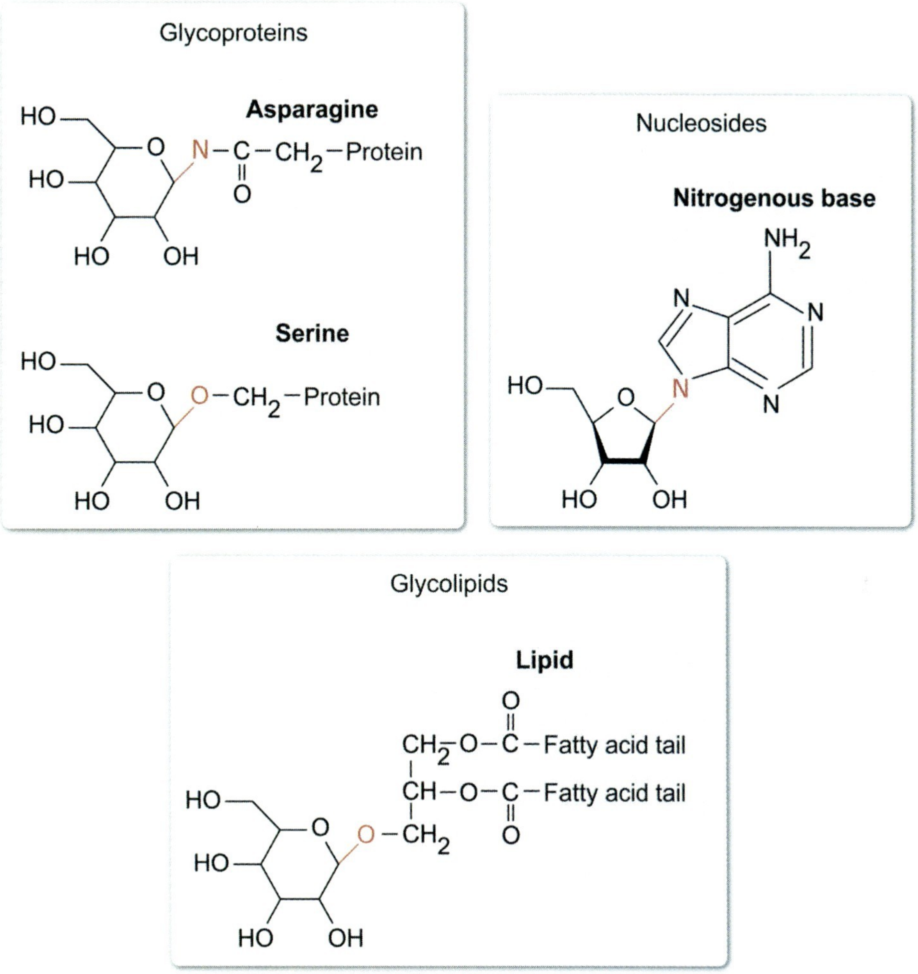

Figure 7.33 Examples of glycosidic linkages between carbohydrates and noncarbohydrates.

Protein glycosylation yields glycoproteins, which facilitate many interactions between cells. In addition, viruses and other pathogens may bind to glycoproteins in the first step of cell infection.

Lipid glycosylation yields glycolipids which, like glycoproteins, are commonly found on the extracellular side of cell membranes. ABO blood types stem from different carbohydrates linked to lipids in red blood cell membranes. Defects in glycolipid metabolism can lead to various disease states such as Tay-Sachs disease.

Nucleoside synthesis results in a glycosidic bond between the anomeric carbon of ribose (or deoxyribose) and one of the nitrogenous bases found in DNA or RNA.

7.2.02 Disaccharides

Disaccharides are the smallest, simplest type of complex carbohydrate. They consist of two monosaccharide units linked to each other through a single glycosidic bond. Several disaccharides play biologically important roles, but for the exam, the most important disaccharides are maltose, sucrose, and lactose.

Maltose is a disaccharide consisting of two glucose units linked by an α-1,4-glycosidic bond. Maltose is derived from the digestion of starch (see Concept 7.2.03) and can be further digested to yield individual glucose units.

The linkage between the glucose units is critical. When the linkage is changed to a β-1,4-glycosidic bond, the sugar becomes cellobiose. Although cellobiose is identical to maltose in every way *except* the orientation of the linkage, animals can digest maltose but they lack the enzymes needed to digest cellobiose. Maltose and cellobiose are shown in Figure 7.34.

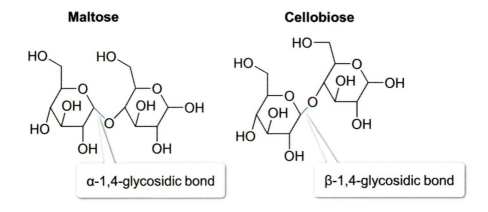

Figure 7.34 Maltose and cellobiose differ only by the configuration of the glycosidic linkage.

Sucrose is a nonreducing sugar consisting of one glucose unit and one fructose unit linked to each other through their anomeric carbons. The glucose unit is in the α configuration, and the fructose unit is in the β configuration. In addition, glucose is in the pyranose form, whereas fructose is in the furanose form.

Because the anomeric carbon of glucose is carbon 1 and that of fructose is carbon 2, the glycosidic linkage in sucrose is an α-1,β-2-glycosidic bond. Sucrose can be digested to its fructose and glucose components, which can then be metabolized for energy or other purposes.

Because the anomeric carbons of each monosaccharide in sucrose link to each other, the Haworth projection of one of the monosaccharides (usually fructose) must be flipped along its vertical axis to align the anomeric carbons. Every group that pointed up in the typical depiction points down, and vice versa, in the flipped depiction. The structures of fructose, glucose, and sucrose are shown in Figure 7.35.

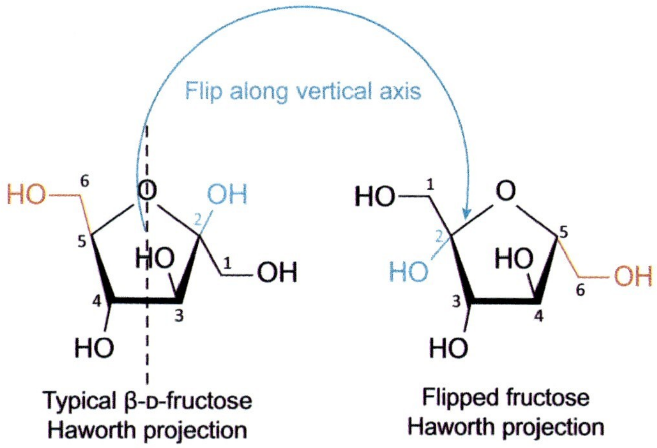

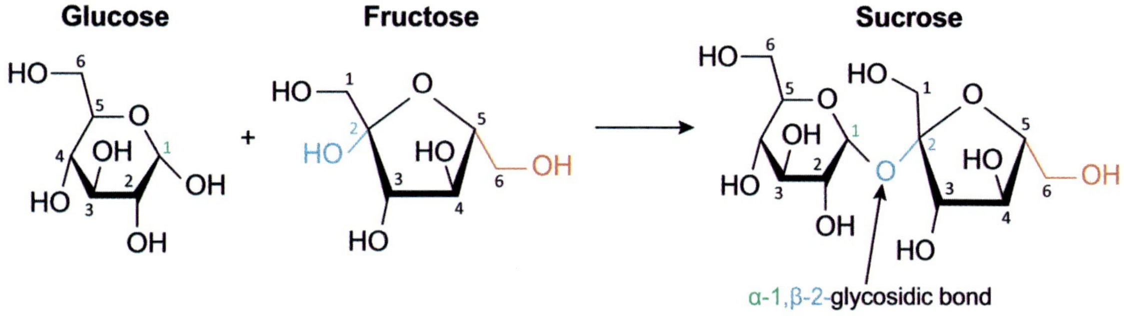

Figure 7.35 Structure of sucrose, assembled from glucose and fructose.

Lactose consists of a galactose unit linked through its anomeric carbon to carbon 4 of a glucose unit. The galactose unit is in the β configuration, and therefore the linkage is a β-1,4-glycosidic bond. Consequently, lactose is identical to cellobiose except for the orientation of carbon 4 at the nonreducing end.

Although humans do not have enzymes that break down cellobiose, most humans can digest the highly similar lactose molecule using the enzyme lactase. This fact illustrates the specificity of some enzymes: a change in the orientation of a single chiral center can significantly impact interactions between enzymes and substrates. Individuals with lactase deficiency may experience lactose intolerance, which can be treated with lactase enzyme supplements. The structure of lactose is shown in Figure 7.36.

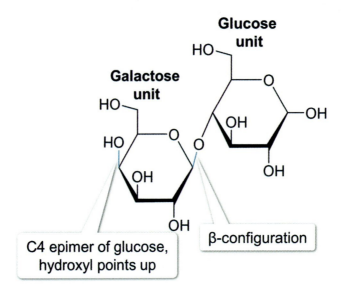

Figure 7.36 Structure of lactose.

 Concept Check 7.10

Which monosaccharide unit in lactose is the reducing end?

Solution

Note: The appendix contains the answer.

7.2.03 Polysaccharides

Polysaccharides are carbohydrates that consist of multiple monosaccharides. All eukaryotes and many prokaryotes use polysaccharides both for structure and energy storage. Storing carbohydrates in this way decreases the glucose molarity (and therefore total cellular osmolarity) within the cell as several small glucose molecules become one much larger molecule.

Because polysaccharides can include branch points, the diversity of ways to assemble a polysaccharide with a given number of monosaccharide units can be enormous. For example, accounting for all possible positions, anomeric carbon orientations (α or β), and ring forms (furanose or pyranose), just two glucose units can link to each other in at least 30 different ways.

The number of possible linkages increases exponentially as more monosaccharide units are included. Consequently, polysaccharides can perform a wide variety of functions, many of which are not currently understood. General features of a few important polysaccharides should be remembered for the exam.

Starch

Starches are polysaccharides consisting entirely of glucose units. The simplest starch is amylose, in which all glucose units are linked to each other through α-1,4-glycosidic bonds in a linear chain. These chains can be broken down into maltose units, which can be further degraded to their glucose constituents.

Amylose is a major component of many staple foods and an important part of human diets. Note that amylose does not have a specific number of glucose units. Instead, amylose is any linear chain of α-1,4-

linked glucose units, and different amylose molecules can differ in size. Figure 7.37 shows the structure of amylose.

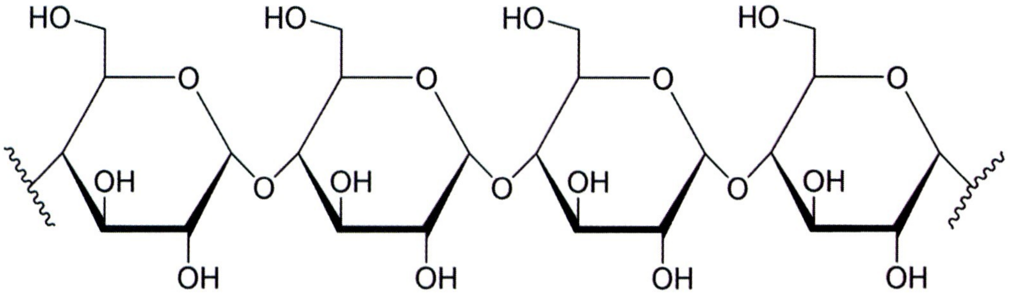

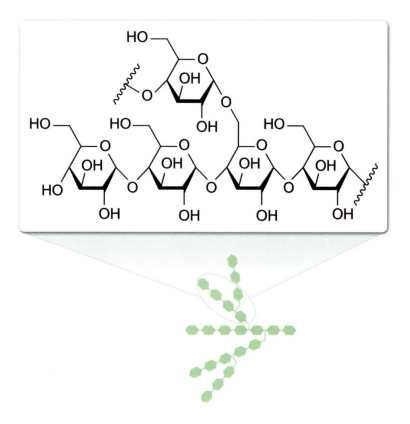

Amylose

Figure 7.37 Structure of amylose, the simplest form of starch.

Plants also contain a form of starch called amylopectin (see Figure 7.38). Like amylose, amylopectin contains glucose units in α-1,4-glycosidic linkages; however, amylopectin also contains branch points that involve α-1,6-glycosidic linkages. On average, amylopectin contains a branch point approximately every 25 glucose units. Animals can digest amylopectin into glucose units.

Figure 7.38 General structure of amylopectin.

Glycogen

Glycogen is the primary form of glucose storage in animals. Like amylopectin, glycogen consists of glucose units linked by α-1,4-glycosidic bonds within linear chains and α-1,6-glycosidic bonds at branch points. The most notable difference between amylopectin and glycogen is that glycogen is more branched, with branch points approximately every 10 to 15 glucose units (Figure 7.39). In addition, glycogen chains in living cells are linked to the protein glycogenin by a glycosidic bond to a tyrosine residue. Therefore, glycogen does not have a free reducing end, whereas amylopectin does.

Note that due to branching, both molecules have *multiple* nonreducing ends. Having multiple nonreducing ends facilitates adding glucose units to the polymer (as in glycogenesis; see Concept 11.4.01) or removing glucose units from the polymer (as in glycogenolysis; see Concept 11.4.02).

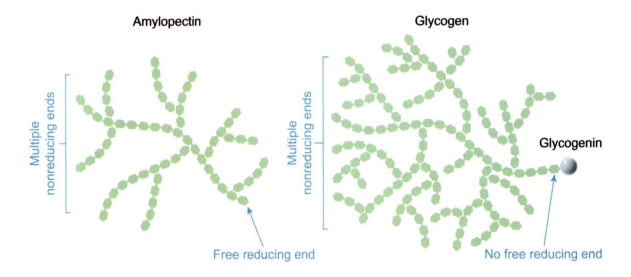

Figure 7.39 Comparison of amylopectin and glycogen.

Cellulose

Cellulose is related to amylose in the same way that cellobiose is related to maltose: they are identical in every way except the orientations of the glycosidic bonds. Cellulose is a linear chain of glucose units linked by β-1,4-glycosidic bonds and is a major component of plant cell walls. Because animals lack enzymes that hydrolyze β-1,4-glycosidic bonds between glucose units, they cannot digest cellulose. Animals that rely on diets high in cellulose (eg, ruminants) require a symbiotic relationship with certain microflora (eg, bacteria), which *do* have the necessary enzymes, to digest their food.

The glucose units in cellulose are linked by β-1,4-glycosidic bonds; as such, the bond from each anomeric carbon to the oxygen of the glycosidic bond might be expected to point upward in a Haworth projection. Similarly, the bond from carbon 4 to the glycosidic bond oxygen would be expected to be shown pointing downward. However, connecting the anomeric carbon of one unit to carbon 4 of another *without* altering the Haworth projection would lead to visual depiction of a chain of monosaccharides moving diagonally up the page. To convert the projection to display a horizontal chain, it is convenient to draw cellulose by flipping the Haworth projection of every other glucose unit in the chain (see Figure 7.40).

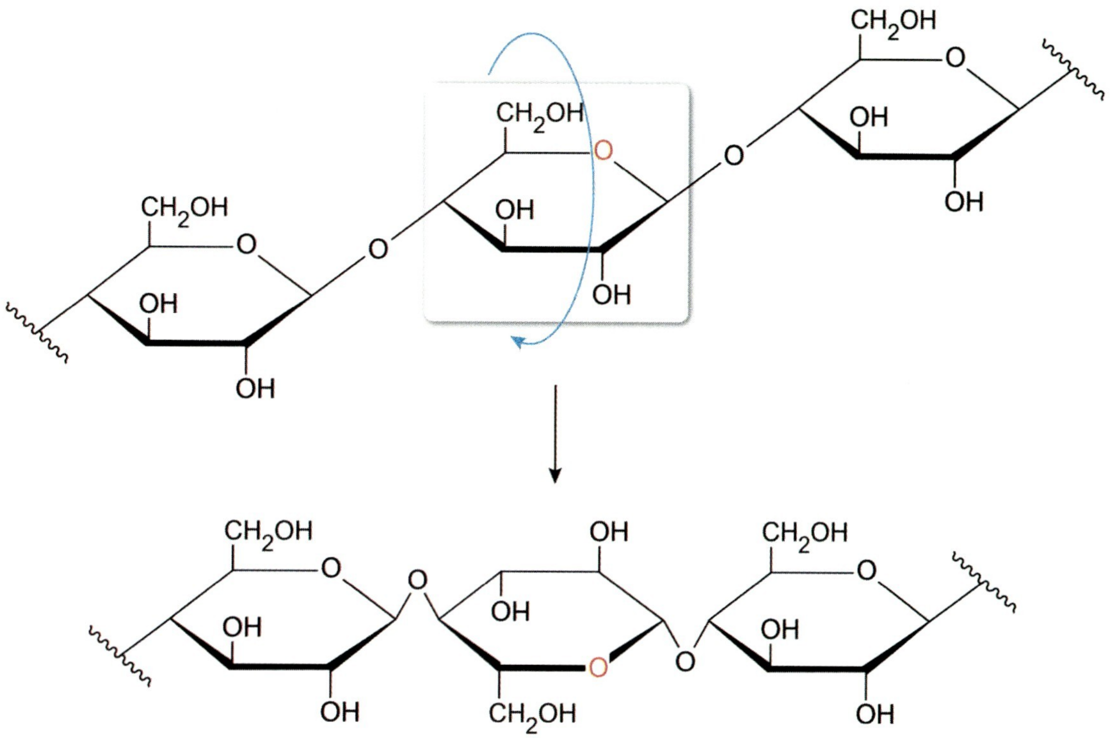

Figure 7.40 Two depictions of cellulose. Flipping the Haworth projection of every other glucose unit facilitates a horizontal depiction.

Other Polysaccharides

Although other polysaccharides do not need to be memorized, it is important to be aware of them. These polysaccharides primarily include the glycan chains found on glycoproteins and glycolipids. They commonly consist of several types of monosaccharide units, including various sugar derivatives, and often contain multiple branch points with various types of glycosidic linkages.

An example of such a polysaccharide is shown in Figure 7.41. Note that this molecule serves only to illustrate the complexity of carbohydrates and need not be memorized.

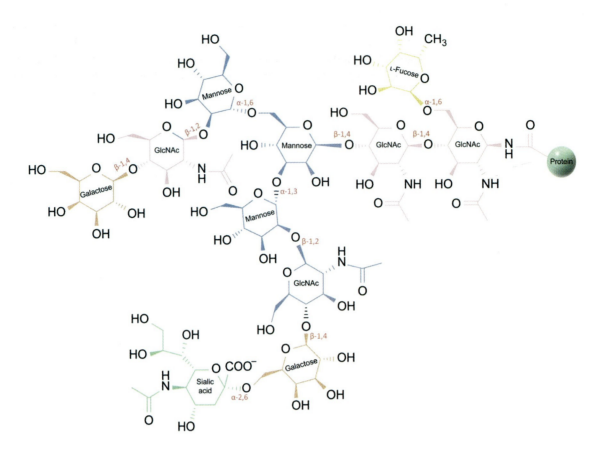

Figure 7.41 A typical glycan found on the proteins of many human cells. This glycan includes *N*-acetyl derivatives, an L-deoxysugar, and a carboxylic acid derivative.

Lesson 8.1

Nucleotides

Introduction

Nucleotides are a class of molecules that store and transmit genetic information, provide energy for thermodynamically unfavorable reactions, and facilitate various biologically important oxidation-reduction reactions. These molecules consist of a nitrogenous base (ie, a system of one or more nitrogen-containing aromatic rings) glycosidically linked to a sugar (usually ribose or deoxyribose). The sugar contains at least one phosphate group, typically linked to the sugar's highest-numbered carbon atom. Figure 8.1 shows the general structure of a nucleotide.

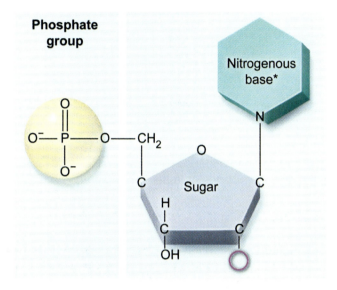

◯ = H in DNA (deoxyribose sugar) and OH in RNA (ribose sugar).

Figure 8.1 General structure of a nucleotide.

This lesson explains the structures of the nitrogenous bases used in DNA and RNA, the arrangement of the components in a nucleotide, nucleotides other than those in DNA and RNA, and the biological roles that nucleotides perform.

8.1.01 Nitrogenous Bases

Just as amino acids are identified by their side chains, nucleotides are identified by their nitrogenous bases. The most abundant bases are those found in nucleic acids (ie, DNA, RNA). The structures of these bases are commonly tested on the exam.

The nitrogenous bases found in nucleic acids are divided into purines and pyrimidines. Of the five most common nitrogenous bases, two are purines (adenine and guanine) and three are pyrimidines (cytosine, thymine, and uracil). This concept explains the structure of each base and provides mnemonics to help remember each structure.

The **pyrimidine bases** consist of a single six-membered aromatic ring that contains two nitrogen atoms. The structure of pyrimidine, from which these bases can be derived, is shown in Figure 8.2.

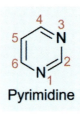

Figure 8.2 The structure of pyrimidine, from which pyrimidine bases are derived.

In the orientation shown in Figure 8.2, the members of the pyrimidine ring are numbered starting with the nitrogen atom at the bottom of the ring and ascending in counterclockwise order. This numbering scheme assigns the nitrogen atoms the smallest numbers possible (ie, 1 and 3).

The **purine bases** consist of two fused aromatic rings, one with five members and one with six members. The structure of purine, shown in Figure 8.3, resembles that of pyrimidine with a five-membered ring attached. Note that because the five- and six-membered rings share two atoms, only three *additional* atoms (two nitrogens and one carbon) are needed to form the second ring of purine.

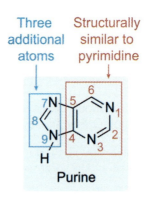

Figure 8.3 Structure of purine with similarities and differences to pyrimidine.

The numbering of the atoms in purine is different from that of pyrimidine. As shown in Figure 8.3, the six-membered ring has its constituents numbered in *clockwise* order, with position 1 assigned to the nitrogen atom that is the farthest away from the five-membered ring. The remaining three atoms of the five-membered ring are numbered 7 to 9 from top to bottom.

This historical numbering scheme also gives the pyrimidine nitrogens, as well as the shared carbons, the lowest numbers possible (ie, 1 and 3 for the nitrogens, 4 and 5 for the carbons). The numbering of the atoms in purine and pyrimidine *do not need to be memorized* but are helpful for discussion of the relative positions of the substituents in each nitrogenous base.

To remember that purines have two rings, it may be helpful to use the mnemonic, "Purines are pure rings." In addition, the mnemonic that these rings are "**A**s **G**ood **A**s **G**old," which contains two As and two Gs, may help in remembering that **A**denine and **G**uanine are purines and have two rings each. The other bases (cytosine, uracil, and thymine) are pyrimidines with one ring each.

The nitrogenous bases in DNA and RNA interact with each other through hydrogen bonds, and pyrimidine and purine can be modified to yield these bases by placing hydrogen bond donors and acceptors in certain positions on the rings.

In general, hydrogen bond *donors* are oxygen, nitrogen, or fluorine atoms covalently linked to at least one hydrogen atom. Hydrogen bond *acceptors* are oxygen, nitrogen, or fluorine atoms with at least one lone pair of electrons.

The mnemonic "**ADD**itionally, **G**uanine is **G**enerous" may be used to remember the structure of guanine: being generous, as guanine has two donors (and one acceptor). The order of these hydrogen bond participants is, from top to bottom, **A**cceptor, **D**onor, **D**onor (**ADD**).

To satisfy these hydrogen bond requirements, guanine contains a carbonyl at position 6 (ie, at the top of the six-membered ring). The carbonyl oxygen acts as a hydrogen bond acceptor. The nitrogen at position 1 is bonded to a hydrogen atom and therefore acts as a donor. The carbon atom at position 2 is linked to a primary amino group ($-NH_2$), which acts as another hydrogen bond donor. Figure 8.4 shows the structure of guanine compared to purine.

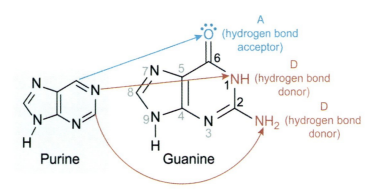

Figure 8.4 The structure of purine can be converted to guanine by adding a hydrogen bond acceptor and two donors in the correct positions.

Once the structure of guanine is known, the other bases can be derived. Guanine pairs with cytosine (see Lesson 8.2), so cytosine must have acceptors that align with guanine's donors and vice versa. To achieve this alignment, cytosine has an amino group as a donor at position 4 (ie, at the top of the ring). The nitrogen atom at position 3 has a double bond to carbon 4 and a *free lone pair* outside the aromatic system, making it a hydrogen bond acceptor. The carbon at position 2 is a carbonyl, and the carbonyl oxygen atom is also an acceptor.

Cytosine's interaction with guanine is shown in Figure 8.5, with the pyrimidine ring flipped relative to its orientation in Figure 8.2 to facilitate hydrogen bond pair alignment.

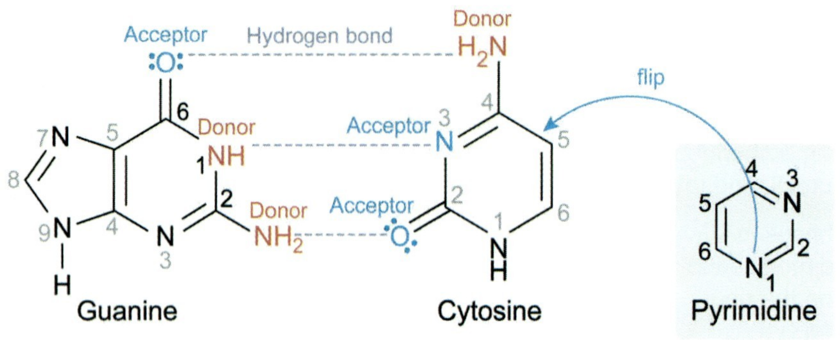

Figure 8.5 Interactions between guanine and cytosine. The relationship of cytosine to pyrimidine is also shown.

Adenine's ring structure can be derived from that of guanine in that adenine must pair with a *different* set of pyrimidines (Lesson 8.2). Therefore, adenine has opposite donor and acceptor configurations at positions 1 and 6. At carbon 6 adenine has an amino group (a donor). Position 1 (a ring nitrogen) must be an acceptor, and therefore it has a lone pair and a double bond to the carbon at position 6. Adenine has only two hydrogen bond participants in its base pairing, so position 2 is neither an acceptor nor a donor. Figure 8.6 shows the relationship between guanine and adenine.

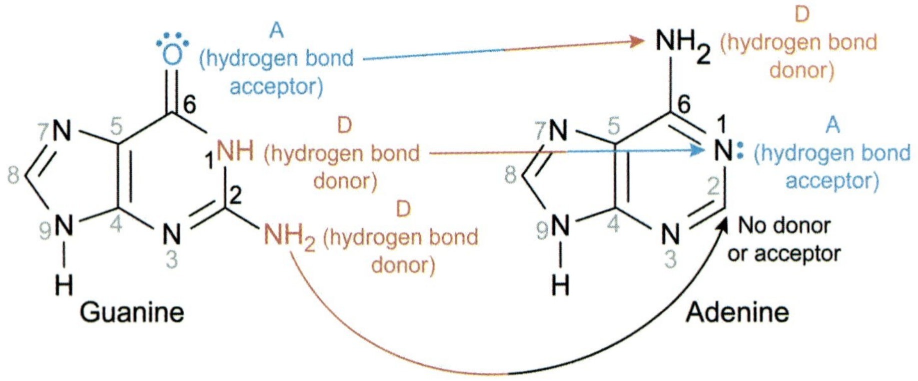

Figure 8.6 Relationship between the structures of guanine and adenine.

Thymine and uracil both base pair with adenine. Therefore, the adenine donor must align with an acceptor in thymine and uracil, and vice versa. Consequently, uracil has a carbonyl at position 4 (making it an acceptor), and the nitrogen atom at position 3 is bonded to hydrogen (making it a donor). This bonding pattern is shown in Figure 8.7, with the pyrimidine ring of uracil flipped relative to the orientation shown in Figure 8.2.

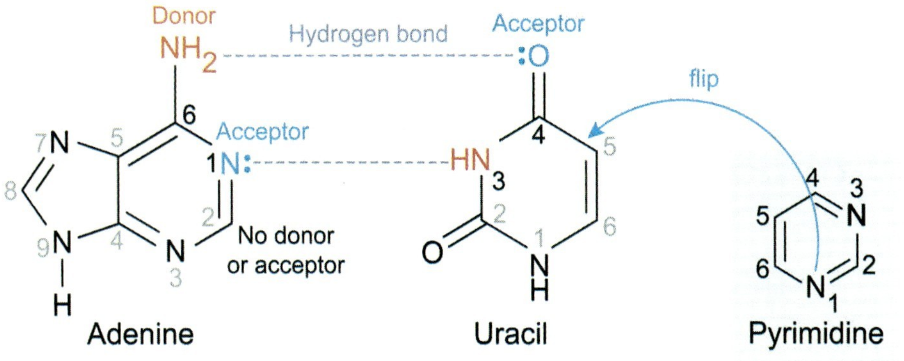

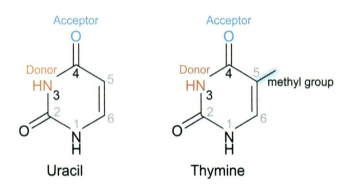

Figure 8.7 Interaction between adenine and uracil. The relation between uracil and pyrimidine is also shown.

Note that carbon 2 of uracil is a carbonyl. This *could* act as an acceptor in other contexts but because adenine has no corresponding donor, this carbonyl oxygen does *not* act as an acceptor when interacting with adenine. Thymine is identical to uracil except for a methyl group at thymine's position 5, as shown in Figure 8.8.

Figure 8.8 Comparison of uracil and thymine.

8.1.02 Chemical Modifications of Nitrogenous Bases

The nitrogenous bases in the nucleotides of DNA and RNA are susceptible to certain chemical modifications. These modifications can cause incorrect base pairing (see Lesson 8.2), which can lead to mutations. This concept provides information about several of the most common nucleotide modifications and their biological relevance.

Tautomerization

Tautomerization is the transfer of a proton from one site within a molecule to another site in the same molecule. This process also involves movement of a double bond.

Each of the nitrogenous bases found in DNA and RNA can undergo tautomerization. Adenine and cytosine both contain amino groups that can transfer a proton to the ring nitrogen while shifting a double bond to generate an imino group (see Figure 8.9).

Guanine, thymine, and uracil contain carbonyl groups linked to nitrogen, which is in turn linked to hydrogen. This functional group as a whole is a cross between a *lactone* and an *amide* (ie, a lactam). Transfer of the hydrogen atom to the carbonyl oxygen produces a hydroxyl group and results in a double bond shift. The result is that the nitrogen now carries a double bond (becoming more imine-like) and is a *lactim*. Tautomerization of each nitrogenous base is shown in Figure 8.9.

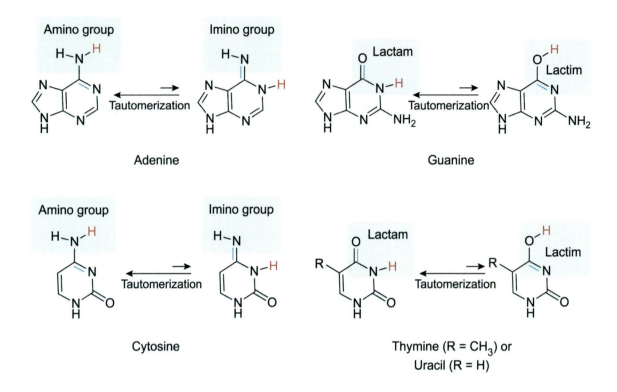

Figure 8.9 Tautomerization of the nitrogenous bases.

Tautomerization occurs naturally in the nitrogenous bases and interferes with normal hydrogen bonding patterns by converting hydrogen bond donors to acceptors and acceptors to donors. However, at physiological pH the amino and lactam forms of the nitrogenous bases are much more stable than the imino and lactim forms and predominate in nature by a large margin.

Concept Check 8.1

The lactim form of guanine is shown in the diagram with the ring positions that participate in hydrogen bonding numbered. Identify each position as a hydrogen bond donor or acceptor.

Solution
Note: The appendix contains the answer.

Deamination

Adenine, guanine, and cytosine each contain one primary amino group (ie, a nitrogen atom bonded to two hydrogen atoms and one carbon atom). These amino groups can be removed and replaced by carbonyls in a process called deamination.

Like tautomerization, deamination can alter hydrogen bonding patterns by converting a hydrogen bond donor (C–NH$_2$) into a hydrogen bond acceptor (C=O). Deamination of nitrogenous bases is spontaneous and can occur even without an enzyme. Figure 8.10 depicts deamination of adenine, guanine, and cytosine.

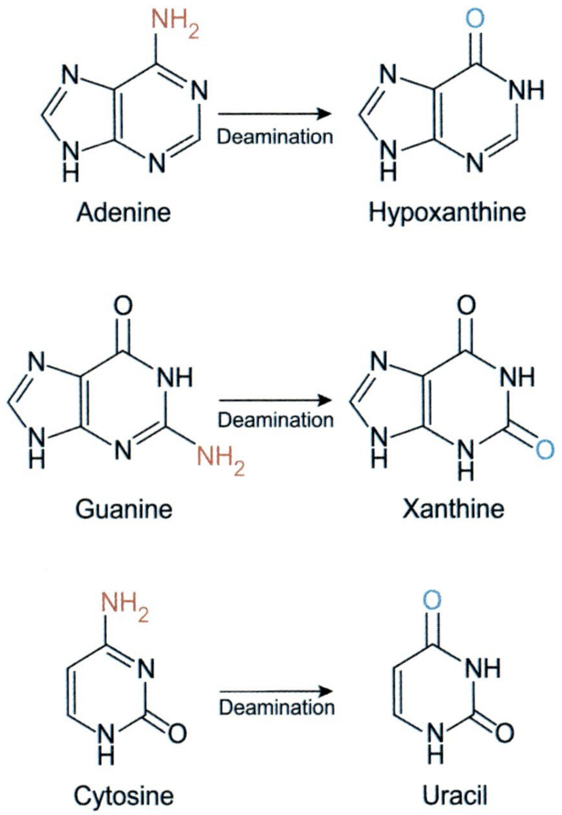

Figure 8.10 Deamination of several nitrogenous bases.

Importantly, deamination of cytosine produces uracil. Uracil is normally absent from DNA, which uses thymine in its place. When cytosine within a DNA strand undergoes deamination, enzymes easily recognize its transformation into uracil as an error because DNA does not normally contain uracil; because of this, the error can then be corrected. In RNA, such a change is *unlikely* to be detected because uracil is a normal component of RNA. However, this is less harmful because RNA is usually degraded quickly.

Deamination of adenine or guanine produces nitrogenous bases that are not often found in DNA *or* RNA, so these changes are also easily detected.

Methylation

Cytosine can become methylated at position 5. Unlike the other chemical modifications, cytosine methylation is commonly used for regulation of DNA expression (see Biology Lesson 2.4). Therefore, many organisms have enzymes that intentionally cause cytosine methylation. Cytosine is shown before and after methylation in Figure 8.11.

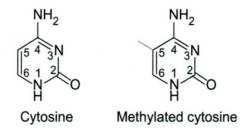

Cytosine   Methylated cytosine

Figure 8.11 Cytosine can be methylated at position 5.

Note that if cytosine is methylated and subsequently deaminated, it becomes thymine, as shown in Figure 8.12. Thymine in DNA is *not* recognized as an error, and therefore sites where cytosine is methylated often become especially susceptible to mutation.

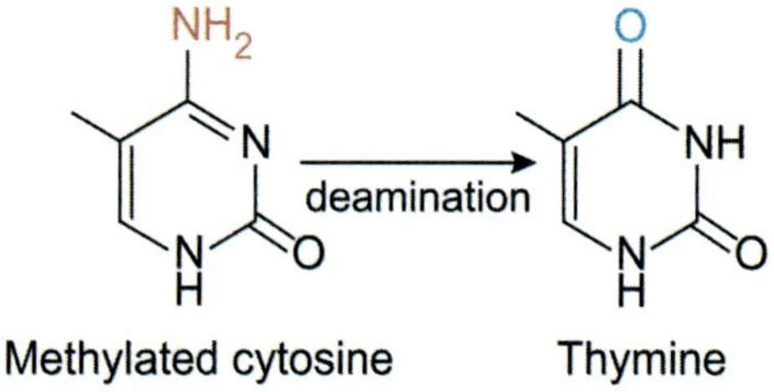

Methylated cytosine   Thymine

Figure 8.12 Deamination of methylated cytosine produces thymine.

8.1.03 Nucleosides and Nucleotides

When a carbohydrate (or deoxy carbohydrate derivative) forms a bond with a nitrogenous base, the result is a molecule that belongs to a class of compounds called **nucleosides**. In biological systems, the carbohydrate is typically ribose or deoxyribose. Just as amino acids have a constant backbone structure, the ribose or deoxyribose sugar is the constant backbone of a nucleoside. Figure 8.13 shows the furanose forms of ribose and deoxyribose.

Ribose   Deoxyribose

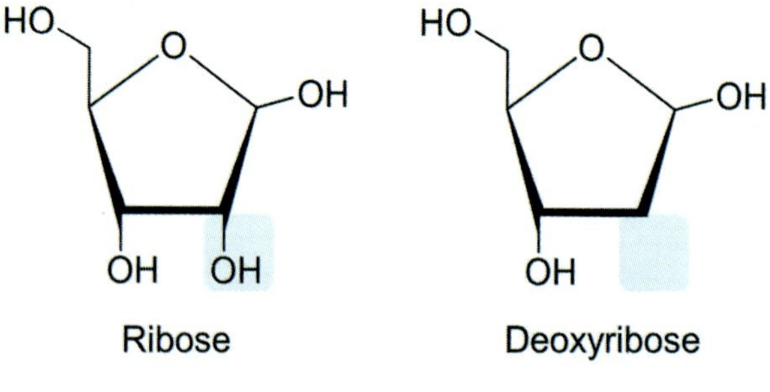

Figure 8.13 Structures of ribose and deoxyribose in their furanose forms, which they adopt in nucleosides and nucleotides.

The ring atoms in each nitrogenous base are numbered separately from the carbons in the sugar backbone. To avoid confusion with the numbering of the carbons in the nitrogenous base, the backbone carbons are numbered with a prime (') symbol added.

For instance, the anomeric carbon is the 1' carbon, the next carbon is the 2' carbon, and so on. The glycosidic bond between the sugar and the base connects the 1' carbon to nitrogen at position 1 in pyrimidines and position 9 in purines. The structures of pyrimidine and purine nucleosides are shown in Figure 8.14.

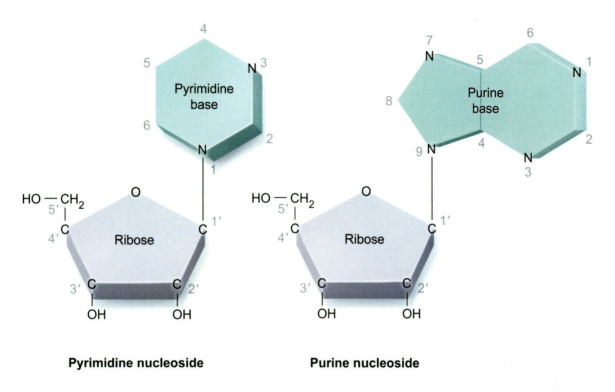

Pyrimidine nucleoside   **Purine nucleoside**

Figure 8.14 General structures of pyrimidine and purine nucleosides.

Nucleosides are given names derived from the nitrogenous base involved in the glycosidic bond. When adenine or guanine are attached to ribose, the molecule formed is called **adenosine** or **guanosine**, respectively. Similarly, cytosine, thymine, and uracil become **cytidine**, **thymidine**, and **uridine**. When the carbohydrate is deoxyribose, the names become deoxyadenosine, deoxyguanosine, and so forth. Nucleosides found in DNA and RNA are shown in Figure 8.15.

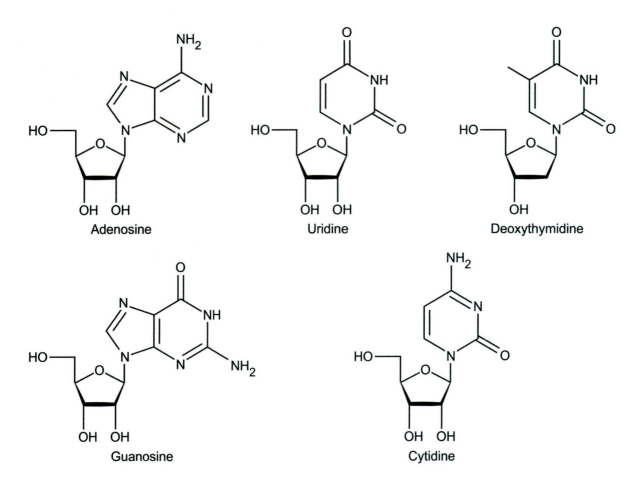

Figure 8.15 Nucleosides found in DNA and RNA. Thymine is generally only present in DNA, so deoxythymidine is shown.

Recall that Lesson 7.1 introduces the phosphorylation of carbohydrates. Ribose (as well as deoxyribose) is a pentose (ie, a five-carbon sugar) and is commonly phosphorylated on carbon 5, or in the case of a nucleoside, 5'. A *nucleoside* (or deoxynucleoside) phosphorylated at the 5' carbon is called a *nucleotide* (or deoxynucleotide).

The 5' carbon of a nucleotide is attached to a single phosphate group or to a chain of phosphate groups that are themselves linked to each other through phosphoanhydride bonds. Nucleotides are classified by the number of phosphates in the chain. A nucleotide with a single phosphate group is a nucleoside monophosphate (NMP), a nucleotide with two phosphates is a nucleoside diphosphate (NDP), and a nucleotide with three phosphates is a nucleoside triphosphate (NTP), as shown in Figure 8.16. For deoxynucleotides, these are designated as dNMPs, dNDPs, and dNTPs, respectively.

Nucleoside monophosphate (NMP)

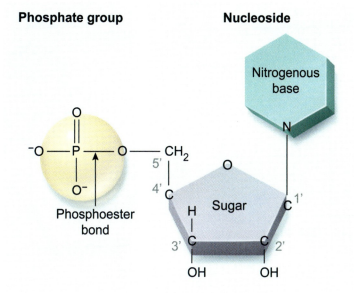

Nucleoside diphosphate (NDP)

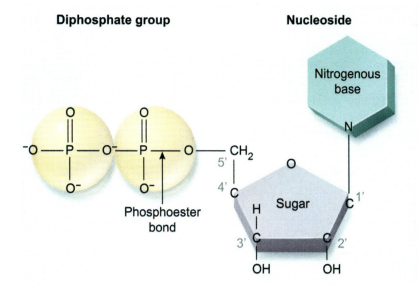

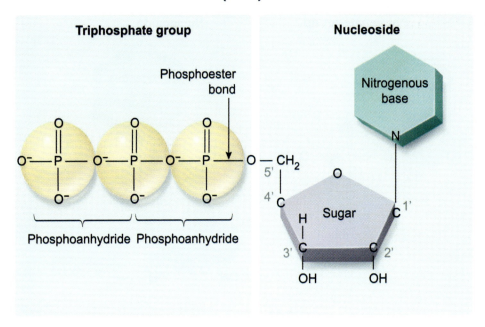

Figure 8.16 General structures of NMP, NDP, and NTP molecules.

These designations can also be applied to specific nucleotides. For example, adenosine linked to three phosphate groups is adenosine triphosphate (ATP), and deoxycytidine linked to two phosphates is deoxycytidine diphosphate (dCDP).

Different nucleotides play different roles in various metabolic pathways, as described briefly in Concept 8.1.04. An alternate nomenclature simply names the nucleotides as the salt of their phosphoric acid functional group. For example, AMP may be called adenylate.

Chapter 8: Nucleotides and Nucleic Acids

Concept Check 8.2

Identify each of the nucleotides shown.

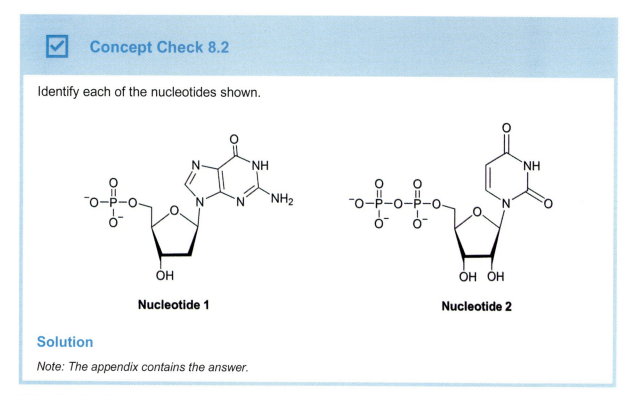

Solution

Note: The appendix contains the answer.

Other Nucleotides

Certain nucleotides other than those involved in DNA and RNA play important biological roles. The structures of these nucleotides do not need to be memorized, but an awareness of their structural features may be beneficial in understanding their biological importance.

For example, nicotinamide adenine dinucleotide (NAD$^+$/NADH) is a common oxidoreductase cofactor (see Concept 4.2.02) and plays major roles in glycolysis, the citric acid cycle, and the electron transport chain. This molecule includes a nucleotide that uses nicotinamide as its nitrogenous base. The nicotinamide nucleotide and the adenosine nucleotide are linked through their two phosphate groups (see Figure 8.17). The molecule NADPH (nicotinamide adenine dinucleotide phosphate) is identical to NADH but contains a phosphate group linked to the 2′ carbon of the adenosine component.

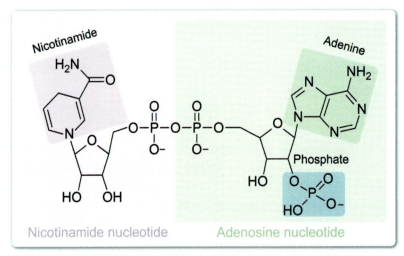

Figure 8.17 Structures of NADH and NADPH. The reduced form of the nicotinamide base is shown.

Flavin adenine dinucleotide (FAD) is similar to NAD but differs in two important ways. First, instead of having nicotinamide as its nitrogenous base, FAD contains a three-ring heterocycle called a flavin group. Second, the flavin component of FAD is linked to ribitol (a sugar alcohol) instead of ribose. Consequently, the bond between the flavin group and ribitol is not a true glycosidic bond.

A related molecule, flavin mononucleotide (FMN) contains only the flavin component (ie, it does not contain adenosine) and is linked to a single phosphate group. Figure 8.18 shows the structures of FAD and FMN.

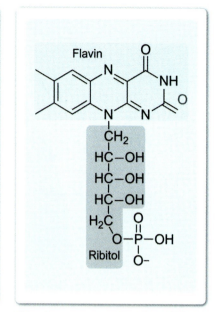

Flavin adenine dinucleotide (FAD) **Flavin mononucleotide (FMN)**

Figure 8.18 Structures of FAD and FMN. The oxidized form of the flavin base is shown.

8.1.04 Overview of Nucleotide Function

Nucleotides are believed to be among the first molecules involved in giving rise to life. As such, these molecules perform many critical biological functions. This concept describes several aspects of nucleotide function, including genetic and nongenetic roles.

Nucleotides in Genetics

As discussed further in Lesson 8.2, nucleotides combine to form DNA and RNA, which store and transmit genetic information. DNA contains all the information necessary to produce the enzymes and proteins that carry out metabolism and other necessary functions. Portions of DNA that encode proteins are read to produce mRNA (ie, transcribed; see Biology Lesson 2.2), which ribosomes then translate into proteins.

The path from DNA to RNA to protein is called the **central dogma of molecular biology** and is depicted in Figure 8.19.

DNA gives rise to RNA via transcription, and RNA gives rise to proteins via translation.

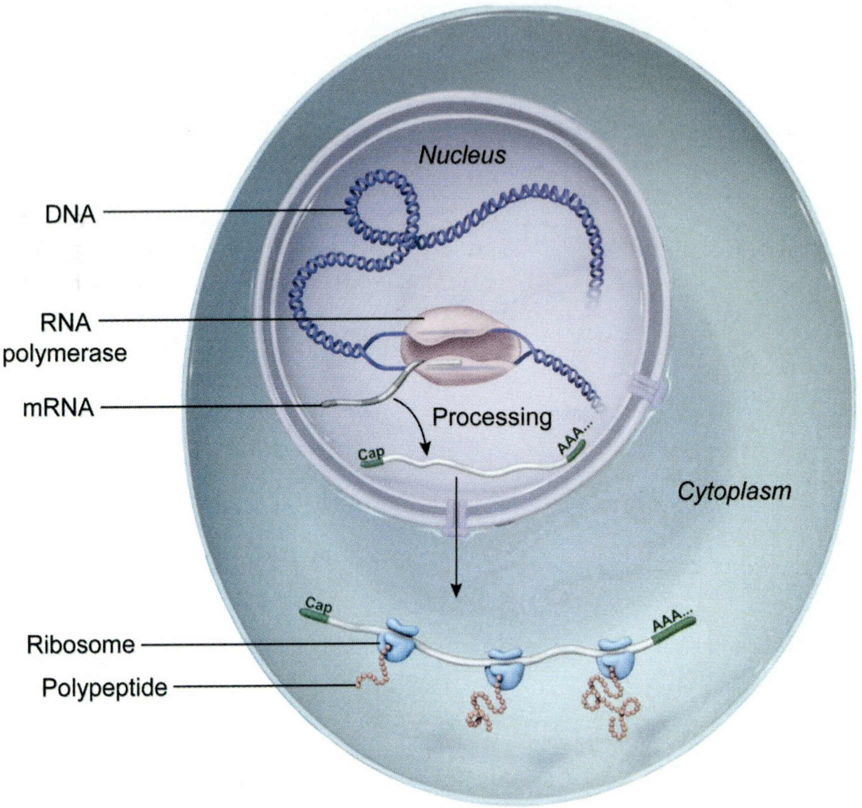

Figure 8.19 The central dogma of molecular biology.

Forms of RNA other than mRNA are themselves the end product of gene expression and serve various biological roles other than coding for proteins. For instance, ribosomes consist of both RNA and proteins, and the RNA component (ie, rRNA) is responsible for the catalytic steps in protein synthesis. Transfer RNA (tRNA) carries amino acids to the ribosome for incorporation into proteins. Other RNA molecules such as small interfering RNA (siRNA) help regulate protein synthesis.

Metabolic Nucleotide Functions

Nucleotides also play biological roles outside storage and transmission of genetic information. Adenosine triphosphate (ATP), discussed briefly in Concept 8.1.03, is a nucleotide that serves as the primary energy currency for all known life. For example, it is the fuel source involved in muscle contractions.

Various metabolic pathways involve nucleotides. Catabolic pathways (discussed in Unit 4) center on the production of ATP and, to a lesser extent, energetically equivalent nucleoside triphosphates such as GTP. ATP, GTP, CTP, and UTP all provide the energy necessary for various anabolic (and thermodynamically unfavorable) processes. For example, UTP is required for glycogen synthesis (see Chapter 11), and each of the nucleoside triphosphates play roles in the synthesis of glycan chains. Select metabolic roles of certain nucleotides are shown in Figure 8.20.

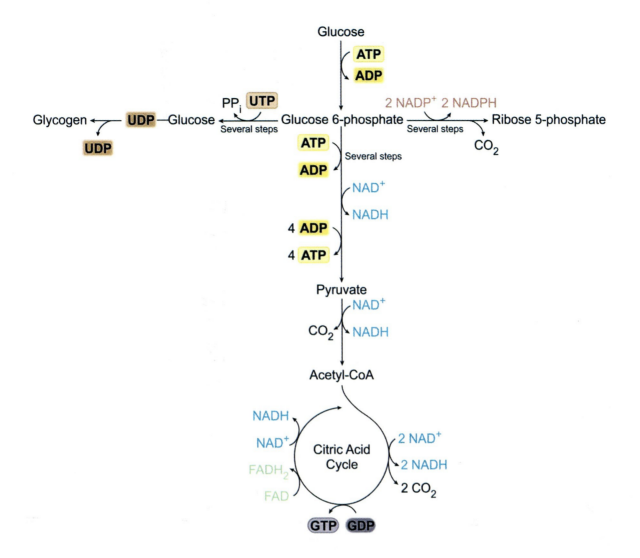

Figure 8.20 Examples of nucleotide involvement in several metabolic pathways.

Note that various steps in these pathways involve the nucleotides NAD, NADP, and FAD. The nicotinamide portions of NAD or NADP undergo redox reactions that interconvert them between the reduced form (NADH or NADPH) and the oxidized form (NAD$^+$ or NADP$^+$). NAD and FAD reduction and oxidation are shown in Figure 8.21.

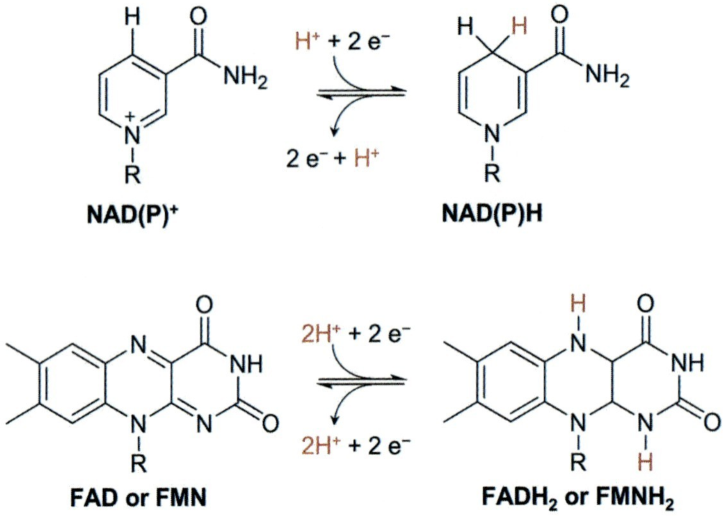

Figure 8.21 Reduction and oxidation of NAD (or NADP) and FAD (or FMN).

Regulatory Nucleotide Functions

As discussed in Unit 2, many enzymes are tightly regulated to ensure that they are active only when the cell needs them. Nucleotides are highly involved in enzyme regulation. For example, the glycolysis enzyme phosphofructokinase-1 (PFK-1) is inhibited by ATP. Both ADP and AMP counter this effect.

A modified form of AMP, cyclic AMP (cAMP), is involved in regulating many pathways and is itself regulated by certain G protein–coupled receptors (GPCRs) (discussed in Lesson 3.2). Figure 8.22 shows the involvement of several nucleotides in a G protein–coupled receptor pathway.

Chapter 8: Nucleotides and Nucleic Acids

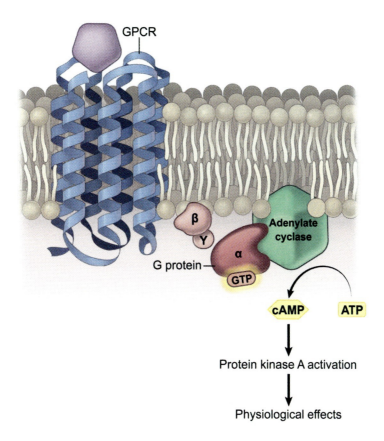

Figure 8.22 G protein–coupled receptor pathways require the nucleotide GTP and often include conversion of ATP into cAMP.

Nucleotides are also involved in regulation of proteins through phosphorylation. Lesson 2.4 and Lesson 6.2 discuss how phosphorylation can activate or inactivate a protein. In general, the phosphorylation of proteins involves transfer of a phosphate group from ATP or another nucleoside triphosphate, as shown in Figure 8.23.

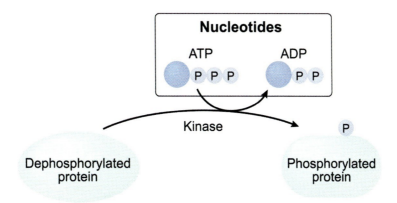

Figure 8.23 Protein phosphorylation typically requires the nucleotide ATP, which becomes ADP after donating a phosphate group.

The examples given in this concept are not comprehensive. However, they demonstrate how nucleotides play critical roles in almost every biological pathway.

Lesson 8.2
Nucleic Acids

Introduction

Just as amino acids polymerize to form peptides and monosaccharides polymerize to form polysaccharides, nucleotides polymerize to form nucleic acids. DNA and RNA are two distinct types of nucleic acids used by all known living organisms. They consist of deoxyribonucleotides and ribonucleotides, respectively. This lesson explains the process of nucleotide polymerization and describes interactions between and within nucleic acids.

8.2.01 Nucleic Acid Formation

Concept 8.1.03 explains that the carbons in the sugar component of a nucleotide are numbered using the prime symbol ('). This numbering system is important in understanding the structure of nucleic acids. Just as peptides form when the carboxyl group of one amino acid reacts with the amino group of another, nucleic acids form when the hydroxyl group on the 3' carbon of one nucleotide reacts with a phosphate group on the 5' carbon of another nucleotide. The result is a phosphodiester bond linking the two nucleotides.

When the 3' hydroxyl group of one nucleoside triphosphate (NTP or dNTP) attacks the α phosphate at the 5' end of the other NTP, the β and γ phosphates are released together as pyrophosphate (PP_i). Therefore, nucleic acid synthesis is a **condensation reaction** in which pyrophosphate is released instead of water. A condensation reaction between two nucleotides is shown in Figure 8.24.

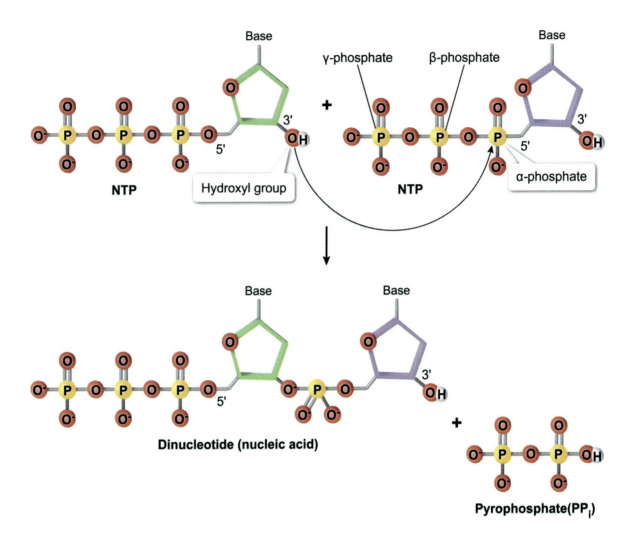

Figure 8.24 A condensation reaction between two nucleotides to form a dinucleotide.

The resulting dinucleotide has one free 5' triphosphate and one free 3' hydroxyl. The 5' and 3' groups that were involved in the reaction, however, are no longer free (ie, they are unavailable to react with other nucleotides). The end of the nucleic acid corresponding to the free 5' phosphate is called the **5' end**, and the end corresponding to the free 3' hydroxyl is called the **3' end**. This is analogous to the N- and C-termini of a peptide or the nonreducing and reducing ends of a complex carbohydrate (see Figure 8.25). By convention, the sequence of a nucleic acid is always written *from the 5' end to the 3' end*, unless specified otherwise.

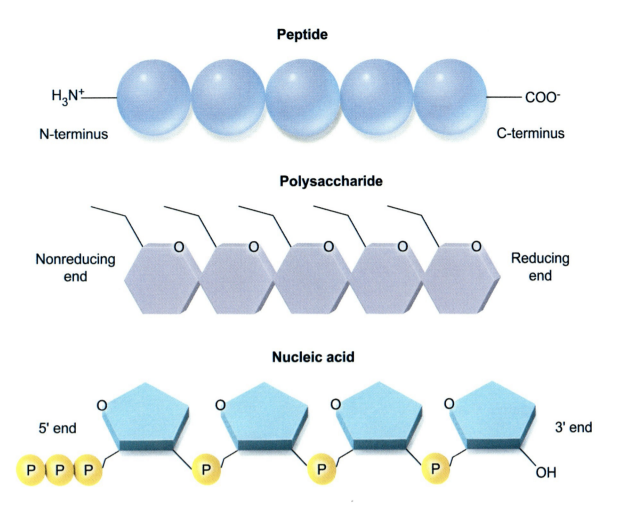

Figure 8.25 Polarity in peptides, polysaccharides, and nucleic acids.

The 3' end of a dinucleotide can react with the 5' end of another NTP to form a trinucleotide. This process can continue to form nucleic acids of any length. Nucleic acids containing a few nucleotides are often called **oligonucleotides**. However, the length of a nucleic acid can be extensive. For example, RNA molecules commonly contain hundreds or even thousands of individual nucleotide components. DNA is commonly even longer: each DNA strand of the human X chromosome consists of approximately 155 million individual nucleotides. These large molecules may be called **polynucleotides** but are often simply called **nucleic acids**.

 Concept Check 8.3

Describe the reaction that must occur for ATP and CTP to form a dinucleotide with the sequence CA. Then describe how a reaction between CA and UTP could form the trinucleotide CAU.

Solution

Note: The appendix contains the answer.

Each nucleic acid has a repeating **backbone structure** consisting of a sugar (eg, ribose, deoxyribose) linked to a phosphate, which is linked to another sugar, which is linked to another phosphate, and so on. This is known as a **sugar-phosphate backbone**. At physiological pH, each phosphate has a negative charge, and therefore the backbones of DNA and RNA are highly **negatively charged**. Eukaryotes use positively charged histone proteins to neutralize and store DNA (see Biology Lesson 1.4). Figure 8.26 shows the repeating sugar-phosphate structure and negative charges in a nucleic acid backbone.

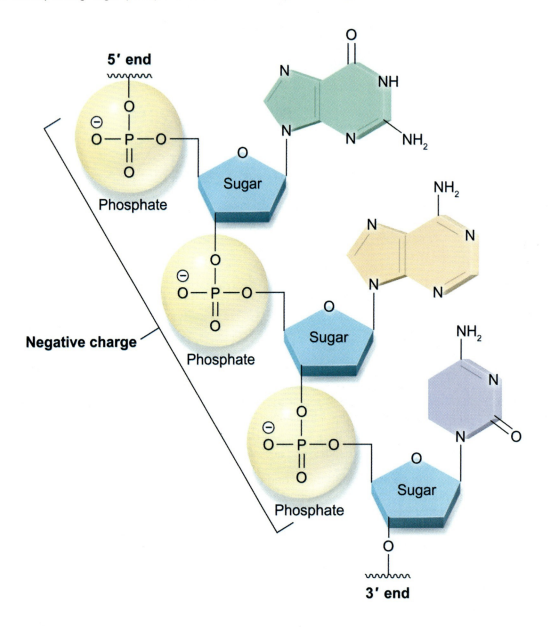

Figure 8.26 Example of a nucleic acid, with important aspects of the sugar-phosphate backbone emphasized.

Nucleic acid synthesis is facilitated by the enzymes DNA polymerase (for DNA synthesis) and RNA polymerase (for RNA synthesis). Different types of RNA are synthesized by different RNA polymerases, as detailed in Biology Chapter 2. The nucleotide at the 5' end of a growing nucleic acid is the first to have been incorporated into the chain, and the nucleotide at the 3' end is incorporated last (Figure 8.27). As the nucleic acid grows, the *3' end of the growing strand* attacks the *5' end of the next nucleotide* to be added.

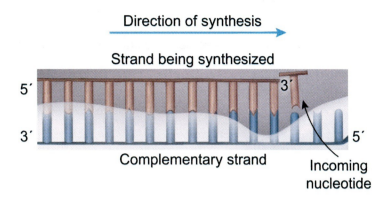

Figure 8.27 Nucleic acids are synthesized starting at the 5′ end and moving toward the 3′ end.

Once a nucleic acid is synthesized, it may be modified. For example, DNA polymerase occasionally inserts the wrong nucleotide. Various mechanisms exist to detect such errors, and many DNA polymerase enzymes include 3′-5′ exonuclease activity. Exonucleases are hydrolases that remove nucleotides from the *ends* of a nucleic acid strand (in this case the 3′ end) by hydrolysis.

Lesson 8.1 discusses how the nitrogenous bases in nucleotides may undergo chemical changes such as deamination. When this occurs, the affected base may be removed by a glycosylase enzyme, which hydrolyzes specific glycosidic bonds. This leaves a ribose or deoxyribose sugar with no nitrogenous base. The sugar can then be removed by an **endonuclease**, which hydrolyzes both phosphodiester bonds in the middle of a nucleic acid.

DNA polymerase can then insert a new nucleotide into the DNA strand in place of the removed one. However, this process leaves a gap, called a nick, at the 3′ end of the repaired nucleotide. DNA ligase uses ATP to provide the energy needed to repair the nick. Figure 8.28 shows the mechanism of nucleotide removal and substitution *within* a nucleic acid.

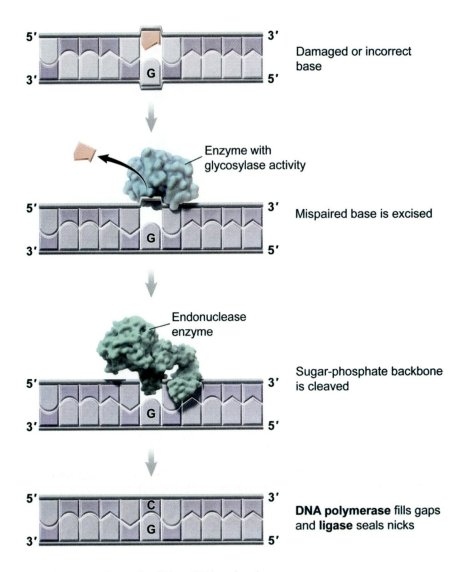

Figure 8.28 Repair of a base mismatch within a DNA molecule.

8.2.02 Base Pairing and Secondary Structure in Nucleic Acids

The sequence of nucleotides in a nucleic acid constitutes the **primary structure** of that nucleic acid. Just as peptides can form secondary structures, so can nucleic acids. This concept outlines the **secondary structures** commonly found in both DNA and RNA and the forces that hold them together.

Perhaps the best-known secondary structure in nucleic acids is the double helix. In living cells, DNA exists as a double-stranded molecule (ie, two distinct DNA strands held together noncovalently). These two strands twist around each other to form the helical structure. The process of double-strand formation is called annealing. The annealed strands align in an **antiparallel orientation** (see Figure 8.29), meaning that the 3' end of one strand aligns with the 5' end of the other. Several forces help hold the two strands together.

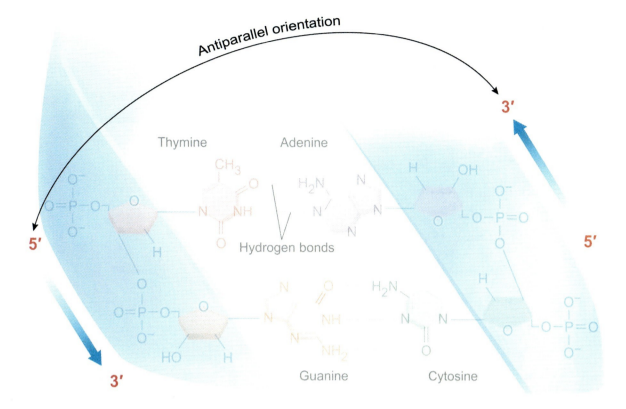

Figure 8.29 The strands in a double helix are aligned in an antiparallel manner.

Lesson 8.1 introduces the concept of hydrogen bonding between particular nitrogenous bases to form base pairs. When aligned properly, *guanine pairs with cytosine* and *adenine pairs with thymine* in DNA. Note that each base pair consists of *one purine and one pyrimidine*. A purine and a pyrimidine that pair with each other in this way are said to be **complementary**. Combined with the antiparallel structure of a double helix, the two strands of a double-stranded DNA molecule are said to be the **reverse complements** of each other.

The hydrogen bonds between complementary bases on opposite DNA strands provide a significant portion of the force that drives double helix formation. DNA strands that do not have complementary bases do not anneal, and even a single mismatch (ie, alignment of noncomplementary bases) can significantly reduce the favorability of annealing.

In addition to hydrogen bonding between bases within a double helix, the structure is also stabilized by a **hydrophobic effect** similar to that seen in protein folding. Compared to the negatively charged backbone, the bases in DNA are relatively hydrophobic. Therefore, when these bases are exposed to an aqueous environment, water must form a more organized solvation layer around them than around the more hydrophilic backbone groups.

Double helix formation allows the nitrogenous bases to "hide" from the surrounding water, while the water can interact favorably with the exposed sugar and phosphate groups. Several interactions that stabilize double helices are shown in Figure 8.30.

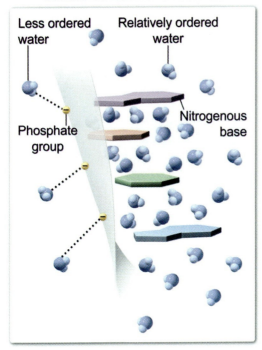 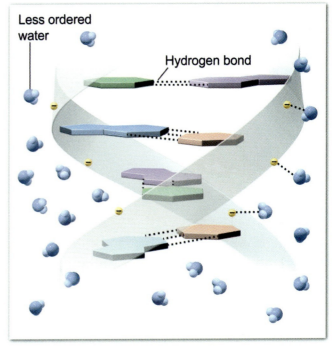

Figure 8.30 Various forces help stabilize double helix structure.

At the same time, within a double helix, the base pairs of each strand are stacked almost on top of each other. These bases are aromatic and have pi orbitals above and below their planes. In this arrangement, each base can therefore participate in a noncovalent interaction called pi stacking with the bases that are adjacent to it. Pi stacking helps stabilize the specific DNA conformation.

RNA can also form secondary structures. During transcription, for example, the RNA molecule being synthesized temporarily pairs with the DNA strand that serves as the template and may briefly form a DNA-RNA double helix.

More importantly, many RNA molecules perform biological functions that require specific structures (eg, tRNA, ribosomal RNA). These molecules tend to be single-stranded but nevertheless form secondary *and* tertiary structures. Commonly, the bases in one portion of a functional RNA strand are complementary to another portion of the *same strand*. In such a case, the strand may fold on itself and allow the complementary bases to pair with each other. This pairing gives rise to a helical structure *within* a single RNA strand. Like proteins, RNA may further fold into complex three-dimensional shapes that correspond to their functions.

RNA structure is often depicted in two-dimensional form, with phosphodiester bonds between adjacent nucleotides shown as solid lines and hydrogen bonds between complementary bases indicated by dashed lines. These depictions often result in structures that resemble hairpins and are consequently named hairpin structures. Figure 8.31 shows both two-dimensional and three-dimensional depictions of tRNA. Note that the 2D depiction clearly shows complementary base pairing in a three-leafed clover shape; however, the true 3D structure of tRNA is more L-shaped.

2D and 3D structures of tRNA

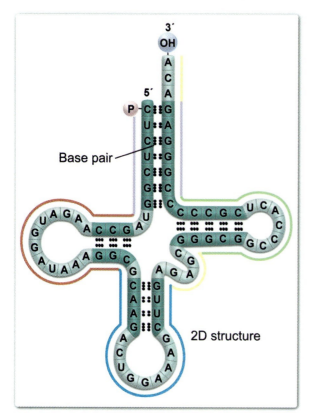
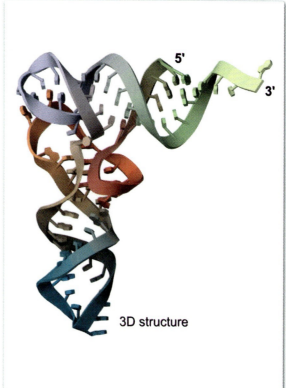

Figure 8.31 Two- and three-dimensional structures of tRNA.

8.2.03 Chargaff's Rules

Base pairing in nucleic acids gives rise to a set of rules known as Chargaff's rules, which dictate certain aspects of the composition of paired nucleic acid strands. Chargaff's rules state the following:

- The number of adenine bases in double-stranded DNA equals the number of thymine bases, and the number of guanine bases equals the number of cytosine bases. This is because every adenine base in one strand is paired with a thymine base in the complementary strand, and every guanine base is paired with a cytosine base.
- The sum of purines equals the sum of pyrimidines because every purine is paired with a pyrimidine.

Note that Chargaff's rules apply only to *double-stranded* nucleic acids, specifically DNA. Each individual strand may have any number of adenine, guanine, cytosine, and thymine bases. Only when a single strand is paired with its complementary strand do Chargaff's rules apply. Figure 8.32 shows an example of Chargaff's rules applied to a small, double-stranded DNA molecule.

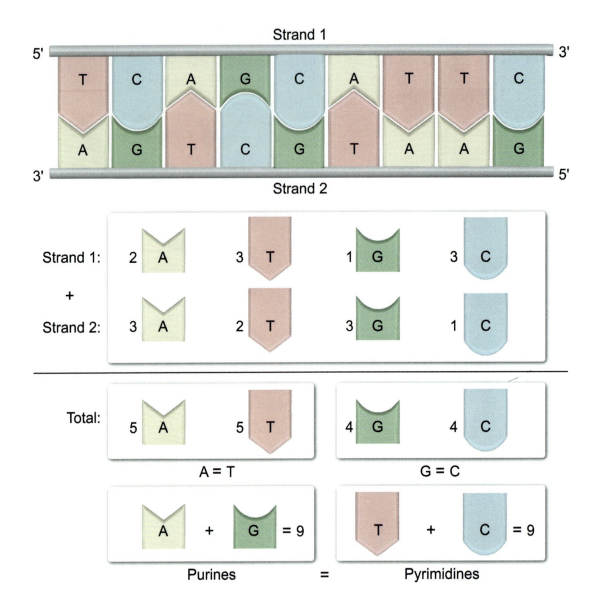

Figure 8.32 Example of a double-stranded DNA molecule adhering to Chargaff's rules. Neither individual strand conforms to the rules, but the two strands together do.

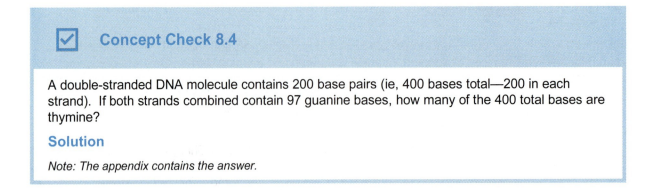

Concept Check 8.4

A double-stranded DNA molecule contains 200 base pairs (ie, 400 bases total—200 in each strand). If both strands combined contain 97 guanine bases, how many of the 400 total bases are thymine?

Solution

Note: The appendix contains the answer.

Base pairing within an RNA strand follows a variant of Chargaff's rules, in which the number of adenine bases equals the number of *uracil* bases instead of thymine. However, functional RNA molecules commonly exhibit some unusual base pairing called "wobble pairing" (eg, G paired with U), so Chargaff's rules do not *universally* apply to RNA. A tRNA molecule with a wobble pair *and* sections that obey Chargaff's rules is shown in Figure 8.33.

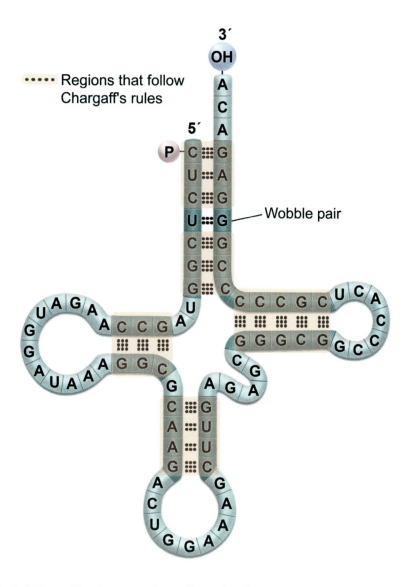

Figure 8.33 A variant of Chargaff's rules generally applies to functional RNA, although unusual base pairs (ie, wobble pairs) are sometimes present.

As mentioned in Concept 8.2.02, the bases in a DNA strand can pair with the bases in an RNA strand. This occurs, for example, when RNA polymerase uses DNA as a template to synthesize RNA or when reverse transcriptase uses RNA as a template to synthesize DNA.

When a DNA strand is paired with an RNA strand, yet another variation of Chargaff's rules applies. In this case, the RNA strand contains uracil, which pairs with adenine bases in the DNA strand. In contrast, the DNA strand contains thymine, which pairs with adenine in the RNA strand. Therefore, the total number of adenine bases in the complex is equal to the *sum* of the thymine and uracil bases in the complex. Figure 8.34 shows Chargaff's rules applied to a DNA-RNA complex.

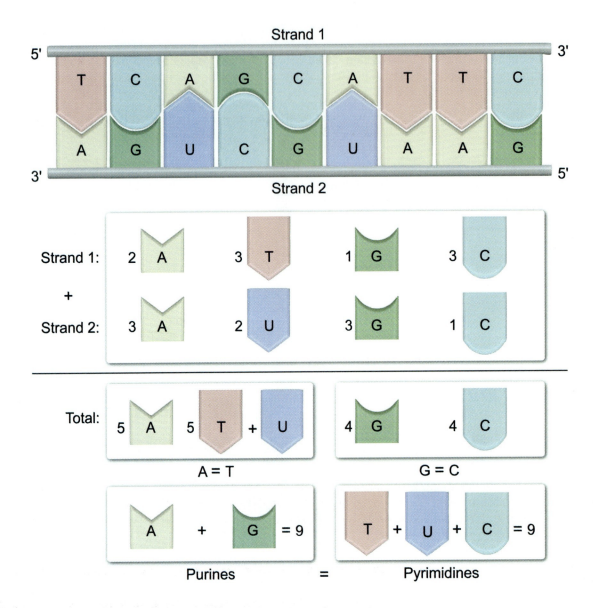

Figure 8.34 Chargaff's rules as applied to a DNA-RNA complex.

8.2.04 Double Helix Stability

As with higher levels of protein structure, nucleic acid secondary structure may be disrupted by environmental conditions. These conditions include pH, salt concentrations, and temperature.

The optimal pH range for DNA annealing is near 7.4, because at this pH the hydrogen bond participants in each base are neutral and can form the expected hydrogen bonds with their complementary bases. If the pH is brought too far below or above this level, functional groups in bases can become protonated or deprotonated, making them positively or negatively charged, and acid- or base-catalyzed tautomerism of the bases (Concept 8.1.02) may also occur. This alters interactions between bases (see Figure 8.35).

Chapter 8: Nucleotides and Nucleic Acids

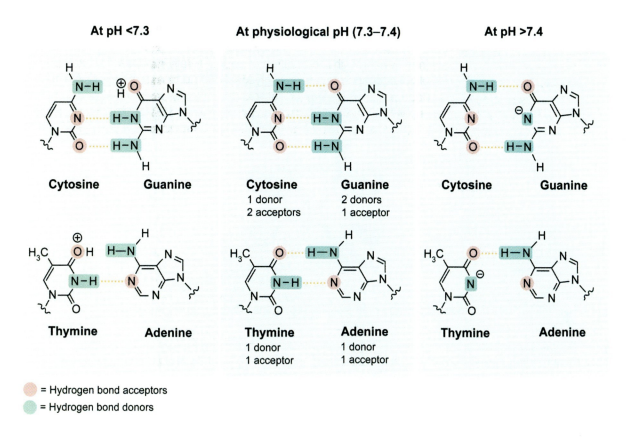

Figure 8.35 Effects of pH on hydrogen bonding between nitrogenous bases.

The sugar-phosphate backbone is negatively charged, so the phosphate groups repel each other, which can destabilize helix formation. Eukaryotes largely resolve this by providing DNA with positively charged histones. The phosphate groups may also be stabilized by the presence of salts because the cations from the salt (eg, Mg^{2+}) interact well with the phosphate groups. The interaction between a cation and a phosphate group tends to be stronger than the interaction between water and a phosphate group. Consequently, *higher salt concentrations* tend to result in *greater double helix stability* (see Figure 8.36).

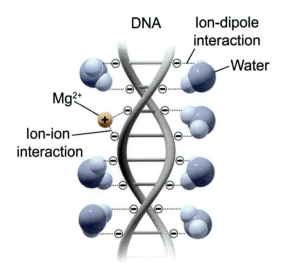

Figure 8.36 Interactions between cations and DNA stabilize repulsions between phosphate groups.

Higher temperatures cause the bonds within a double helix to move more rapidly and allow the interactions holding strands together to break. This causes the two strands in a double helix to **denature** (ie, to come apart), as shown in Figure 8.37.

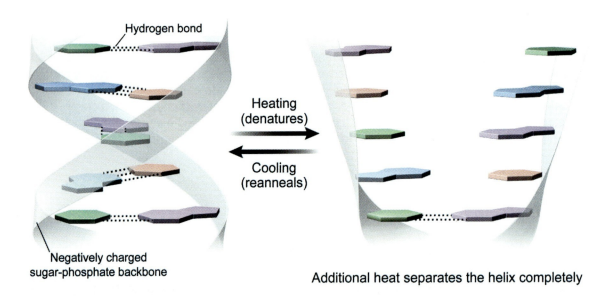

Figure 8.37 DNA denatures as temperatures increase and reanneals as temperatures decrease.

The temperature that causes 50% of double helices in solution to denature is called the **melting temperature**, T_m, for that double helix.

The composition of a double helix can significantly impact its T_m. A longer helix contains more hydrogen bonds, each of which must break for denaturation to occur. Consequently, a higher temperature is required to melt a long strand than to melt a short strand of similar composition.

Guanine and cytosine pair through three hydrogen bonds, whereas adenine and thymine (or adenine and uracil) pair through only two hydrogen bonds. Correspondingly, a helix with a *high percentage of guanine and cytosine* (ie, high **GC content**) has a *higher* T_m than a helix of similar length with lower GC content. The effects of length and GC content on melting temperature are shown in Figure 8.38.

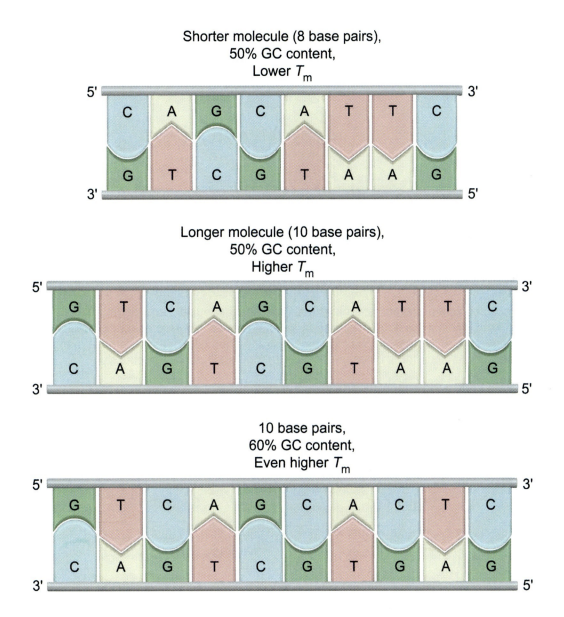

Figure 8.38 Effects of length and GC content on melting temperature T_m.

The melting temperature of a double helix is particularly important in biotechnology contexts such as the polymerase chain reaction (PCR) (see Biology Lesson 4.1). In PCR, it is important to ensure that primers anneal at certain temperatures and denature at others. Consequently, PCR primers are typically designed to optimize T_m by manipulating GC content, and salt and pH conditions are carefully controlled.

The melting temperature of a given double helix can be determined experimentally by generating a melting curve. A sample of DNA or RNA with at least some double helical portions is gradually heated and monitored for denaturation. Often this is done by including a fluorescent dye that binds to double-stranded DNA (or annealed RNA) but *not* to single-stranded nucleic acids. Upon binding, fluorescence intensity increases. As the sample is heated, the dye unbinds from the denatured nucleic acid and fluorescence decreases. An example of the resulting melting curve is shown in Figure 8.39.

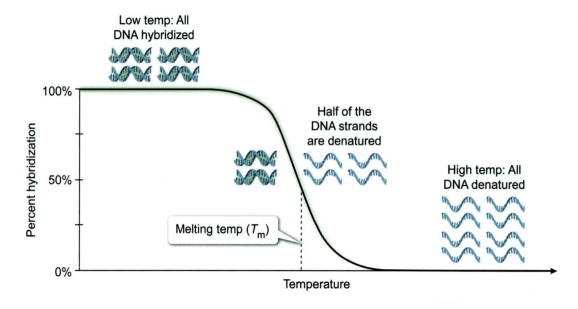

Figure 8.39 A typical DNA melting curve.

Note the sigmoidal shape of the curve. This shape indicates that nucleic acid denaturation is **positively cooperative**: as portions of the double helix separate, it becomes easier to separate additional portions. Similarly, as temperature decreases, complementary strands reanneal in a cooperative manner: as interactions between strands form, it becomes easier to form additional interactions.

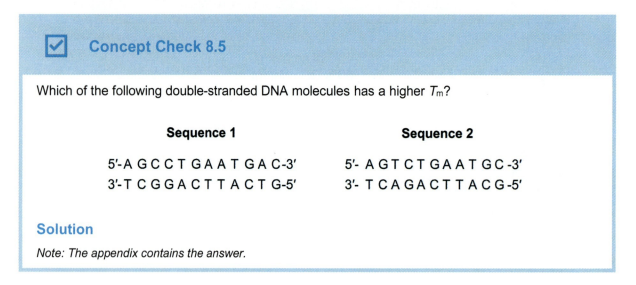

Concept Check 8.5

Which of the following double-stranded DNA molecules has a higher T_m?

Sequence 1	Sequence 2
5'-A G C C T G A A T G A C-3'	5'- A G T C T G A A T G C -3'
3'-T C G G A C T T A C T G-5'	3'- T C A G A C T T A C G -5'

Solution

Note: The appendix contains the answer.

Lesson 9.1

Energy Storage Lipids

Introduction

After proteins (Units 1 and 2), carbohydrates (Chapter 7), and nucleic acids (Chapter 8), lipids are the fourth and final major class of biological macromolecules. Compared with other macromolecules, the basic units of lipids are much more varied in structure, which reflects their diverse functional roles. This lesson covers the general properties of lipids, with a focus on energy storage lipids. Structural and signaling lipids are discussed in the other lessons of this chapter.

9.1.01 General Properties of Lipids

As a class of molecules, lipids are defined by their **hydrophobicity** and their **insolubility** in water. Lipid molecules contain only a few, if any, polar functional groups relative to their size. Because of this, the water molecules in an aqueous solution can only interact with the hydrophobic portions of a lipid molecule through London dispersion forces. Because these forces are weak, water molecules instead form strong hydrogen bonds with other nearby water molecules, forming a solvation layer around the lipid. The inability of a water molecule to strongly interact with a lipid *restricts* the water molecules' rotational freedom, leading to low entropy.

Because systems with *higher* entropy are favored, lipids and other hydrophobic molecules tend to aggregate in aqueous solutions, as shown in Figure 9.1. The forces leading to this hydrophobic aggregation in aqueous systems are collectively called **hydrophobic interactions**. As discussed in Lesson 2.3, hydrophobic interactions are a crucial driver of protein folding. They are also the driving force of the organization of structural lipids into membranes and of storage lipids into lipid droplets.

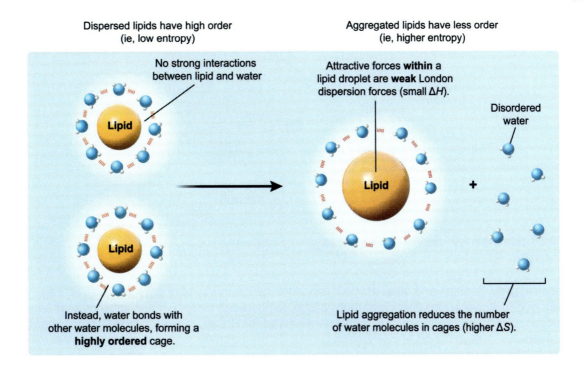

Figure 9.1 Lipids aggregate in aqueous solutions, which maximizes the entropy of the water molecules.

As their name implies, energy storage lipids can be used as an energy source for most cells. Within a cell, the aggregation of these lipids pulls them out of solution (ie, the cytosol) in a manner similar to the precipitation of ionic salts with a low K_{sp}. A small amount of lipid remains solvated to maintain the equilibrium constant, but the bulk of the lipid is removed from the aqueous phase. However, the lipids do not enter the solid phase as ionic precipitates do; they instead separate into a distinct *liquid* phase.

This separation minimizes the effect that individual lipid molecules have on the osmolarity of the aqueous cytosol, which allows cells to maintain their osmolarity while also maintaining a large energy store. This benefit is similar to the benefit of storing glucose monosaccharides as a glycogen polymer (see Concept 7.2.03), but lipid aggregates provide an even greater density of chemical energy per volume.

However, the aggregation of lipids does affect the *speed* at which the energy can be accessed. Cytosolic proteins and second messengers can only interact with molecules at the *surface* of the lipid droplet, as shown in Figure 9.2. This limits the rate at which lipids in the droplet can be **mobilized** (ie, released from storage and used for energy).

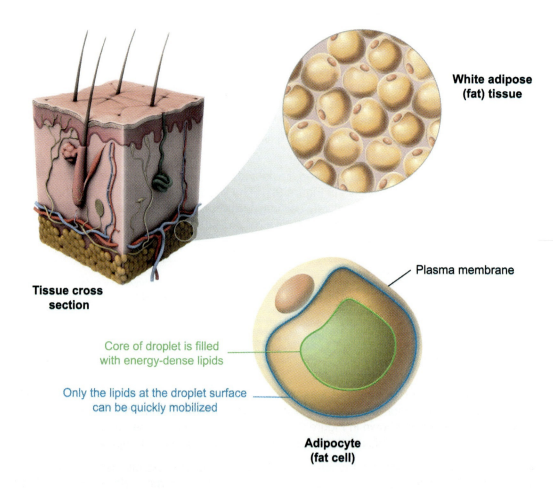

Figure 9.2 Lipids within lipid droplets provide a large, energy-dense reservoir for ATP production, but lipids in these droplets are slow to mobilize.

For these reasons, humans and other animals can store vast amounts of energy in energy storage lipids. However, due to slower accessibility and mobilization compared to more hydrophilic storage molecules (eg, glucose, glycogen), lipid stores are generally reserved for slower, lower-intensity, and long-term energy needs.

9.1.02 Fatty Acids

The main energy storage lipid is the triglyceride, which is composed of three fatty acids esterified onto a glycerol backbone. The features of the glycerol backbone and the triglyceride are discussed in Concept 9.1.03. This concept focuses on the general properties of **fatty acids** as they apply to energy storage lipids.

Fatty acids are carboxylic acids with an R group that is an aliphatic (ie, nonaromatic) hydrocarbon. The carboxylic acid (carboxylate at physiological pH) is a hydrophilic functional group; however, the hydrocarbon "tail" of longer fatty acids provides significant hydrophobic character. Because they have both hydrophilic *and* hydrophobic character, fatty acids can be classified as an **amphipathic**, or **amphiphilic**, molecule. An example of a typical fatty acid is shown in Figure 9.3.

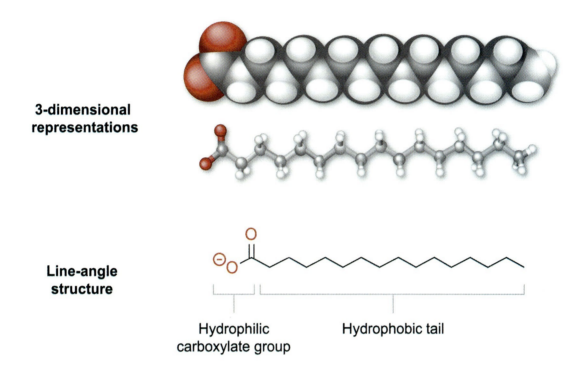

Figure 9.3 Different representations of a typical fatty acid.

A common mistake when counting carbons in a line-angle structure of a fatty acid is to forget to include the carboxylate carbon. It is important to include *both* the carboxylate carbon *and* the carbons in the hydrocarbon tail. The fatty acid shown in Figure 9.3 has a 16-carbon chain; the numbering starts with the carboxylate carbon as carbon 1. This numbering scheme is shown explicitly in Figure 9.4.

Figure 9.4 also shows Greek letter designations of the fatty acid carbons. As discussed in Organic Chemistry Lesson 4.7, carbons *next* to a carbonyl are also called the α-carbon (alpha-carbon). In a fatty acid, carbon 2 is the α-carbon. The next few carbons along the chain are given the subsequent Greek letter designations (ie, carbon 3 = β [beta], carbon 4 = γ [gamma]). Regardless of chain length, the final carbon in a fatty acid is called the ω-carbon (omega-carbon). Another commonly used numbering system that starts from the ω end of the molecule is described later in this concept.

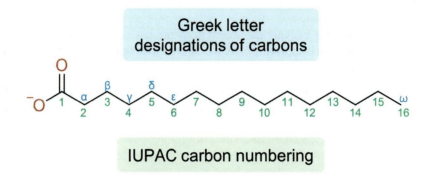

Figure 9.4 Designations of carbon position in a fatty acid.

Fatty acids can vary greatly in structure, including in number of carbons and in degrees of unsaturation. The fatty acid shown in Figure 9.3 and Figure 9.4 is a saturated, long-chain fatty acid. Fatty acids with a chain length of 7 to 12 carbons are medium-chain fatty acids, and those with a chain length of 6 carbons or fewer are considered short-chain fatty acids.

Most important fatty acids in human metabolism have an even number of carbons, although odd-chain fatty acids exist as well. The hydrocarbon tail of the fatty acid shown in Figure 9.3 and Figure 9.4 is saturated (ie, it has no carbon-carbon double bonds), resulting in its classification as a **saturated fatty acid**. In contrast, **unsaturated fatty acids** have at least one carbon-carbon double bond in the hydrocarbon tail. In natural fatty acids, these double bonds are typically in the *cis* configuration, which provides a "kink" in the fatty acid tail, as shown in Figure 9.5.

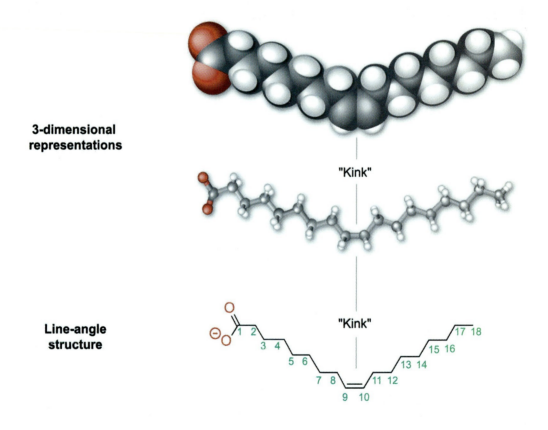

Figure 9.5 An example of an unsaturated fatty acid.

Fatty acids may have multiple double bonds in their chain, but these polyunsaturated fatty acids are typically *not* conjugated. In other words, there is typically at least one sp^3 hybridized methylene group ($-CH_2-$) between any pair of double bonds. Because many double bonds are closer to a fatty acid's ω end than they are to its carboxylate end, many fatty acids are classified based on ω numbering of the double bond nearest the ω end.

The polyunsaturated fatty acid shown in Figure 9.6 is an example of a fatty acid with two double bonds. The first double bond is at position 9 (IUPAC numbering) or ω−9 by ω numbering. The second double bond is at position 12 (IUPAC numbering) or ω−6 by ω numbering. Because the double bond nearest the ω end is at the ω−6 position, the fatty acid shown in Figure 9.6 is an ω−6 fatty acid.

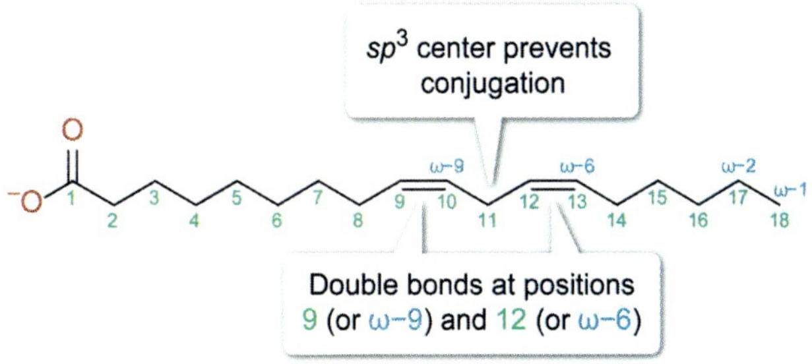

Figure 9.6 An example of a polyunsaturated fatty acid.

Despite the presence of double bonds, fatty acids are highly reduced molecules compared to other energy molecules, such as carbohydrates. Consequently, fatty acids store a lot of energy that can be released upon oxidation. Specifically, a fatty acyl group undergoes β-oxidation in the mitochondria or peroxisome of a eukaryotic cell. During β-oxidation, carbons 1 and 2 form acetyl-CoA and the β-carbon is oxidized, as shown in Figure 9.7. Acetyl-CoA and β-oxidation are discussed further in Lesson 13.1.

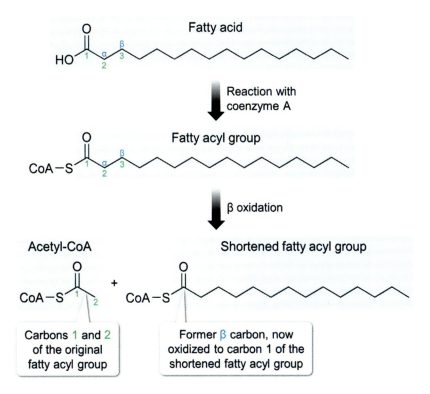

Figure 9.7 Some of the energy stored in fatty acids is liberated through β-oxidation.

Although fatty acids are typically stored in triglycerides, the triglycerides in adipose tissue (ie, fat tissue) are hydrolyzed when needed, and **free fatty acids (FFA)** are released into the bloodstream. Here, the insoluble FFAs must bind to carrier proteins, such as albumin, which can carry lipid molecules until they reach their target tissue for further processing.

9.1.03 Triacylglycerides

Fatty acids are amphipathic molecules because they have a hydrophobic tail and a hydrophilic carboxylate group. Free fatty acids, therefore, do not aggregate into lipid droplets, but instead form structures called micelles. However, micelles are much smaller than lipid droplets, and an individual micelle cannot hold a large reservoir of fatty acids. To store fatty acids in a large lipid droplet, the carboxylate group must be made hydrophobic, which is typically done by **esterifying** the carboxylate group with an alcohol.

In storage lipids, fatty acids are typically esterified onto **glycerol**. Glycerol is a three-carbon compound with an –OH group on each of its three carbons. If all three hydroxyl groups are esterified with fatty acids, the resulting molecule is a **triacylglycerol**, also known as a **triglyceride**. The structures of glycerol and a triglyceride are shown in Figure 9.8.

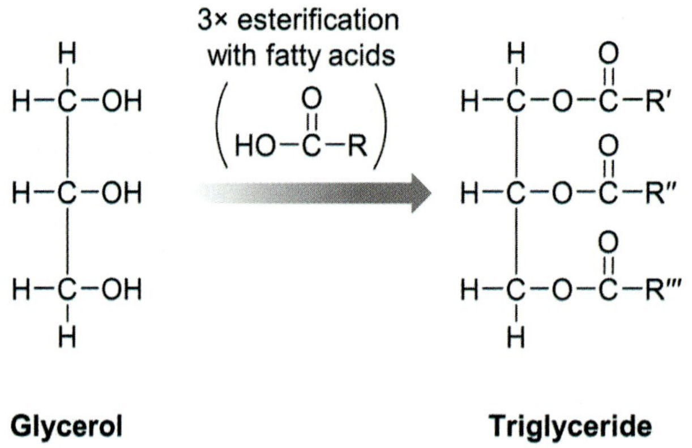

Figure 9.8 Structures of glycerol and a triglyceride.

Esterification results in a much less polar (and more hydrophobic) molecule because the ester group lacks the carboxylate's charge and the alcohol's ability to serve as a hydrogen bond donor. Consequently, triglycerides are much more easily stored in large lipid droplets.

The glycerol molecule that forms the backbone of a triglyceride is a *prochiral* molecule. In other words, glycerol is achiral itself but can *become chiral* through a chemical reaction. Specifically, if the fatty acids esterified onto carbon 1 and carbon 3 are different, then carbon 2 becomes a chiral center (Figure 9.9).

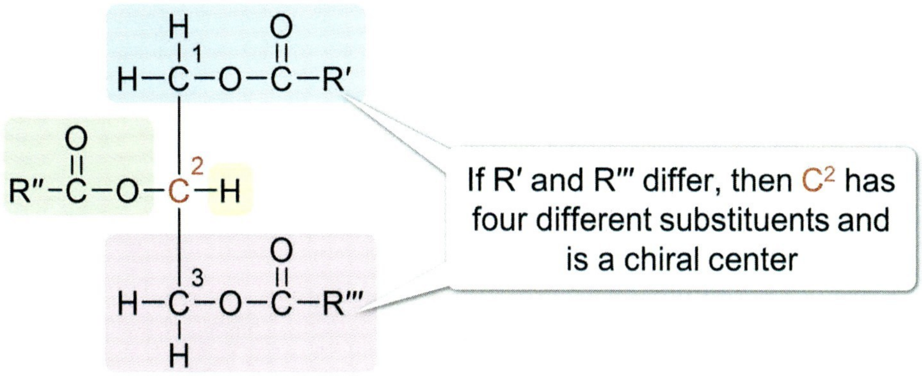

Figure 9.9 Carbon 2 of a triglyceride is a chiral center if the acyl groups on carbons 1 and 3 differ.

Upon mobilization of lipids in adipose tissue (ie, fat tissue), the ester linkages in a triglyceride are enzymatically **hydrolyzed**, yielding free fatty acids and free glycerol. As discussed in Concept 9.1.02, the free fatty acids enter the bloodstream, bind to albumin, and are oxidized in target tissue through β-oxidation. The glycerol backbone can enter glycolysis or gluconeogenesis for further processing (see Chapter 11).

Although albumin can transport free fatty acids throughout the circulatory system, it cannot transport intact triglycerides. Instead, intact triglycerides are transported via lipoprotein particles (eg, chylomicrons, HDL, LDL). Lipoprotein particles are similar to large lipid droplets and have characteristic proteins embedded in their surface, which target each particle to the correct tissue.

Lesson 9.2
Structural Lipids

Introduction

The same hydrophobic interactions that drive lipid droplet formation in triglycerides also drive certain lipids to form biochemically important structures, including micelles, lipid monolayers, and lipid bilayers. Lipid bilayers are especially important as major components of biological **membranes**.

The lipids found in membranes and related structures are **amphiphilic**, meaning they contain both a hydrophobic region and a hydrophilic region. This property allows these lipids to interact with both aqueous and hydrophobic environments by orienting themselves accordingly. Figure 9.10 shows examples of amphiphilic lipids forming membranes and membrane-like structures.

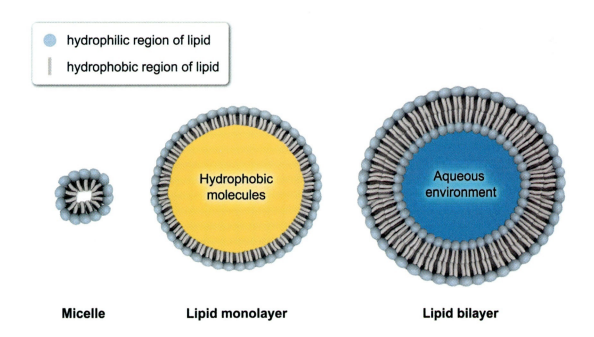

Figure 9.10 In aqueous solutions, amphiphilic lipids can arrange themselves into one of various biochemically important structures.

This lesson provides an overview of the biochemical properties of membrane-forming lipids, as well as the properties of the membranes they form. Lipids that serve structural roles apart from forming membranes are also discussed.

9.2.01 Glycerophospholipids

The most common lipid component of lipid bilayer membranes is the phospholipid. There are two major types of phospholipids: glycerophospholipids and sphingophospholipids. Of these types, glycerophospholipids are more common, and the term "phospholipids" is often used to refer to glycerophospholipids in particular. This concept covers the structure and functions of glycerophospholipids; sphingolipids are discussed in Concept 9.2.02.

General Structure of Phospholipids

The name **glycerophospholipid** hints at the molecule's structure—glycerophospholipids are lipids with a **glycerol backbone** attached to a **phosphate group**. Lesson 9.1 discussed how a triglyceride is composed of a glycerol backbone with three fatty acids esterified onto its three hydroxyl groups. A glycerophospholipid has a structure similar to a triglyceride, but it is esterified with only *two* fatty acids; the third hydroxyl group of glycerol is attached to a phosphate group. Figure 9.11 compares the structures of a triglyceride and a glycerophospholipid.

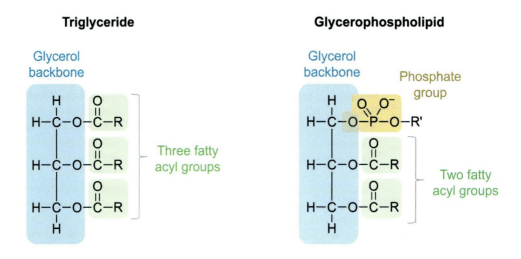

Figure 9.11 Comparison of triglyceride and glycerophospholipid structures.

Phospholipid Tails

Glycerophospholipids consist of a head group and two tail groups. The tail groups come from the hydrophobic tails of the fatty acyl groups esterified onto the glycerol backbone.

In membrane lipids, the tails may consist of any number of carbons but are commonly **long-chain fatty acids** (ie, 14 to 24 carbon atoms). The tails may have *no* double bonds (which makes them saturated) or *at least one* double bond (which makes them unsaturated). The saturation of the fatty acid tails has important implications for membrane fluidity (Concept 9.2.08). Note that the tails on a glycerophospholipid do not need to be the same type of fatty acid. Figure 9.12 shows a common depiction of a glycerophospholipid with two different tail groups.

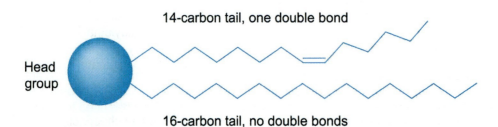

Figure 9.12 A glycerophospholipid with two different tails.

Phospholipid Head Groups

The head group of a glycerophospholipid consists of the phosphate group and anything to which the phosphate is attached (ie, the R′ group shown in Figure 9.11). If the R′ group is simply an H atom, then the molecule is phosphatidic acid (<u>phospha</u>t<u>e</u> + lip<u>id</u> + –<u>ic acid</u>).

The R′ position can be replaced by a variety of different groups. For example, if R′ comes from ethanolamine (ie, R′ = –OCH$_2$CH$_2$NH$_3^+$), the molecule is phosphatidylethanolamine (ie, ethanolamine with a phosphatidyl substituent). These R′ groups are typically polar molecules and contribute to the hydrophilic character of the phospholipid head group. Figure 9.13 shows examples of common head groups that appear on glycerophospholipids.

	Phosphatidyl-				
	-ethanolamine	-choline	-glycerol	-serine	-inositol
Head group charge	Zwitterionic (neutral)	Zwitterionic (neutral)	Negative	Negative	Negative
Polar head group	(structure)	(structure)	(structure)	(structure)	(structure)
Backbone	Glycerol	Glycerol	Glycerol	Glycerol	Glycerol
Hydrophobic tails					

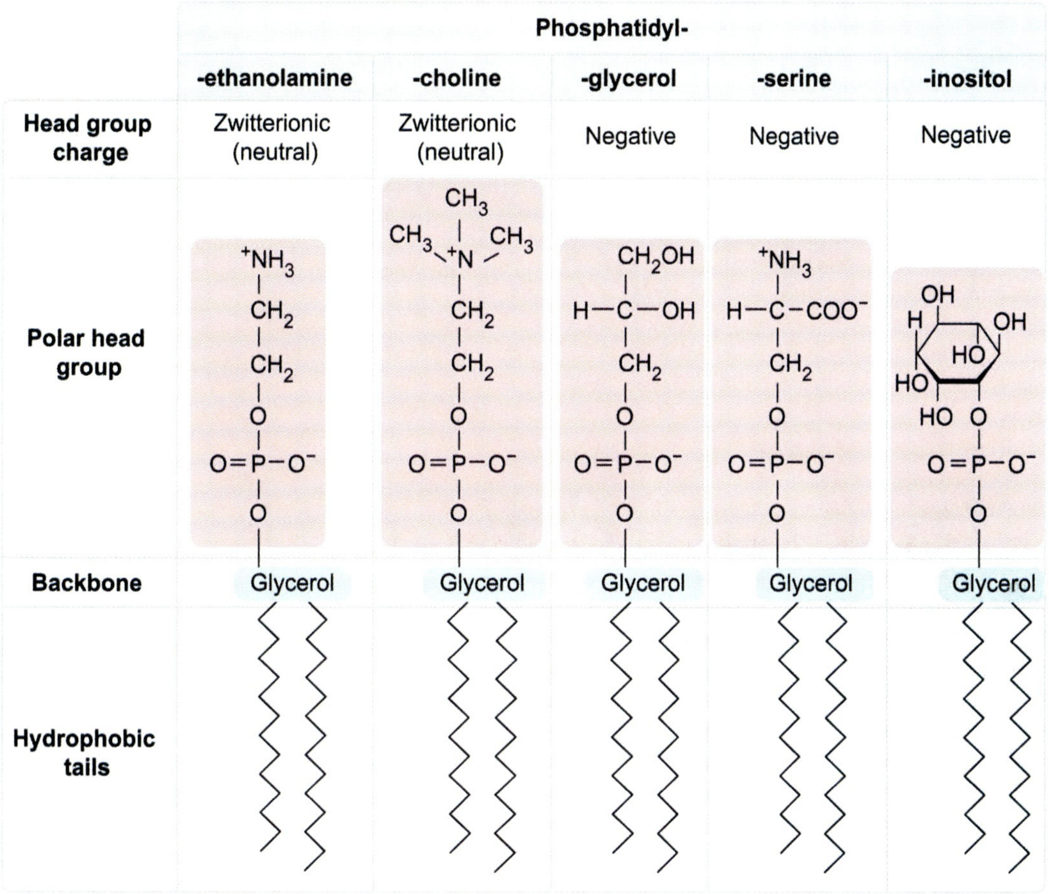

Figure 9.13 Glycerophospholipids with different head groups.

In Figure 9.13, the phospholipids are divided into two groups: those with a zwitterionic head group and those with a negative head group. **Zwitterions** are compounds with an equal number of positive and negative charges and therefore have a net neutral charge. Both phosphatidylethanolamine (PE) and phosphatidylcholine (PC) have positively charged amine groups (a primary amine and a quaternary amine, respectively), which cancel the remaining negative charge of the phosphate group. Therefore, PE and PC have zwitterionic head groups.

In contrast, phosphatidylglycerol (PG) and phosphatidylinositol (PI) have neutral R′ substituents. With no positive charge to offset the negative charge of the phosphate group, these phospholipids have negatively charged head groups. The zwitterionic R′ substituent of phosphatidylserine (PS) combined with the negative phosphate charge gives the PS head group an overall negative charge as well.

Chapter 9: Lipids

Head Groups Can Be Glycosylated to Form a Glycolipid

In addition to the small molecule head groups shown in Figure 9.13, the head groups can be further modified through the addition of carbohydrate groups. For example, the reducing end of an oligosaccharide can form a glycosidic bond to one of the –OH groups of the phosphatidylinositol (PI) headgroup, as shown in Figure 9.14. The resulting molecule, comprising both a lipid and a carbohydrate, is called a **glycolipid**. The glycolipid glycosylphosphatidylinositol (GPI) can be further attached to certain proteins to form a **lipid anchor** that permanently attaches the protein to the membrane (Figure 9.14).

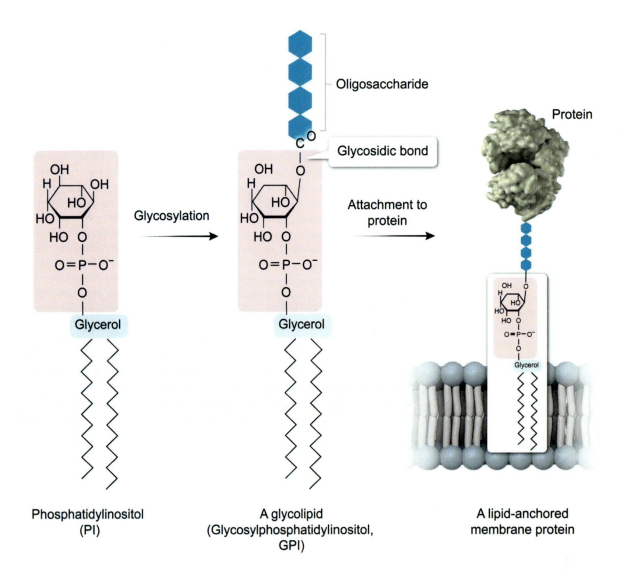

Figure 9.14 Glycolipids are composed of a lipid attached to a carbohydrate. Membrane glycolipids can act as lipid anchors for membrane proteins.

9.2.02 Sphingolipids

Sphingolipids—both in their phosphorylated *and* nonphosphorylated variants—are another major component of amphipathic membranes. Unlike glycerophospholipids, which have a glycerol backbone, sphingolipids have the molecule **sphingosine** as their backbone. Sphingosine is compared to glycerol in Figure 9.15.

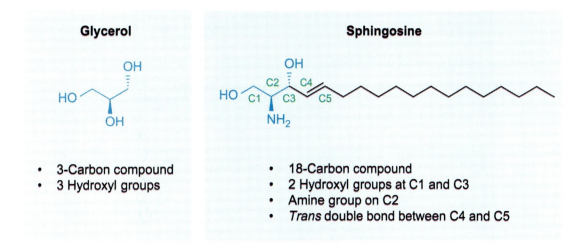

Figure 9.15 A comparison of glycerol (the backbone molecule of glycerophospholipids) and sphingosine (the backbone molecule of sphingolipids).

Figure 9.15 shows several interesting similarities and differences between the two backbone molecules. Whereas glycerol is only three carbons long, sphingosine is much longer at 18 carbons. Each of the three carbons of glycerol forms a single bond to an electronegative O atom (as part of a hydroxyl group). Similarly, the first three carbons of sphingosine each form a single bond to an electronegative atom; however, C2 is bonded to an N instead of an O.

Just as glycerol can esterify with fatty acids to become a lipid, sphingosine can amidify with a fatty acid through the C2 amine to become a sphingolipid. A sphingolipid is compared to a glycerophospholipid in Figure 9.16.

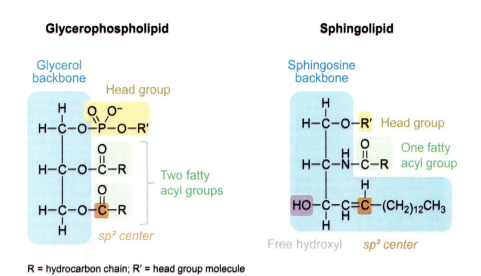

Figure 9.16 A comparison of glycerophospholipids and sphingolipids.

Figure 9.16 shows that glycerophospholipids are bonded to two fatty acyl groups through ester linkages, whereas sphingolipids are bound to only *one* fatty acyl group through an amide linkage on C2. Despite having a free hydroxyl group on C3, sphingolipids do *not* react with a second fatty acid through the C3 hydroxyl; instead, the extra carbons of the sphingosine backbone essentially act as the second tail of the sphingolipid. The remaining hydroxyl group is bonded to a head group.

Sphingolipids Can Have Various Head Groups

Like glycerophospholipids, sphingolipids can have various head groups. Figure 9.16 shows that the head group is attached to the C1 hydroxyl position of sphingosine in a position analogous to that of the phosphate head group of glycerophospholipids.

Sphingolipids that have an –H group as the head group are called **ceramides**, highlighting the amide bond connecting sphingosine to its attached fatty acid. **Sphingophospholipids** are similar to glycerophospholipids and have a phosphate group as the head group. The phosphate may be further substituted, just as with glycerophospholipids. Sphingophospholipids with a phosphocholine or phosphoethanolamine head group are classified as the zwitterionic **sphingomyelins**. Sphingomyelins are particularly common in the myelin sheath surrounding the axons of neurons (see Figure 9.17).

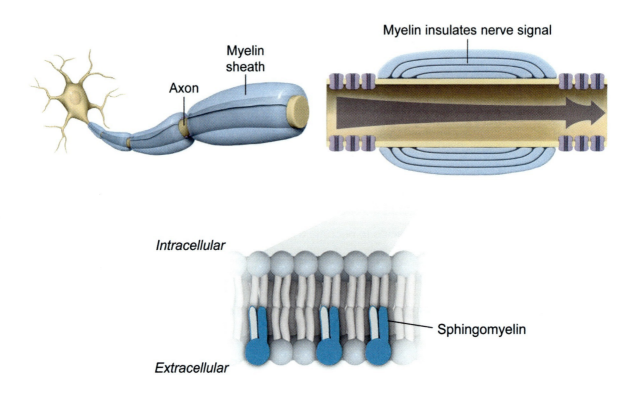

Figure 9.17 Myelin sheaths are enriched in sphingomyelin.

Glycosphingolipids have carbohydrates as the sphingolipid head group and are connected by glycosidic bonds between the C1 hydroxyl and the anomeric carbon of the carbohydrate. The carbohydrate head group of these glycolipids can be a simple monosaccharide (eg, the **cerebrosides**), a longer oligosaccharide (eg, the **globosides**), or more complex polysaccharides that include sialic acid derivatives (eg, the **gangliosides**). Importantly, these glycosphingolipids are *not* phospholipids because they do not have phosphate groups; however, they still are important components of cell membranes.

The classifications of sphingolipids based on head group are summarized in Figure 9.18.

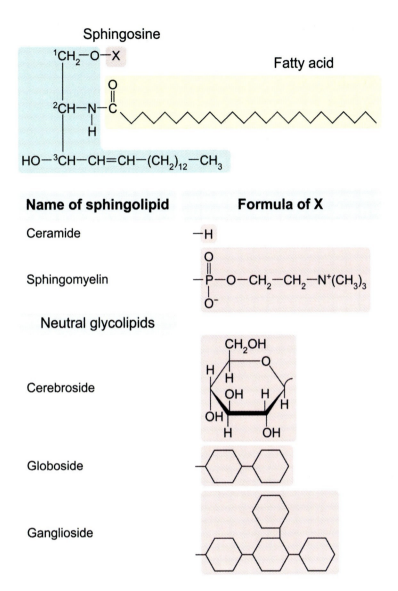

Figure 9.18 Sphingolipids may be classified by their head group.

> **✓ Concept Check 9.1**
>
> Determine whether each of the following statements applies to glycerophospholipids, sphingolipids, or both:
>
> A. Can be a membrane lipid
> B. Attached to only one fatty acid
> C. Attached to two fatty acids
> D. Esterified to its fatty acid(s)
> E. Amidified to its fatty acid(s)
>
> **Solution**
>
> *Note: The appendix contains the answer.*

9.2.03 Cholesterol

Cholesterol is the final major membrane lipid discussed in this lesson. Several representations of the structure of cholesterol are shown in Figure 9.19.

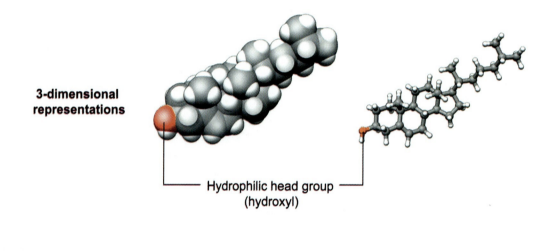

Figure 9.19 Structure of cholesterol.

Unlike the lipids previously described, the hydrophobic portion of cholesterol contains a system composed of four fused rings: three six-membered rings and one five-membered ring. The rigidity imparted by the fused ring system has implications for membrane fluidity, which is discussed in Concept 9.2.08. A single equatorial hydroxyl group serves as the hydrophilic portion of the amphiphilic membrane lipid.

The hydroxyl group of membrane cholesterol is unmodified. It is not phosphorylated to become a phospholipid, nor is it esterified with a fatty acid. However, cholesterol *can* be esterified to become a cholesteryl ester, but this typically occurs only if cholesterol is being transported in the hydrophobic *interior* of a lipoprotein particle. This is because the ester group is much less polar than a hydroxyl due to its inability to act as a hydrogen bond donor. Consequently, only free, unesterified cholesterol can serve as an amphiphilic membrane lipid (Figure 9.20).

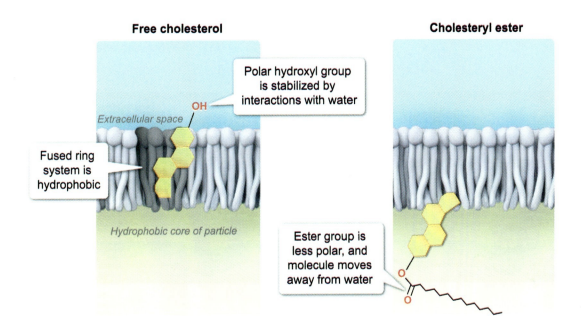

Figure 9.20 Free cholesterol is amphiphilic and is stabilized in membrane monolayers or bilayers. Cholesteryl esters are hydrophobic and are stable in a fully hydrophobic environment.

Importantly, the lack of an ester group in free cholesterol means that there is *no hydrolyzable functional group* in free cholesterol. In other words, cholesterol is a **nonhydrolyzable lipid**. In contrast, triglycerides, glycerophospholipids, and sphingolipids, as well as waxes (Concept 9.2.04), all have hydrolyzable ester or amide bonds and therefore are classified as hydrolyzable lipids.

Cholesterol Metabolism

The pathways of cholesterol metabolism are complex. Although these reaction pathways are not required knowledge for the exam, it is important to know general features of the metabolites used to form cholesterol, as well as those of other biomolecules that are formed *from* cholesterol.

Like other lipids, cholesterol is synthesized from acetyl-CoA, using ATP and NADPH as energy and redox cofactors, respectively (see Lesson 13.1 for more on lipid synthesis). The rate-limiting enzyme of cholesterol synthesis is the enzyme **HMG-CoA reductase**. Consequently, HMG-CoA reductase is a common target for inhibition by drugs such as statins, which lower cholesterol levels.

Several acetyl-CoA molecules undergo a series of reactions (including one catalyzed by HMG-CoA reductase) to produce activated (ie, phosphorylated) **isoprene** units, which are branched five-carbon molecules (see Figure 9.21).

Two isoprene units join to make a **terpene**. Two terpenes (ie, four isoprenes) join to form a diterpene. Six isoprene units can combine to make a **triterpene**. Cyclization of a triterpene is an important precursor step in cholesterol synthesis (Figure 9.21). Cholesterol and its derivatives may be classified as isoprenoids, terpenoids, or triterpenoids.

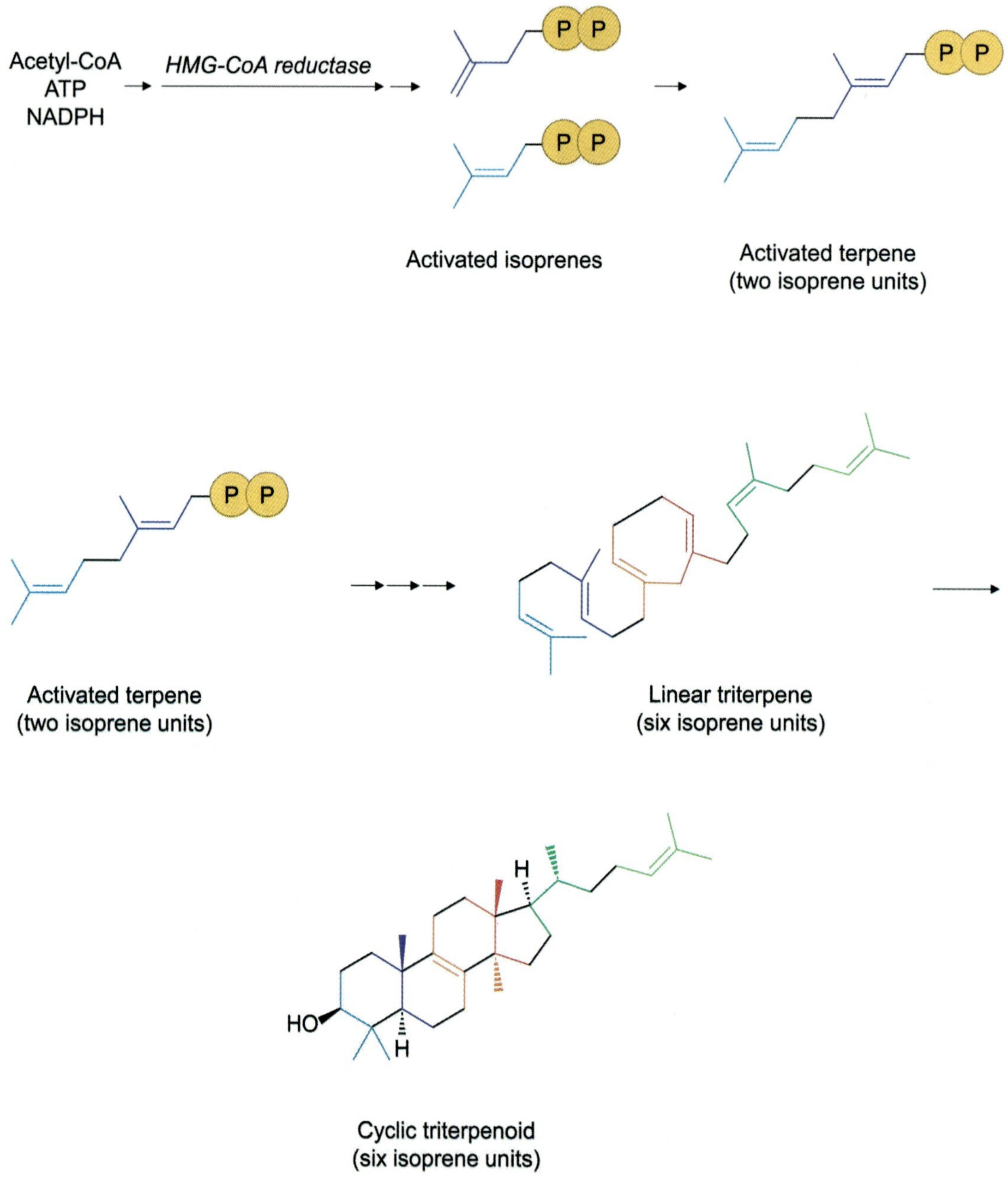

Figure 9.21 Cholesterol synthesis involves isoprene and terpene units.

Once synthesized, cholesterol can be used as a precursor molecule for many other important biomolecules such as steroid hormones and bile salts. Because **steroid hormones** are derived from cholesterol, they are generally hydrophobic enough to cross a cell's lipid bilayer membrane without a membrane receptor. Figure 9.22 compares the structure of a selection of steroid hormones to the structure of cholesterol.

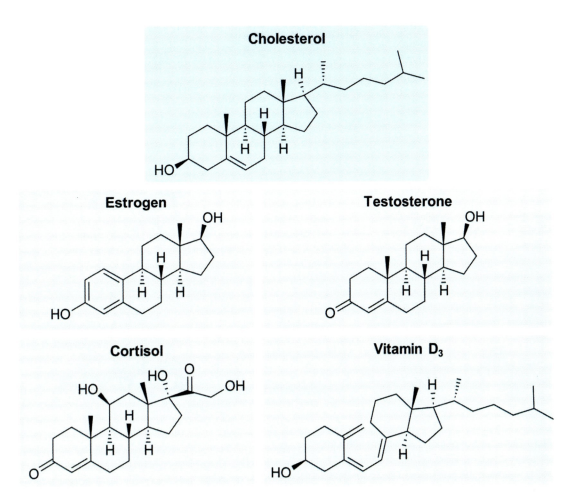

Figure 9.22 Examples of steroid hormones, which are synthesized using cholesterol as a precursor.

Bile salts are emulsifying agents used during digestion to break down large lipid aggregates into smaller, more accessible droplets. Like unmodified cholesterol, bile salts are amphiphilic structures with both a hydrophilic portion and a hydrophobic portion. However, the hydrophilic groups of bile salts are more numerous and arranged differently. Figure 9.23 shows the structure of cholate, the salt of a primary bile acid, which has three axial hydroxyl groups, as well as a negatively charged carboxylate group, on its hydrophilic face.

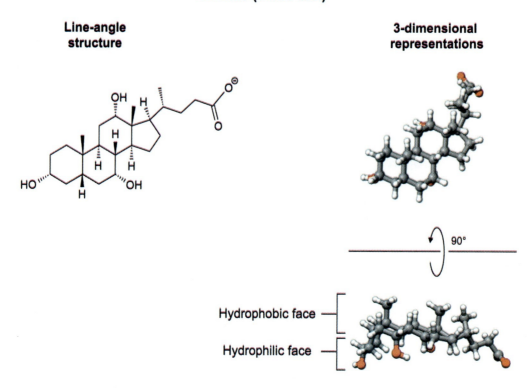

Figure 9.23 Structure of cholate, a primary bile salt.

During digestion, the hydrophobic face of a bile salt adheres to large, hydrophobic lipid aggregates. Meanwhile the hydroxyls and the negative carboxylate group maintain an interaction with the aqueous intestinal lumen.

Unlike membrane lipids, which can stabilize large lipid aggregates, bile salts serve to *break down the aggregate* and stabilize the ingested lipids in smaller droplets. The contents of these smaller droplets can then be more easily accessed by digestive lipases and other enzymes (Figure 9.24).

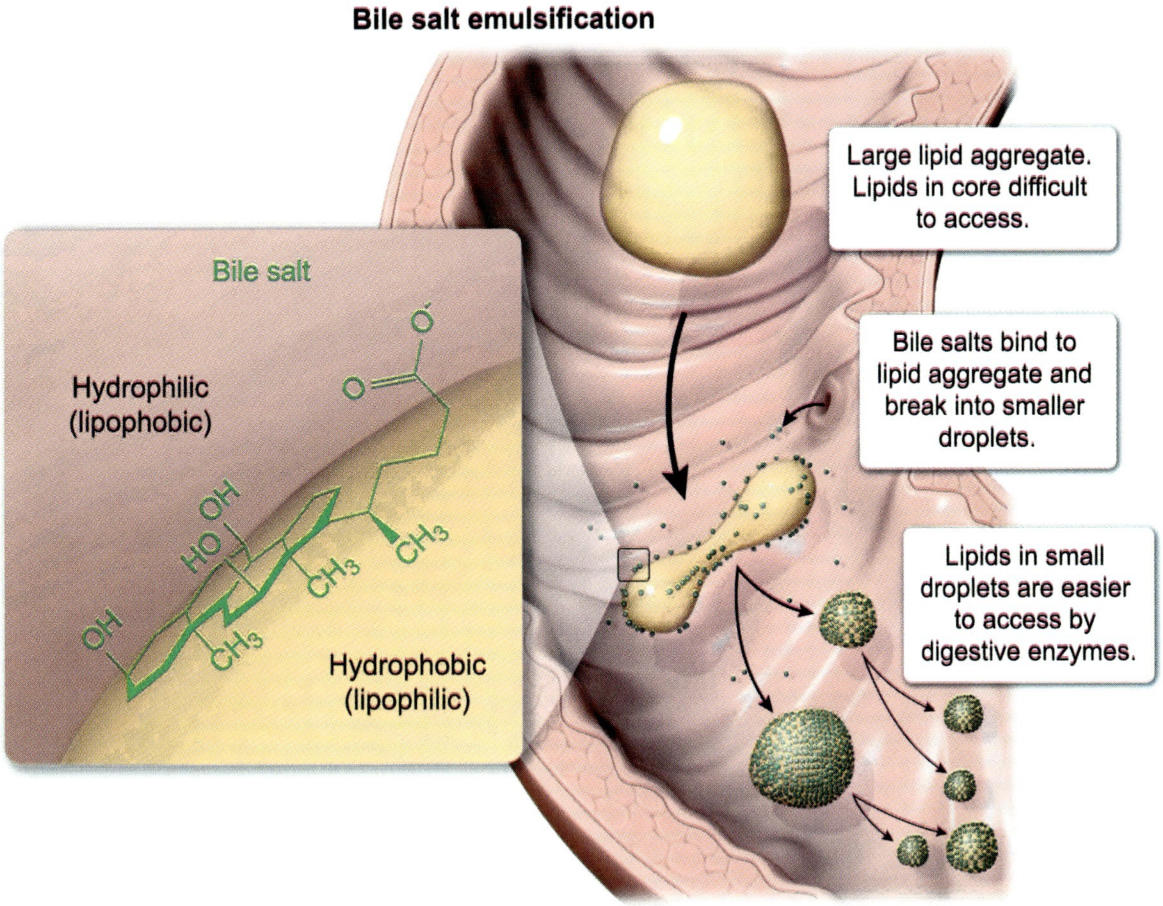

Figure 9.24 Bile salts are cholesterol-derived emulsifying agents that emulsify lipids during digestion.

9.2.04 Waxes

Unlike most of the structural lipids described in this lesson, **waxes** are *not* membrane lipids. However, waxes serve a structurally supportive role as a protective coating for hair and skin in humans, and they also serve an energy storage role in other organisms. Structurally, a biological wax is the ester of a long-chain fatty acid and a long-chain alcohol. Unlike glycerol, the alcohol component of a wax has only a single primary alcohol group with no other functional group. Figure 9.25 shows the structure of a biological wax.

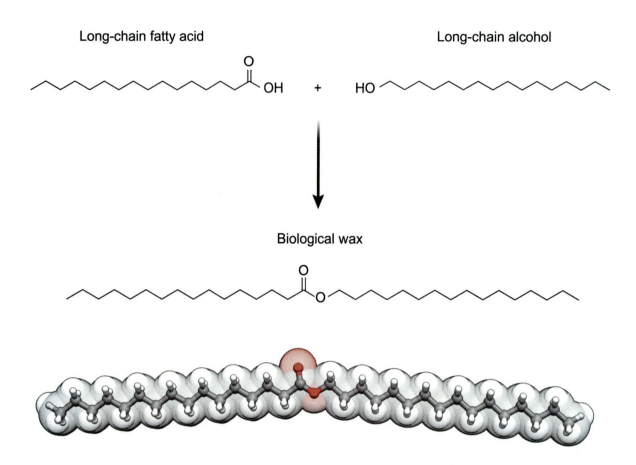

Figure 9.25 Structure of a biological wax.

Figure 9.25 shows that biological waxes are long, fairly linear molecules. These features allow for extensive intermolecular London forces between individual molecules, making waxes solid at room temperature. When used to coat hair, skin, or other surfaces, the hydrophobic nature of waxes makes them effective waterproofing agents.

9.2.05 Organization of Structural Lipids

Some Amphiphiles Organize as Micelles and Act as Detergents

Many **amphiphilic** molecules, such as fatty acids, form spherical structures called **micelles** when placed in an aqueous solution. Micelles are organized with the hydrophobic portions of each molecule oriented toward the center of the sphere and the hydrophilic portions oriented outward. The general structure of a micelle is shown in Figure 9.26.

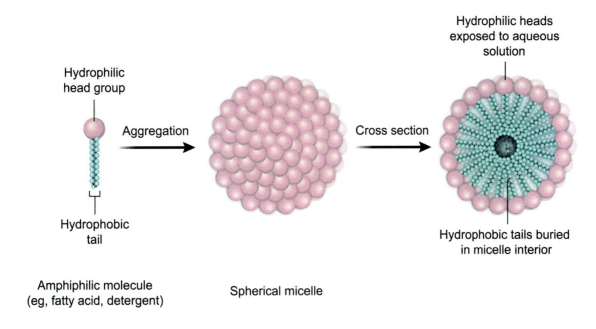

Figure 9.26 The general structure of a micelle.

The name for the chemical hydrolysis (ie, base-catalyzed hydrolysis) of a triglyceride to liberate free fatty acids is *saponification*, which is derived from the use of the reaction's products as soap. Just like soap, fatty acids and other micelle-forming amphiphiles are **detergents** that can be used for various biochemical purposes, including disruption of hydrophobic aggregates, cell lysis, and destruction of enveloped viruses. Sodium dodecyl sulfate is a common detergent that denatures proteins, and it is discussed further in Chapter 14.

Other Amphiphiles Organize into Layers

Not all amphiphilic molecules organize into spherical micelles. Phospholipids, for example, are amphiphilic molecules that organize into flatter layers. An individual lipid layer (ie, a monolayer) organizes with the hydrophilic components on the aqueous side and the hydrophobic components on the nonaqueous side.

Monolayers can serve as the barrier between an aqueous solution and a hydrophobic aggregate. For example, lipid droplets in a cell and lipoprotein particles in the blood are typically enclosed in a phospholipid monolayer. By enclosing lipid aggregates, phospholipid monolayers eliminate the need for a highly ordered solvation layer, as shown in Figure 9.27.

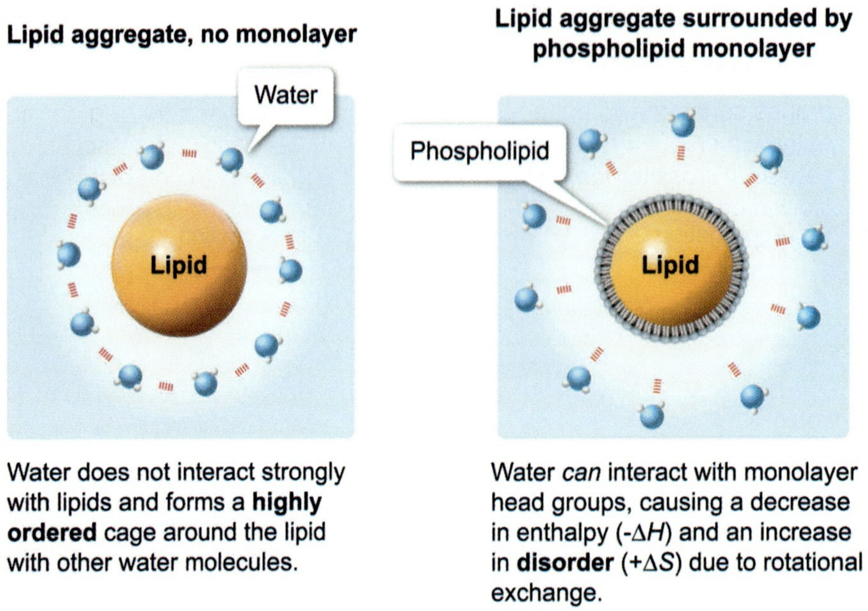

Lipid aggregate, no monolayer

Water

Lipid

Water does not interact strongly with lipids and forms a **highly ordered** cage around the lipid with other water molecules.

Lipid aggregate surrounded by phospholipid monolayer

Phospholipid

Lipid

Water *can* interact with monolayer head groups, causing a decrease in enthalpy ($-\Delta H$) and an increase in **disorder** ($+\Delta S$) due to rotational exchange.

Figure 9.27 A phospholipid monolayer barrier eliminates the high order of the water cage surrounding lipids.

Layers of phospholipids can also act as barriers between two different aqueous solutions. In these cases, however, a monolayer is not sufficient because the hydrophobic tails would be forced to interact with one of the aqueous solutions; therefore, a second layer is needed. The two layers, sometimes called **leaflets**, are organized such that the hydrophobic portions of both layers face each other, and the hydrophilic portions face the two aqueous solutions. These **phospholipid bilayers** (sometimes simply called "lipid bilayers") serve as the basis for the membranes that enclose cells and organelles. Figure 9.28 compares a phospholipid monolayer with a phospholipid bilayer.

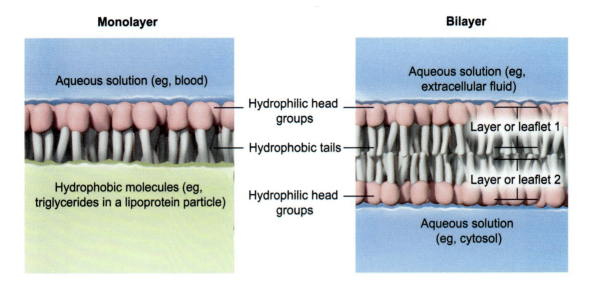

Figure 9.28 Phospholipids can organize as either monolayers (left) or bilayers (right).

9.2.06 Properties of Lipid Bilayers

Lipid Bilayer Membranes Are Semipermeable

Lipid bilayer membranes separate two aqueous compartments from each other (eg, the cytosol from the extracellular space, the lumen of an organelle from the cytosol). Most solutes dissolved in an aqueous compartment *cannot easily cross* through a lipid bilayer by themselves. In contrast, small hydrophobic molecules (eg, dissolved gases) *can* cross a lipid bilayer because they interact more favorably with the lipid tails (see Figure 9.29).

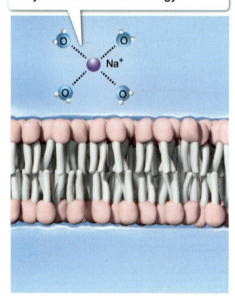
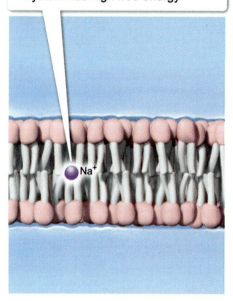

Figure 9.29 Hydrophilic solutes cannot easily cross a lipid membrane, whereas hydrophobic solutes can.

However, membrane-enclosed cells and organelles need both hydrophobic *and* hydrophilic molecules to enter and exit. To allow hydrophilic molecules to cross, cells use channel and carrier proteins (see Lesson 3.3). Cellular membranes allow only certain solutes to cross while blocking most others. Because of this, cellular lipid bilayer membranes are often described as **semipermeable**. Figure 9.30 provides a summary overview of common biochemically relevant solutes and their cellular membrane permeability.

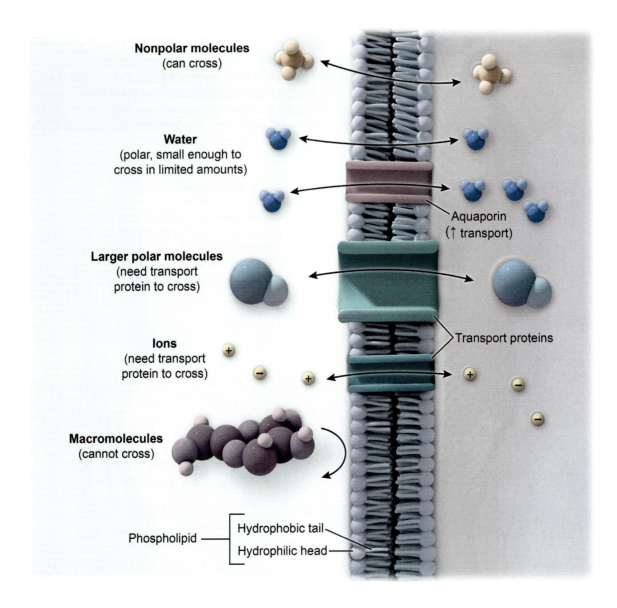

Figure 9.30 Cellular lipid bilayer membranes are semipermeable.

Membranes Are Asymmetric

Although the individual molecules in a membrane have free *lateral* diffusion (see Concept 9.2.07), they do *not* experience free *transverse* diffusion. In other words, a plasma membrane phospholipid in the outer leaflet (where its hydrophilic head group faces the extracellular solution) *cannot* freely diffuse into the inner leaflet (where its hydrophilic head group would face the cytosol) (Figure 9.31). Similarly, membrane proteins cannot flip the orientations of their extracellular and intracellular domains.

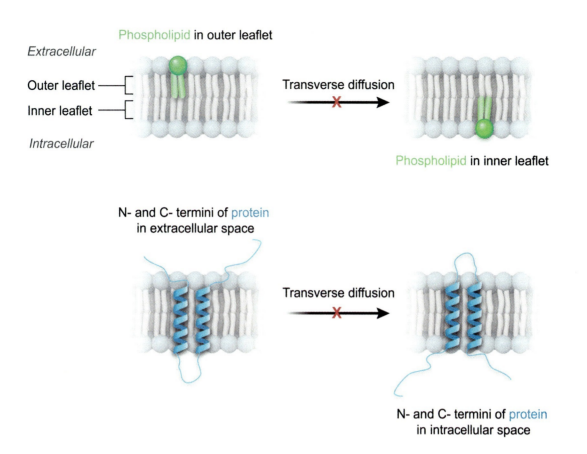

Figure 9.31 Transverse diffusion, or the movement of membrane components across leaflets, generally does not occur freely.

Restricted transverse diffusion results in an **asymmetric phospholipid distribution** in most cell membranes. For example, phosphatidylcholine is predominantly found in the outer leaflet, but phosphatidylethanolamine and phosphatidylserine are predominantly found in the inner leaflet. Maintenance of lipid asymmetry keeps signaling lipids on one side of the membrane, whereas disruption of lipid asymmetry can itself be a signal for other processes (eg, apoptosis, see Biology Lesson 5.4).

9.2.07 The Fluid Mosaic Model

The Fluid Mosaic Model

Cell membranes consist of amphiphilic lipids and membrane proteins. The amphiphilic lipids include both phospholipids and cholesterol, and the membrane proteins include both transporters (Lesson 3.3) and nontransport membrane proteins (eg, transmembrane receptors, Concept 3.2.02).

Although lipids are often classified as macromolecules, they are *not* polymers. Unlike proteins (amino acids joined by peptide bonds), polysaccharides (monosaccharides joined by glycosidic bonds), and nucleic acids (nucleotides joined by phosphodiester linkages), the individual lipids in a lipid aggregate such as a membrane are *not* covalently bonded to each other.

Because of this, membrane lipids and membrane proteins have free **lateral diffusion** throughout the bilayer. In other words, individual lipid molecules or membrane proteins can move side-to-side or back-to-front across a lipid bilayer, as shown in Figure 9.32.

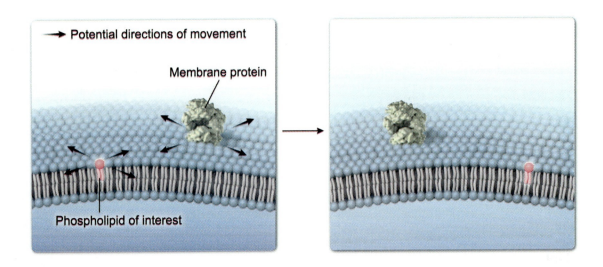

Figure 9.32 Components in a lipid bilayer have free lateral diffusion.

Free diffusion of molecules is a property of the *fluid* states of matter (eg, liquids, gases). Additionally, the heterogeneous composition of membranes, which includes membrane proteins *and* membrane lipids, reminded early scientists of a tile mosaic. Consequently, the model used by scientists to describe a lipid bilayer cell membrane is called the fluid mosaic model. Figure 9.33 shows a representation of a cell membrane according to the fluid mosaic model.

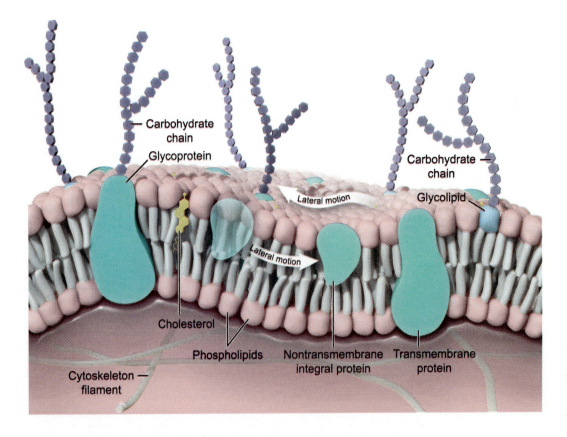

Figure 9.33 The fluid mosaic model of a membrane. The membrane contains both membrane lipids and membrane proteins, and each molecule can diffuse laterally.

Figure 9.33 also shows that the fluidity of the molecules within the membrane allows the entire membrane to be flexible. Rather than existing as a flat plane or a perfectly uniform sphere, membranes can bend, curve, and mold themselves into a variety of shapes, depending on the environment and the cell's needs.

The shape of the cell membrane can be influenced by interactions with the cytoskeleton, with elements in the extracellular matrix, or with membrane components of nearby cells. Figure 9.34 shows examples of various structural arrangements that membranes can adopt due to their fluidity and their interactions with other molecules.

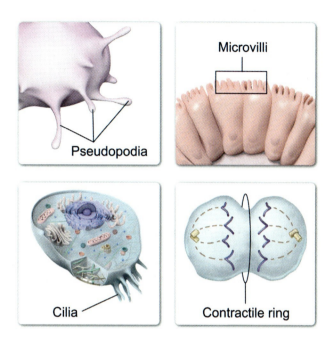

Figure 9.34 Membranes can form a variety of structures depending on their fluidity and their interactions with other molecules.

Interactions with the cytoskeleton, extracellular matrix, and cell-cell junctions can also act as a **scaffold** to restrict the movement of some membrane components. Therefore, despite the free lateral diffusion described by the fluid mosaic model, cells can have regions of specialization (eg, the synaptic junctions of neurons) or show polarization (eg, the apical and basolateral ends of absorptive intestinal cells) Figure 9.35 shows examples of cells with regions of membrane specialization.

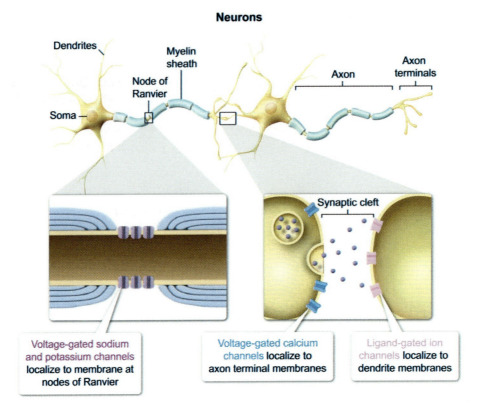
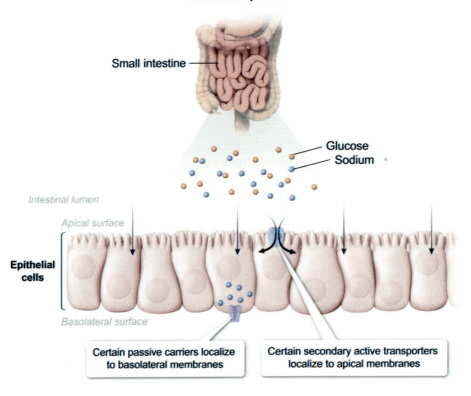

Figure 9.35 Despite the lateral diffusion of the fluid mosaic model, certain membrane regions can be enriched in specific membrane components.

Concept Check 9.2

Two samples of vesicles enclosed by phospholipid bilayers were prepared. One sample was prepared such that some of its inner leaflet phospholipids were tagged with a red fluorophore label. The other sample was prepared such that the outer leaflet phospholipids were tagged with a green fluorophore label. Two vesicles—one from each sample—were induced to fuse and allowed to incubate and achieve equilibrium. Colored lights combine in the way described by the following table:

Label combination	Final color
Red fluorophore only	Red
Green fluorophore only	Green
Both red and green fluorophores.	Yellow

1) What color will the fused vesicle be? (Choose among Options A through D.)
2) How will the labeled phospholipids in the fused vesicle be arranged? (Choose among Options E through H.)

Solution

Note: The appendix contains the answer.

Lipid Rafts Are Localized Microdomains in Biological Membranes

Despite lateral diffusion, a given leaflet of a biological membrane does not necessarily have a homogeneous (ie, evenly spread) distribution of its membrane components. Some membranous components self-assemble into **microdomains** known as **lipid rafts**.

Lipid rafts are localized regions of the membrane that are rich in the *less fluid* components of the membrane bilayer. For example, compared to the surrounding membrane, lipid rafts tend to be enriched in sphingolipids, cholesterol, and glycerophospholipids with long, saturated fatty acid tails. Figure 9.36 shows a visual depiction of a lipid raft. Lipid rafts can serve to group membrane proteins together, increasing the activity of the metabolic pathway they are involved in.

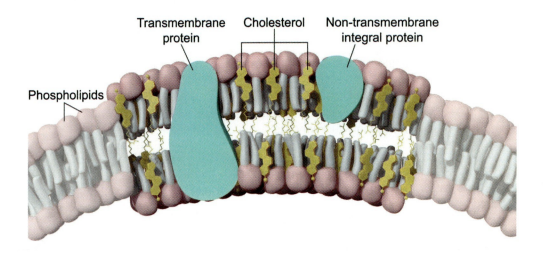

Figure 9.36 Lipid rafts are membrane microdomains enriched in sphingolipids, cholesterol, and long, saturated fatty acids.

9.2.08 Determinants of Fluidity

The fluidity of a membrane is essential to its function in cell biology. A membrane that is too rigid (ie, solid-like) is prone to breakage upon deformation, whereas a membrane that is too flexible and fluid is prone to increased permeability and leakage. External temperature is one factor that influences membrane fluidity. Like all matter, as the temperature of the membrane increases, its fluidity also increases.

Individual cells cannot control external temperature. However, cells *can* influence membrane fluidity by controlling the *composition* of a membrane in response to temperature changes. By altering the types of lipids included in a membrane, membrane fluidity can increase, decrease, or be buffered against fluidity changes. Broadly speaking, membrane fluidity is controlled by the strength of the intermolecular forces (IMFs) between the membrane components.

Influence of Phospholipid Fatty Acyl Tails

Membrane phospholipids may be attached to a variety of fatty acids to form the phospholipid tails. Fatty acid tails vary in length, but commonly range between 14 to 24 carbons (Concept 9.2.01). Furthermore, fatty acids can be saturated or unsaturated, depending on the presence of carbon-carbon double bonds in the hydrocarbon tail. Both factors affect the intermolecular **London dispersion forces** holding the membrane together; therefore, the properties of the phospholipid fatty acyl groups have a significant impact on membrane fluidity.

Increasing tail length causes stronger IMFs because there are more atoms available to interact. Consequently, phospholipids with *longer tails* are more viscous and therefore *less fluid*. In contrast, shorter tails lead to more fluidity (Figure 9.37).

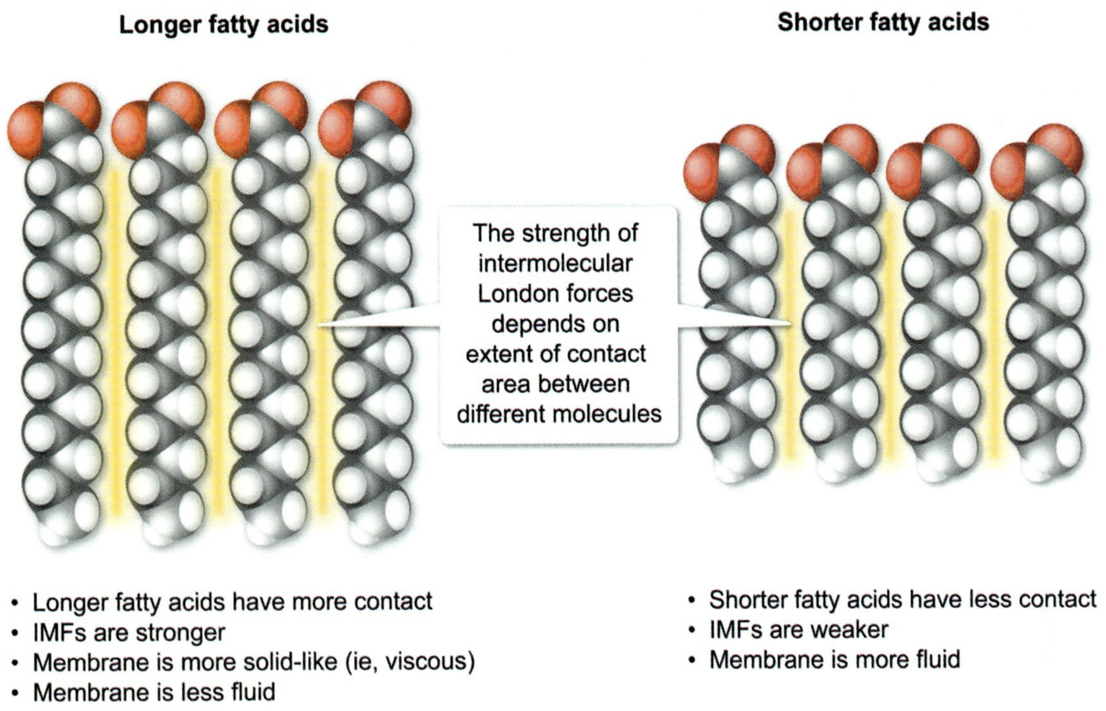

Longer fatty acids　　　　　　　　　　**Shorter fatty acids**

The strength of intermolecular London forces depends on extent of contact area between different molecules

- Longer fatty acids have more contact
- IMFs are stronger
- Membrane is more solid-like (ie, viscous)
- Membrane is less fluid

- Shorter fatty acids have less contact
- IMFs are weaker
- Membrane is more fluid

Figure 9.37 Glycerophospholipids with longer fatty acid tails tend to decrease membrane fluidity.

Similarly, the **saturation** of fatty acid tails also affects membrane fluidity. Saturated fatty acids contain no double bonds in the hydrocarbon tail and are fairly linear. Saturated fatty acids can therefore pack tightly together, resulting in a more viscous, less fluid membrane. In contrast, unsaturated fatty acids have at least one carbon-carbon double bond that results in a "kinked" tail structure. This kink prevents tight packing of the fatty acid tails and results in a *more* fluid membrane (Figure 9.38).

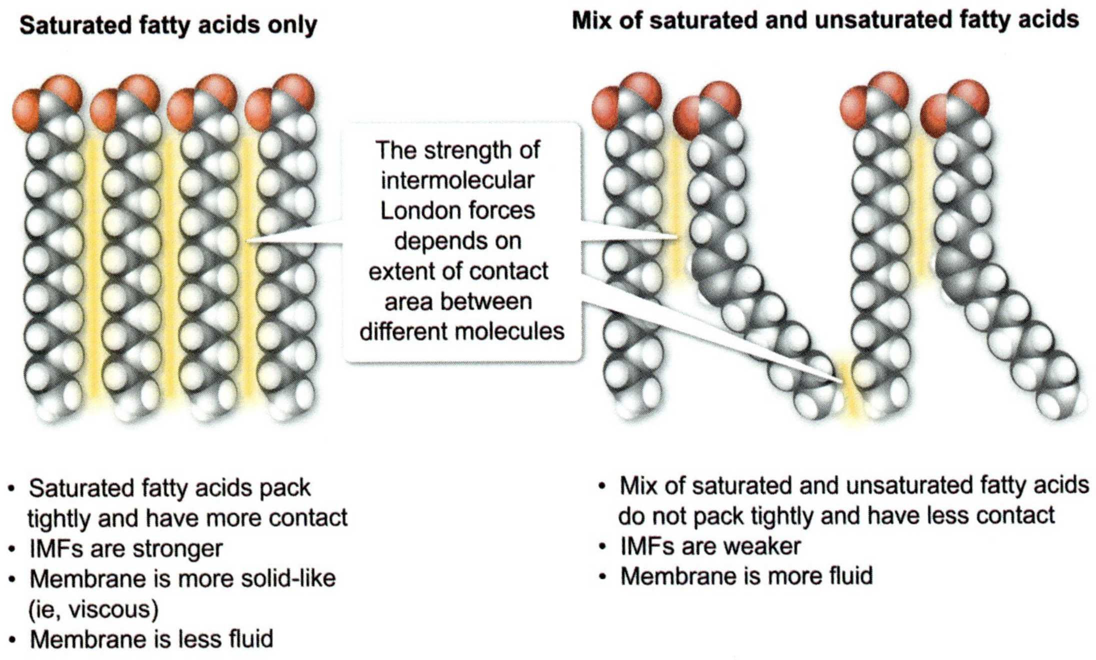

Saturated fatty acids only　　　　**Mix of saturated and unsaturated fatty acids**

The strength of intermolecular London forces depends on extent of contact area between different molecules

- Saturated fatty acids pack tightly and have more contact
- IMFs are stronger
- Membrane is more solid-like (ie, viscous)
- Membrane is less fluid

- Mix of saturated and unsaturated fatty acids do not pack tightly and have less contact
- IMFs are weaker
- Membrane is more fluid

Figure 9.38 Glycerophospholipids with saturated fatty acid tails tend to decrease membrane fluidity.

Cholesterol Is a Membrane Fluidity Buffer

Cholesterol has an interesting effect on membrane fluidity. The fused ring structure causes cholesterol to be relatively rigid, inflexible, and conformationally locked. Because of this, cholesterol's presence can make the membrane itself more rigid and inflexible and can therefore *decrease* the fluidity of *overly fluid* membranes.

However, cholesterol's unique structure does not pack tightly with most lipid components in the membrane. Because cholesterol interrupts the packing interactions between the phospholipid tails of other membrane lipids, cholesterol can also *increase* the fluidity of *overly rigid* membranes.

Because cholesterol can both increase *and* decrease membrane fluidity, cholesterol can be thought of as a *buffer* of membrane fluidity that prevents changes away from an optimal level (Figure 9.39).

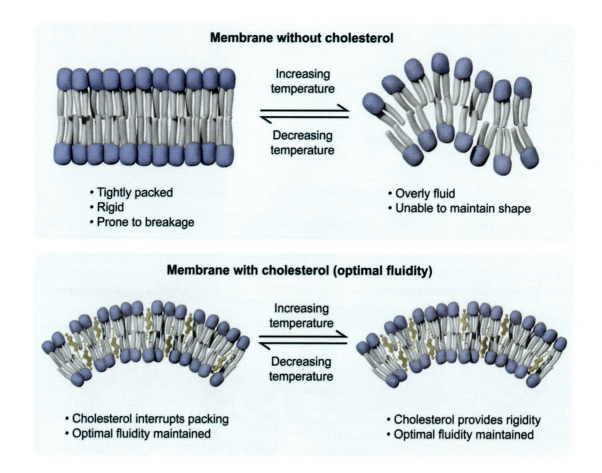

Figure 9.39 Cholesterol acts as a membrane fluidity buffer to maintain an optimal fluidity level.

Membrane Composition Can Change in Response to Temperature

Membrane composition is not static. If external temperatures change a cell's membrane fluidity, the cell can alter its membrane composition in a homeostatic response to *restore* its preferred level of fluidity. For example, if the temperature rises and a cell's membrane fluidity increases, the cell can increase the average length and proportion of saturated fatty acids in the membrane to bring fluidity down (see Figure 9.40).

Chapter 9: Lipids

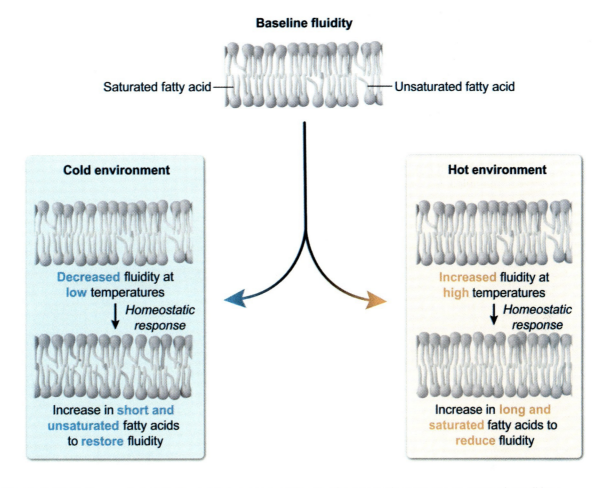

Figure 9.40 Cells can change their membrane composition in a homeostatic response to external conditions.

☑ Concept Check 9.3

A particular bacterial species infects a patient who subsequently undergoes a fever response (ie, increased body temperature). How would the membrane fatty acid composition of the bacterial population change in the post-fever population compared to the pre-fever population?

Solution
Note: The appendix contains the answer.

Lesson 9.3

Signaling Lipids

Introduction

In addition to their roles as energy-storage molecules (Lesson 9.1) and the primary structural component of membranes (Lesson 9.2), lipids can also serve metabolic roles as signaling molecules, regulatory molecules, or coenzymes. This lesson provides a brief introduction to some of these additional roles. Note that details of pathways involving signaling lipids are discussed in this lesson only to provide *examples* of lipids' signaling role. For details of these pathways as they may appear on the exam, see Unit 4.

9.3.01 Signaling Lipids Derived from Hydrolyzable Membrane Lipids

Signaling Lipids Derived from Glycerophospholipids

Membrane lipids surround every cell and every membrane-bound organelle. Thus, they are abundant throughout an organism and can serve as an important reservoir of signaling molecules. Recall that lipid bilayer membranes are asymmetric—certain lipids are more common in one leaflet than the other.

Phosphatidylserine (PS) is an example of a glycerophospholipid that is found only on the inner (ie, cytosolic) leaflet of the plasma membrane of healthy cells. However, in apoptotic cells, this asymmetry is disrupted, and PS moves to the outer leaflet. Outer leaflet PS is a recognition signal for macrophages to phagocytose apoptotic bodies, as shown in Figure 9.41.

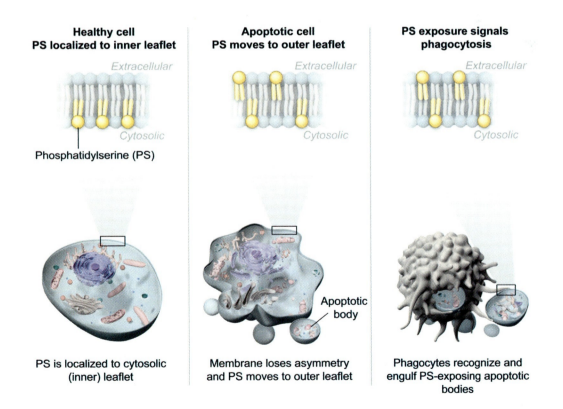

Figure 9.41 Phosphatidylserine (PS) is an example of a phospholipid that acts as a signal when localized to a certain region.

Phosphatidylinositol (PI) and its derivatives are another example of glycerophospholipids that can serve as both structural lipids and signaling lipids. For example, phosphatidylinositol 4,5-bisphosphate (PIP$_2$) can be phosphorylated by phosphatidylinositol 3-kinase (PI3K) to make phosphatidylinositol 3,4,5-trisphosphate (PIP$_3$). Membrane PIP$_3$ can then allosterically activate downstream effector molecules in various pathways, such as cell growth or neuronal plasticity (Figure 9.42).

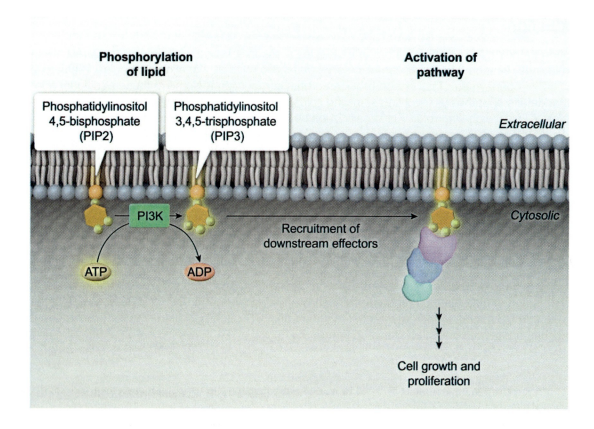

Figure 9.42 Phosphatidylinositol (PI) is an example of a membrane lipid that can act in a signaling role based on its phosphorylation status.

Alternatively, PIP$_2$ can be hydrolyzed to form two different molecules that both have signaling properties (Figure 9.43). The enzyme phospholipase C (PLC) hydrolyzes the phosphoester bond between the glycerol backbone and the polar head group to yield inositol 1,4,5-trisphosphate (IP$_3$) and diacylglycerol (DAG). IP$_3$ is a second messenger that activates the IP$_3$ receptor to release calcium into the cytosol, and DAG is an allosteric activator of the enzyme protein kinase C (PKC).

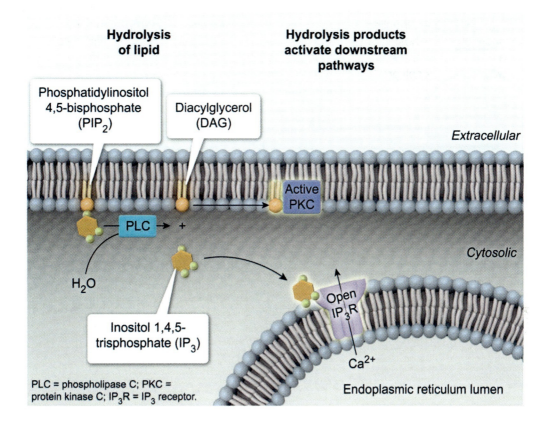

Figure 9.43 PIP$_2$ is an example of a membrane lipid that is hydrolyzed to release products involved in signaling pathways.

Whereas PLC hydrolyzes a phospholipid's polar head group from the rest of the phospholipid molecule, the phospholipase A (PLA) enzymes hydrolyze a fatty acid group from the rest of the molecule. In this way, hydrolyzable membrane lipids serve as a reservoir of fatty acids, which can also serve as signaling molecules, as shown in Figure 9.44.

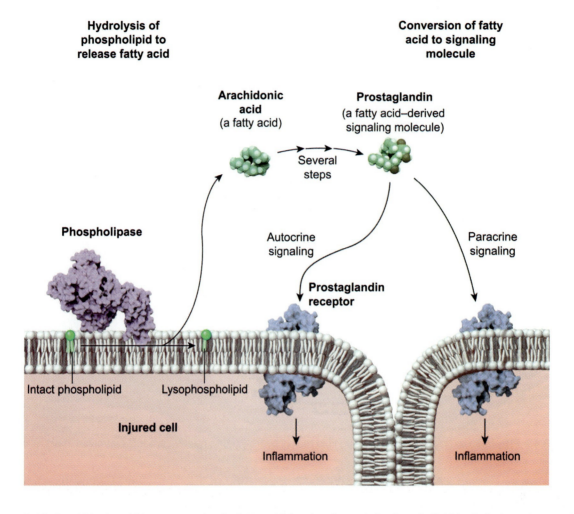

Figure 9.44 Arachidonic acid is an example of a fatty acid that is released via phospholipid hydrolysis and converted into a signaling molecule.

Arachidonic acid is a 20-carbon polyunsaturated fatty acid that serves as a precursor molecule to the various **eicosanoid** signaling molecules. Eicosanoids include the prostaglandins, the leukotrienes, and the thromboxanes. Eicosanoids have various effects but tend to act as paracrine or autocrine signals. Because of this, eicosanoids often function in local processes such as injury response or **inflammation**.

Prostaglandins are eicosanoids that contain a *cyclic* five-membered ring as well as several *oxidized* functional groups. Many nonsteroidal anti-inflammatory drugs (NSAIDs) decrease prostaglandin synthesis by targeting the *cyclooxygenase* (COX) enzymes that produce them.

Figure 9.45 shows the structure of arachidonate (ie, the salt of arachidonic acid), a prostaglandin, and examples of several other eicosanoids.

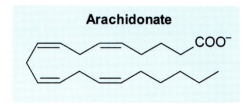

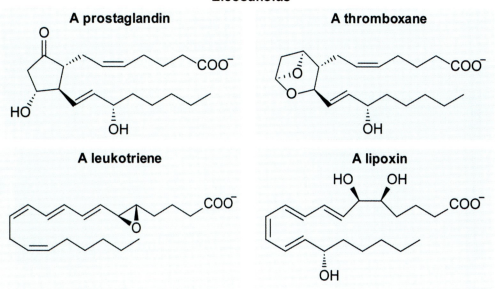

Figure 9.45 Arachidonate and examples of eicosanoids.

Finally, fatty acids can also serve as signaling molecules by acting as lipid modifications of other molecules. Concept 2.4.02 discusses lipidation as an example of a protein post-translational modification. The 16-carbon saturated fatty acid palmitic acid (palmitate in its deprotonated form) is commonly used to post-translationally lipidate a protein and anchor it to the membrane.

Unlike the permanent GPI lipid anchors discussed in Concept 9.2.01, palmitoylation (ie, the process of adding a palmitate group substituent) is a dynamically reversible process, similar to phosphorylation. Palmitoylation of a protein can traffic that protein toward the membrane or toward lipid rafts within a membrane, whereas removal of the palmitate group results in diffusion away from rafts or the membrane. Figure 9.46 shows an example of this type of regulation by lipid modification.

By regulating the localization of target proteins, fatty acids can act in a regulatory role to bring certain proteins or enzymes close together to activate a pathway or disperse them to inhibit a pathway.

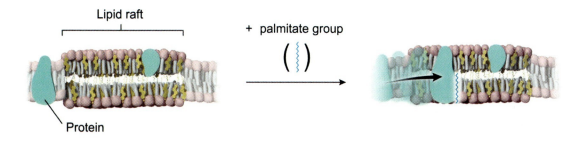

Figure 9.46 Palmitoylation is an example of a post-translational addition of a fatty acid to a protein and of fatty acids acting as regulatory signals.

Chapter 9: Lipids

Sphingolipids as Signaling Molecules

Like the glycerophospholipids, sphingolipids are also important membrane components and serve as an easily accessible reservoir of signaling molecules. Membrane sphingolipids can be hydrolyzed to form signaling molecules (eg, ceramide, sphingosine 1-phosphate), or they can serve as signals in their intact form.

Glycosphingolipids are the most common type of glycolipid in mammals. Glycosphingolipids are commonly found in the outer leaflet of the plasma membrane, where their carbohydrate moiety (ie, attached carbohydrate group) is exposed to the external environment. Other cells or proteins can then recognize the exposed carbohydrate group of the glycosphingolipid.

For example, the ABO blood types arise from the variety of carbohydrates present on membrane glycosphingolipids of human red blood cells (Figure 9.47). Blood transfusions of an incompatible blood type are recognized as foreign antigens, and the immune system develops antibodies against them. In contrast, the glycosphingolipid carbohydrates present on *compatible* blood types are recognized as a familiar component of the patient's blood and do *not* trigger an immune response.

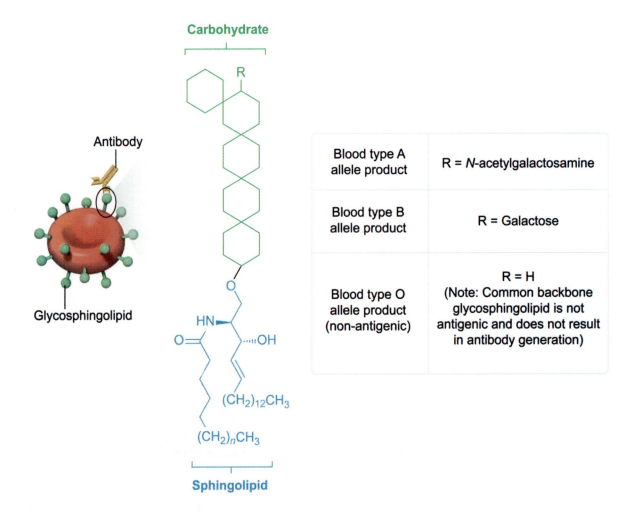

Figure 9.47 ABO blood types are an example of glycolipids that can send a signal (eg, native versus foreign body) based on recognition of the attached carbohydrate.

9.3.02 Terpenoids as Signaling Molecules

Steroid Hormones

Terpenoids (also called isoprenoids) are molecules derived from the combination of two or more isoprene units. Recall from Concept 9.2.03 that **cholesterol** is derived from a triterpene and serves as the precursor of several **steroid hormones**.

Steroid hormones can be used to coordinate a large variety of processes in a multicellular organism. For example, the **sex steroid** hormones (eg, estrogen, testosterone) affect a variety of cells and tissues to regulate development of secondary sex characteristics and other metabolic processes (Figure 9.48). The sex steroid hormones are sometimes called gonadocorticoids because they are produced by the gonads (ie, ovaries, testes).

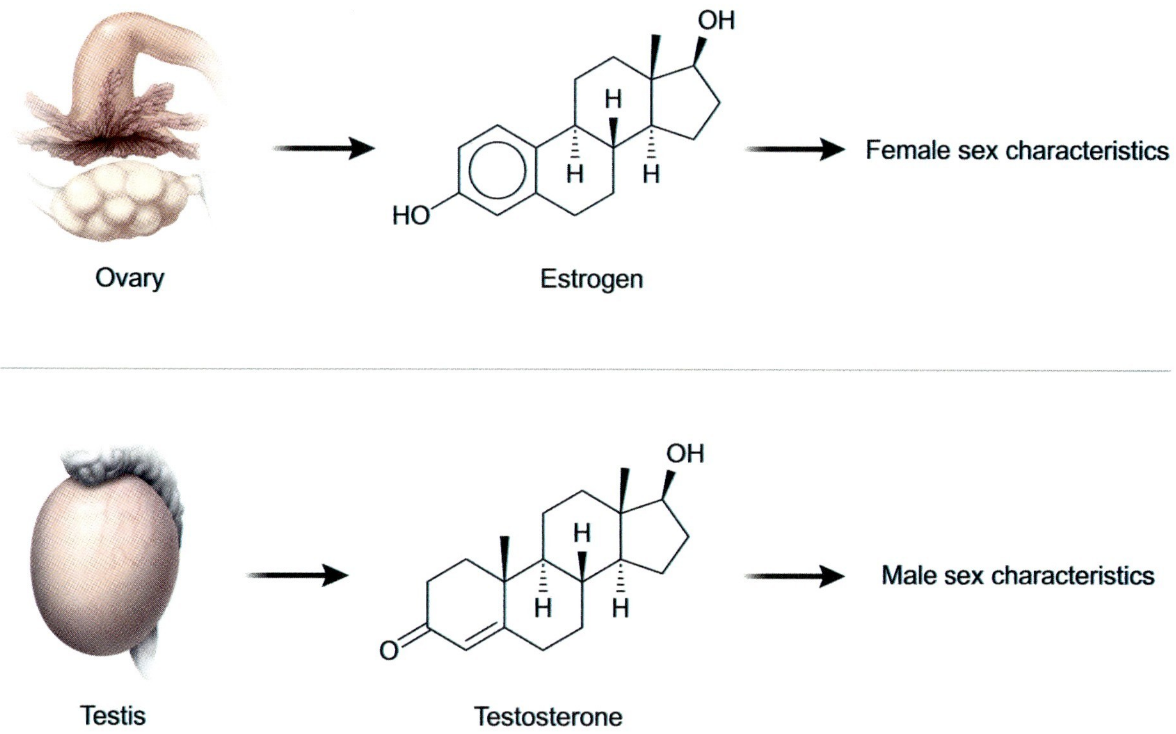

Figure 9.48 Examples of sex steroids produced by the gonads (ie, ovaries, testes).

Glucocorticoids such as cortisol are especially important in the stress response and consequently have effects on energy metabolism, inflammation, the immune response, and other processes. In contrast, mineralocorticoids such as aldosterone regulate mineral excretion and reabsorption in the kidney and thereby regulate blood volume and blood pressure. Figure 9.49 shows examples of these two corticosteroids.

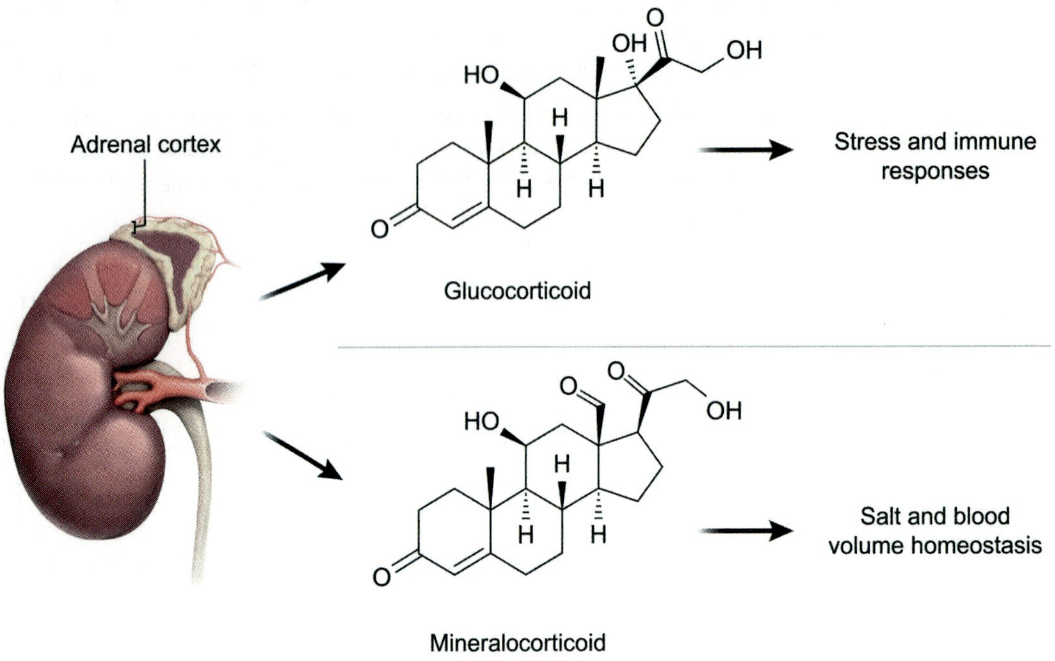

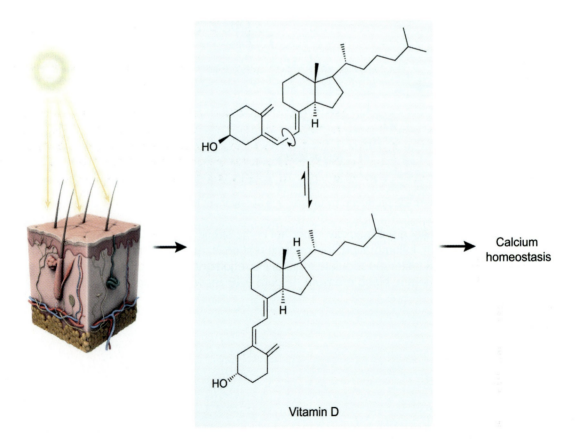

Figure 9.49 Corticosteroids produced by the adrenal cortex.

Vitamin D (Figure 9.50) is another steroid hormone that can be made in limited quantities in the skin through an ultraviolet light–catalyzed reaction. Vitamin D is involved in calcium homeostasis.

Figure 9.50 Example of a form of vitamin D, a signaling lipid.

Most steroid hormones can be recognized from the backbone of *four fused rings* they share with their cholesterol precursor. Vitamin D is an exception because the UV light–catalyzed reaction breaks one of the rings open. Nevertheless, all steroid hormones are generally hydrophobic like cholesterol and have limited solubility in the blood. Therefore, steroid hormones must be bound to a protein carrier such as albumin during transport through the circulatory system.

Once they arrive at their target tissue, the hydrophobic character of steroid hormones allows them to *passively diffuse* through the lipid bilayer plasma membrane. Once inside their target cell, steroid hormones then bind an *intracellular* receptor, which can be either cytosolic or already in the nucleus. Note that this contrasts with soluble peptide hormones, which typically must bind membrane receptors. Figure 9.51 shows an example of steroid hormones traveling through the blood on a protein carrier before passively diffusing through the plasma membrane of its target cell.

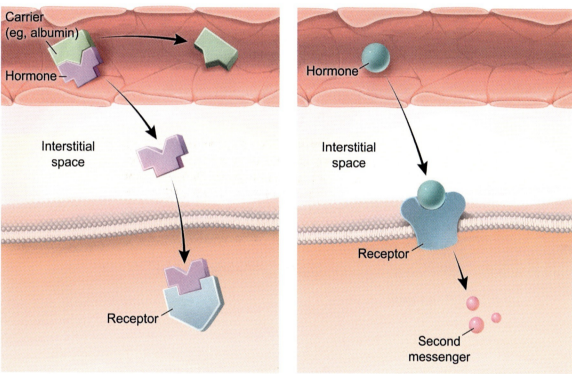

Figure 9.51 Lipid hormones (eg, steroids) travel in the blood, bound to a carrier protein, and can passively diffuse through the plasma membrane into their target cell.

The intracellular receptor–steroid complex then traffics to the nucleus, where it regulates the expression of various target genes (Figure 9.52).

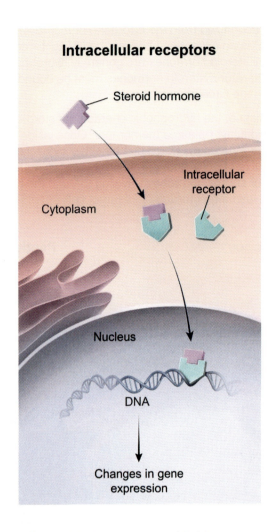

Figure 9.52 Steroid hormones typically regulate the expression of their target genes.

Other Terpenoid Signaling Lipids

Retinal, derived from vitamin A, is a terpenoid that acts as a signaling lipid in the visual system, covalently binding as a ligand to a G protein–coupled receptor known as rhodopsin. Retinal has a long, conjugated hydrocarbon tail, which allows it to absorb specific wavelengths of light. The absorbed light energy can then induce *cis-trans* isomerization of one of its double bonds, allowing it to act as an activating agonist for rhodopsin, as shown in Figure 9.53.

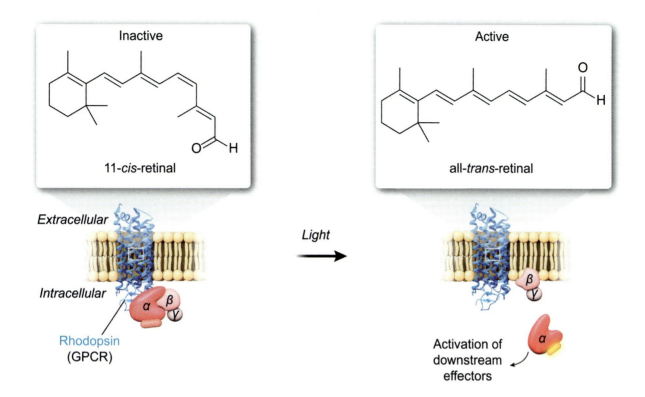

Figure 9.53 Retinal is an example of a lipid ligand that can activate a receptor based on its *cis-trans* configuration.

Retinoic acid, another form of vitamin A, acts like a steroid hormone in that it binds a nuclear receptor to induce expression of genes related to growth and development.

Another lipid with a signaling role is ubiquinone (ubiquinol in its reduced form). Ubiquinone, also called coenzyme Q (CoQ) is a nonsteroid terpenoid that acts as a redox cofactor in the electron transport chain (ETC). Ubiquinone (Q) receives a pair of electrons (and two H^+) at either Complex I or Complex II of the ETC, which reduces the lipid to ubiquinol (QH_2). QH_2 then diffuses through the inner mitochondrial membrane to Complex III, where it deposits its electrons one at a time (see Lesson 12.2). Ubiquinone and ubiquinol are shown in Figure 9.54.

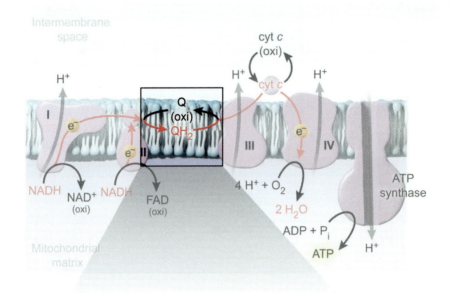

Figure 9.54 Ubiquinone and ubiquinol are examples of lipids acting as redox coenzymes.

Vitamins E and K are also lipids with redox capability and with terpenoid tails. Vitamin E acts as a general antioxidant, whereas vitamin K functions as a redox coenzyme in the clotting pathway.

Collectively, vitamins A, D, E, and K are known as the lipid-soluble vitamins. Lipid-soluble vitamins (also called fat-soluble vitamins) are nonhydrolyzable lipids that can be stored in adipose tissue (ie, fat tissue). These vitamins are shown in Table 9.1.

Table 9.1 The lipid-soluble vitamins.

Vitamin	Structure	Biological function
A		Cell maintenance and repair, immune system maintenance, eyesight
D		Calcium absorption and utilization in the development and growth of bones and teeth
E		Antioxidant, protects cell membranes from oxidative damage, particularly red blood cells
K		Required for blood clot formation

☑ Concept Check 9.4

Which of the following describe a role of lipids? Select all that apply.

    A. An intracellular signaling molecule
    B. A barrier around cells and organelles
    C. A specific recognition signal for other cells or antibodies
    D. An extracellular signaling molecule
    E. A fuel that can be broken down for energy
    F. A waterproof coating of hair and skin
    G. A means of storing excess energy
    H. An allosteric regulator
    I. A redox coenzyme

Solution

Note: The appendix contains the answer.

Chapter 9: Lipids

END-OF-UNIT MCAT PRACTICE

Congratulations on completing **Unit 3: Carbohydrates, Nucleotides, and Lipids**.

Now you are ready to dive into MCAT-level practice tests. At UWorld, we believe students will be fully prepared to ace the MCAT when they practice with high-quality questions in a realistic testing environment.

The UWorld Qbank will test you on questions that are fully representative of the AAMC MCAT syllabus. In addition, our MCAT-like questions are accompanied by in-depth explanations with exceptional visual aids that will help you better retain difficult MCAT concepts.

TO START YOUR MCAT PRACTICE, PROCEED AS FOLLOWS:

1) Sign up to purchase the UWorld MCAT Qbank
   IMPORTANT: You already have access if you purchased a bundled subscription.
2) Log in to your UWorld MCAT account
3) Access the MCAT Qbank section
4) Select this unit in the Qbank
5) Create a custom practice test

Unit 4 Metabolic Reactions

Chapter 10 Catabolism and Anabolism

10.1 Catabolism

10.1.01	ATP and Other Cellular Sources of Energy
10.1.02	Energy of ATP Hydrolysis
10.1.03	Coupling of ATP Hydrolysis to Other Reactions

10.2 Anabolism

10.2.01	Anabolism Uses Energy to Produce Structural and Storage Molecules

Chapter 11 Carbohydrate Metabolism

11.1 Glycolysis and Fermentation

11.1.01	Glycolysis: Energy-Investment Phase
11.1.02	Glycolysis: Energy-Payoff Phase
11.1.03	Alternate Entries into Glycolysis
11.1.04	Fermentation

11.2 Gluconeogenesis

11.2.01	Gluconeogenesis
11.2.02	Alternate Entry Points of Gluconeogenesis
11.2.03	The Cori Cycle
11.2.04	Regulation of Glucose Metabolism

11.3 The Pentose Phosphate Pathway

11.3.01	The Oxidative Phase
11.3.02	The Nonoxidative Phase

11.4 Glycogen Metabolism

11.4.01	Glycogenesis
11.4.02	Glycogenolysis
11.4.03	Regulation of Glycogen Metabolism

Chapter 12 Aerobic Respiration

12.1 The Citric Acid Cycle

12.1.01	Pyruvate Entry into the Citric Acid Cycle
12.1.02	Citric Acid Cycle Reactions
12.1.03	Citric Acid Cycle Products

12.2 The Electron Transport Chain

12.2.01	Complex I
12.2.02	Complex II
12.2.03	Complex III
12.2.04	Complex IV

12.3 Oxidative Phosphorylation

12.3.01	The Proton Motive Force
12.3.02	ATP Synthase
12.3.03	Uncoupling of ATP Synthase

Chapter 13 Noncarbohydrate Metabolism

13.1 Fatty Acid Metabolism

- 13.1.01 Lipid Transport and Catabolism
- 13.1.02 Fatty Acid Oxidation
- 13.1.03 Ketone Bodies
- 13.1.04 Fatty Acid Synthesis
- 13.1.05 Regulation of Fatty Acid Metabolism

13.2 Protein Catabolism

- 13.2.01 Protein Degradation
- 13.2.02 Transamination, Deamination, and Deamidation
- 13.2.03 The Urea Cycle
- 13.2.04 Glucogenic and Ketogenic Amino Acids

Lesson 10.1

Catabolism

Introduction

Metabolism encompasses all chemical reactions that an organism performs to sustain its life. Metabolic processes that *break molecules down* are called **catabolic processes** and typically *release* energy, which can be used to power various cellular processes. One use of that energy is to *build other types of molecules*—such processes are called **anabolic processes** (Figure 10.1). Nearly all metabolic reactions are catalyzed by enzymes (Unit 2) and often involve the breaking down or building up of proteins, carbohydrates, nucleic acids, and lipids.

Unit 4 focuses on **metabolism**, with a specific focus on the catabolic pathways that release energy from carbohydrates or lipids and the anabolic pathways that store energy in carbohydrates or lipids. This lesson describes the forms that the energy released through catabolism may take between its release from molecules and its later use by the cell. One such form is adenosine triphosphate (ATP), the primary chemical unit of energy storage in most cells.

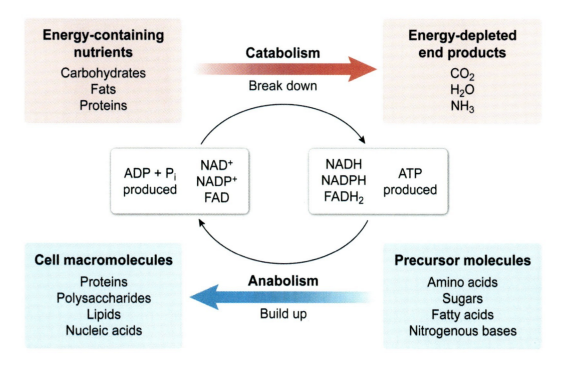

Figure 10.1 A comparison of catabolic and anabolic metabolism.

10.1.01 ATP and Other Cellular Sources of Energy

Glucose is commonly considered the preferred fuel source for cells, yet the oxidative combustion of glucose yields approximately 2,840 kJ/mol. This amount of energy is much more than can be coupled to any individual endergonic biochemical reaction. Consequently, biochemical catabolism of biomolecules does not occur as a single combustion step, but over several enzyme-catalyzed steps (ie, a **metabolic pathway**). This facilitates the release of energy in *smaller portions* that can be stored for *later use* to power *one reaction at a time*.

ATP as Energy Currency

Storing energy in molecules of ATP is one of the most common ways of managing the massive amount of energy released by glucose oxidation and other catabolic processes. As discussed in Concept 8.1.04, ATP is a nucleoside triphosphate that contains a chain of three phosphate groups connected by phosphoanhydride linkages. Phosphoanhydride hydrolysis has a negative ΔG, as shown in Figure 10.2, but this ΔG is much smaller than that of glucose combustion. This smaller amount of energy is *much more appropriate* for coupling to biochemical processes.

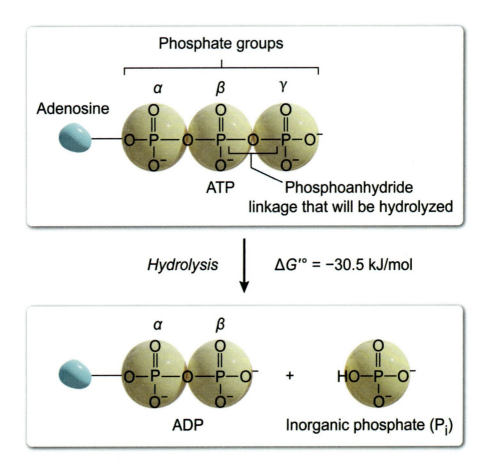

Figure 10.2 ATP hydrolysis releases energy that cells can use to power endergonic biochemical processes.

For example, ATP hydrolysis is used to power reactions catalyzed by many ligases (Concept 4.2.07) and translocases (Concept 4.2.08). Because of the common use of ATP by many enzymes, either as a source of energy or as a way to store energy, ATP is considered the primary *energy currency* of the cell. Hydrolysis of the bond connecting the γ-phosphate to the β-phosphate in ATP liberates ADP and inorganic phosphate (P_i).

$$\text{ATP} \xrightarrow{H_2O} \text{ADP} + P_i \qquad \Delta G = -30.5 \, \text{kJ}/\text{mol}$$

Table 10.1 compares the energy of the complete oxidative combustion of glucose to the energy captured by ATP through the complete *biochemical* oxidation of glucose using aerobic respiration. Note that the energy is *not* captured with 100% efficiency; nevertheless, the ability to store energy in smaller, more usable amounts has proven to be evolutionarily more beneficial than the ability to capture *more* energy in a *less* usable amount.

Chapter 10: Catabolism and Anabolism

Table 10.1 Energy yield per mole of glucose combustion compared to aerobic respiration.

Molecule	$\Delta G'^\circ$ (kJ/mol)	ATP yield after aerobic respiration	Total energy (kJ)	Efficiency of energy capture
Glucose	−2,840 (Combustion)	—	−2,840	—
ATP	−30.5 (Hydrolysis to ADP)	32	−976	34.4%

The Nucleotides Are Energetically Equivalent

ATP is the nucleotide most often used as energy currency, but the other common ribonucleotides (ie, GTP, CTP, UTP) are also used to power reactions. Although the nitrogenous base differs among the four nucleotides, the energy released by hydrolysis of their phosphoanhydride linkages is essentially the same. Furthermore, ATP can be used to regenerate the other nucleoside triphosphates, making the γ-phosphate hydrolysis of *any* nucleoside triphosphate essentially equivalent to one unit of ATP.

Figure 10.3 provides an example of this equivalence using a step of protein translation. The ribosome uses one GTP to power peptide bond formation between an aminoacyl-tRNA and a growing peptide. The γ-phosphate of GTP is hydrolyzed to produce GDP + P_i, which releases an amount of energy essentially equivalent to the energy of ATP hydrolysis. ATP can be used to regenerate GTP; therefore, the ribosome can be said to use the equivalent of one unit of ATP to power peptide bond formation.

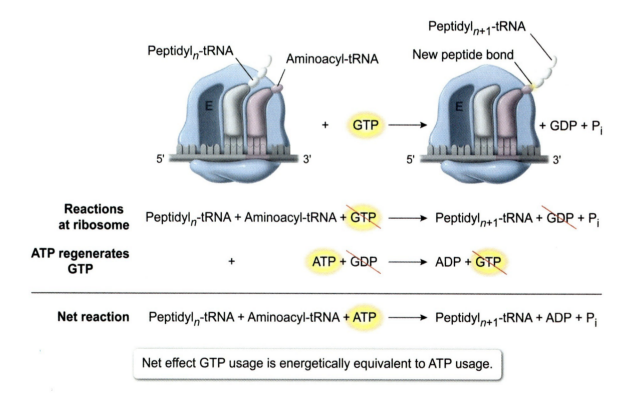

Figure 10.3 Hydrolysis of GTP (or any other NTP) is energetically equivalent to hydrolysis of ATP.

Some biochemical reactions require more energy input than hydrolysis of the γ-phosphate can provide. Rather than bind multiple NTP molecules, enzymes can bind a single NTP molecule but hydrolyze it at the phosphoanhydride linkage between the α- and β-phosphates instead of at the linkage between the β- and γ-phosphates. This results in the production of an NMP (instead of an NDP) and an inorganic *pyrophosphate* (PP_i) (instead of a single phosphate). PP_i is then hydrolyzed to two P_i molecules in another exergonic reaction, releasing more energy.

Ultimately, the hydrolysis of an NTP to NMP + PP_i costs two units (or equivalents) of ATP. This is because ATP can be used to regenerate the used NTP by adding one phosphate back at a time. Figure 10.4 shows the net effect of this kind of hydrolysis.

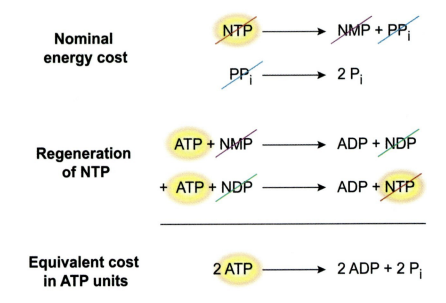

Figure 10.4 Hydrolysis of NTP to NMP + PP_i is energetically equivalent to two ATP units.

Other High-Energy Molecules Can Regenerate ATP by Substrate-Level Phosphorylation

Other molecules may facilitate ATP synthesis from ADP. For example, creatine phosphate is an important molecule in skeletal muscle. Under conditions of high energy demand, its guanidinyl phosphate group can be hydrolyzed to quickly replenish ATP stores under low-oxygen conditions. Transfer of a phosphate from one molecule to ADP to form ATP is called **substrate-level phosphorylation**.

Other examples of high-energy substrates include acyl phosphates, enol phosphates, and thioesters. Table 10.2 lists several molecules that can act as high-energy substrates in biochemical pathways commonly tested on the exam.

Table 10.2 Examples of high-energy substrates.

High-energy substrate	Hydrolysis products	Example substrate/ pathway
Phosphoric acid anhydride		ATP/ various
Acyl phosphate		1,3-bisphosphoglycerate/ glycolysis
Enol phosphate		Phosphoenolpyruvate/ glycolysis
Guanidinyl phosphate		Creatine phosphate/ anaerobic muscle metabolism
Thioester		Succinyl-CoA/ citric acid cycle

Other Forms of Energy Currency

Aside from storing energy in ATP or ATP equivalents, the potential energy of high-energy electrons can also be stored via the redox potential ($E°$) of cofactors such as NADH or $FADH_2$. For example, the high-energy electrons of NADH enter Complex I of the electron transport chain, and the oxidation of NADH to NAD^+ provides the energy to pump protons across the mitochondrial inner membrane, as shown in Figure 10.5. (See Lesson 12.2 for more on the electron transport chain and proton pumping.)

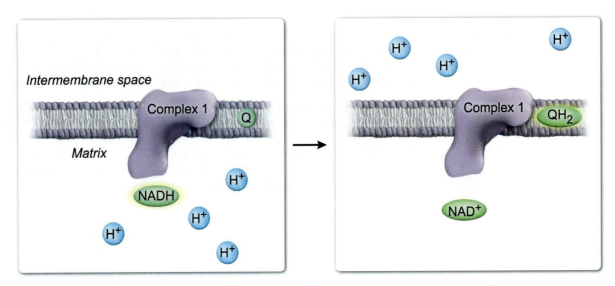

High-energy electrons of NADH are used to pump protons to the intermembrane space.
Q = ubiquinone; QH$_2$ = ubiquinol

Figure 10.5 The energy of electrons transferred during redox reactions can be used to power biochemical processes.

The electrochemical gradient of charged particles (ie, ions) contributes to the membrane potential (Lesson 3.3), which is another form of potential energy that cells can use to power endergonic processes. The energy stored in the membrane potential can be used to synthesize chemical bonds. For example, the protons pumped across the membrane by the electron transport chain create an electrochemical gradient across the mitochondrial inner membrane called the proton motive force, which powers ATP synthase. Consequently, generation of ATP in this way—that is, through a proton gradient that itself comes from oxidation of redox coenzymes—is known as oxidative phosphorylation (see Chapter 12).

Figure 10.6 illustrates the use of electrochemical gradients and the membrane potential as a cellular source of energy.

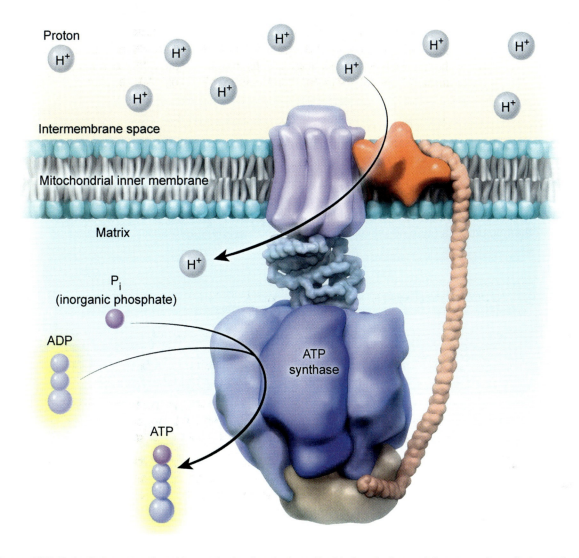

Figure 10.6 Potential energy stored in an electrochemical gradient is freed when solutes move down their gradient. This energy can be used to power biochemical processes such as ATP synthesis.

☑ Concept Check 10.1

Protein synthesis (ie, translation) at the ribosome involves numerous energy-requiring processes. One round of elongation involves:

- "Charging" of a tRNA with a free amino acid (coupled to $ATP \xrightarrow{H_2O} AMP + PP_i$)
- Peptide bond condensation (coupled to $GTP \xrightarrow{H_2O} GDP + P_i$)
- Translocation of peptidyl-tRNA from the A site to the P site (coupled to $GTP \xrightarrow{H_2O} GDP + P_i$)

Given these steps, what is the total net cost (in units of ATP) for adding one free amino acid to a growing peptide?

Solution

Note: The print book appendix contains the short-form answer.

10.1.02 Energy of ATP Hydrolysis

Concept 10.1.01 explains that the phosphoanhydride linkages of NTPs, as well as some other bonds and linkages (see Table 10.2), are considered high energy. Hydrolysis of these linkages is highly exergonic (−ΔG) and often highly exothermic (−ΔH), meaning it produces heat. This concept discusses the chemistry of biochemically important high-energy linkages and their hydrolysis products.

Importantly, hydrolysis of a biochemical bond involves *more* than the simple breaking of a *single* covalent bond. As discussed in General Chemistry Lesson 3.5, breaking bonds always requires enthalpy input (ie, breaking bonds is endothermic [+ΔH]). However, hydrolysis *also* involves the *formation* of bonds to the atoms of the water molecule, which *releases* energy (ie, forming bonds is exothermic [−ΔH]).

For example, in ATP hydrolysis (Figure 10.7), a phosphoanhydride bond and an O–H bond break, but another O–H bond forms (in the released inorganic phosphate) and a P–OH bond forms on the β-phosphate of ADP. The energy of the bonds *formed* outweighs that of the bonds *broken*, yielding an overall exothermic process (ie, a negative ΔH) and contributing to the favorability of the reaction.

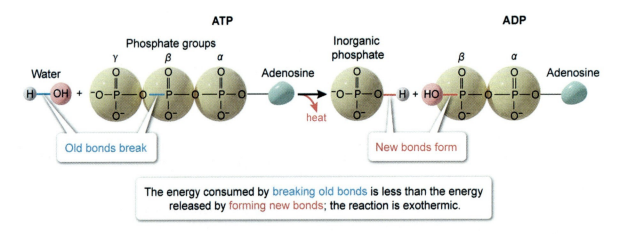

Figure 10.7 Exothermic hydrolysis of ATP.

Other Factors That Contribute to the Free Energy Change of Hydrolysis

However, not all exothermic reactions are exergonic, and not all hydrolysis reactions are as exergonic as ATP hydrolysis. Exergonic reactions are considered energy releasing because their products have *less* free energy than their reactants. The magnitude of this change can be influenced by two general factors: factors that contribute to the high energy of the reactants (ie, factors that make the reactants *unstable*) and factors that contribute to the low energy of the products (ie, factors that make the products *more stable*). Common contributing factors include charge-charge repulsion, ionization, resonance, and tautomerization.

Charge-charge repulsion is an important feature of nucleoside triphosphates that contributes to their high energy. Each of the phosphate groups of the nucleotide holds at least one negative charge at physiological pH. Because like charges repel each other, this close arrangement of negative charges is relatively unstable and therefore high energy.

Hydrolysis of the phosphoanhydride bond between the β- and γ-phosphate groups allows those two groups and their negative charges to separate. Consequently, hydrolysis of an NTP is an exergonic process because it partially relieves charge-charge repulsion, thereby releasing energy (Figure 10.8).

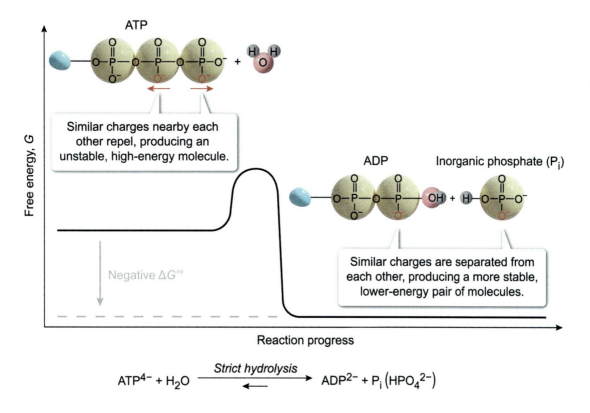

Figure 10.8 Hydrolysis of a phosphoanhydride bond allows nearby negative charges to separate, stabilizing the hydrolyzed products.

Ionization of a molecule *after* hydrolysis is another factor that increases the energy released. For example, Figure 10.9 shows that hydrolysis of ATP^{4-} yields ADP^{2-} and HPO_4^{2-} as its direct products. However, the predominant forms of ATP and ADP under physiological conditions have formal charges of −4 and −3, respectively.

After ATP hydrolysis, the β-phosphate of ADP is quickly deprotonated by water. This spontaneous deprotonation contributes to the overall negative ΔG of hydrolysis (Figure 10.9).

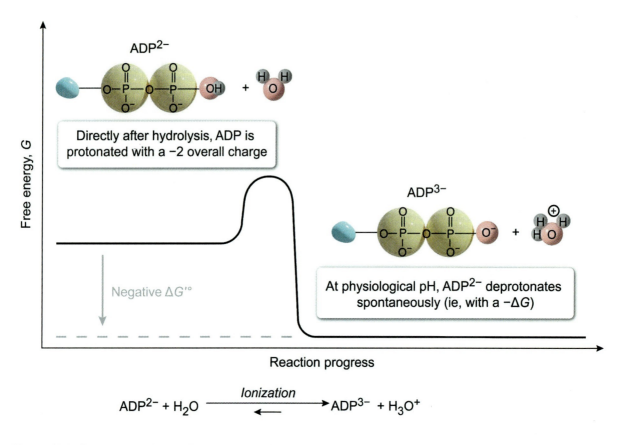

Figure 10.9 Spontaneous (ie, $-\Delta G$) ionization at physiological pH contributes to the exergonic nature of ATP hydrolysis.

Resonance (see Organic Chemistry Lesson 2.7) is another stabilizing factor that contributes to the negative ΔG of ATP hydrolysis. In general, the more resonance forms a molecule has, the more stable it is. Although ATP has several resonance forms *before* hydrolysis, additional resonance forms are possible *after* hydrolysis (see Figure 10.10).

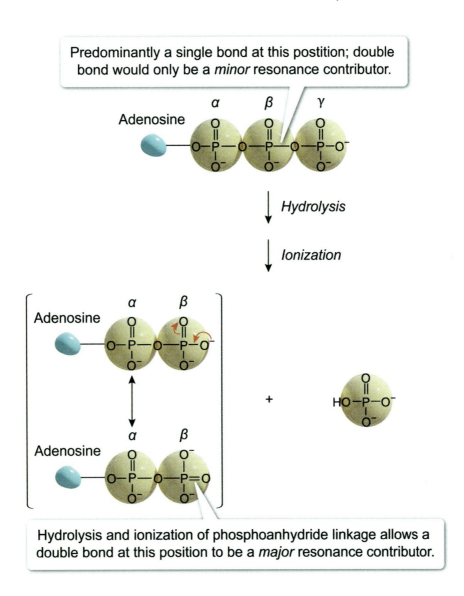

Figure 10.10 The β-phosphate of ADP is more resonance stabilized than the β-phosphate of ATP.

Other high energy molecules such as acyl phosphates (eg, 1,3-bisphosphoglycerate) and thioesters (eg, succinyl-CoA) are also stabilized by charge separation, ionization, and resonance.

Tautomerization is another stabilizing factor that can contribute to the exergonic nature of high-energy molecule hydrolysis. Tautomerization does not play a large role in stabilizing the products of ATP hydrolysis; however, it does explain why phosphoenolpyruvate (PEP) can serve as a high-energy molecule to power a substrate-level phosphorylation reaction at the end of glycolysis (see Lesson 11.1).

This is because the phosphate group of PEP traps the molecule in its enol form. However, keto tautomers are generally much more stable than enol tautomers (see Organic Chemistry Chapter 9.4). Consequently, once PEP's phosphate group is removed, the resulting enol form of pyruvate spontaneously tautomerizes to its keto form, stabilizing it (Figure 10.11).

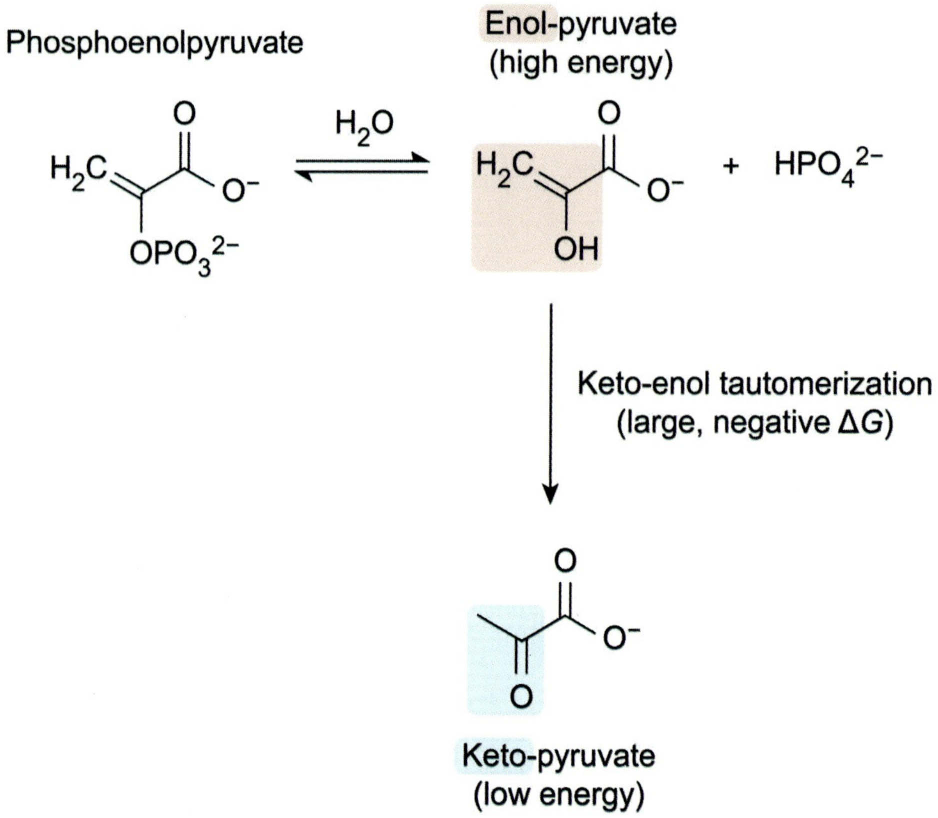

Figure 10.11 The spontaneous tautomerization of pyruvate to its keto form contributes to the negative ΔG of phosphoenolpyruvate hydrolysis.

10.1.03 Coupling of ATP Hydrolysis to Other Reactions

Concepts 10.1.01 and 10.1.02 discuss the energetics of ATP hydrolysis (and hydrolysis of other high-energy substrates) as being exergonic and therefore able to power endergonic processes. Importantly, an actual hydrolysis reaction (ie, breaking bonds by addition of water) *does not need to occur* to allow the potential energy stored within those molecules to be utilized. However, the *energy* of hydrolysis is relevant to the thermodynamics of reactions that use ATP (or its equivalents), even if hydrolysis does not occur in the reaction mechanism.

Concept 4.1.03 discusses that free energy *G* is a state function that depends only on a system's state and *not* on the path taken to achieve that state. Kinases "couple" ATP hydrolysis to substrate phosphorylation, which forms a phosphorylated substrate and ADP. However, for many kinases the actual mechanism is simply the *transfer* of a phosphoryl group—no hydrolysis actually occurs. Nevertheless, the thermodynamic energy of phosphoryl transfer is the same as the *sum* of the individual energies of ATP hydrolysis and phosphosubstrate formation (Figure 10.12).

Chapter 10: Catabolism and Anabolism

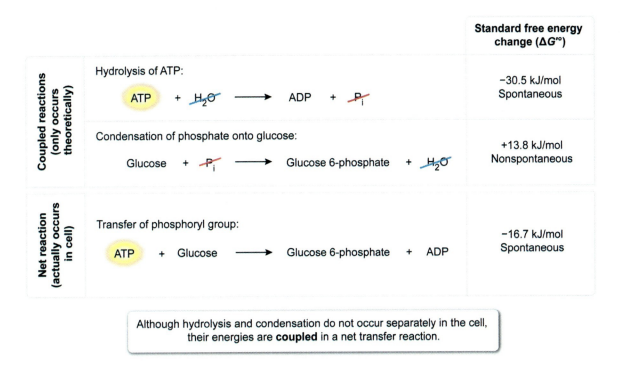

Figure 10.12 Because free energy is a state function, the free energy of hydrolysis can be considered to power a reaction, even if hydrolysis does not actually occur.

Some enzymes couple ATP hydrolysis to an endergonic reaction through a series of phosphate transfers either to the substrate or to the enzyme, with the phosphate group "activating" the molecule it is attached to. Temporary phosphorylation of the enzyme can trigger a **conformational change** that facilitates enzyme function. For example, phosphorylation of a translocase often induces a conformational change that moves ions across a membrane against their electrochemical gradient. Figure 10.13 shows the sodium-potassium pump, a translocase that couples ATP hydrolysis to ion movement by temporarily phosphorylating the translocase.

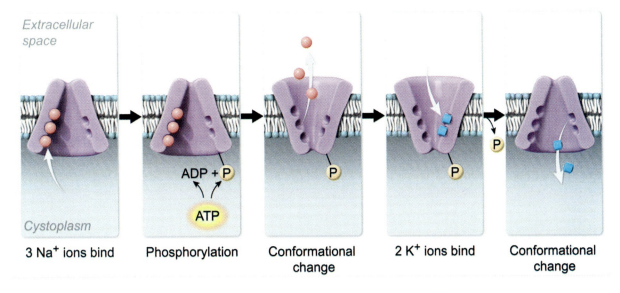

Figure 10.13 ATP can power reactions by temporarily phosphorylating an enzyme to trigger a conformational change.

ATP hydrolysis can also be coupled to endergonic processes through conformational changes induced by protein-ligand interactions. During muscle contraction, for example, ATP *is* hydrolyzed and the muscle filaments slide along each other based on their binding to ATP, ADP, or P_i (muscle contraction is discussed in more detail in Biology Chapter 17).

ATP hydrolysis can also be *indirectly* coupled to an endergonic process. Electrochemical gradients and membrane potentials are a source of potential energy for powering the cell and are often generated and maintained using ATP hydrolysis. Secondary active transporters, which couple the spontaneous movement of one ion *down* its gradient while cotransporting a second solute *against* its gradient, are the most common example of this (Figure 10.14).

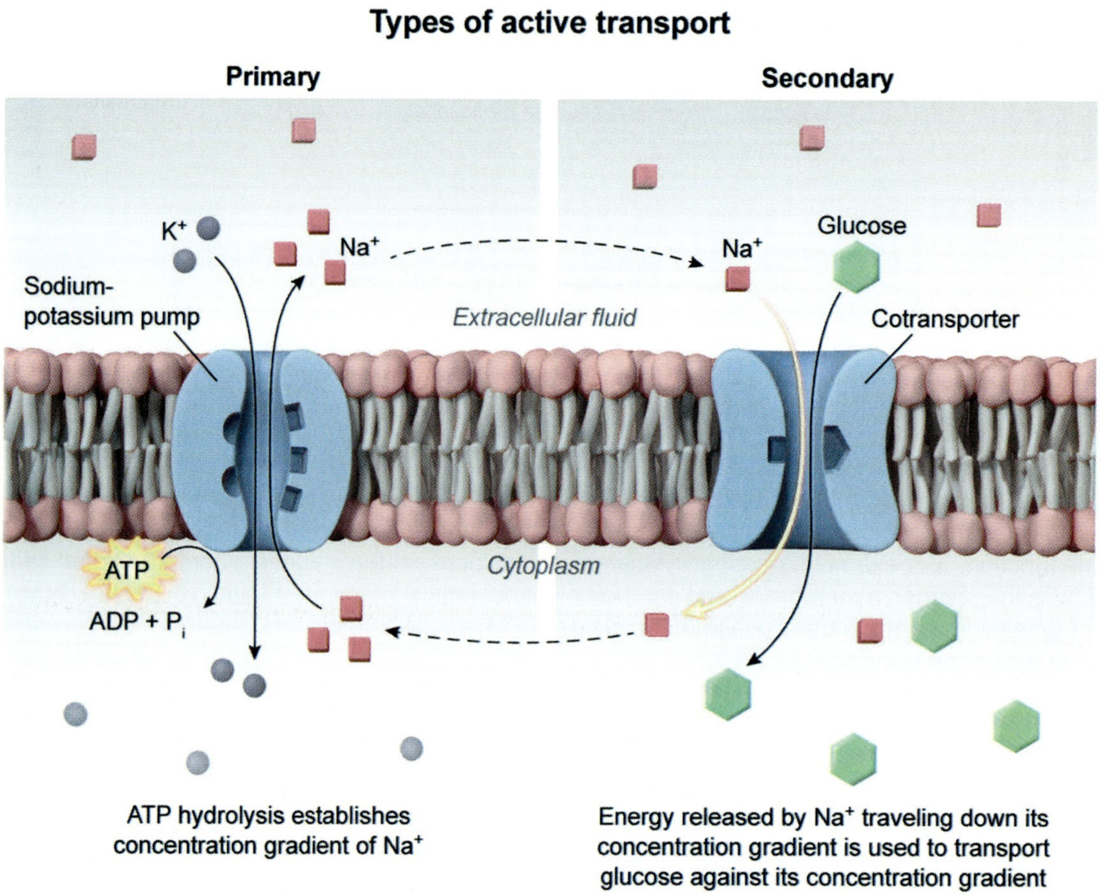

Types of active transport

Figure 10.14 ATP can indirectly power reactions by converting its chemical energy into the potential energy of an electrochemical gradient.

Lesson 10.2

Anabolism

10.2.01 Anabolism Uses Energy to Produce Structural and Storage Molecules

Lesson 10.1 describes catabolism as the breakdown of large biomolecules, which permits storage of potential energy in various forms, including the high-energy chemical bonds of ATP, the redox potential of NADH, and the membrane potential that results from electrochemical gradients of ions. In contrast, this lesson discusses **anabolism**, which *uses* those different forms of energy currency to *build* larger biomolecules (Figure 10.15).

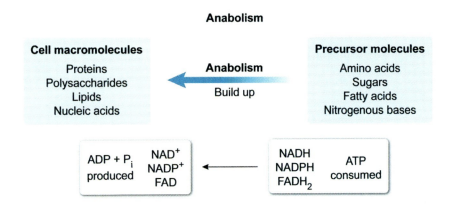

Figure 10.15 Anabolic metabolism involves the use of energy to build larger biomolecules from smaller precursors.

Concept 10.1.03 explains that the potential energy stored in energy currencies (particularly in ATP or ATP equivalents) can be coupled to endergonic processes. Anabolic applications of these coupled reactions can be used to build larger structural molecules or molecules that store excess energy.

For example, protein translation couples GTP hydrolysis to peptide bond formation. Polysaccharide and glycerophospholipid synthesis use an NTP to "activate" monomeric precursors before formation of the glycosidic bond or ester linkage, respectively. Nucleic acid synthesis transfers an incoming nucleotide onto a growing strand, breaking the phosphoanhydride linkages of the incoming NTP or dNTP in the process. In all these cases, energy from an NTP is liberated to build larger molecules with distinct structural and functional purposes.

Alternately, anabolism can be used to store excess energy. Organic fuels that have *not* been completely oxidized to CO_2 can instead be converted to glucose or fatty acids. Although energy is required to build these molecules, their complete oxidation later liberates *much more energy* (mostly in the form of ATP) than the amount of energy invested into building them. For example, the energy required to link glucose molecules to form glycogen is less than the energy released by converting glycogen back to glucose *and then continuing* to oxidize glucose to CO_2 through glycolysis and aerobic respiration.

By storing energy in molecules such as glycogen or triglycerides, organisms can conserve fuel obtained during a feast and use them during periods of fasting or famine when fuel is less abundant, as shown in Figure 10.16.

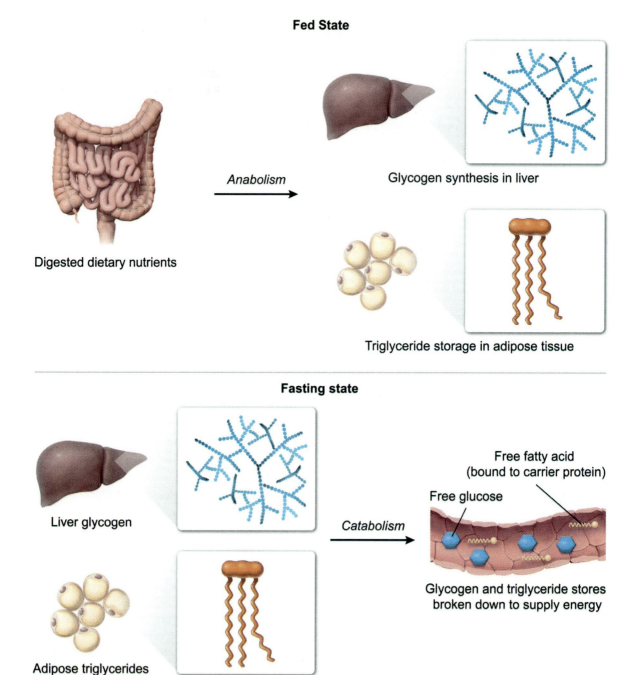

Figure 10.16 Glycogen and triglycerides are energy-storage molecules that are synthesized through anabolic metabolism in the fed state. They are broken down to supply energy for the body during fasting.

Just as catabolism can produce different forms of energy currency (see Concept 10.1.01), anabolism can utilize different forms of energy currency. NTPs are the most common form of energy currency; however, energy stored in the electrochemical potential of the proton gradient is used to synthesize ATP from ADP + P_i in the mitochondria. Redox cofactors such as NADPH are used in the synthesis of fatty acids and other lipids from acetyl-CoA.

Catabolic and Anabolic Processes Must Be Regulated to Prevent Wasting Energy

Energy capture from biomolecule catabolism is rarely 100% efficient. In addition, the anabolic building of molecules can waste energy. Thermodynamically, these inefficient reactions can be identified as those with large, negative ΔG values for the *net* reaction. A large, negative ΔG value means two things: 1) that at least some energy was *not* captured or stored in a usable form, and 2) that the reaction is biochemically irreversible.

For example, glycolysis (see Chapter 11) includes two ATP-producing steps. The first, catalyzed by phosphoglycerate kinase, has a small, near-equilibrium ΔG ($\Delta G \approx 0.09$ kJ/mol), indicating that the energy of its acyl phosphate hydrolysis is efficiently captured in ATP condensation. However, the second ATP-producing step, catalyzed by pyruvate kinase, has a ΔG of much *larger* magnitude ($\Delta G \approx -23.0$ kJ/mol). Although some of the energy of the enol phosphate hydrolysis is captured in ATP condensation, approximately 23 kilojoules per mole is *not* captured (Figure 10.17).

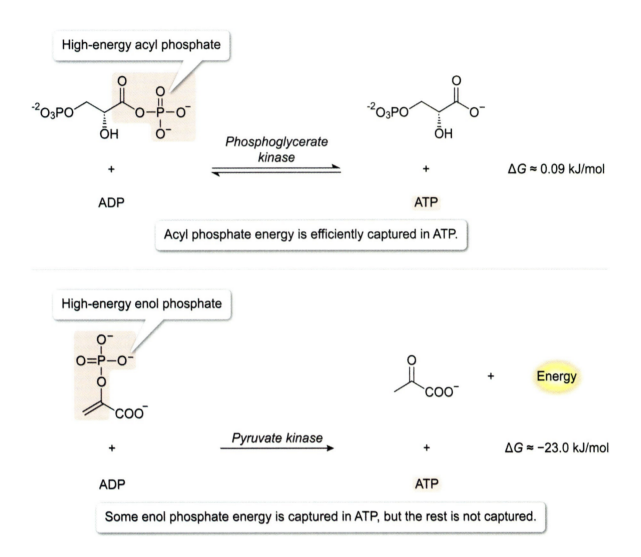

Figure 10.17 Exergonic reactions release energy.

The anabolic counterpart of glycolysis is gluconeogenesis (see Chapter 11). Gluconeogenesis uses many of the same enzymes as glycolysis, but because certain enzymes catalyze irreversible reactions, enzymes that catalyze **bypass reactions** must be used instead.

For example, gluconeogenesis cannot use pyruvate kinase to help build glucose because the pyruvate kinase reaction is irreversible. Instead, gluconeogenesis uses a pair of bypass reactions, catalyzed by pyruvate carboxylase and phosphoenolpyruvate carboxykinase (PEPCK), to convert pyruvate to phosphoenolpyruvate (Figure 10.18). Both bypass reactions are also highly exergonic (ie, large, negative ΔG).

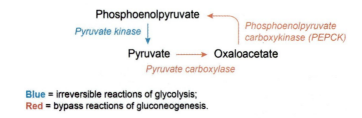

Figure 10.18 To circumvent the irreversible reaction catalyzed by pyruvate kinase, gluconeogenesis uses pyruvate carboxylase and phosphoenolpyruvate carboxykinase to catalyze bypass reactions.

Therefore, both the catabolic process of glycolysis and the anabolic process of gluconeogenesis involve reactions that result in uncaptured energy. This is indicated by the energetic yield and cost of each process, respectively. One molecule of glucose undergoing glycolysis yields 2 pyruvate, 2 NADH, and 2 net molecules of ATP. In contrast, *synthesis* of one molecule of glucose through gluconeogenesis consumes 2 pyruvate, 2 NADH, and *6* molecules of ATP.

Although the amount of pyruvate and NADH produced by glycolysis is the same as the amount consumed by gluconeogenesis, the difference in ATP amounts means energy is lost every time gluconeogenesis occurs. This also means that cells must **regulate** these processes to prevent a catabolic process and its anabolic counterpart from happening at the same time.

A cell undergoing both glycolysis and gluconeogenesis at the same time would cycle glucose between both processes with a net cost of 4 ATP per cycle and no gain (Figure 10.19). This unregulated cycling between a catabolic process and its anabolic counterpart is known as a **futile cycle** or a substrate cycle.

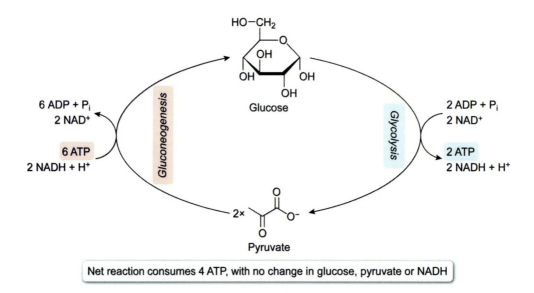

Figure 10.19 If glycolysis and gluconeogenesis took place simultaneously, the result would be a futile cycle that wastes energy without any material gain.

In the subsequent chapters of this unit, a common theme in the regulation of metabolic processes is the reciprocal regulation of catabolic and anabolic counterparts to prevent futile cycles. This regulation can be accomplished through allosteric means, with the metabolites themselves acting as allosteric effectors. Regulation can also be the result of external, hormonal signals. Keeping in mind the biological logic of preventing futile cycles can help to simplify the understanding of the complex interplay of metabolic regulation.

Concept Check 10.2

The hormone insulin is considered an anabolic hormone that stimulates fatty acid synthesis. Based on this, what is the expected effect (stimulation, inhibition, or no effect) of insulin on β-oxidation (ie, the catabolism of fatty acids)?

Solution

Note: The appendix contains the answer.

Lesson 11.1

Glycolysis and Fermentation

Introduction

Glucose is often considered the primary fuel source for cells. Not only can glucose be broken down to release energy, but its carbon skeleton can also be used to form other molecules. In addition, byproducts of glucose oxidation can be used as reducing agents to build other molecules or to protect the cell against oxidative damage. This lesson focuses on the catabolism of glucose monosaccharides and how glucose is broken down into two pyruvate molecules.

11.1.01 Glycolysis: Energy-Investment Phase

The process of glucose catabolism begins with **glycolysis**. Unlike some other catabolic pathways, which take place in the mitochondria, glycolysis is a metabolic pathway that occurs completely in the **cytosol** of cells. Glycolysis converts one molecule of glucose into two molecules of a three-carbon compound known as pyruvate while reducing two NAD^+ molecules to NADH and producing two *net* ATP. The net result of glycolysis is shown in Figure 11.1.

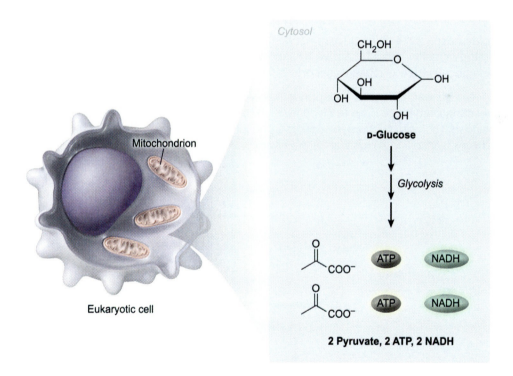

Figure 11.1 The net result of glycolysis is the conversion of one molecule of glucose into two molecules of pyruvate and the production of two molecules of ATP and two molecules of NADH.

Glycolysis occurs over a series of *10 enzyme-catalyzed reactions*, which can be divided into *two phases*. Although glycolysis ultimately liberates energy in the form of ATP and the reduced cofactor NADH, this can only occur *after* some energy is invested during the first phase of glycolysis. This concept describes the first five steps of glycolysis, known collectively as the **energy-investment phase**.

Step 1: Phosphorylation of Glucose by Hexokinase

The first step of glycolysis is the irreversible phosphorylation of glucose to form glucose 6-phosphate. Condensation of glucose and *inorganic phosphate* is thermodynamically unfavorable. Therefore, this phosphorylation reaction requires ATP as the phosphate donor and is catalyzed by the enzyme **hexokinase**. The substrates and products of the hexokinase reaction are shown in Figure 11.2.

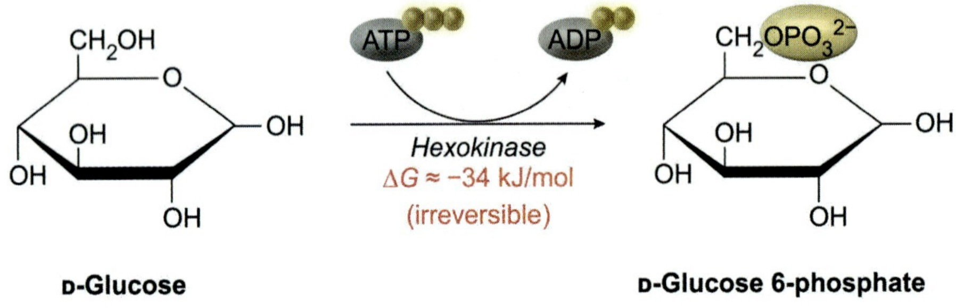

D-Glucose + ATP → D-Glucose 6-phosphate + ADP
Hexokinase, $\Delta G \approx -34$ kJ/mol (irreversible)

Figure 11.2 Hexokinase catalyzes the first reaction of glycolysis.

The phosphorylation of glucose by ATP (instead of inorganic phosphate) is highly *exergonic* and is therefore considered biochemically *irreversible*. Much of the free energy change can be explained by the energetic coupling to ATP hydrolysis. This is the *first step in which energy is invested* into glycolysis by converting ATP to ADP.

The irreversible phosphorylation of glucose serves to "trap" glucose in the cell. Glucose typically enters cells by facilitated diffusion through passive transporters. Phosphorylation prevents passive exit from the cell by diffusion because the glucose transporters cannot transport the phosphorylated substrate.

Consequently, hexokinase-mediated phosphorylation of glucose commits glucose to the *cell*; however, it does *not* necessarily commit glucose to *glycolysis*. This is because glucose 6-phosphate can enter pathways besides glycolysis, such as the pentose phosphate pathway (Lesson 11.3) or the glycogen synthesis pathway (Lesson 11.4), as shown in Figure 11.3.

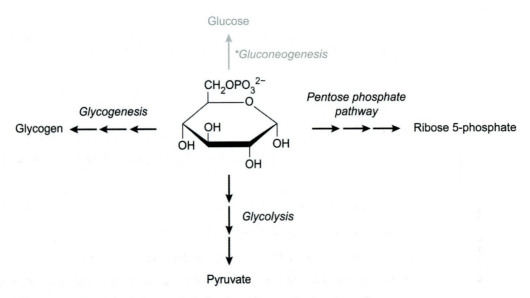

*Gluconeogenesis and glycolysis are not typically active at the same time in a given cell.

Figure 11.3 Several possible fates of glucose 6-phosphate.

The exergonic, irreversible nature of the hexokinase reaction means that hexokinase does experience some regulation (particularly in tissues that can also perform gluconeogenesis, see Lesson 11.2); however, because its products are not committed to glycolysis, it is not the *major* regulated step of glycolysis.

Step 2: Isomerization of Glucose 6-Phosphate by Phosphoglucose Isomerase

The second step of glycolysis is the conversion of glucose 6-phosphate (an aldohexose) into fructose 6-phosphate (a ketohexose, see Figure 11.4). No atoms are gained or lost during this conversion, so this reaction is therefore considered an **isomerization** reaction. The enzyme that catalyzes this reaction is called **phosphoglucose isomerase**. Alternate names for the enzyme include phosphohexose isomerase and glucose-6-phosphate isomerase (GPI).

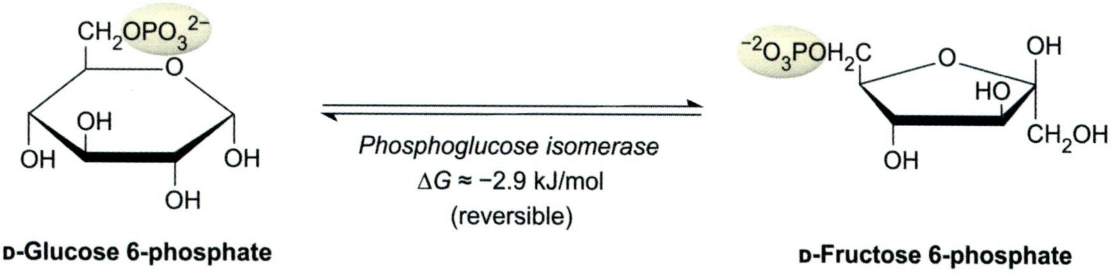

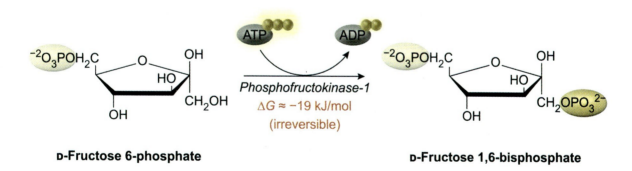

Figure 11.4 Phosphoglucose isomerase catalyzes the second step of glycolysis.

This isomerization reaction has a relatively small ΔG under physiological conditions and is therefore considered a reversible reaction.

Step 3: Phosphorylation of Fructose 6-Phosphate to Fructose 1,6-Bisphosphate by Phosphofructokinase-1

The third step of glycolysis is another phosphorylation step. In this case, the phosphate group is added to the carbon at position 1 of fructose 6-phosphate, resulting in the molecule fructose 1,6-bisphosphate (F1,6BP). The enzyme that catalyzes this reaction is called **phosphofructokinase-1 (PFK-1)**. This reaction is the *second energy-investment step* in the energy-investment phase of glycolysis. The PFK-1 reaction is shown in Figure 11.5.

Figure 11.5 Phosphofrucktokinase-1 catalyzes the third step of glycolysis.

Like step 1, the phosphoryl transfer reaction of step 3 is energetically coupled to ATP hydrolysis and is both *highly exergonic* and *irreversible*. Unlike step 1, the product of step 3 (ie, fructose 1,6-bisphosphate) *is committed to glycolysis*. In other words, in most cells the F1,6BP and the metabolites that appear after it have no alternate pathway to enter until they are converted to pyruvate at the end of glycolysis.

Because PFK-1 is the enzyme that catalyzes the *earliest committed step* of glycolysis, it is also the *most regulated enzyme* of glycolysis.

Step 4: Cleavage of Fructose 1,6-Bisphosphate by Aldolase

The fourth step of glycolysis results in the cleavage of fructose 1,6-bisphosphate. This converts the 6-carbon compound to *two 3-carbon compounds*: dihydroxyacetone phosphate (DHAP) and glyceraldehyde 3-phosphate (GAP or G3P). This reaction, catalyzed by the enzyme **aldolase**, is shown in Figure 11.6.

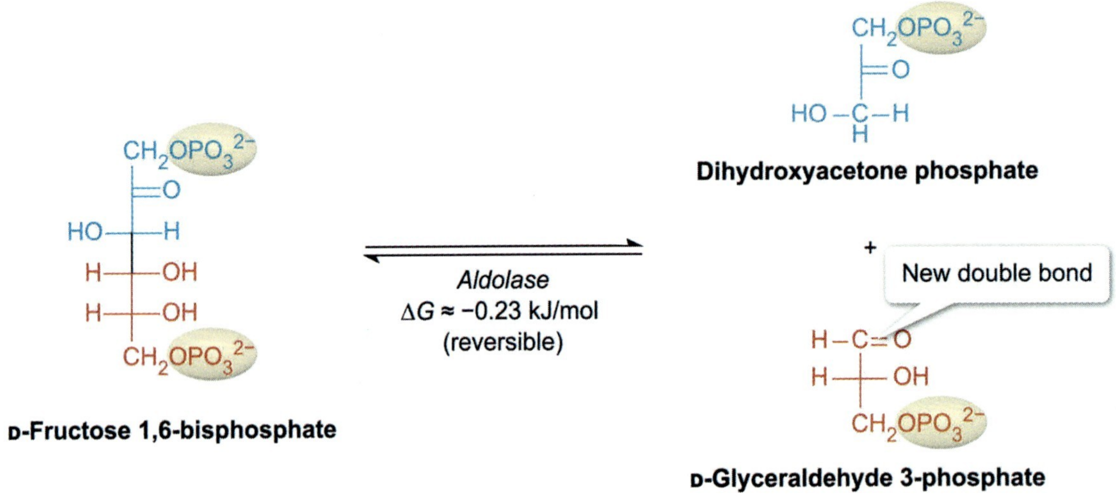

Figure 11.6 Aldolase catalyzes the fourth step of glycolysis.

Aldolase is an example of a lyase enzyme. Lyases split molecules by catalyzing *elimination* reactions and leave behind a double bond or a ring. Aldolase splits the linear form of F1,6BP and forms a double bond between carbon 3 and its hydroxyl oxygen, resulting in a new aldose phosphate.

Step 5: Isomerization of DHAP to GAP by Triose Phosphate Isomerase

The fifth step of glycolysis—and the final step of the energy-investment phase—is the isomerization of dihydroxyacetone phosphate to glyceraldehyde 3-phosphate. This reaction is shown in Figure 11.7.

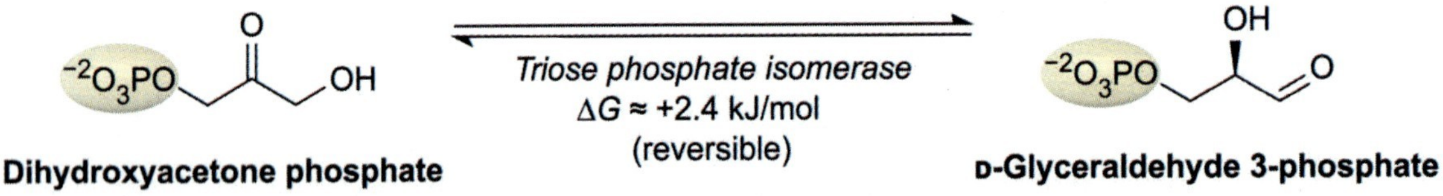

Figure 11.7 Triose phosphate isomerase catalyzes the fifth step of glycolysis.

Recall that the aldolase reaction produces one molecule each of DHAP and G3P. **Triose phosphate isomerase** only acts on DHAP, converting it into a second copy of G3P.

Summary of the Energy-Investment Phase of Glycolysis

The energy-investment phase of glycolysis consists of its first five steps and involves the conversion of one molecule of glucose into two molecules of glyceraldehyde 3-phosphate. During this phase, the energy of two ATP molecules is invested into the pathway at steps 1 and 3. These two steps are highly exergonic (and therefore irreversible). Step 3, catalyzed by phosphofructokinase-1, is the major point of regulation for glycolysis because it is the first *committed* step of glycolysis.

The two G3P molecules can then each enter the second phase of glycolysis separately, during which the invested energy pays off as even *more* energy is released from the fuel to produce ATP and NADH. The energy-investment phase of glycolysis is summarized in Figure 11.8.

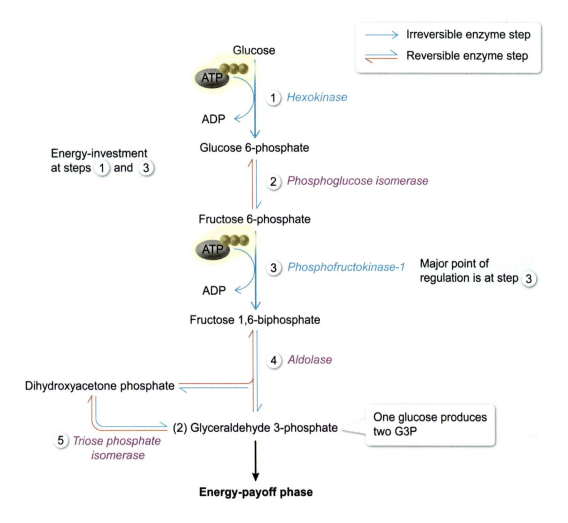

Figure 11.8 A summary of the energy-investment phase of glycolysis.

11.1.02 Glycolysis: Energy-Payoff Phase

The energy-payoff phase comprises the second half of glycolysis. During these five enzyme-catalyzed reactions, the glyceraldehyde 3-phosphate (G3P) molecules produced at the end of the energy-investment phase undergo a series of oxidation and phosphoryl transfer reactions that produce more than enough ATP and NADH to offset the energy invested in the first five steps.

This concept discusses each of the reactions of the energy-payoff phase as a *single* molecule of G3P is metabolized through it. Importantly, *two* molecules of G3P are produced at the end of the energy-investment phase. Therefore, the yield of each reaction should be multiplied by two to give the yield *per glucose* molecule metabolized.

Step 6: Oxidative Phosphorylation of G3P by Glyceraldehyde-3-Phosphate Dehydrogenase

The sixth step of glycolysis, and the first step of the energy-payoff phase, is the reversible oxidation of glyceraldehyde 3-phosphate. The oxygen of an inorganic phosphate group (P_i) replaces the aldehyde hydrogen, thereby oxidizing the aldehyde to the **acyl phosphate** group of the product **1,3-bisphosphoglycerate**. The hydrogen and its electrons, meanwhile, are transferred to NAD^+ in the form

Chapter 11: Carbohydrate Metabolism

of a hydride ion, reducing the NAD⁺ to **NADH**. This reaction is catalyzed by the enzyme **glyceraldehyde-3-phosphate dehydrogenase (GAPDH)** and is shown in Figure 11.9.

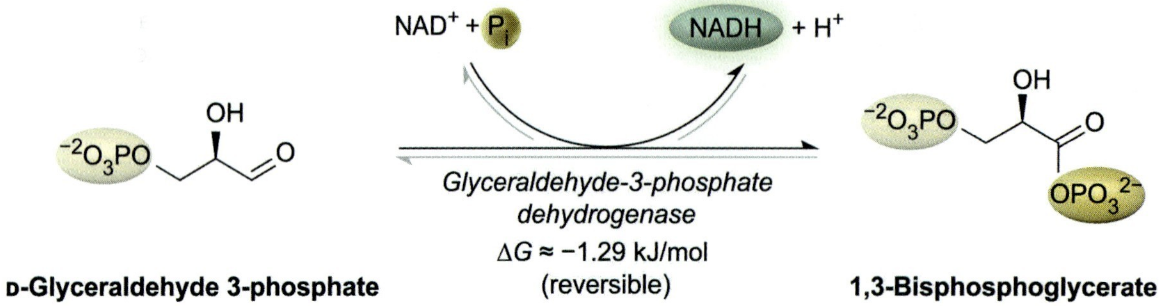

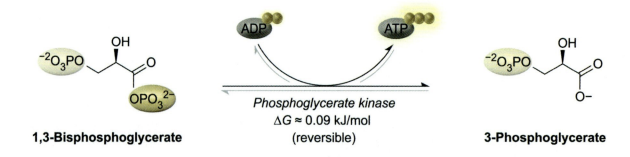

Figure 11.9 Glyceraldehyde-3-phosphate dehydrogenase catalyzes the sixth step of glycolysis.

NADH is the first unit of energy currency produced during glycolysis. Although NADH does not directly replace the ATP spent in the energy-investment phase, the high-energy electrons transferred to NADH can later enter the electron transport chain and contribute to the production of ATP, as discussed in Chapter 12.

Note that *inorganic* phosphate is used to phosphorylate G3P. Because each phosphate group will be used to produce ATP by substrate-level phosphorylation, this allows glycolysis to produce *more* ATP than was invested in the first phase of the pathway. In other words, the phosphate groups eventually transferred to ATP come from *both* the ATP used in the investment phase *and* the inorganic phosphate added by GAPDH.

Step 7: ATP Production through Substrate-Level Phosphorylation by Phosphoglycerate Kinase

The seventh step produces the first molecule of ATP in the pathway. Acyl phosphate hydrolysis is coupled to ATP synthesis through a phosphoryl *transfer* reaction catalyzed by **phosphoglycerate kinase**. The resulting products are **3-phosphoglycerate** and ATP. This reaction is shown in Figure 11.10.

Figure 11.10 Phosphoglycerate kinase catalyzes the seventh step of glycolysis.

Note that the enzyme is named for the *reverse* reaction—the enzyme-catalyzed transfer of a phosphoryl group *from* ATP onto 3-phosphoglycerate. Nevertheless, enzymes catalyze both directions of any given reaction (see Concept 4.1.04); therefore, the name "phosphoglycerate kinase" *also* accurately describes the enzyme that transfers a phosphoryl group from 1,3-bisphosphoglycerate *onto* ADP.

The phosphoglycerate kinase reaction produces one ATP per G3P molecule that enters the energy-payoff phase; therefore, it produces *two* ATP molecules per glucose that enters glycolysis. Because two ATP were invested in the energy-investment phase, by this point the ATP investment has been fully recouped (ie, the *net* amount of ATP produced and consumed at this point is zero). All ATP made by future steps contribute to the *net* gain of energy currency molecules.

Step 8: Functional Group Movement by Phosphoglycerate Mutase

The eighth step of glycolysis is a reversible isomerization reaction. Specifically, the phosphate group on carbon 3 moves to carbon 2, forming **2-phosphoglycerate**. This reaction is catalyzed by the enzyme **phosphoglycerate mutase**, as shown in Figure 11.11.

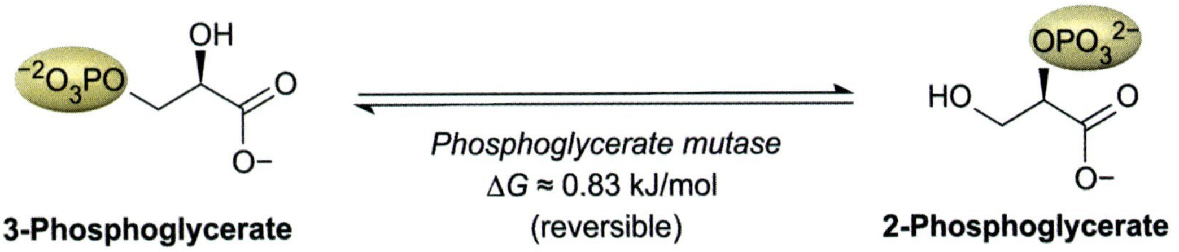

Figure 11.11 Phosphoglycerate mutase catalyzes the eighth step of glycolysis.

Step 9: Dehydration of 2-Phosphoglycerate to Form Phosphoenolpyruvate by Enolase

The ninth step of glycolysis is a reversible dehydration reaction. The hydroxyl on carbon 3 and a hydrogen on carbon 2 are eliminated, yielding water and the molecule **phosphoenolpyruvate**. This reaction is catalyzed by the lyase enzyme **enolase**, as shown in Figure 11.12.

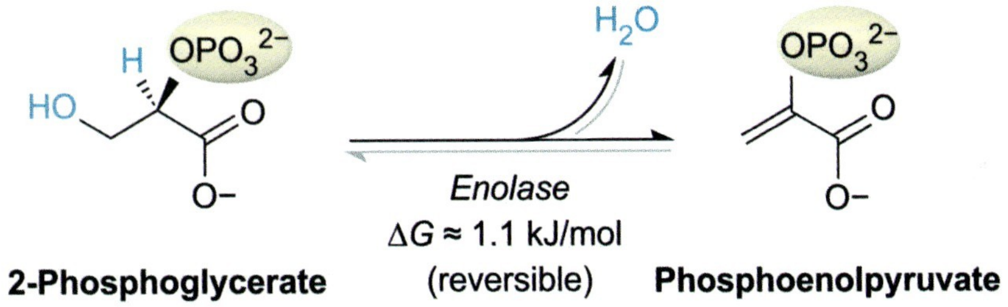

Figure 11.12 Enolase catalyzes the ninth step of glycolysis.

Phosphoenolpyruvate is a high-energy *enol phosphate*. The phosphate group prevents the thermodynamically favorable tautomerization of the enol to the keto form, making transfer of the phosphate group favorable.

Step 10: ATP Production through Substrate-Level Phosphorylation by Pyruvate Kinase

The final step of glycolysis is catalyzed by the enzyme **pyruvate kinase**, which catalyzes the irreversible transfer of the phosphate from phosphoenolpyruvate (PEP) to ADP. This forms ATP by substrate-level phosphorylation. This reaction is shown in Figure 11.13.

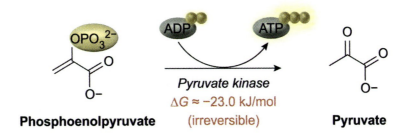

Figure 11.13 Pyruvate kinase catalyzes the final step of glycolysis.

Like phosphoglycerate kinase, pyruvate kinase is named for the *reverse* reaction. Unlike phosphoglycerate kinase, however, the large free energy change ($\Delta G \approx -23.0$ kJ/mol) means that pyruvate kinase facilitates phosphate transfer in *only one direction* under physiological conditions. In other words, pyruvate kinase is only active in glycolysis; it is *not* active in gluconeogenesis.

Pyruvate kinase produces one ATP molecule per PEP molecule. Therefore, it produces *two* ATP molecules per glucose that enters glycolysis. Because the ATP investment was already recouped by step 7 (phosphoglycerate kinase), the two ATP produced by pyruvate kinase represent the *net gain* of ATP by the end of glycolysis.

Summary of Glycolysis

The energy-payoff phase begins at step 6 of glycolysis and uses the G3P molecules produced at the end of the energy-investment phase (ie, steps 1–5). For each G3P molecule that enters, the energy-payoff phase produces one NADH molecule and two ATP molecules, and converts the carbon skeleton from G3P to pyruvate. Therefore, per glucose molecule, the energy-payoff phase produces *two* NADH, *four* ATP, and *two* pyruvate, as shown in Figure 11.14.

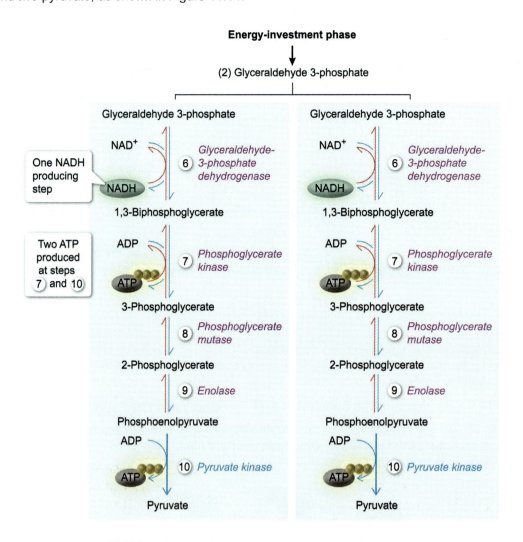

Figure 11.14 A summary of the energy-payoff phase of glycolysis. This phase occurs twice per glucose because two molecules of glyceraldehyde 3-phosphate are produced during the energy-investment phase.

However, when considering the *net* gain of glycolysis as a whole, the ATP consumed during the energy-investment phase must also be taken into account. These two ATP consumed are subtracted from the four ATP produced, resulting in the *net* production of only *two ATP per glucose*. The complete net outcome of glycolysis is summarized in Figure 11.15.

Energy-investment phase

Glucose + 2 ATP → 2 G3P + 2 ADP + 2 P$_i$

+

Energy-payoff phase

2 G3P + 2 NAD$^+$ + 4 ADP + 4 P$_i$ → 2 Pyruvate + 2 NADH + 2 H$^+$ + 4 ATP

Net effect of glycolysis

Glucose + 2 NAD$^+$ + 2 ADP + 2 P$_i$ → 2 Pyruvate + 2 NADH + 2 H$^+$ + 2 ATP

G3P = Glyceraldehyde 3-phosphate.

Figure 11.15 The net outcome of glycolysis is the production of two pyruvate, two ATP, and two NADH.

Of the 10 enzyme-catalyzed reactions that compose glycolysis, *three* are highly exergonic and therefore irreversible: step 1, step 3, and step 10. Of those three, phosphofructokinase-1 (PFK-1, step 3) catalyzes the earliest committed step of glycolysis and is the most tightly regulated.

☑ Concept Check 11.1

Three molecules of free glucose are catabolized by glycolysis.

1) How many ATP molecules are produced by the enzymes phosphoglycerate kinase and pyruvate kinase?

2) How many *net* ATP molecules are produced by the entire glycolytic pathway?

Solution

Note: The appendix contains the answer.

11.1.03 Alternate Entries into Glycolysis

In addition to glucose, other metabolites can also be processed by the glycolytic pathway. When this occurs, the metabolites can enter or exit at different points of the pathway. In other words, not all enzymes need to be used. Furthermore, additional enzymes are typically needed to prepare for entry into glycolysis. Because they use some, but not all, enzymes of glycolysis, these alternate substrates have *yields* or *regulatory mechanisms* that *may differ* from the classic pathway.

The catabolism of nonglucose monosaccharides provides several examples of metabolites with alternate entry points into glycolysis. The details of nonglucose metabolism are not a high priority for memorization for the exam, but they provide a deeper understanding of the interrelatedness of biochemical pathways and the effect on net yield and regulation of alternate entry points.

In addition to glucose, fructose and galactose are the main monosaccharides obtained from the diet. Mannose is another important monosaccharide often incorporated into glycoproteins or glycolipids. These monosaccharides have six carbons each (ie, they are hexoses, Figure 11.16).

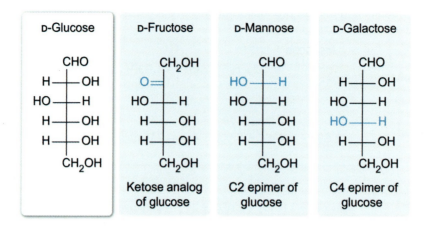

Figure 11.16 Various hexoses can be processed through the glycolytic pathway.

Mannose Catabolism

Hexokinase, the enzyme that catalyzes the first step of glycolysis, has broad substrate specificity and can also phosphorylate mannose and fructose on carbon 6. The product of mannose phosphorylation by hexokinase is mannose 6-phosphate.

In contrast, the second enzyme of glycolysis, phosphoglucose isomerase, does *not* act on mannose 6-phosphate. Instead, the enzyme **phosphomannose isomerase** catalyzes the analogous reaction that converts mannose 6-phosphate to a ketose. Despite acting on different starting substrates, phosphoglucose isomerase and phosphomannose isomerase produce the same product: fructose 6-phosphate.

This is because glucose 6-phosphate and mannose 6-phosphate are C2 epimers of each other. Isomerization of an aldose to a ketose changes only the stereochemistry of carbons 1 and 2, making both carbons achiral. Carbons 3–6 all have the same stereochemistry in D-glucose, D-mannose, and D-fructose (and their derivatives), as shown in Figure 11.17.

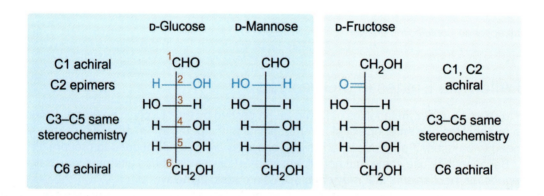

Figure 11.17 Glucose, mannose, and fructose differ only at positions C1 and C2.

After conversion to fructose 6-phosphate, mannose catabolism uses the same enzymes as the traditional glycolysis pathway, as shown in Figure 11.18.

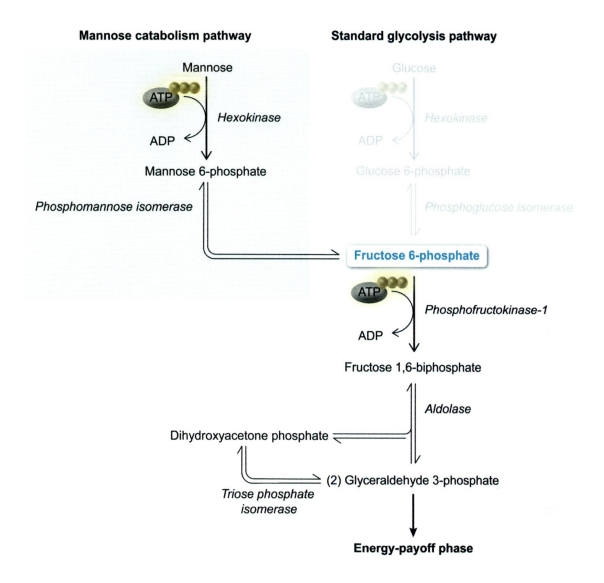

Figure 11.18 Mannose catabolism enters the traditional glycolysis pathway after conversion to fructose 6-phosphate.

Fructose Catabolism

As mentioned previously, hexokinase can act on fructose, converting it to fructose 6-phosphate directly. Fructose 6-phosphate can then proceed through glycolysis without the need for an additional isomerization step. However, glucose competes with fructose for the hexokinase enzyme, so processing of fructose by hexokinase happens only in limited amounts.

The liver metabolizes the majority of dietary fructose using the enzyme fructokinase. Unlike hexokinase, fructokinase adds a phosphate group to carbon *1*, forming fructose 1-phosphate. Fructose 1-phosphate can be acted upon by aldolase, which splits the molecule to form dihydroxyacetone phosphate (DHAP) and an unphosphorylated glyceraldehyde. The enzyme triokinase then phosphorylates glyceraldehyde to glyceraldehyde 3-phosphate (G3P), and both DHAP and G3P enter glycolysis and proceed along the classic pathway (Figure 11.19).

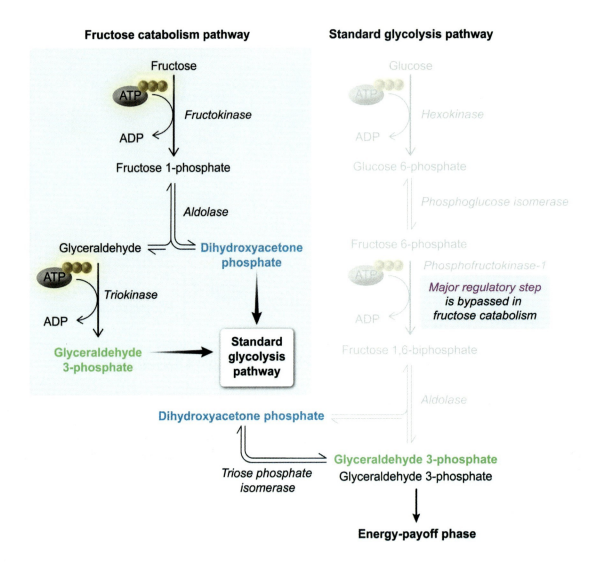

Figure 11.19 Fructose is converted into dihydroxyacetone phosphate and glyceraldehyde 3-phosphate, both of which can then proceed down the typical glycolysis pathway into the energy-payoff phase.

This fructose catabolism pathway allows fructose to be metabolized into two molecules of G3P, but it *bypasses the critical regulatory enzyme phosphofructokinase-1 (PFK-1)*. Consequently, excessive fructose intake can lead to unregulated hepatic glycolysis and is associated with nonalcoholic fatty liver disease (NAFLD).

Galactose Catabolism

Galactose is the C4 epimer of glucose and one of the monomers that compose the disaccharide lactose, which is found in milk products. Like fructose, galactose has its own kinase, galactokinase, which phosphorylates the sugar on carbon 1. The resultant galactose 1-phosphate then participates in a transfer reaction with the molecule UDP-glucose. This reaction, catalyzed by galactose-1-phosphate uridylyltransferase (GALT), transfers the UDP group from glucose to galactose, producing UDP-galactose and glucose 1-phosphate (Figure 11.20).

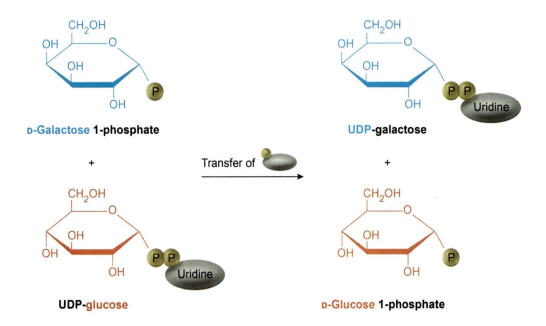

Figure 11.20 In galactose catabolism, galactose 1-phosphate receives a UDP group from UDP-glucose, producing UDP-galactose and glucose 1-phosphate.

The enzyme phosphoglucomutase moves the glucose phosphate group from position 1 to position 6, and the newly produced glucose 6-phosphate molecule can then enter glycolysis as normal. Meanwhile, the UDP-galactose molecule is used to regenerate the consumed UDP-glucose. The enzyme UDP-galactose 4-epimerase (GALE) accomplishes this by inverting the stereochemistry of carbon 4, as shown in Figure 11.21.

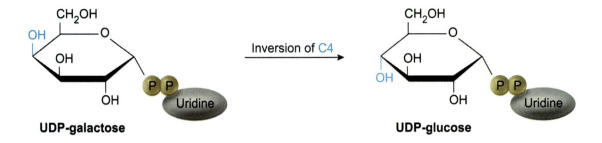

Figure 11.21 UDP-galactose is converted to UDP-glucose by inversion of the stereochemistry of carbon 4.

The regenerated UDP-glucose can then react with more galactose 1-phosphate to be catabolized further, or it can be used in other metabolic processes such as glycogenesis (see Lesson 11.4). Galactose catabolism is summarized in Figure 11.22.

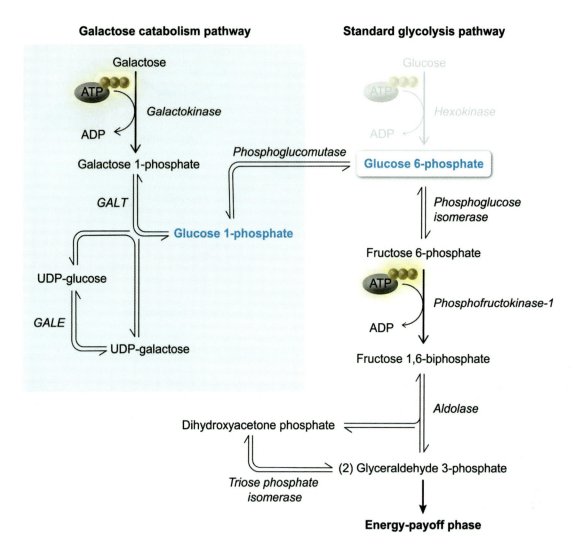

Figure 11.22 Galactose catabolism.

11.1.04 Fermentation

The net reaction of glycolysis is:

$$\text{Glucose} + 2\ \text{NAD}^+ + 2\ \text{ADP} + 2\ P_i \rightarrow 2\ \text{Pyruvate} + 2\ \text{NADH} + 2\ \text{H}^+ + 2\ \text{ATP}$$

Under aerobic (ie, abundant oxygen) conditions, pyruvate and NADH can proceed with aerobic respiration in the mitochondria (see Chapter 12), which converts the pyruvate to CO_2 and the NADH molecules back into NAD^+. Cells and tissues that either do not possess mitochondria (eg, red blood cells) or have limited oxygen availability (eg, muscles under vigorous exercise conditions) are unable to perform aerobic respiration. Without an alternative to aerobic respiration, pyruvate and NADH accumulate, and NAD^+ levels deplete, eventually preventing glycolysis from continuing.

Fermentation is an alternate pathway to respiration through which cells can handle pyruvate molecules and regenerate NAD^+ from NADH. Mammalian cells in particular use a type of fermentation called **lactic acid fermentation**. In this process, NADH is converted to NAD^+ by the enzyme **lactate dehydrogenase**. To power this oxidation reaction, the enzyme also reduces the pyruvate molecule to a lactate molecule. Like glycolysis, this reaction occurs completely in the **cytosol** and therefore is available even to cells without mitochondria. The lactic acid fermentation reaction is shown in Figure 11.23.

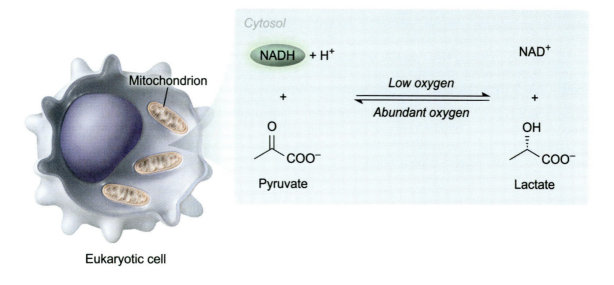

Figure 11.23 The lactic acid fermentation reaction, catalyzed by lactate dehydrogenase.

The lactic acid fermentation reaction is **reversible** and can proceed in either direction, depending on the relative concentrations of the reactants and products (ie, the reaction quotient Q). In cells that cannot perform respiration, NADH levels are high and the reaction tends toward reduction of pyruvate to lactate. In tissues that *can* perform respiration, NADH levels are relatively low because NADH is consumed by respiration pathways, and NAD^+ levels are higher. This shifts the equilibrium toward oxidation of lactate back into pyruvate.

Multicellular organisms such as mammals can take advantage of this reversibility by exporting lactate from anaerobic cells to be processed by well-oxygenated cells. The well-oxygenated cells can then process lactate back to pyruvate due to the abundance of oxygen (and therefore NAD^+) available. This cycle is elaborated further as part of the **Cori cycle**, discussed in Lesson 11.2.

Alcoholic Fermentation

In addition to lactic acid fermentation, many microbial organisms are capable of another type of fermentation known as **alcoholic fermentation**. In alcoholic fermentation, pyruvate is first decarboxylated to form carbon dioxide (CO_2) and acetaldehyde. Reduction of acetaldehyde is then coupled to the oxidation of NADH, which regenerates NAD^+ and produces ethanol. This pathway involves two enzymes: pyruvate decarboxylase catalyzes the first, irreversible step and alcohol dehydrogenase catalyzes the second, reversible step. The alcoholic fermentation pathway is shown in Figure 11.24.

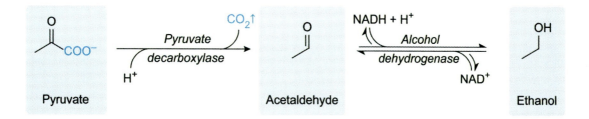

Figure 11.24 Alcoholic fermentation consists of two steps: decarboxylation and oxidation-reduction.

Microbial organisms use alcoholic fermentation to regenerate NAD^+ if they cannot use respiration to do so. Unlike the lactate produced during lactic acid fermentation, the CO_2 and ethanol produced during alcoholic fermentation are both volatile (ie, they easily evaporate into the gas phase); therefore, alcoholic fermentation also allows for elimination of the excess carbons from pyruvate.

Humans and other higher vertebrates do not perform alcoholic fermentation because they lack the pyruvate decarboxylase enzyme needed for the first step of the pathway. However, humans do possess the alcohol dehydrogenase enzyme, which operates in the reverse direction (ie, oxidizing ethanol to acetaldehyde and reducing NAD^+ to NADH) following ethanol consumption. Acetaldehyde can be further oxidized to acetate, which can then be converted to acetyl coenzyme A. This molecule, often abbreviated as acetyl-CoA, is metabolized by the citric acid cycle (Lesson 12.1).

Acetaldehyde is highly toxic and responsible for many of the unpleasant effects of alcohol consumption. Consequently, low doses of inhibitors that prevent conversion of acetaldehyde to acetate are sometimes used to treat alcoholism; the painful effects increase as acetaldehyde builds up, which provides a disincentive for future alcohol consumption.

Lesson 11.2
Gluconeogenesis

Introduction

Lesson 11.1 discusses the process of glycolysis, which catabolizes (ie, breaks down) glucose into two pyruvate molecules while liberating energy in the form of two net ATP and two NADH. This lesson begins with a discussion of gluconeogenesis, the process by which cells *build* glucose from two pyruvate molecules.

In addition to the two pyruvate molecules, gluconeogenesis consumes two NADH and *six* ATP equivalents, meaning it costs more energy to *make* glucose than is released by glycolysis. Although gluconeogenesis is useful to provide glucose during times of fasting or to recycle glycolysis end products, gluconeogenesis and glycolysis must be regulated to prevent the wasting of energy. In addition to the process of gluconeogenesis, this lesson also discusses the regulation of glucose metabolism.

11.2.01 Gluconeogenesis

Gluconeogenesis builds up glucose from two pyruvate molecules. Unlike glycolysis, which can occur in the cytosol of *all* cells, gluconeogenesis in mammals mainly occurs in the **liver**, although a few other cell types can support gluconeogenesis in limited amounts. This compartmentalization of processes reflects the liver's role in providing fuel, such as glucose, for the rest of the body during times of fasting.

Although glycolysis and gluconeogenesis are highly related and share several enzymes, gluconeogenesis is *not* simply glycolysis in reverse. Glycolysis uses three **biochemically irreversible** enzyme-catalyzed reactions, namely the reactions catalyzed by hexokinase, phosphofructokinase-1, and pyruvate kinase (see Concept 11.1.01 and Concept 11.1.02). Although glycolysis uses these enzymes in the *catabolism* of glucose, gluconeogenesis *cannot* use the same enzymes in the anabolism of two pyruvate molecules back to glucose.

Instead, for each of these three irreversible steps of glycolysis, the process of gluconeogenesis must use one or more **bypass reactions**, which are themselves irreversible and are each catalyzed by unique enzymes *not used in glycolysis*. Figure 11.25 depicts an overview of glycolysis with its irreversible enzymes and the bypass reactions used in gluconeogenesis. The other seven enzymes of glycolysis, which catalyze *reversible* reactions, are shared with gluconeogenesis.

Chapter 11: Carbohydrate Metabolism

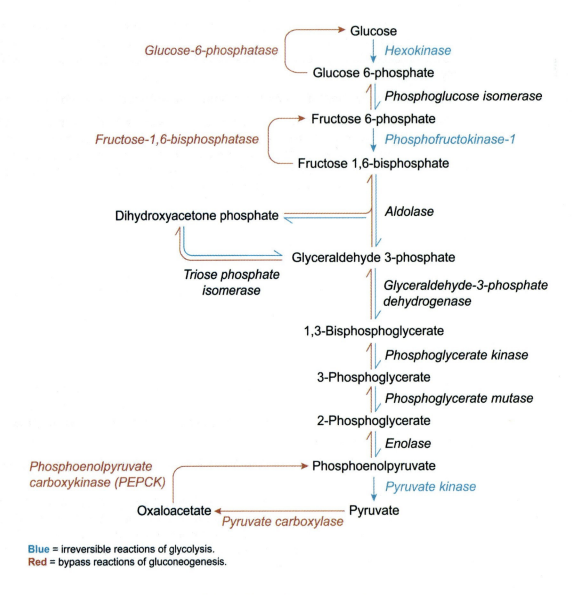

Figure 11.25 The irreversible enzymes of glycolysis and gluconeogenesis.

The net reaction for gluconeogenesis is

$$2\ \text{Pyruvate} + 6\ \text{ATP} + 2\ \text{NADH} + 2\ \text{H}^+ \rightarrow \text{Glucose} + 6\ \text{ADP} + 6\ \text{P}_i + 2\ \text{NAD}^+$$

Gluconeogenesis: The First Set of Bypass Reactions

Because glycolysis ends with the production of pyruvate, this discussion of gluconeogenesis begins with pyruvate. However, it is worth noting that other starting points are possible and several of these are discussed in Concept 11.2.02.

Pyruvate is produced by the final enzyme of glycolysis, pyruvate kinase, which transfers a phosphate from phosphoenolpyruvate to ADP, thereby forming pyruvate and ATP. Because the pyruvate kinase reaction is biochemically irreversible, pyruvate kinase cannot convert pyruvate back to phosphoenolpyruvate for gluconeogenesis. Instead, this single reaction of glycolysis requires *two bypass reactions*—and therefore *two enzymes*—to convert pyruvate to phosphoenolpyruvate in gluconeogenesis.

The first reaction is the carboxylation of pyruvate to form oxaloacetate, catalyzed by the enzyme **pyruvate carboxylase**. The carboxyl group comes from a bicarbonate ion, which itself comes from the reaction of carbon dioxide with water. The pyruvate carboxylase reaction *consumes one molecule of ATP per molecule of pyruvate.*

Unlike glycolysis, in which every enzyme is cytosolic, pyruvate carboxylase is a mitochondrial enzyme. Therefore, the pyruvate molecule must first be transported to the mitochondria before gluconeogenesis can occur. The pyruvate carboxylase reaction is shown in Figure 11.26.

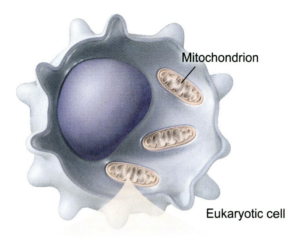

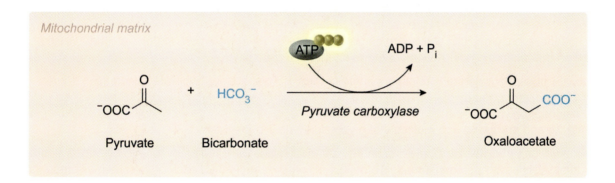

Figure 11.26 The pyruvate carboxylase reaction, the first of two reactions needed to bypass the pyruvate kinase reaction.

The second bypass reaction is catalyzed by the enzyme **phosphoenolpyruvate carboxykinase (PEPCK)**. In gluconeogenesis, the PEPCK reaction accomplishes two sub-reactions: first, the oxaloacetate is decarboxylated to form an enolate; second, a phosphate is transferred from GTP to the substrate to form phosphoenolpyruvate. The same carboxyl group that was *added* by pyruvate carboxylase is *removed* by PEPCK.

Humans possess both a mitochondrial and a cytosolic isoform of PEPCK. Therefore, the PEPCK reaction can occur in either compartment, depending on the levels of other metabolites and other metabolic needs (see Concept 11.2.02). The PEPCK reaction is shown in Figure 11.27.

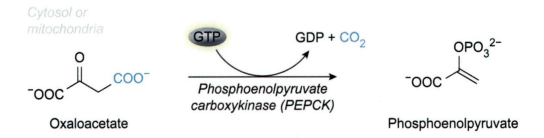

Figure 11.27 The phosphoenolpyruvate carboxykinase (PEPCK) reaction.

Importantly, although the pyruvate kinase reaction produces *one* ATP molecule per pyruvate formed, the bypass reactions together *consume two ATP equivalents* (one ATP and one GTP) to convert one pyruvate to phosphoenolpyruvate.

If both glycolysis and gluconeogenesis were active at the same time, pyruvate and phosphoenolpyruvate would cycle between the two processes, resulting in a futile cycle. Because the net effect of a futile cycle is simply to hydrolyze ATP (ie, no other metabolite levels are changed), futile cycles are said to "waste" energy. To prevent wasting of ATP, the irreversible enzymes of glycolysis and gluconeogenesis are tightly regulated, as discussed in Concept 11.2.04.

Gluconeogenesis: Most Enzymes Are Shared with Glycolysis

After phosphoenolpyruvate is synthesized by PEPCK, it is trafficked to the cytosol if not already there. At this point, the next six reactions are catalyzed by the *same enzymes used in glycolysis.* This can occur because these enzymes all catalyze reversible reactions and therefore can operate in either direction based on the relative concentrations of the reactants and products.

Notably, 3-phosphoglycerate is phosphorylated to form 1,3-bisphosphoglycerate during this process. This reaction, catalyzed by phosphoglycerate kinase, consumes *one molecule of ATP* per reaction. 1,3-Bisphosphoglycerate is then reduced by glyceraldehyde-3-phosphate dehydrogenase, *consuming one NADH* and producing one molecule of glyceraldehyde 3-phosphate (G3P).

At this point one pyruvate molecule has been converted to one G3P molecule, and 3 ATP equivalents and one NADH have been consumed. To build a full glucose molecule, each of the previous steps must occur twice, consuming a total of *6 ATP equivalents*, *2 NADH*, and 2 pyruvates to make two G3P molecules.

One of the G3P molecules is isomerized to dihydroxyacetone phosphate (DHAP) by triose phosphate isomerase. DHAP and the remaining molecule of G3P then combine to make fructose 1,6-bisphosphate in a reaction catalyzed by aldolase.

Gluconeogenesis: The Final Pair of Bypass Reactions

In glycolysis, fructose 1,6-bisphosphate is synthesized in an irreversible reaction by the enzyme phosphofructokinase-1. In gluconeogenesis, the enzyme **fructose-1,6-bisphosphatase (FBPase-1)** catalyzes the bypass reaction. Whereas the glycolysis reaction is a phosphate transfer reaction that consumes an ATP molecule, the bypass reaction in gluconeogenesis is a **hydrolysis** reaction that *does not involve ATP* at all. In this reaction, the phosphate on carbon 1 is hydrolyzed, producing fructose 6-phosphate and inorganic phosphate (P_i, Figure 11.28).

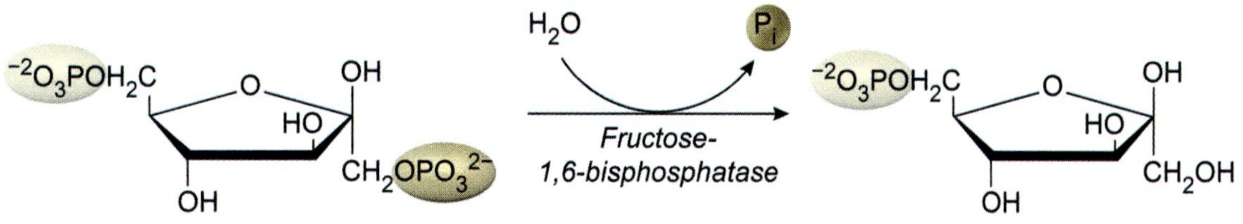

D-Fructose 1,6-bisphosphate → (Fructose-1,6-bisphosphatase, +H₂O, −Pᵢ) → **D-Fructose 6-phosphate**

Figure 11.28 The fructose-1,6-bisphosphatase reaction of gluconeogenesis is used to bypass the phosphofructokinase-1 reaction of glycolysis.

Fructose 6-phosphate is then isomerized to glucose 6-phosphate by the same reversible enzyme used in glycolysis, phosphoglucose isomerase.

The final step of gluconeogenesis is another bypass reaction to oppose the irreversible reaction catalyzed by hexokinase. The bypass reaction is catalyzed by **glucose-6-phosphatase**, as shown in Figure 11.29.

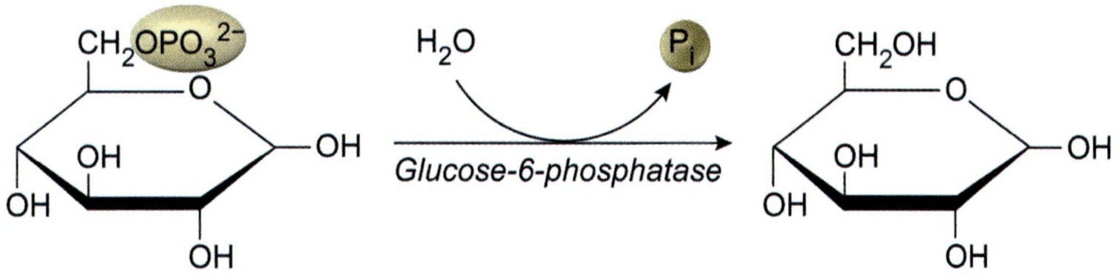

D-Glucose 6-phosphate → (Glucose-6-phosphatase, +H₂O, −Pᵢ) → **D-Glucose**

Figure 11.29 The glucose-6-phosphatase reaction of gluconeogenesis bypasses the hexokinase reaction of glycolysis.

Without the phosphate group, the free glucose molecules are no longer trapped in the cell. As gluconeogenesis proceeds, the intracellular glucose concentration in the liver cell eventually surpasses the blood glucose concentration level and glucose passively leaves the liver cell by facilitated diffusion through glucose transporters. The glucose molecules are then carried through the blood to target tissues.

The enzymes that differ between glycolysis and gluconeogenesis are summarized in Table 11.1.

Table 11.1 Irreversible glycolysis enzymes and corresponding bypass enzymes of gluconeogenesis. Gluconeogenesis steps are numbered starting from pyruvate.

Step of glycolysis	Enzyme of glycolysis	Corresponding enzyme(s) of gluconeogenesis	Step of gluconeogenesis
1	Hexokinase	Glucose-6-phosphatase	11
3	Phosphofructokinase-1	Fructose-1,6-bisphosphatase	9
10	Pyruvate kinase	Phosphoenolpyruvate carboxykinase (PEPCK)	2
		Pyruvate carboxylase	1

☑ **Concept Check 11.2**

One molecule of glucose is broken down through glycolysis. The resulting pyruvate molecules are then immediately sent through gluconeogenesis to rebuild glucose. What is the net outcome of these processes?

Solution

Note: The appendix contains the answer.

11.2.02 Alternate Entry Points of Gluconeogenesis

This discussion of gluconeogenesis has so far focused on the production of glucose from pyruvate molecules. However, many other molecules can also be used as starting substrates for gluconeogenesis.

For example, alanine can be converted to pyruvate through deamination. **Lactate** can similarly enter gluconeogenesis. As discussed in Concept 11.1.04, pyruvate is reversibly reduced to lactate during the process of lactic acid fermentation, which occurs in oxygen-poor (or mitochondria-lacking) tissue. In the oxygen-rich liver, lactate can be oxidized back into pyruvate, which can then enter gluconeogenesis.

In addition to molecules that are converted to pyruvate, many gluconeogenic substrates are metabolized directly into **oxaloacetate** *without* becoming pyruvate first. Although oxaloacetate appears only after the first step of the *classical* gluconeogenesis pathway, substrates that can be metabolized directly to oxaloacetate can bypass the pyruvate carboxylase step and therefore consume less ATP. In addition, because oxaloacetate is also a member of the citric acid cycle (see Chapter 12), molecules that are metabolized into other citric acid cycle members can be further metabolized into oxaloacetate to act as a gluconeogenic substrate.

Examples of molecules that produce oxaloacetate—either directly or through the citric acid cycle—include many amino acids and odd-chain fatty acids. The conversion of these molecules into gluconeogenic precursors is shown in Figure 11.30. For details of the biochemical pathways involved in producing these gluconeogenic precursors, see Chapters 12 and 13. Notably, acetyl-CoA *cannot* act as a precursor for gluconeogenesis.

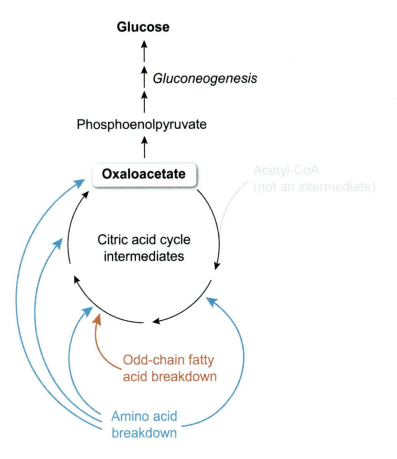

Figure 11.30 Many molecules can serve as gluconeogenesis substrates by being converted to oxaloacetate.

As discussed in Concept 11.2.01, oxaloacetate can be converted to phosphoenolpyruvate (PEP) by a *cytosolic isoform* of PEPCK. To enter the cytosol, however, oxaloacetate must first be reduced to malate in the mitochondrial matrix and re-oxidized to oxaloacetate in the cytosol. Like the malate-aspartate shuttle (discussed in Chapter 12), this process also results in the effective transport of NADH from the mitochondria to the cytosol, and the cytosolic NADH can then be used to power the GAPDH reaction.

Interestingly, the oxidation of lactate to pyruvate produces *cytosolic NADH* directly. In this case, the malate shuttle is no longer needed to transport NADH, so the *mitochondrial isoform* of PEPCK can be used to convert oxaloacetate to phosphoenolpyruvate (PEP). PEP can then be exported to the cytosol directly to proceed with gluconeogenesis. The difference in NADH production, and consequently the difference in PEPCK isoform localization, is shown in Figure 11.31.

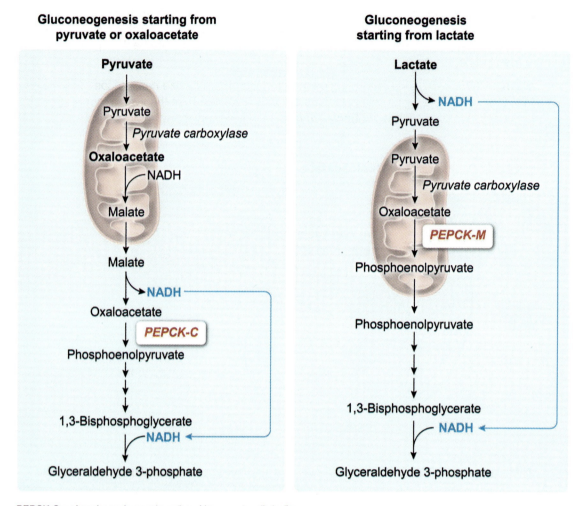

Figure 11.31 In humans, the choice of PEPCK isoform during gluconeogenesis is influenced by the need to shuttle NADH to the cytosol.

Glycerol, produced as a byproduct of triglyceride catabolism, can also serve as a gluconeogenic precursor. Glycerol can be phosphorylated into glycerol 3-phosphate before being oxidized to dihydroxyacetone phosphate (DHAP). DHAP can then enter gluconeogenesis directly as one of the metabolites found along the pathway. In the fasting state, the liver breaks down fat to power its ability to provide glucose for the rest of the body. Therefore, the glycerol backbones of triglycerides produced by this process commonly proceed through gluconeogenesis rather than glycolysis.

11.2.03 The Cori Cycle

The **Cori cycle** is an example of coordination across tissues and organ systems that allows multicellular organisms to adapt and thrive in a variety of habitats and conditions. Whereas most cells in aerobic organisms can use aerobic respiration to provide ample ATP production (see Chapter 12), cells without mitochondria (eg, red blood cells) and hypoxic cells (eg, vigorously contracting muscle) are limited to glycolysis for energy production.

Without aerobic respiration, fermentation is required to convert the NADH product of glycolysis back into NAD^+, which can then be used to power further rounds of glycolysis (see Concept 11.1.04). Unlike the volatile ethanol and CO_2 produced during alcoholic fermentation, the lactate produced during lactic acid fermentation remains in solution. The cells that produce lactate can export it and, in multicellular organisms, send it through the blood to the oxygen-rich liver, which *can* use aerobic respiration to produce an abundant supply of NAD^+. Therefore, the liver can oxidize lactate back to pyruvate, which can then be metabolized further.

The pyruvate molecules in the liver undergo gluconeogenesis to synthesize new glucose molecules (Concept 11.2.01). These new glucose molecules are then exported from the liver through the blood to the target tissues that need glucose. An overview of the Cori cycle is shown in Figure 11.32.

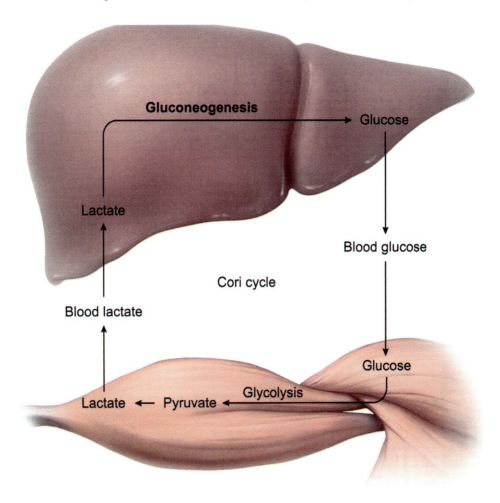

Figure 11.32 The Cori cycle.

Although cycling between glycolysis and gluconeogenesis *within a single cell* would be a wasteful futile cycle, the separation of processes across different tissues is an efficient means of resource management by multicellular organisms. It costs the liver more ATP to synthesize glucose than can be produced from glycolysis, but the oxygen-rich liver has alternate sources of energy that produce much higher ATP yields (eg, aerobic respiration of stored glycogen and fat).

Red blood cells and hypoxic muscle, on the other hand, are dependent on anaerobic glycolysis for ATP production. Rather than allow those tissues to run out of glucose, the liver spends ATP to power gluconeogenesis and to provide the whole body with a consistent supply of fuel molecules.

11.2.04 Regulation of Glucose Metabolism

Glycolysis, like most metabolic pathways, must be regulated to maintain relatively constant metabolite levels. Levels of ATP that are too high, for example, may shift the equilibrium of ATP-consuming reactions, whereas levels that are too low would not allow energy-requiring coupled reactions to occur. In tissues that express gluconeogenesis enzymes (eg, liver), the regulation of glycolysis must be even more controlled to prevent the occurrence of a futile cycle. This concept provides an overview of the regulation of glycolysis and gluconeogenesis.

Allosteric Feedback Mechanisms Regulate Glycolytic Flux

In tissues that do *not* perform gluconeogenesis (eg, muscle, red blood cells), the flux—or overall rate—of glycolysis is largely controlled by feedback inhibition. In other words, excessive production of products causes the rate of glycolysis to slow down, whereas low levels of products allow glycolysis to speed up. As discussed in Lesson 6.1, these regulatory mechanisms are focused mainly on the enzymes that catalyze *irreversible reactions*, namely the enzymes hexokinase, phosphofructokinase-1, and pyruvate kinase.

In most cells, hexokinase, which catalyzes the first step of glycolysis, is inhibited by its product, glucose 6-phosphate. Glucokinase, the hexokinase isoform in the liver and some other tissues, lacks this regulation so that it can serve as a glucose sensor. In other words, high levels of glucose result in high levels of glucose 6-phosphate in the liver so that the liver can "sense" the blood glucose level, whereas other tissues limit glucose 6-phosphate production to the amount those tissues need for their own metabolism.

Pyruvate kinase, which catalyzes the final step of glycolysis, is allosterically inhibited by its product (ATP), as well as by acetyl-CoA and long-chain fatty acids. Acetyl-CoA is a downstream metabolite of glucose catabolism (see Lesson 12.1). Long-chain free fatty acids are an alternate energy source for tissues such as muscle, and their presence indicates glycolysis is not needed for energy.

These mechanisms are summarized in Figure 11.33.

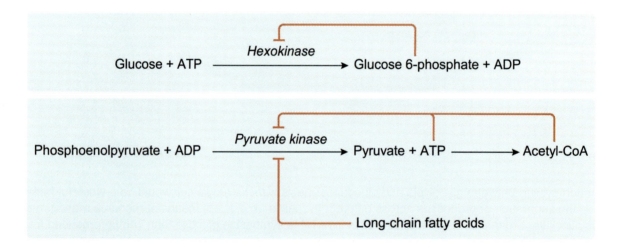

Figure 11.33 Allosteric regulatory mechanisms of hexokinase and pyruvate kinase.

Phosphofructokinase-1 (PFK-1) catalyzes the third step of glycolysis. This reaction is also the *earliest committed step* of glycolysis (see Concept 11.1.01); therefore, PFK-1 is the *most highly regulated* enzyme of glycolysis.

Like hexokinase and pyruvate kinase, PFK-1 is affected by feedback inhibition. In this case, the allosteric inhibitor is **ATP**, which is a product of the glycolysis pathway rather than a product of the reaction catalyzed by PFK-1. PFK-1 can also be *activated* by the molecules **ADP** and **AMP**. Therefore, when

ATP levels drop (and ADP and AMP levels rise), not only is feedback inhibition relieved, but enzyme activity is also *stimulated* to facilitate more ATP production (Figure 11.34).

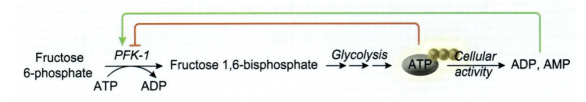

PFK-1 = phosphofructokinase-1.

Figure 11.34 ATP decreases PFK-1 activity by feedback inhibition. High levels of ADP or AMP stimulate PFK-1 activity by feedback activation.

In addition, PFK-1 is regulated by the levels of the metabolite **citrate**. Citrate is a member of the citric acid cycle, which occurs in the mitochondrial matrix, whereas PFK-1 (like the other enzymes of glycolysis) is cytosolic. However, citrate is trafficked to the cytosol during the process of fatty acid synthesis (see Concept 13.1.04), which occurs during periods of energy excess. Therefore, citrate inhibits PFK-1 because its presence in the cytosol is a sign that glycolysis does *not* need to occur to meet a cell's energy needs.

One of the most potent allosteric stimulators of PFK-1 activity is the molecule fructose 2,6-bisphosphate (F2,6BP). F2,6BP is synthesized by phosphorylation of fructose 6-phosphate by the enzyme phosphofructokinase-2 (PFK-2) and is degraded back to fructose 6-phosphate by the enzyme fructose-2,6-bisphosphatase (FBPase-2). Note that the position of the phosphate group on the molecule is very important in distinguishing the molecules: F*2*,6BP, PFK-*2*, and FBPase-*2* are involved in regulating glycolysis (and gluconeogenesis), whereas F*1*,6BP, PFK-*1*, and FBPase-*1* are direct *members* of glycolysis (or gluconeogenesis, Figure 11.35).

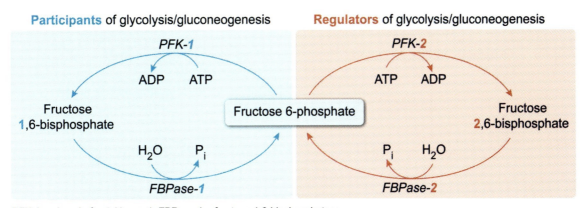

PFK-1 = phosphofructokinase-1; FBPase-1 = fructose-1,6-bisphosphatase;
PFK-2 = phosphofructokinase-2; FBPase-2 = fructose-2,6-bisphosphatase.

Figure 11.35 Fructose 2,6-bisphosphate, PFK-2, and FBPase-2 all regulate glycolysis (and gluconeogenesis) and are not to be confused with fructose 1,6-bisphosphate, PFK-1, or FBPase-1.

F2,6BP and its associated enzymes are especially important in the liver, where hormones induce covalent regulation of PFK-2/FBPase-2 to help coordinate glycolysis and gluconeogenesis. However, F2,6BP also plays an important role in regulating glycolysis in nongluconeogenic tissues, such as muscle. In muscle, PFK-2 and F2,6BP levels are regulated by substrate levels (ie, fructose 6-phosphate concentration) rather than covalent modifications to the enzyme.

To understand the role of F2,6BP in glycolysis regulation, consider a muscle cell that has received a signal that causes a sudden increase in glucose 6-phosphate levels. Phosphoglucose isomerase, which is reversible and unregulated, converts glucose 6-phosphate to fructose 6-phosphate (F6P); however, PFK-1 activity may not be able to keep up with the sudden influx of F6P. The accumulating F6P then proceeds down a side pathway, catalyzed by PFK-2, to make F2,6BP.

F2,6BP then plays its role as a potent **allosteric activator** of PFK-1 and therefore of glycolysis, which relieves the buildup of F6P and allows the cell to come back to homeostasis. This scenario is depicted in Figure 11.36.

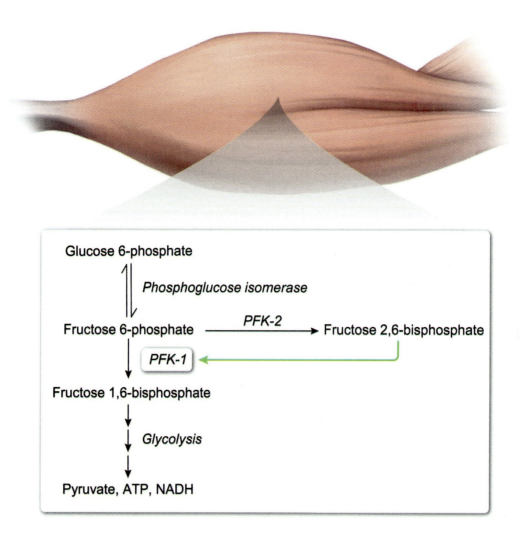

Figure 11.36 In muscle, a buildup of fructose 6-phosphate leads to production of fructose 2,6-bisphosphate. Fructose 2,6-bisphosphate then activates PFK-1 to stimulate glycolysis.

A summary of the regulatory mechanisms controlling glycolysis in nongluconeogenic cells is given in Table 11.2.

Table 11.2 Glycolysis regulatory mechanisms in nongluconeogenic tissue.

Enzyme	Inhibitor	Activator	Rationale for regulation
Hexokinase	• Glucose 6-phosphate		• Feedback inhibition
Phosphofructokinase-1	• Citrate • ATP	• ADP • AMP • Fructose 2,6-bisphospate	• Sign of fatty acid synthesis • Feedback inhibition • Sign of low energy stores • Sign of low energy stores • Sign of slow flux
Pyruvate kinase	• ATP • Acetyl-CoA • Long-chain fatty acids	• Fructose 1,6-bisphospate	• Feedback inhibition • Feedback inhibition • Alternate energy source • Feedforward activation

Covalent Regulation Affects Glycolysis in the Liver

The liver, which must coordinate *both* glycolysis *and* gluconeogenesis together, has more complex regulatory mechanisms that allow for the control of both processes. Because the liver must take into account signals from other tissues and organs, much of liver glycolysis regulation is controlled by the action of hormones that induce covalent regulation on certain enzymes.

Two important hormones in the regulation of glycolysis and gluconeogenesis are insulin and glucagon. Insulin is a peptide hormone released when blood glucose is high (eg, during the fed state), and it binds to the insulin receptor. In contrast, the peptide hormone glucagon is released when blood glucose is low (eg, during the fasting state), and it binds to the glucagon receptor. Epinephrine (also known as adrenaline) is another hormone that binds to a $G_{s\alpha}$-coupled G protein–coupled receptor (GPCR) and has similar action to glucagon.

Both insulin and glucagon have many and varied effects. One of the most prominent effects of **glucagon** is the eventual activation of cAMP-dependent protein kinase A (PKA), which **phosphorylates** many enzymes involved in glycolysis and gluconeogenesis. In the context of glycolysis and gluconeogenesis, **insulin** activates **phosphoprotein phosphatase-1 (PP1)**, which **dephosphorylates** the enzymes acted upon by PKA (Figure 11.37).

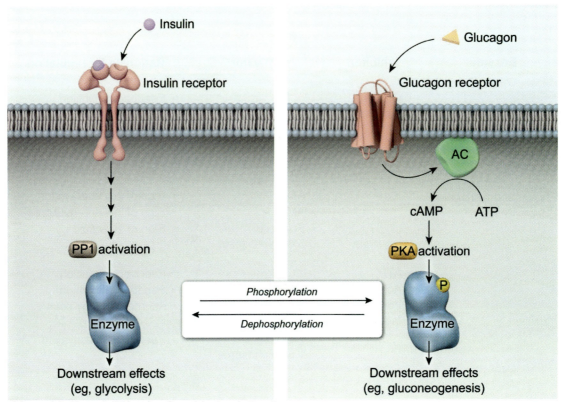

Figure 11.37 Glucagon binding results in the phosphorylation of many enzymes involved in energy metabolism. Insulin binding results in the dephosphorylation of many of those same enzymes.

Although glucose is the preferred fuel for most cells, *high* levels of blood glucose are damaging. Consequently, insulin signals for many tissues to take up glucose and signals the liver to *upregulate glycolysis* to decrease blood glucose levels. To do so, insulin triggers the *dephosphorylation of PFK-2*. Dephosphorylation of PFK-2 *activates* the enzyme, resulting in an increase in liver F2,6BP levels. As discussed in the previous section, F2,6BP is a potent allosteric activator of PFK-1 and therefore of glycolysis.

Because the liver can also perform gluconeogenesis, insulin must also *inhibit gluconeogenesis* to prevent a futile cycle. Not only does F2,6BP stimulate PFK-1, but it also *inhibits FBPase-1*, the bypass enzyme that corresponds to PFK-1 (Figure 11.38).

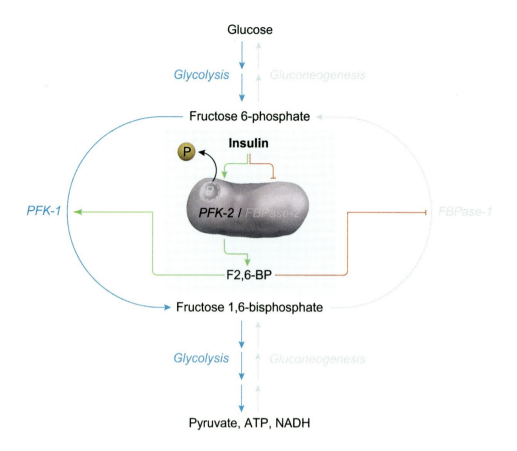

Figure 11.38 Insulin dephosphorylates and activates PFK-2, which produces F2,6BP. F2,6BP stimulates glycolysis by activating PFK-1 and inhibits gluconeogenesis by inhibiting FBPase-1.

Note that insulin-mediated stimulation of glycolysis (and inhibition of gluconeogenesis) continues *despite the liver cell having an excess of energy* (ie, high ATP levels). This is because F2,6BP overrides the feedback regulation that would normally inhibit PFK-1 (ie, high [ATP], high [citrate]). To help the liver cell manage the influx of glucose and increased glycolytic flux, insulin also stimulates glycogen synthesis (Lesson 11.4) and fatty acid synthesis (Lesson 13.1).

Glucagon, released when blood sugar levels are low, signals the liver to *upregulate gluconeogenesis* so that it can synthesize and release glucose for the rest of the body. To prevent a futile cycle, glucagon must therefore also *inhibit glycolysis* in the liver cell.

To accomplish both goals, glucagon signaling leads to the stimulation of PKA, which *phosphorylates and inactivates* the enzyme *PFK-2*. Interestingly, the enzymes PFK-2 and FBPase-2 exist as two domains on the same protein; therefore, the same event that leads to phosphorylation and inhibition of PFK-2 also causes *activation* of *FBPase-2*. This covalent regulation leads to a *decrease in F2,6BP levels*, increased gluconeogenesis, and decreased glycolysis (Figure 11.39). PKA also phosphorylates pyruvate kinase, inhibiting it and decreasing glycolysis even further.

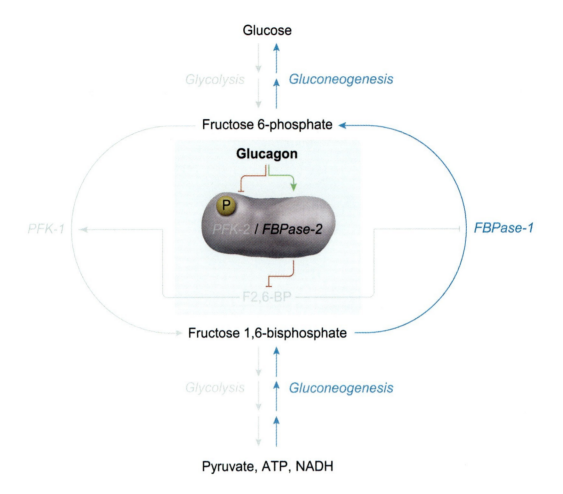

Figure 11.39 Glucagon causes the phosphorylation and activation of FBPase-2 to degrade F2,6BP. Low F2,6BP levels promote gluconeogenesis and prevent glycolysis.

Glucagon is normally secreted during the fasting state, when stores of readily available fuel for energy are low. Nevertheless, the liver expends any energy it *does* have to create fuel for the rest of the body. To help the liver cell power gluconeogenesis despite the low stores of fuel in the blood, glucagon also stimulates glycogenolysis (Lesson 11.4) and fatty acid oxidation (Chapter 13).

Like glucagon, the hormone epinephrine activates PKA, and therefore also stimulates gluconeogenesis in the liver cell. Yet epinephrine stimulates *glycolysis* in the muscle cell. This phenomenon illustrates the differing effects a signaling molecule may have on different tissues, depending on the genes and proteins expressed. Liver tissue expresses a phosphorylatable isoform of PFK-2 and therefore downregulates glycolysis upon epinephrine exposure. Muscle cell PFK-2 is *not* regulated by phosphorylation, and therefore epinephrine-induced increases in fructose 6-phosphate cause an increase in glycolysis.

Concept Check 11.3

Forskolin is a compound that stimulates production of cyclic AMP (cAMP), which activates protein kinase A (PKA).

1) If forskolin is applied to liver cells, what is the expected effect on glycolysis and gluconeogenesis?
2) If forskolin is applied to muscle cells, what is the expected effect on glycolysis and gluconeogenesis?

Solution

Note: The appendix contains the answer.

Lesson 11.3
The Pentose Phosphate Pathway

Introduction

The **pentose phosphate pathway**, sometimes called the hexose monophosphate shunt, acts on intermediates of the glycolysis pathway to ultimately convert glucose molecules into pentose phosphate molecules (ie, phosphorylated five-carbon sugars). The pentose phosphates can then be used to form nucleosides and nucleotides (Lesson 8.1). In addition, part of the pentose phosphate pathway can be used to produce NADPH, which can then be used by the cell as a reducing agent.

The pentose phosphate pathway consists of *two* related but separate sub-pathways. These sub-pathways are sometimes called the different "phases" of the pentose phosphate pathway. Each phase can operate independently, and *both* can create pentose phosphates. Alternatively, they can work together in sequence, with one phase *creating* pentose phosphates and the second phase *consuming* them (Figure 11.40).

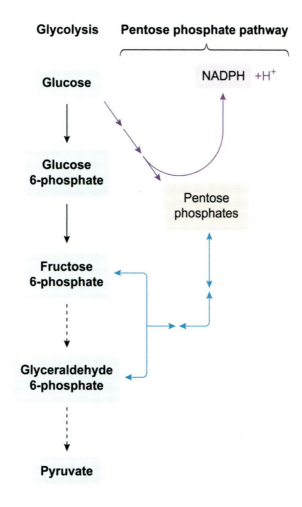

Figure 11.40 Overview of the pentose phosphate pathway.

11.3.01 The Oxidative Phase

The oxidative phase of the pentose phosphate pathway (PPP) is a sequence of three **irreversible** enzyme-catalyzed reactions that ultimately result in the production of a pentose phosphate, two molecules of NADPH, and a CO_2 molecule. The substrate molecule that enters the oxidative phase is glucose 6-phosphate.

Because all three steps of the oxidative phase are irreversible, the first step in the sequence (acting on glucose 6-phosphate) is the earliest committed (and therefore rate-determining) step. The enzyme that catalyzes this reaction is **glucose-6-phosphate dehydrogenase (G6PD)**. As its name implies, G6PD is an oxidoreductase that oxidizes (or dehydrogenates) the anomeric carbon of glucose 6-phosphate into a cyclic ester called 6-phosphoglucono-δ-lactone.

Unlike the enzymes of glycolysis, G6PD does *not* use NAD^+ as a coenzyme; instead, it uses a related but different coenzyme called $NADP^+$, which becomes **NADPH** (Figure 11.41).

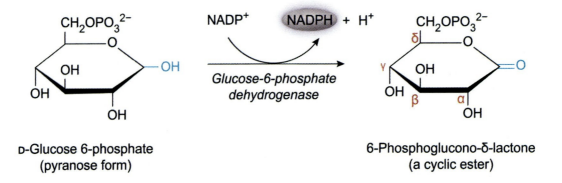

Figure 11.41 Glucose-6-phosphate dehydrogenase (G6PD) catalyzes the first, rate-determining step of the oxidative phase of the PPP and reduces $NADP^+$ to NADPH.

After the first oxidation step, the cyclic lactone is hydrolyzed by the second enzyme (6-phosphogluconolactonase) forming 6-phosphogluconic acid. At physiological pH, this molecule is deprotonated to 6-phosphogluconate (6PG), as shown in Figure 11.42.

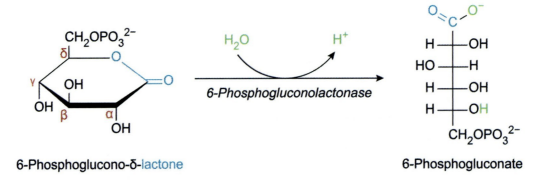

Figure 11.42 The second step of the oxidative phase is hydrolysis of the lactone.

The third and final reaction unique to the oxidative phase is *oxidative decarboxylation* of 6PG, catalyzed by 6-phosphogluconate dehydrogenase. The C1 carboxylate is removed, the rest of the carbon skeleton is oxidized to a pentose phosphate, and another molecule of $NADP^+$ is reduced to NADPH. This reaction is shown in Figure 11.43.

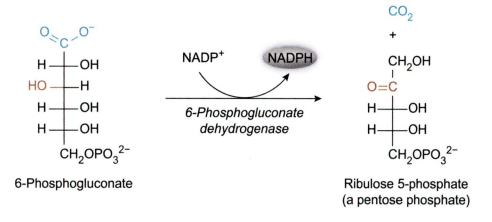

Figure 11.43 The third reaction of the oxidative phase of the PPP is an oxidative decarboxylation, producing CO_2, a pentose phosphate, and NADPH.

The pentose phosphate produced directly by 6-phosphogluconate dehydrogenase is the phosphoketopentose ribulose 5-phosphate (Ru5P). However, this pentose phosphate quickly equilibrates with two other biologically relevant pentose phosphates: ribose 5-phosphate and xylulose 5-phosphate. The enzyme ribose-5-phosphate isomerase converts Ru5P into its corresponding aldose, ribose 5-phosphate, and the enzyme ribulose-5-phosphate 3-epimerase converts Ru5P into its C3 epimer, xylulose 5-phosphate.

Because all three of the pentose phosphates are in equilibrium with each other, the PPP is said to produce pentose phosphates in general, rather than producing a *specific* pentose phosphate. The three biochemically relevant pentose phosphates are shown in Figure 11.44.

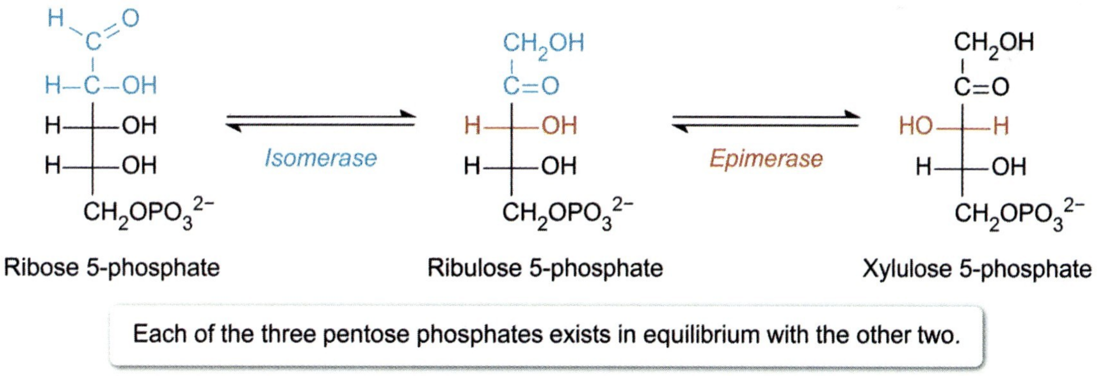

Each of the three pentose phosphates exists in equilibrium with the other two.

Figure 11.44 The three biochemically relevant pentose phosphates.

The net outcome of the oxidative phase of the PPP is

Glucose 6-phosphate + 2 $NADP^+$ + H_2O → Pentose phosphate + CO_2 + 2 NADPH + 2 H^+

The pentose phosphates can be used for various purposes such as nucleotide synthesis, or they can re-enter glycolysis by going through the nonoxidative phase of the PPP (Concept 11.3.02). However, the lost matter (ie, the released CO_2 molecule) means that the ATP yield is always slightly decreased compared to a glucose 6-phosphate molecule that goes through glycolysis directly. This demonstrates that the major outcome of the oxidative phase is *not* the pentose phosphates themselves but rather the *production of cytosolic NADPH.* This is supported by the fact that NADPH feedback inhibits the oxidative (but *not* the nonoxidative) phase by inhibition of G6PD.

One function of cytosolic NADPH is as a reducing factor in lipid synthesis (see Concept 13.1.04 for a discussion of fatty acid synthesis). NADPH is also used to reduce ribonucleotides to 2'-deoxyribonucleotides, which are needed for DNA replication. One of the most clinically relevant roles of cytosolic NADPH is its use as a reducing agent in the protection of cells *against* oxidative damage (ie, its role as an antioxidant).

The oxidative phase of the PPP serves as the *primary source of cytosolic NADPH.* Deficiencies in G6PD—the first and rate-limiting enzyme of the oxidative phase—can lead to hemolytic anemia (ie, red blood cell destruction) because patients' red blood cells cannot produce sufficient NADPH to protect themselves from oxidative damage (Figure 11.45). G6PD deficiency is the most common genetically inherited enzyme deficiency.

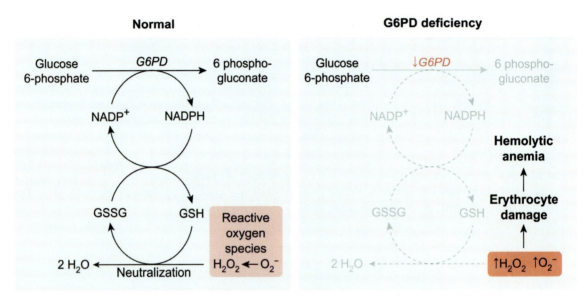

G6PD = glucose-6-phosphate dehydrogenase; GSH = glutathione; GSSG = glutathione disulfide.

Figure 11.45 A deficiency in G6PD and the oxidative phase of the PPP leads to insufficient NADPH production. This leads to increased oxidative damage and symptoms such as hemolytic anemia.

11.3.02 The Nonoxidative Phase

The nonoxidative phase of the pentose phosphate pathway (PPP) consists of a sequence of *reversible* reactions that convert intermediates of glycolysis into pentose phosphates. Unlike the oxidative phase, in which some mass is lost as CO_2 and some reducing power is transferred to NADPH, the nonoxidative phase uses only reversible transfer reactions, so no mass or reducing power is lost to other molecules.

The reactions of the nonoxidative phase are unlikely to be tested on the exam; however, a common misconception is that the nonoxidative phase serves only as a continuation of the oxidative phase to re-enter glycolysis, rather than as a separate sequence of reactions that *can operate independently* of the oxidative phase. To emphasize this important property, the nonoxidative phase is presented here proceeding *from* the intermediates of glycolysis *toward* production of pentose phosphates. Descriptions of the reaction details serve to reinforce mastery of carbohydrate structure (Chapter 7).

Unlike the oxidative phase, which produces one pentose phosphate from one hexose phosphate, the nonoxidative phase requires multiple substrate molecules and produces multiple pentose phosphate molecules. The stoichiometry of the overall reaction can be easily remembered as
5 hexose phosphates ⇌ 6 pentose phosphates, as shown in Figure 11.46. In this reaction, 30 carbons

enter as hexose phosphates (5 hexoses × 6 carbons per hexose) and 30 carbons exit as pentose phosphates (6 pentoses × 5 carbons per pentose).

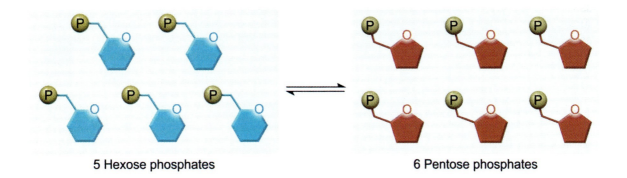

Figure 11.46 In the nonoxidative phase of the PPP, five 6-carbon hexose phosphates are converted to six 5-carbon pentose phosphates.

It may also be helpful to describe the process as three pentose phosphate molecules being produced from two and a half hexose phosphate molecules. In this description, the "half hexose phosphate" molecule is the three-carbon triose molecule **glyceraldehyde 3-phosphate**, produced by aldolase during glycolysis.

The reversible, nonoxidative phase of the PPP uses two different transferase enzymes to catalyze three different reactions. Both enzymes transfer carbons from a ketose phosphate onto an aldose phosphate. Transketolase transfers two carbons (C1 and C2) and acts twice in the nonoxidative phase; transaldolase transfers three carbons (C1–C3) and acts once, in the middle of the sequence. The transketolase and transaldolase reactions are shown schematically in Figure 11.47.

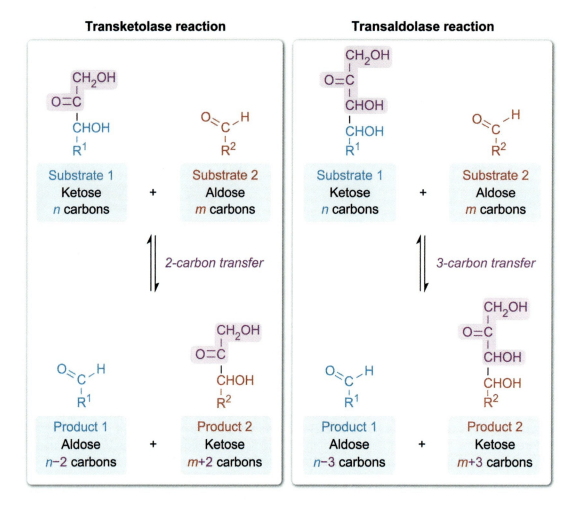

Figure 11.47 The transketolase and transaldolase reactions.

The first reaction of the nonoxidative phase is catalyzed by transketolase acting on **fructose 6-phosphate (F6P)** and **glyceraldehyde 3-phosphate (G3P)**. Transketolase transfers two carbons from F6P to G3P. Consequently, the products are the four-carbon aldose *erythrose 4-phosphate* and the five-carbon ketose **xylulose 5-phosphate**. Xylulose 5-phosphate is one of the desired products (a pentose phosphate) and therefore leaves the pathway. This first reaction of the nonoxidative phase is shown in Figure 11.48.

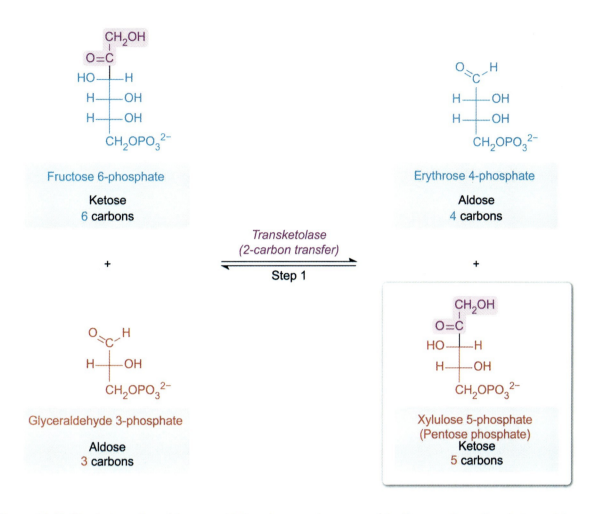

Figure 11.48 The first reaction of the nonoxidative phase produces one of the three pentose phosphates, xylulose 5-phosphate.

The second reaction of the nonoxidative phase is catalyzed by transaldolase. This reaction acts on the erythrose 4-phosphate (E4P) produced in the previous step and on *another molecule of fructose 6-phosphate* (F6P). Transaldolase transfers three carbons from F6P to E4P. Consequently, the products are the three-carbon aldose *glyceraldehyde 3-phosphate* (G3P) and the seven-carbon ketose *sedoheptulose 7-phosphate* (Su7P). The second reaction of the nonoxidative phase is shown in Figure 11.49.

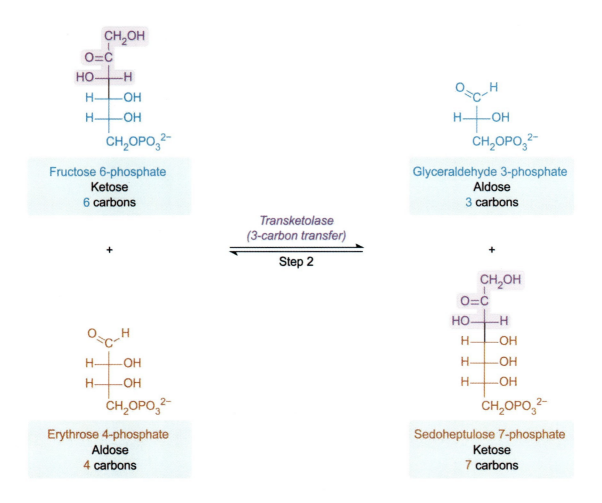

Figure 11.49 The second reaction of the nonoxidative phase.

The final reaction of the nonoxidative phase is again catalyzed by transketolase. This reaction acts on the products of the previous reaction: G3P and Su7P. Transketolase transfers two carbons from Su7P to G3P. Consequently, the products are the five-carbon aldose **ribose 5-phosphate** and the five-carbon ketose **xylulose 5-phosphate**. The third reaction of the nonoxidative phase is shown in Figure 11.50.

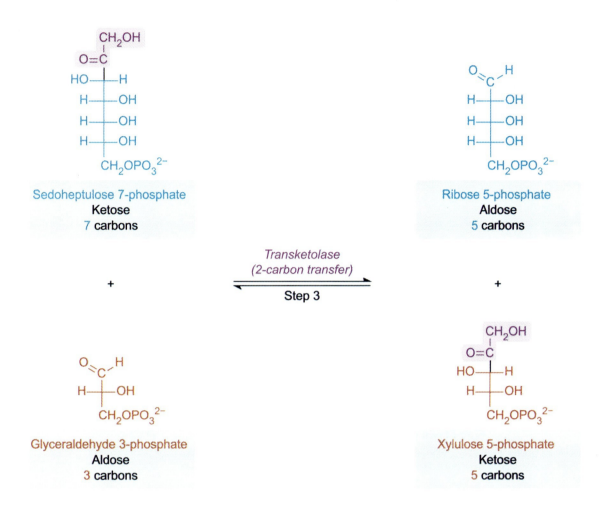

Figure 11.50 The third reaction of the nonoxidative phase produces two pentose phosphates.

By the end of the nonoxidative phase, two molecules of fructose 6-phosphate and one molecule of glyceraldehyde 3-phosphate have reacted together to produce *three pentose phosphate molecules*.

The overall process of the reversible, nonoxidative phase of the PPP is shown in Figure 11.51. Five hexose phosphate molecules become six pentose phosphate molecules through two rounds of the nonoxidative phase.

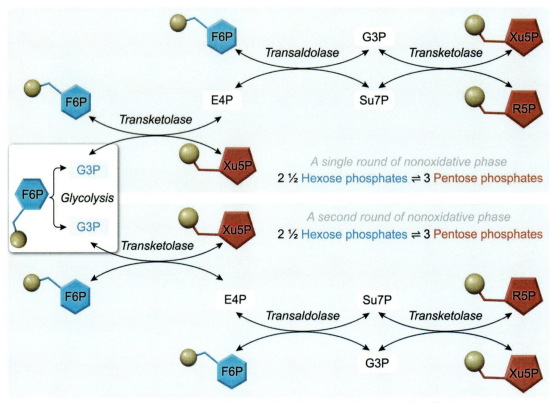

Sum of two rounds: 5 Hexose phosphates ⇌ 6 Pentose phosphates

E4P = erythrose 4-phosphate; F6P = fructose 6-phosphate; G3P = glyceraldehyde 3-phosphate; R5P = ribose 5-phosphate; Su7P = sedoheptulose 7-phosphate; Xu5P = xylulose 5-phosphate.

Figure 11.51 The nonoxidative phase of the PPP.

Putting the Two Phases Together

The irreversible, oxidative phase of the PPP (Concept 11.3.01) produces pentose phosphates *and* NADPH. The nonoxidative phase can *also* produce pentose phosphates but does *not* produce NADPH; however, the nonoxidative phase is *reversible* and can return the carbons of the pentose phosphates to glycolysis. What is the benefit of having two separate phases of pentose phosphate production?

The reversible, nonoxidative phase allows cells to produce pentose phosphates without also producing NADPH if it is not needed (eg, NADPH is already in excess). Conversely, because this phase is reversible, it can also *recycle* pentose phosphates back into glycolysis. This allows the *oxidative* phase to produce NADPH without burdening the cell with excess pentose phosphates (because they are recycled by the nonoxidative phase). If a cell needs both NADPH *and* pentose phosphates, the oxidative phase can provide both products.

Therefore, the different phases of the PPP give cells the flexibility to produce pentose phosphates alone, NADPH alone, or both NADPH *and* pentose phosphates together. An overview of both phases is shown in Figure 11.52.

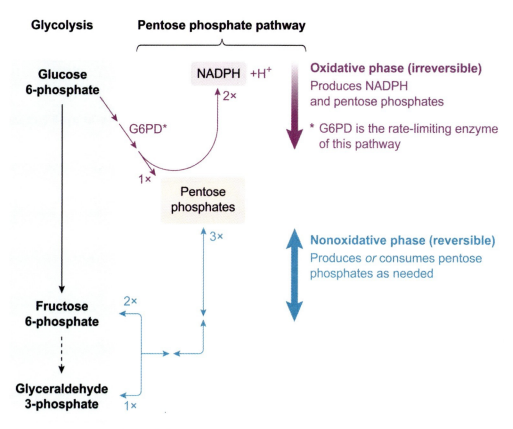

Figure 11.52 The oxidative and nonoxidative phases of the PPP yield different products.

 Concept Check 11.4

A resting cell receives a signal to upregulate one of the following processes. Based on the metabolic needs of the process, will the cell experience increased flux (in either direction, if applicable) through the oxidative phase, the nonoxidative phase, or both phases of the pentose phosphate pathway?

1) Increased DNA replication (requires both pentose phosphates and NADPH)
2) Increased gene transcription (requires pentose phosphates only)
3) Increased fatty acid synthesis (requires NADPH only)

Solution

Note: The appendix contains the answer.

Concept Check 11.5

Three glucose molecules entering a cell are directed through the oxidative phase of the pentose phosphate pathway. The pentose phosphates produced then proceed through the nonoxidative phase to re-enter glycolysis, after which the intermediates are all metabolized to pyruvate. What is the net yield of pyruvate, CO_2, NADH, NADPH, and ATP at the end of this process?

Solution

Note: The appendix contains the answer.

Lesson 11.4
Glycogen Metabolism

Introduction

Glucose, the preferred fuel for most cells, is generally acquired from the diet; however, organisms must continue to fuel their cells *between* feedings. Lesson 11.2 describes the process of gluconeogenesis, in which the liver can produce glucose during times of need. However, liver production of glucose may not be quick enough to satisfy the needs of vigorously contracting muscle, for example. Moreover, gluconeogenesis requires fuel to provide the carbon skeleton for the produced glucose, which may not be available between feedings.

Instead of relying on gluconeogenesis, animal cells can store glucose in the form of a polysaccharide known as **glycogen**. As discussed in Lesson 7.2, storing glucose as glycogen provides cells quick access to large amounts of glucose *without* disrupting the osmolarity of the cell.

This lesson describes the biochemical reactions through which glycogen molecules are synthesized and elongated (**glycogenesis**), the reactions through which glycogen is broken down (**glycogenolysis**), and the ways in which these processes are regulated. Figure 11.53 shows general information about glycogenesis and glycogenolysis.

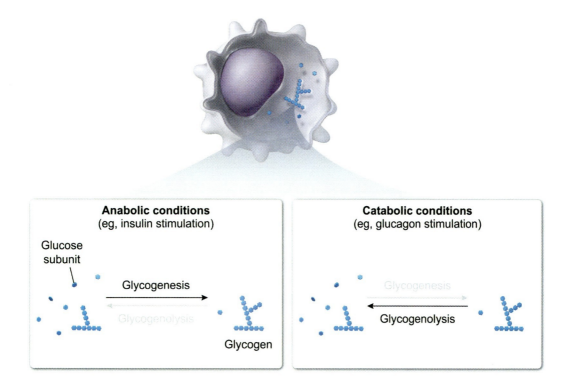

Figure 11.53 Glycogenesis and glycogenolysis.

11.4.01 Glycogenesis

Like any other metabolic process that involves use of glucose by the cell, the first step of glycogenesis is phosphorylation of glucose to glucose 6-phosphate (G6P). This is accomplished by hexokinase, the first enzyme of glycolysis. The G6P produced is not committed to any one pathway but can go down one of several pathways (eg, glycolysis, pentose phosphate pathway). To proceed down glycogenesis, G6P must be further activated before it can be added to an existing glycogen polymer.

Activation of Glucose for Glycogenesis

The glucose derivative that reacts in the glycogen elongation reaction is UDP-glucose, which has a uridylyl functional group (ie, a uridine nucleotide) attached to carbon 1. Therefore, to further activate glucose 6-phosphate for glycogen synthesis, the *phosphate group must be moved* to carbon 1 and a *uridylyl group must be transferred* onto the phosphate. These reactions activate the anomeric carbon—and therefore the reducing end—of the incoming glucose, which allows it to later form a glycosidic bond with a *nonreducing end* of the growing glycogen molecule.

The enzyme that moves the phosphate group from carbon 6 to carbon 1 is **phosphoglucomutase**. The conversion of glucose 6-phosphate into glucose 1-phosphate is *reversible* under physiological conditions and is shown in Figure 11.54.

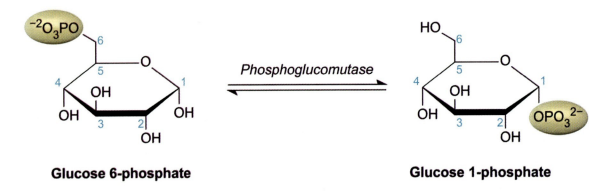

Figure 11.54 Phosphoglucomutase interconverts glucose 6-phosphate with glucose 1-phosphate.

A uridine monophosphate (UMP) group is then transferred from UTP onto glucose 1-phosphate, leaving a pyrophosphate (PP$_i$) behind. The phosphate group of UMP forms a phosphoanhydride bond with the phosphate of glucose 1-phosphate, so the resulting molecule has a *di*phosphate bridge and therefore is called **UDP-glucose** (ie, uridine diphosphate glucose).

This transfer reaction is catalyzed by the enzyme UDP-glucose pyrophosphorylase. Although this enzyme is named for the reverse reaction (in which PP$_i$ breaks up UDP-glucose to form UTP and glucose 1-phosphate), the reaction is **irreversible** under physiological conditions. The uridylyl transfer reaction is shown in Figure 11.55.

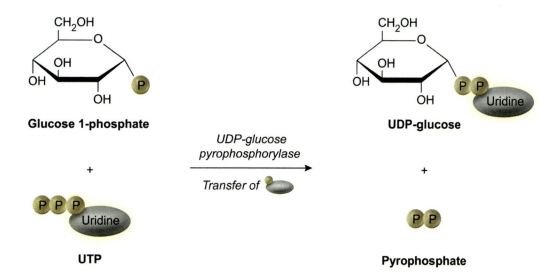

Figure 11.55 To activate glucose for glycogen synthesis, a uridine nucleotide is transferred to glucose 1-phosphate to form UDP-glucose.

Although the uridylyl transfer reaction is an early irreversible step of glycogenesis, it is *not* the rate-limiting step. This is because UDP-glucose can participate in other reactions and pathways in addition to glycogenesis. For example, UDP-glucose appears in galactose catabolism, as discussed in Concept 11.1.03.

Glycogen Elongation

Once an activated UDP-glucose has been formed, it can be added to an existing glycogen chain by the enzyme **glycogen synthase**. Glycogen synthase catalyzes the *exergonic* transfer of glucose from UDP-glucose onto the **nonreducing end** of an existing glycogen molecule. This transfer reaction results in the release of UDP and the formation of a new **α-1,4-glycosidic bond** (see Chapter 7 for a review of carbohydrate glycosidic bonds). The glycogen synthase reaction is shown in Figure 11.56.

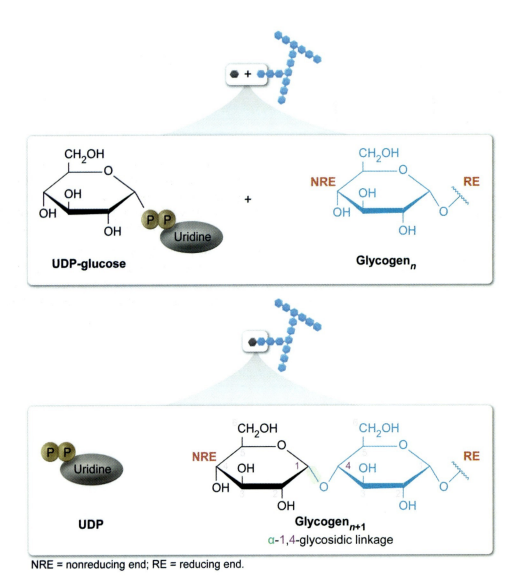

Figure 11.56 Glycogen synthase transfers a glucose unit from UDP-glucose onto the nonreducing end of a growing glycogen chain.

Although UTP loses two phosphate groups as PP$_i$ in the UMP transfer reaction (Figure 11.55), it subsequently *gains* a phosphate group from glucose 1-phosphate. Consequently, a single UTP molecule is invested to activate glucose 6-phosphate, and a single *UDP* molecule is released at the end of the glycogen synthase reaction. Overall, the net cost of incorporating a **glucose 6-phosphate** molecule into glycogen is *one* UTP molecule, which is energetically the same as **one ATP**.

The net cost of incorporating a *free* glucose into glycogen, on the other hand, must account for the ATP used by the hexokinase reaction of glycolysis. Therefore, the energetic cost of *free* glucose incorporation into glycogen is *two* ATP equivalents. The net cost of glucose and G6P incorporation into glycogen is shown in Figure 11.57.

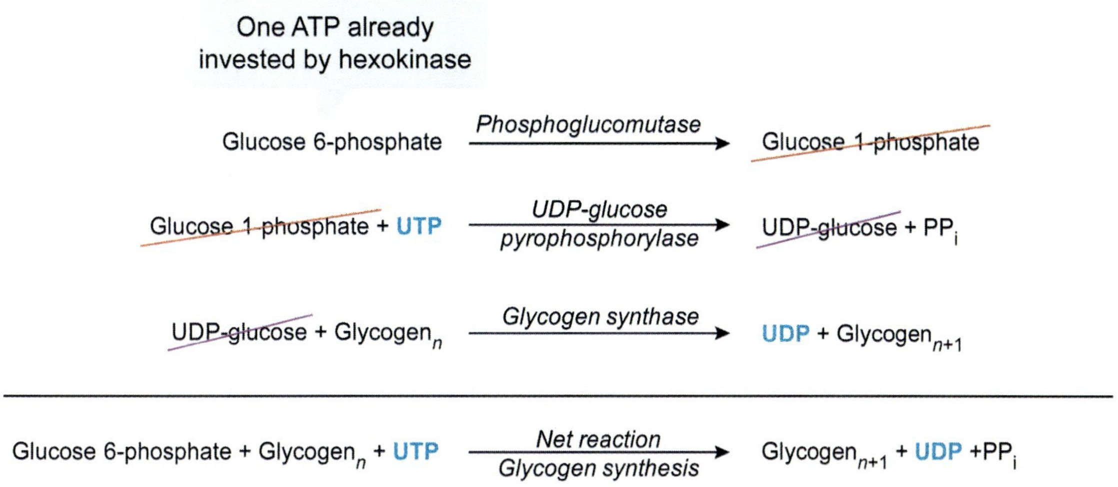

One ATP already invested by hexokinase

Glucose 6-phosphate $\xrightarrow{\text{Phosphoglucomutase}}$ Glucose 1-phosphate

Glucose 1-phosphate + UTP $\xrightarrow{\text{UDP-glucose pyrophosphorylase}}$ UDP-glucose + PP$_i$

UDP-glucose + Glycogen$_n$ $\xrightarrow{\text{Glycogen synthase}}$ UDP + Glycogen$_{n+1}$

Glucose 6-phosphate + Glycogen$_n$ + UTP $\xrightarrow[\text{Glycogen synthesis}]{\text{Net reaction}}$ Glycogen$_{n+1}$ + UDP + PP$_i$

Figure 11.57 Glycogen synthesis involves the investment of one UTP molecule (one ATP equivalent) per glucose 6-phosphate. The incorporation of free glucose requires investment of another ATP by hexokinase.

Priming

Glycogen synthase can only elongate *existing* glycogen polymers. Lesson 7.2 explains that the core of a glycogen molecule contains a protein called **glycogenin**, which is connected to the reducing end of the glycogen polysaccharide by an α-1,Tyr-glycosidic bond. In addition to serving as the structural core, glycogenin is an enzyme that catalyzes the reaction that joins a glucose from UDP-glucose to one of glycogenin's own tyrosine residues. Glycogenin then catalyzes the first few elongation steps, as shown in Figure 11.58.

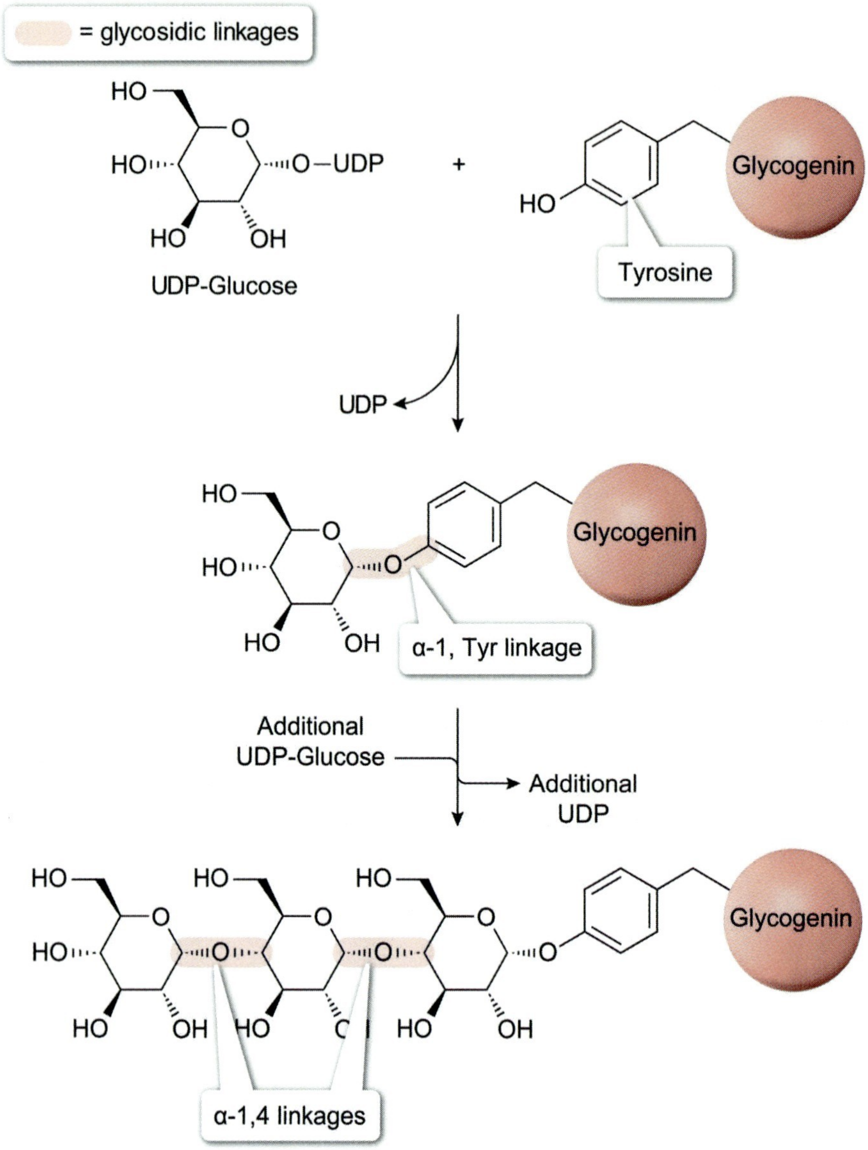

Figure 11.58 Glycogenin catalyzes the formation of an α-1,Tyr linkage to glucose and several α-1,4 linkages between glucose units before glycogen synthase acts on the growing chain.

Once the polysaccharide is around eight units long, glycogen synthase can take over as the main enzyme in glycogen elongation. Figure 11.59 depicts a portion of a glycogen molecule and indicates which glucose subunits are added by glycogenin and which are added by glycogen synthase.

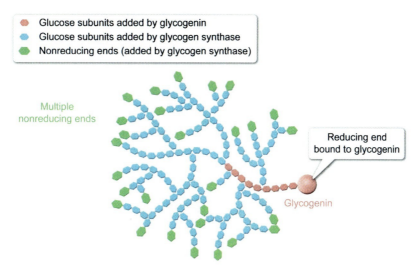

Figure 11.59 After glycogenin catalyzes the incorporation of the first eight glucose subunits into glycogen, glycogen synthase catalyzes the incorporation of the remaining glucose subunits.

Branching

Lesson 7.2 explains that glycogen is characterized as being highly branched. Branching of glycogen is accomplished by the aptly named **glycogen-branching enzyme**. Once glycogen synthase catalyzes formation of a sufficiently long series of glucose subunits, glycogen-branching enzyme can transfer an oligosaccharide from the nonreducing end of a linear chain to carbon 6 of a more interior glucose unit, forming a branch point. An overview of this process is shown in Figure 11.60.

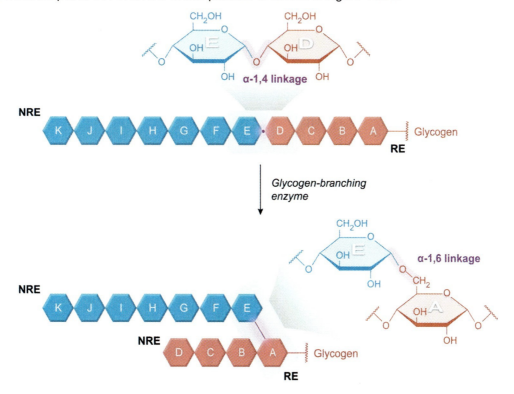

NRE = nonreducing end; RE = reducing end.

Figure 11.60 Glycogen-branching enzyme transfers an oligosaccharide (units E through K) from the nonreducing end of a linear chain to carbon 6 of an interior glucose (unit A), forming a branch point.

Note that the branching process creates *additional nonreducing ends*. Figure 11.60 shows that, in addition to subunit K, which serves as a nonreducing end both before and after branching, subunit D *also* becomes a nonreducing end after branching occurs. Molecules of glycogen synthase can then act on *either* nonreducing end to store glucose in an elongated polysaccharide.

11.4.02 Glycogenolysis

Glycogenolysis is the process by which glycogen is broken down to release stored glucose subunits. Rather than being hydrolyzed, the vast majority of glucose subunits in glycogen are **phosphorolyzed**. Phosphorolysis is like hydrolysis, except an *inorganic phosphate* is used to break apart a functional group rather than a water molecule. Phosphorolysis of a glycogen molecule is shown in Figure 11.61.

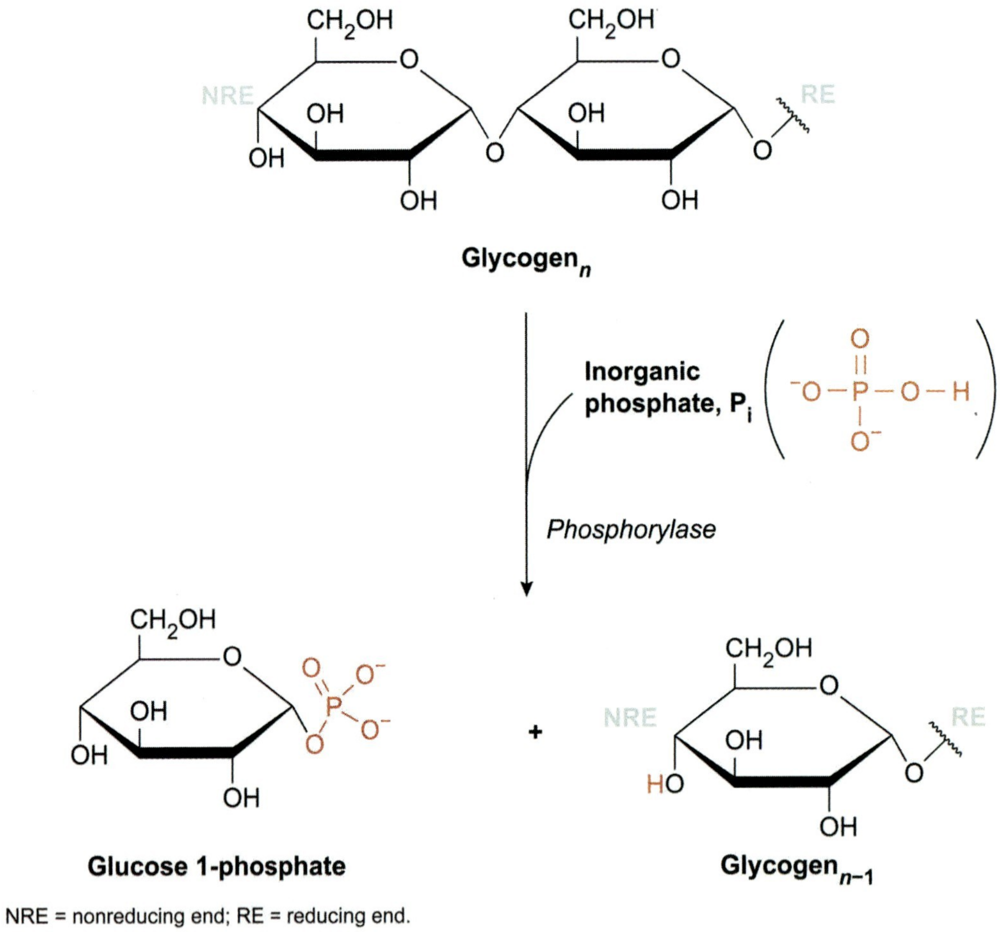

NRE = nonreducing end; RE = reducing end.

Figure 11.61 Phosphorolysis of glycogen produces a shortened glycogen molecule and releases a glucose 1-phosphate.

This reaction is catalyzed by the enzyme **glycogen phosphorylase**, which is often simply referred to as phosphorylase. The active form of the enzyme may also be specified with the name phosphorylase a.

Phosphorylase facilitates the removal of a glucose subunit from a **nonreducing end** of glycogen by catalyzing the phosphorolysis of the α-1,4-glycosidic bond connecting it to the rest of the molecule. Consequently, the products are a glycogen molecule shortened by one subunit and a glucose 1-phosphate molecule.

Glucose 1-phosphate is isomerized to glucose 6-phosphate by the reversible enzyme phosphoglucomutase. When glycogenolysis is occurring, phosphoglucomutase, discussed in Concept 11.4.01, operates in the reverse direction from its action during glycogenesis.

In most tissues (eg, muscle) the glucose 6-phosphate then proceeds through *glycolysis*, through which it is metabolized to produce energy for the cell in the form of ATP. On the other hand, the liver, which expresses the enzymes needed for gluconeogenesis, usually activates glycogenolysis and *gluconeogenesis* together. Consequently, the glucose 6-phosphate produced after glycogenolysis is typically hydrolyzed by glucose-6-phosphatase. This produces free glucose molecules that are then released into the bloodstream to fuel other tissues. These possibilities are illustrated in Figure 11.62.

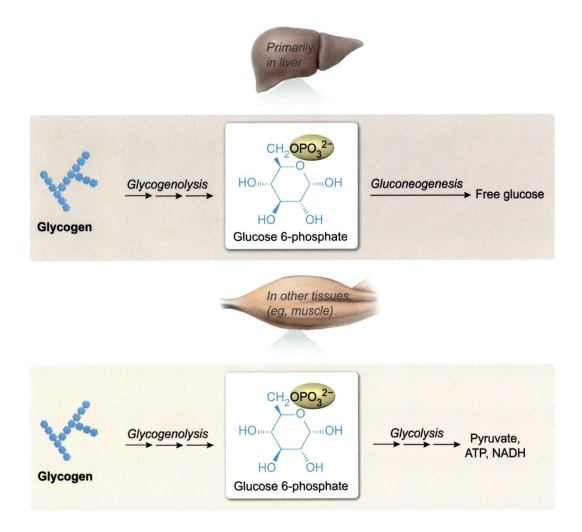

Figure 11.62 Possible fates of the products of glycogenolysis in different tissues.

Debranching

The phosphorylase enzyme acts on only the α-1,4-glycosidic linkages of linear chains, *not* the α-1,6 linkages found at branches. In addition, the enzyme cannot accommodate glucose subunits that are very close to (ie, within four subunits of) a glycogen branch site. After phosphorylase has removed all of the nonreducing end subunits further than four subunits away from the branching glucose, glycogenolysis requires additional enzymes to assist with debranching of glycogen molecules. An overview of the glycogen debranching process is shown in Figure 11.63.

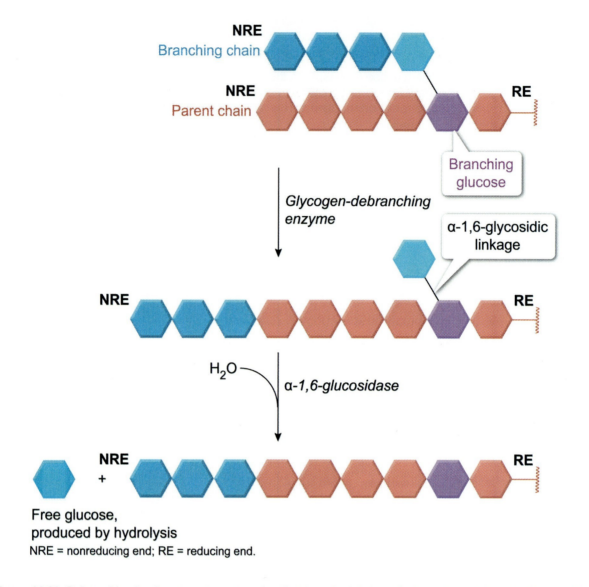

Figure 11.63 Debranching involves two steps: transfer of a trisaccharide from the branching chain onto the parent chain and hydrolysis of the α-1,6-glycosidic linkage connecting the final branching-chain subunit.

The first additional enzyme is the aptly named **debranching enzyme**. The debranching enzyme transfers a three-subunit-long trisaccharide from the branching chain to the parent chain. This leaves behind the final branching-chain subunit, which is connected by an α-1,6-glycosidic linkage to the parent chain. The enzyme α-1,6-glucosidase **hydrolyzes** this glycosidic linkage, resulting in a **free glucose** molecule and a debranched glycogen molecule.

Note the difference in products between phosphorylase releasing a subunit and the debranching pathway releasing a subunit. Whereas phosphorylase releases a glucose 1-phosphate, debranching releases *free* glucose. In cells that direct the products of glycogenolysis to glycolysis, subunits released by phosphorylase *already have* a phosphate group and can proceed to step 2 of glycolysis. In contrast, subunits released by the debranching process *must be phosphorylated by hexokinase* before they can proceed through glycolysis.

Concept Check 11.6

Stimulation of glycogenolysis in a cell results in the release of 100 subunits from a particular glycogen molecule. If eight branches are fully clipped during this process:

1) How many units are released as free glucose?

2) How many units are released as glucose 1-phosphate?

Solution

Note: The appendix contains the answer.

 Concept Check 11.7

A muscle cell receives a free glucose molecule from the bloodstream. That glucose molecule is temporarily incorporated into glycogen through an α-1,4-glycosidic linkage before being released from a linear chain by glycogenolysis. The released molecule is then processed through glycolysis.

1) At the end of glycolysis, what is the net yield of pyruvate, NADH, and ATP equivalents?

2) What is the net yield if the glucose unit is incorporated into a branch point in the glycogen molecule (ie, the subunit is involved in an α-1,6-glycosidic linkage)?

Solution

Note: The appendix contains the answer.

11.4.03 Regulation of Glycogen Metabolism

Regulation of Glycogenolysis

In glycogenolysis, phosphorolysis serves as the rate-determining step. Therefore, glycogen phosphorylase is the major point of regulation for glycogenolysis and is regulated by both covalent (ie, hormonal) and allosteric mechanisms.

Hormonally, glycogen phosphorylase (also simply called phosphorylase) is controlled in part by glucagon. When glucagon binds a G protein–coupled receptor (GPCR, see Concept 3.2.02), the agonist-bound GPCR facilitates exchange of GDP with GTP, turning its associated heterotrimeric G protein "on." The G protein uncouples from the GPCR, and the α subunit dissociates from the β and γ subunits, which remain together as a βγ complex. For the glucagon receptor, the $G_α$ subunit is *stimulatory* ($G_{sα}$), resulting in *increased* activity of adenylyl cyclase and the conversion of ATP to 3′,5′-cyclic AMP (cAMP) and PP_i. The cAMP can then allosterically activate the cAMP-dependent kinase **protein kinase A (PKA)**, which phosphorylates many enzymes involved in energy metabolism.

For pathways commonly tested on the exam, a general trend is that phosphorylation activates enzymes involved in glycogenolysis (ie, glycogen degradation) but *inhibits* enzymes involved in glycogenesis (eg, glycogen synthesis). By regulating both processes in this way, PKA regulation prevents the occurrence of a futile cycle, in which glycogenesis and glycogenolysis would both happen at the same time in a single cell, leading to wasting of energy.

For example, PKA is one of several enzymes that directly phosphorylates and *inhibits* glycogen synthase, thereby directly inhibiting the rate-determining step of glycogenesis. In contrast, PKA causes the phosphorylation and *activation* of glycogen phosphorylase, which catalyzes the rate-determining step of glycogenolysis. However, PKA does *not* phosphorylate phosphorylase directly; instead, it phosphorylates a regulatory protein known as *phosphorylase kinase*. Phosphorylase kinase *then* phosphorylates phosphorylase. This process, from glucagon through phosphorylase activation, is shown in Figure 11.64.

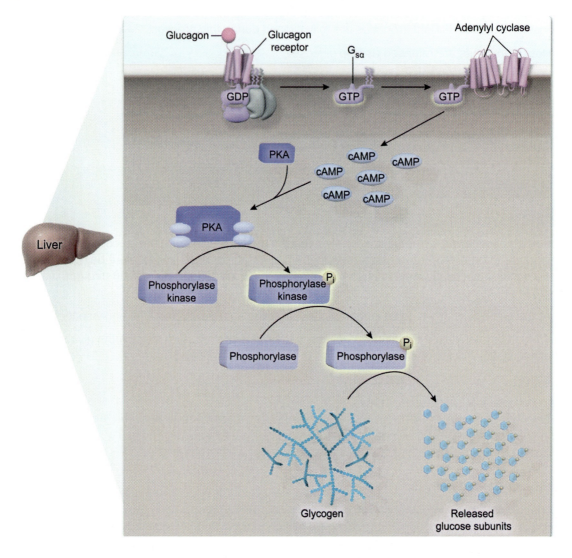

Figure 11.64 Glucagon activates a GPCR, which leads to activation of PKA, phosphorylation of phosphorylase kinase, phosphorylation of phosphorylase, and, finally, glycogenolysis.

Why go through this roundabout way of phosphorylating phosphorylases? One of the benefits of having several intermediary enzymes is that the signal from a single glucagon agonist can be amplified manyfold. Each instance of $G_{s\alpha}$-mediated activation of adenylyl cyclase induces a signaling cascade (see Lesson 3.2) that can ultimately activate thousands of phosphorylase enzymes.

Note that although glucagon activates a catabolic pathway in this case, it may also activate anabolic pathways in other contexts. Recall from Lesson 11.2 that in the liver, glucagon acts through PKA to activate the anabolic process of gluconeogenesis and to inhibit the catabolic process of glycolysis. Therefore, when studying regulatory pathways, it is important to remember not only general trends of signaling pathways but also tissue-specific roles in physiology (Figure 11.65).

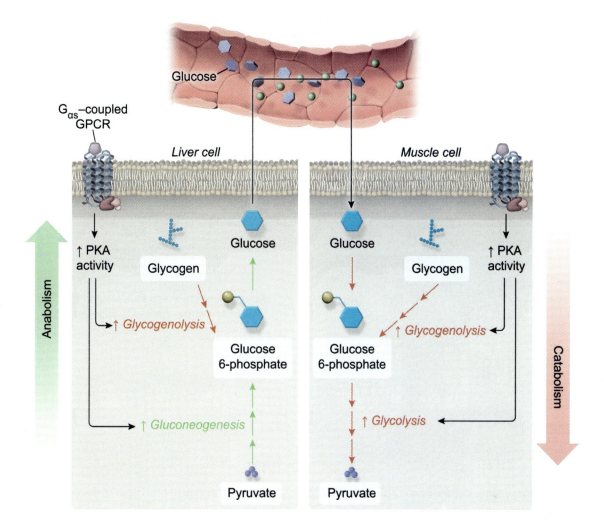

Figure 11.65 A given hormone can activate different pathways in different cells, or different types of pathways in the same cell, depending on the needs of the organism and the tissue.

The fight-or-flight hormone epinephrine (also known as adrenaline) also regulates phosphorylase, in this case by binding the β-adrenergic receptor. Like the glucagon receptor, the β-adrenergic receptor is a $G_{s\alpha}$-coupled GPCR whose activation leads to activation of PKA, and ultimately the phosphorylation (and activation) of both phosphorylase kinase and phosphorylase. This allows muscle cells to have sufficient glucose 6-phosphate released in anticipation of a fight-or-flight event.

In addition, allosteric effectors interact with the covalent phosphorylation of phosphorylase in tissue-specific ways to affect regulation. For example, liver tissue, which undergoes glucagon-stimulated glycogenolysis to produce glucose for *other* tissues, contains an isoform of phosphorylase that is feedback-inhibited by free glucose.

Muscle tissue, which undergoes glycogenolysis to provide fuel for *itself*, contains an isoform of phosphorylase that is stimulated by AMP, which accumulates when energy stores are low. Muscle phosphorylase kinase is also allosterically stimulated by calcium ions (Ca^2), which are released during stimulation of muscle contraction (see Biology Lesson 17.1).

In this way, allosteric and covalent regulation work together to ensure tissues have the amount of glucose they need. Endocrine hormones such as glucagon and epinephrine travel through the blood, activating glycogenolysis in anticipation of glucose need and therefore priming muscles *throughout the body*. Calcium and AMP can further enhance glycogenolysis in the *specific muscle cells* that need the most energy (eg, actively contracting muscles, Figure 11.66).

Figure 11.66 Allosteric and hormone-initiated covalent mechanisms work together to ensure specific tissues are supplied with the appropriate energy for their needs.

Regulation of Glycogenesis

Glucagon and epinephrine can both bind a GPCR that leads to activation of PKA and the eventual stimulation of glycogenolysis (degradation), whereas insulin binds a receptor tyrosine kinase (see Concept 4.2.09) that leads to the eventual stimulation of glycogenesis (synthesis). To prevent a futile cycle, the phosphorylation signals that stimulate glycogenolysis must be removed whenever insulin stimulates glycogenesis. Therefore, one of the most important downstream effects of insulin signaling in terms of energy metabolism is the activation of **protein phosphatases** (eg, phosphoprotein phosphatase 1, PP1).

The specific pathway leading from insulin receptor activation to phosphatase activation is more complex and less straightforward than GPCR pathways and is therefore unlikely to be directly tested on the exam.

In brief, the binding of insulin to its dimeric receptor causes **autophosphorylation**—in other words, each monomer is activated to phosphorylate tyrosine residues on the other monomer of the protein. This activates the receptor so that it can more efficiently phosphorylate tyrosine residues on other substrates. Specific regulatory proteins recognize and bind to phosphotyrosine residues, which causes the protein to undergo a conformational change. The signaling pathway eventually activates several protein phosphatases.

The rate-limiting enzyme of glycogen synthesis is **glycogen synthase**. For pathways commonly tested on the exam, a general trend is that dephosphorylation *activates* enzymes involved in glycogenesis and *inhibits* enzymes involved in glycogenolysis. Therefore, insulin stimulates phosphatases to dephosphorylate and *activate* glycogen synthase and subsequently activate glycogenesis. Meanwhile, the phosphatases also dephosphorylate and *inhibit* phosphorylase and phosphorylase kinase, thereby inhibiting glycogenolysis and preventing a futile cycle (Figure 11.67).

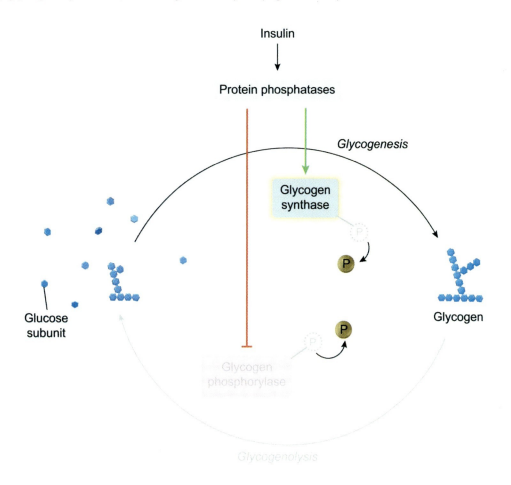

Figure 11.67 To prevent a futile cycle, phosphatases activated by insulin lead to the stimulation of glycogenesis and the inhibition of glycogenolysis.

The phosphate group on glycogen synthase, which insulin removes, can be added back by PKA but typically only after glucagon or epinephrine stimulation. At rest (ie, without hormone stimulation) glycogen synthase is phosphorylated by its regulatory enzyme glycogen synthase kinase (GSK). The action of GSK prevents glycogenesis until it is signaled for.

Insulin signaling therefore must not only dephosphorylate and activate glycogen synthase, but also inhibit GSK. Given the general trend that enzymes of glycogenesis are inhibited by phosphorylation, one of the insulin-dependent kinases phosphorylates and inhibits GSK. Inhibition of GSK, combined with dephosphorylation of glycogen synthase, allows glycogenesis to occur following insulin stimulation (Figure 11.68).

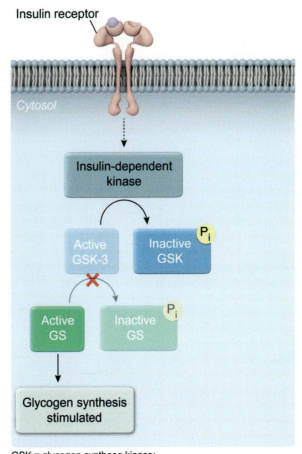

Figure 11.68 Glycogen synthase kinase (GSK) is a regulatory enzyme that normally inhibits glycogen synthesis. Insulin inactivates GSK, which allows glycogenesis to proceed.

☑ Concept Check 11.8

What is the phosphorylation state and activation state of each of the following enzymes in the liver following 1) glucagon stimulation, and 2) insulin stimulation?

a. Phosphorylase
b. Phosphorylase kinase
c. Glycogen synthase
d. Glycogen synthase kinase
e. Phosphofructokinase-2
f. Fructose-2,6-bisphosphatase

Solution

Note: The appendix contains the answer.

Glucose 6-phosphate, which accumulates during periods of fuel excess, allosterically activates glycogen synthase and makes any phosphate groups present more accessible to phosphatases. This makes the enzyme more prone to *de*phosphorylation and an even more robust activation.

Lesson 12.1
The Citric Acid Cycle

Introduction

Lesson 11.1 examines glucose catabolism, including glycolysis, in which glucose is converted to two pyruvate molecules. Glycolysis requires NAD⁺, and under anaerobic conditions (ie, when oxygen is absent or scarce) NAD⁺ is regenerated by fermentation. Under aerobic conditions, however, NAD⁺ is regenerated through **aerobic respiration** (Figure 12.1). After glycolysis, aerobic respiration continues with the citric acid cycle, which produces NADH and FADH₂. These molecules then enter the electron transport chain, where they ultimately pass their electrons to molecular oxygen (O₂). These reactions provide the energy needed for ATP synthesis.

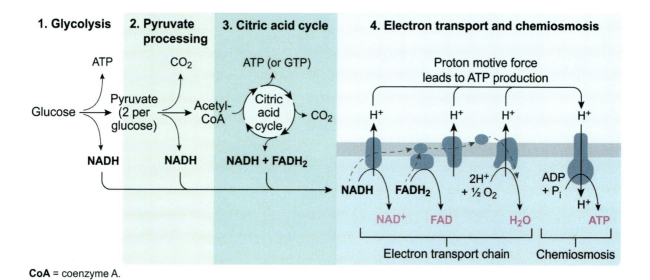

CoA = coenzyme A.

Figure 12.1 Overview of glucose catabolism through aerobic respiration.

This lesson focuses on the **citric acid cycle**, also known as the **Krebs cycle** or the **tricarboxylic acid (TCA) cycle**. Specifically, it focuses on the citric acid cycle using the pyruvate generated by glycolysis as a starting material. Other molecules such as fatty acids and amino acids may also enter the citric acid cycle, and these processes will be covered in Chapter 13.

12.1.01 Pyruvate Entry into the Citric Acid Cycle

The end product of glycolysis is pyruvate. Once formed, pyruvate typically has one of two fates. Under anaerobic conditions, pyruvate is fermented to form either ethanol or lactate, depending on the organism (see Concept 11.1.04). Under aerobic conditions (ie, abundant oxygen), many organisms instead send pyruvate into the citric acid cycle.

In eukaryotes, the citric acid cycle occurs in the mitochondria, whereas glycolysis occurs in the cytosol. Consequently, pyruvate must be transported from the cytosol into the mitochondria prior to use in the citric acid cycle. Mitochondria consist of two membranes that surround the mitochondrial matrix. The outer membrane is porous and permeable to most ions and small molecules. The inner membrane is much more selective but contains transport proteins that carry pyruvate into the matrix, as shown in Figure 12.2.

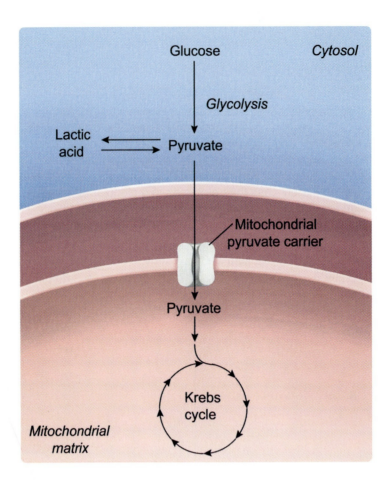

Figure 12.2 Pyruvate transport into the mitochondrial matrix.

The mitochondrial matrix contains multiple proteins involved in catabolism, including the **pyruvate dehydrogenase complex (PDC)** shown in Figure 12.3. This complex consists of three individual enzymes that collectively catalyze the decarboxylation of pyruvate to form acetyl coenzyme A (acetyl-CoA) and CO_2. These enzymes are

1. **Pyruvate dehydrogenase**. This enzyme, often called the E1 component of the complex, uses the molecule thiamine pyrophosphate (TPP) as a coenzyme. TPP nucleophilically attacks the electrophilic carbon at position two (C2) of pyruvate, converting the carboxyl group at position one into CO_2.

2. **Dihydrolipoyl transacetylase**. This enzyme, often called E2, uses a sulfur-containing molecule called lipoic acid (or lipoate) as a cofactor. Lipoic acid attacks the electrophilic carbon and forms a bond with it, forming acyl-lipoate and releasing TPP as a leaving group. Coenzyme A, often represented as CoA-SH, then removes the acyl group (specifically, a 2-carbon acetyl group) from lipoate to form acetyl-CoA.

3. **Dihydrolipoyl dehydrogenase**. This enzyme, called E3, uses the coenzyme FAD to oxidize lipoic acid so that it can repeat the E2 reaction. The FAD is converted to $FADH_2$, which then transfers its electrons to NAD^+, forming NADH and regenerating FAD.

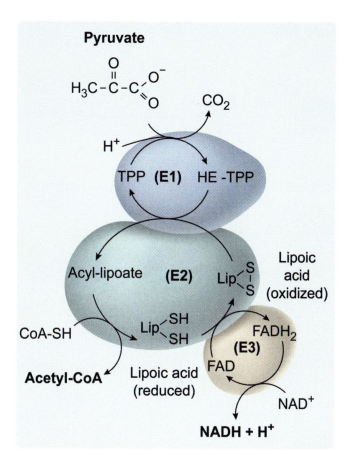

Figure 12.3 Overview of the mechanism of the pyruvate dehydrogenase complex.

The net reaction catalyzed by the PDC, then, is

$$\text{Pyruvate} + NAD^+ + \text{CoA-SH} \rightarrow \text{Acetyl-CoA} + NADH + CO_2$$

The resulting acetyl-CoA enters the citric acid cycle, the NADH enters the electron transport chain (see Lesson 12.2), and the CO_2 is eventually exhaled.

Pyruvate dehydrogenase is inactivated by phosphorylation. The phosphorylation reaction is carried out by pyruvate dehydrogenase kinase (PDK), which is allosterically activated by the products of the PDC: acetyl-CoA and NADH. PDK activation results in the *inactivation* of the PDC, and therefore this pathway works as a form of feedback inhibition.

In contrast, PDK is *downregulated* by the *reactants* of the PDC: pyruvate, coenzyme A, and NAD^+. In this way, buildup of PDC reactants results in the *release of inhibition*—and therefore an indirect *activation*—of the PDC. Interactions between PDK, PDC, and several metabolites are shown in Figure 12.4.

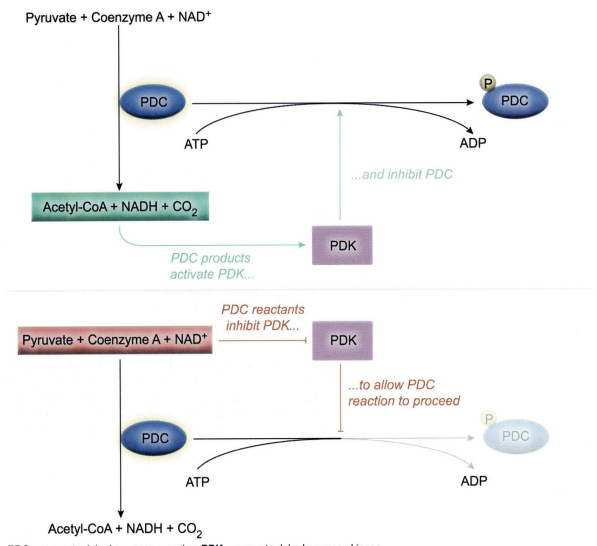

PDC = pyruvate dehydrogenase complex, **PDK** = pyruvate dehydrogenase kinase.

Figure 12.4 Pyruvate dehydrogenase kinase (PDK) action on the PDC, and allosteric regulation of PDK.

The phosphate is removed by pyruvate dehydrogenase phosphatase, which reactivates the PDC. Pyruvate dehydrogenase phosphatase is activated by insulin and by calcium ions. This regulation can help stimulate aerobic respiration during periods of glucose excess and vigorous muscular contraction, respectively.

Chapter 12: Aerobic Respiration

 Concept Check 12.1

Pyruvate is prepared with the radioactive isotope carbon-14 at position 2, as indicated by the asterisk in the following figure.

Pyruvate

When this pyruvate molecule is acted upon by the PDC,

1) Which molecule will contain the radiolabeled carbon after the reaction is complete?
2) Which atom in that molecule will be labeled?

Solution

Note: The appendix contains the answer.

12.1.02 Citric Acid Cycle Reactions

Once pyruvate is converted to acetyl-CoA, it can enter the **citric acid cycle**. The citric acid cycle is an eight-step metabolic pathway that begins when a two-carbon group (the acetyl component of **acetyl-CoA**) combines with a four-carbon molecule (**oxaloacetate**) to form the six-carbon molecule **citrate**.

Over the course of the cycle, two carbon atoms are removed, one at a time, each in the form of CO_2. This converts the six-carbon molecule first into a five-carbon molecule, and then into a four-carbon molecule, which eventually regenerates oxaloacetate. The regenerated oxaloacetate can then react with a *new* acetyl-CoA molecule to repeat the cycle.

The first half of the citric acid cycle (ie, three of the first four steps) is predominantly *irreversible*. Under physiological conditions, the reactions are highly exergonic, and two of them release CO_2, which quickly exits the mitochondria. The three irreversible reactions in the first half of the cycle are the primary points of regulation. The final four reactions, in contrast, are reversible. Figure 12.5 shows an overview of the citric acid cycle.

Chapter 12: Aerobic Respiration

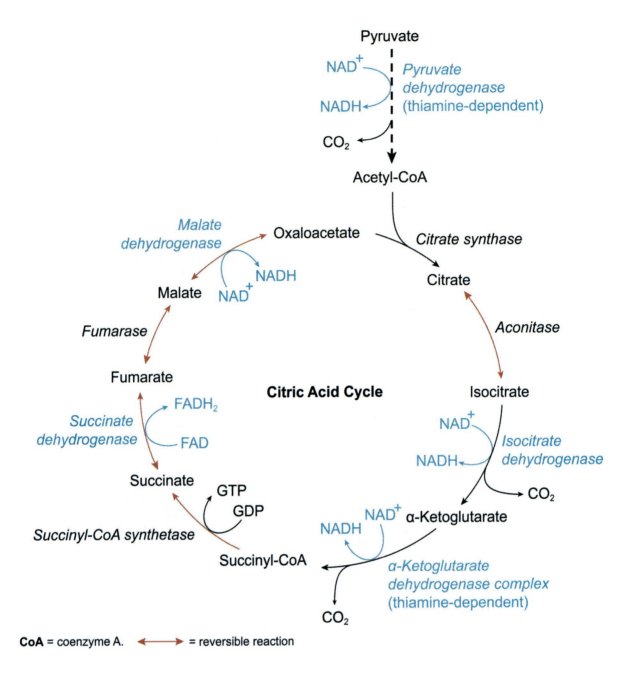

Figure 12.5 Overview of the citric acid cycle.

Step 1: Acetyl-CoA and Oxaloacetate Combine to Become Citrate

The first step of the citric acid cycle is the condensation of the two-carbon acetyl group of **acetyl-CoA** and the four-carbon molecule **oxaloacetate**. A water molecule is also consumed in this reaction, while free coenzyme A (CoA-SH) is produced. The product is the six-carbon molecule **citrate** (the conjugate base of citric acid) from which the cycle takes its name. Citric acid contains three carboxylic acid groups, and accordingly the cycle is also called the tricarboxylic acid (TCA) cycle.

The formation of citrate is catalyzed by the enzyme **citrate synthase**, a transferase that transfers an acetyl group from acetyl-CoA to oxaloacetate. The reaction is shown in Figure 12.6.

Chapter 12: Aerobic Respiration

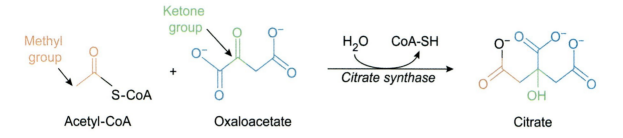

Figure 12.6 Step 1 of the citric acid cycle, catalyzed by citrate synthase.

Importantly, this reaction is irreversible, meaning *acetyl-CoA cannot be regenerated from citrate* by this enzyme. Consequently, this step is carefully regulated. Citrate synthase activity is *downregulated* by citrate, NADH, succinyl-CoA, and ATP, each of which are products of citrate synthase or of downstream enzymes. In other words, these molecules cause feedback inhibition. In contrast, ADP allosterically *upregulates* citrate synthase activity.

Step 2: Citrate Becomes Isocitrate

Citrate is a symmetric molecule with a hydroxyl group on its central carbon. To facilitate formation of α-ketoglutarate in Step 3, the hydroxyl group must move to one of the methylene (–CH_2–) groups of citrate. In other words, *citrate must isomerize* to form isocitrate. This reaction is the only reversible reaction in the first half of the cycle and is facilitated by the enzyme **aconitase**.

Aconitase first removes the hydroxyl group along with a hydrogen atom from a methylene group, forming a water molecule. This dehydration results in the formation of a double bond, yielding the molecule *cis*-aconitate. This classifies aconitase as a lyase.

After the dehydration reaction, water is then added *back* to *cis*-aconitate in a different configuration (ie, it is rehydrated) to form isocitrate. Consequently, the *net* reaction catalyzed by aconitase is an isomerization (the hydroxyl group and hydrogen atom switch places). However, because the enzyme accomplishes isomerization by catalyzing two separate lyase reactions, it is not classified as an isomerase despite its name. Figure 12.7 shows the conversion of citrate to isocitrate.

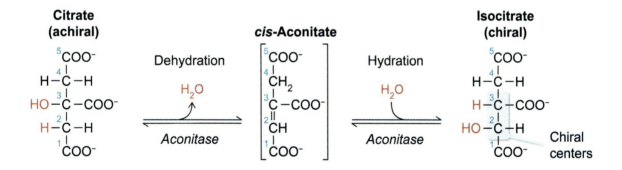

Figure 12.7 Conversion of citrate to isocitrate through a *cis*-aconitate intermediate.

Step 3: Isocitrate Is Decarboxylated to Form α-Ketoglutarate

From an energy standpoint, the primary purpose of the citric acid cycle is to produce the reduced cofactors NADH and $FADH_2$, which later enter the electron transport chain. The third step of the citric acid cycle, in which isocitrate becomes α-ketoglutarate, is the first of three to produce NADH. This step is facilitated by the enzyme **isocitrate dehydrogenase**.

Conversion of **isocitrate** (a six-carbon molecule) to **α-ketoglutarate** (a five-carbon molecule) occurs through **oxidative decarboxylation**. One of the carboxyl groups of isocitrate is oxidized to CO_2 and released. In addition, the hydroxyl group of isocitrate is oxidized to a ketone. Consequently, isocitrate loses two electrons and CO_2 in the reaction. CO_2 cannot readily be added back to α-ketoglutarate, so this reaction is *irreversible*.

When an oxidation reaction occurs, a reduction reaction must occur simultaneously, because the electrons lost by one molecule must be gained by another. In this case, the two electrons lost by isocitrate are transferred to NAD^+ as a hydride ion, forming NADH. Because the reaction is a redox reaction, isocitrate dehydrogenase is an oxidoreductase. Figure 12.8 shows the oxidative decarboxylation of isocitrate.

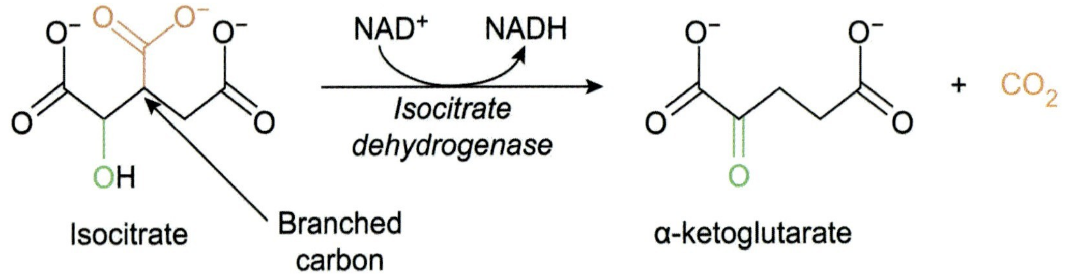

Figure 12.8 Conversion of isocitrate to α-ketoglutarate.

Isocitrate dehydrogenase is closely regulated. Like citrate synthase, isocitrate dehydrogenase is allosterically inhibited by ATP and upregulated by ADP. Interestingly, this enzyme is also upregulated by calcium ions (Ca^{2+}). Calcium is released from the sarcoplasmic reticulum as part of the process of muscle contractions, which consume ATP. Therefore, Ca^{2+} signals that more ATP is needed.

Step 4: α-Ketoglutarate Is Decarboxylated and Becomes Succinyl-CoA

This step, facilitated by the **α-ketoglutarate dehydrogenase** complex, is another irreversible oxidative decarboxylation in which the carboxyl group linked to the ketone carbon of *α-ketoglutarate is oxidized* to CO_2 (Figure 12.9). The ketone carbon is further oxidized by reacting with coenzyme A to form a thioester called **succinyl-CoA**. The succinyl component contains the four carbons remaining after the loss of CO_2. The two electrons lost from α-ketoglutarate are transferred to NAD^+ as a hydride, producing the second of three NADH molecules. The α-ketoglutarate dehydrogenase complex is another *oxidoreductase*.

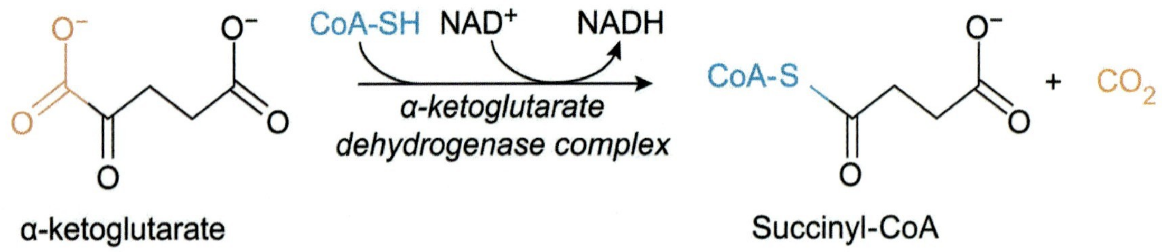

Figure 12.9 Oxidative decarboxylation of α-ketoglutarate to form succinyl-CoA.

α-Ketoglutarate dehydrogenase, like citrate synthase, is allosterically downregulated by succinyl-CoA and NADH (its products). Like isocitrate dehydrogenase, this enzyme is allosterically upregulated by Ca^{2+}.

Note that at this point in the pathway, three CO_2 molecules have been produced in the mitochondria. The first was produced in the step before the cycle began, when pyruvate became acetyl-CoA. The second CO_2 molecule was produced when isocitrate became α-ketoglutarate (Step 3), and the third was produced when α-ketoglutarate became succinyl-CoA (Step 4). Pyruvate is a three-carbon molecule.

Therefore, at this point in the process, the carbon atoms that entered as pyruvate have been offset by carbons that exit as CO₂.

Importantly, the two carbons contributed by acetyl-CoA are still present in succinyl-CoA at this point, while two *other* carbons were removed as CO₂, with the *net effect* being that three carbons went into the mitochondria as pyruvate, and three came out as CO₂. Each glucose molecule that entered glycolysis produced two pyruvate molecules (six carbons), so once both pyruvate molecules enter the mitochondria, all six carbons from glucose are *effectively* converted to CO₂ by these steps. Figure 12.10 highlights the carbons contributed by acetyl-CoA and the carbons that become CO₂.

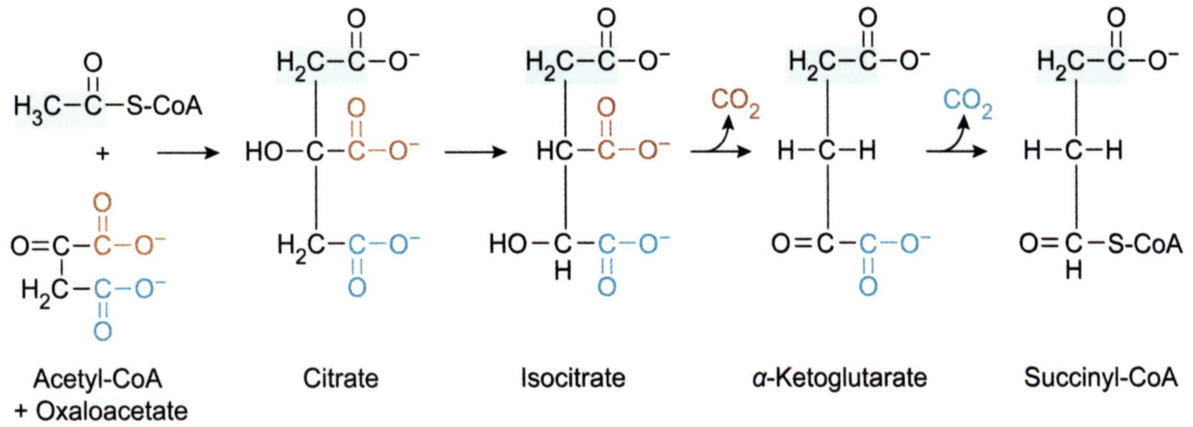

[Figure showing the conversion of Acetyl-CoA + Oxaloacetate → Citrate → Isocitrate → α-Ketoglutarate → Succinyl-CoA, with two CO₂ molecules released in the last two steps]

Figure 12.10 The two carbons that acetyl-CoA contributes to citrate are still present in succinyl-CoA, while two other carbons (contributed by oxaloacetate) become CO₂.

Step 5: Succinyl-CoA Is Converted to Succinate

The thioester bond in succinyl-CoA is a high-energy bond. In humans, the enzyme **succinyl-CoA synthetase** catalyzes formation of GTP by condensing GDP and inorganic phosphate. In the same reaction, succinyl-CoA is broken into the four-carbon molecule **succinate** and **coenzyme A** (Figure 12.11). Interestingly, this reaction is *reversible* because the high-energy phosphoanhydride bond in GTP is energetically similar to that of the thioester in succinyl-CoA.

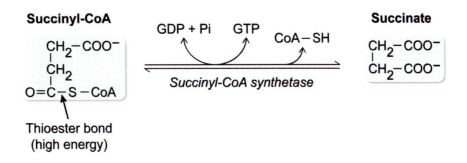

Figure 12.11 Reversible conversion of succinyl-CoA to succinate.

Some organisms transfer the phosphate group to ADP, forming ATP instead of GTP. Note that GTP and ATP are energetically equivalent (see Lesson 10.1). In addition, in organisms that produce GTP in this step, the phosphate group may subsequently be transferred to ADP to produce ATP.

The enzyme in this reaction is named succinyl-CoA synthetase, but in the context of the citric acid cycle, succinyl-CoA is *broken down*; therefore, this enzyme is named for the *reverse* reaction: formation of succinyl-CoA using the energy provided by GTP hydrolysis. Based on this reaction, succinyl-CoA synthetase is a ligase, linking two molecules (succinate and coenzyme A) while separately hydrolyzing a nucleoside triphosphate.

Step 6: Succinate Becomes Fumarate

Succinate contains two methylene (–CH$_2$–) groups linked to each other through a sigma bond. Oxidation of this bond removes one hydrogen atom (including its electrons) from each carbon to form a C=C double bond. The resulting molecule is called **fumarate**. This reversible reaction is facilitated by the *oxidoreductase* enzyme **succinate dehydrogenase**, which effectively transfers the two hydrogen atoms to FAD, forming FADH$_2$. This is the *only* step in the citric acid cycle that produces FADH$_2$. Figure 12.12 shows oxidation of succinate to fumarate.

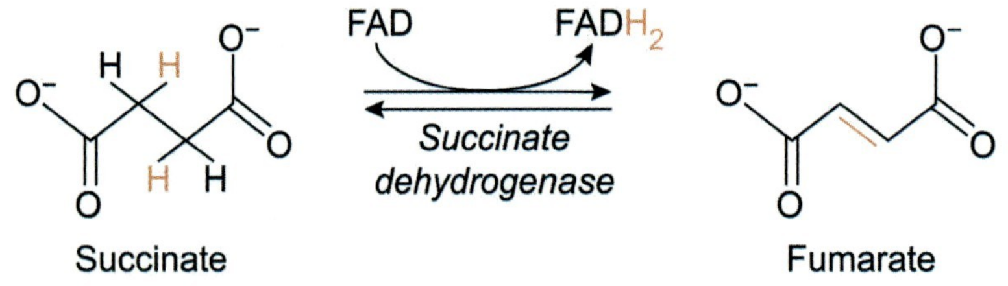

Figure 12.12 Oxidation of succinate to fumarate.

Importantly, succinate dehydrogenase is part of a larger enzyme complex found in the inner mitochondrial membrane: Complex II of the electron transport chain (see Lesson 12.2). The succinate dehydrogenase component of this complex does not use *free* FAD. Instead, FAD is bound to the enzyme as a prosthetic group. As soon as FADH$_2$ is produced, it quickly transfers its electrons through the rest of the complex, regenerating FAD.

Step 7: Fumarate Becomes L-Malate

In this reversible step, the carbon-carbon double bond formed in the previous step is broken by addition of water. This reaction produces L-malate and is catalyzed by the enzyme **fumarase**. Although water is used to break a bond in this reaction, in this case it is a **pi bond**. Therefore, as noted in Lesson 4.2, fumarase is *not* classified as a hydrolase, but is instead classified as a lyase.

Although the product of this reaction is commonly referred to simply as malate, it is important to note that malate is a chiral molecule. Specifically, the hydroxyl carbon is a chiral center, and the reaction catalyzed by fumarase is stereospecific; it always produces the L-form of malate. Figure 12.13 shows the reaction catalyzed by fumarase.

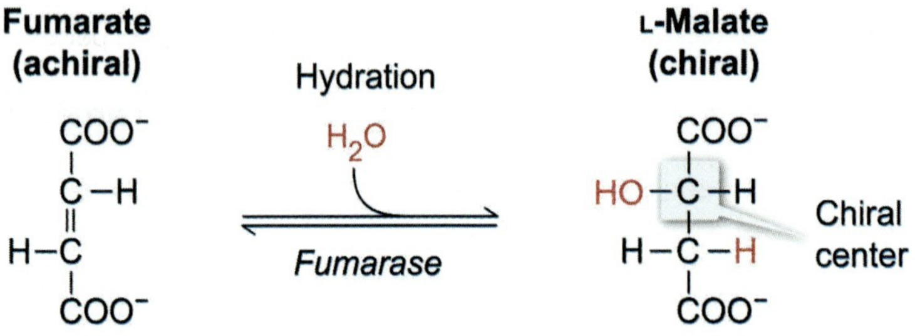

Figure 12.13 Hydration of fumarate to form L-malate.

Step 8: L-Malate Is Oxidized to Oxaloacetate

In the final step of the citric acid cycle, the hydroxyl group of L-malate is oxidized to a ketone to yield oxaloacetate. In this reversible process, which is catalyzed by an **oxidoreductase** called **malate dehydrogenase**, two electrons are transferred to NAD⁺ to form the final NADH molecule of the cycle (Figure 12.14). The resulting oxaloacetate molecule can then react with another acetyl-CoA molecule to begin a new cycle.

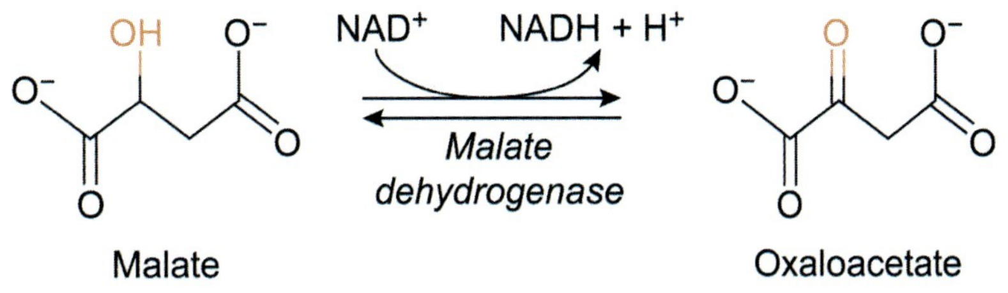

Figure 12.14 Oxidation of malate to oxaloacetate.

12.1.03 Citric Acid Cycle Products

The net reaction for the citric acid cycle is

$$\text{Acetyl-CoA} + 3\,\text{NAD}^+ + \text{FAD} + 2\,\text{H}_2\text{O} + \text{GDP} + \text{P}_i \rightarrow 2\,\text{CO}_2 + 3\,\text{NADH} + \text{FADH}_2 + \text{GTP} + \text{Coenzyme A} + 3\,\text{H}^+$$

Note that because oxaloacetate is consumed and *then regenerated* by the cycle, it is *not* included as a net reactant *or* product. Similarly, none of the intermediates are included because they are consumed immediately after they are produced.

In addition to the reactions of the cycle itself, entry of pyruvate into the cycle follows the reaction

$$\text{Pyruvate} + \text{NAD}^+ + \text{Coenzyme A} \rightarrow \text{Acetyl-CoA} + \text{NADH} + \text{CO}_2$$

Energetically, the most important products are NADH, FADH$_2$, and GTP, each of which can be used to generate ATP or, in the case of GTP, can be used directly to power various biological processes. Table 12.1 shows the numbers of each high-energy product formed per pyruvate molecule that enters the cycle. Because glycolysis produces two pyruvate molecules per glucose, each glucose molecule can produce double the number of high-energy products.

Table 12.1 High-energy products of pyruvate dehydrogenase in the citric acid cycle, per pyruvate or glucose.

Product	Number of products per pyruvate	Number of products per glucose
NADH	4	8
FADH$_2$	1	2
GTP	1	2

This lesson has focused on the citric acid cycle starting with the reaction between acetyl-CoA and oxaloacetate because this is the point of entry for metabolites derived from glucose. However, it is possible for other metabolites to enter the citric acid cycle at different points.

For example, certain amino acids can be converted directly to α-ketoglutarate (discussed further in Lesson 13.2), which can participate in Steps 4–8 of the cycle to become oxaloacetate. In this case, two of the NADH-producing steps are skipped: conversion of pyruvate to acetyl-CoA and conversion of isocitrate to α-ketoglutarate. This is shown in Figure 12.15.

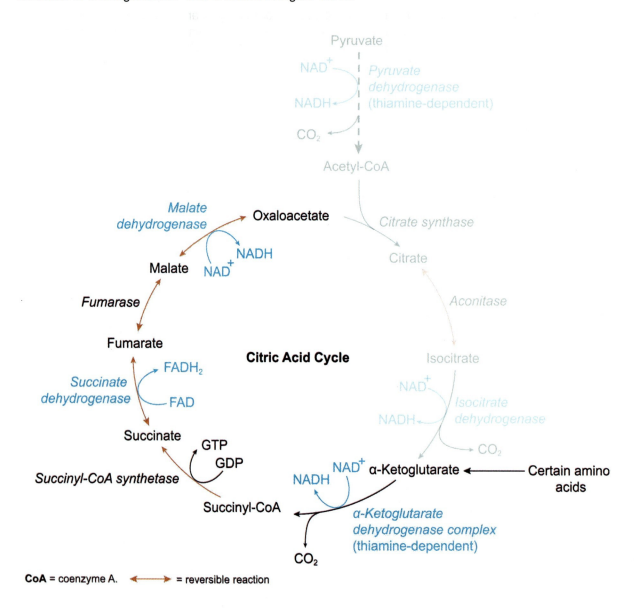

Figure 12.15 Entry into the cycle at a point other than acetyl-CoA alters the amount of NADH, FADH$_2$, and GTP produced.

Once oxaloacetate forms, it cannot continue through another round of the cycle unless it reacts with a new acetyl-CoA. Consequently, when α-ketoglutarate enters the cycle directly, it only provides enough energy to produce two NADH molecules instead of four (FADH$_2$ and GTP production are not affected in this case). The energy for any further NADH production comes from the addition of new acetyl-CoA. In other words, for the citric acid cycle to continue from one round to the next, a constant influx of acetyl-CoA is required.

Concept Check 12.2

A certain molecule is metabolized to produce one acetyl-CoA molecule and one succinyl-CoA molecule. The acetyl-CoA molecule enters the citric acid cycle by reacting with oxaloacetate to become citrate, and the succinyl-CoA molecule enters directly. Once both the citrate and succinyl-CoA molecules are converted to oxaloacetate, what is the final yield of NADH, FADH$_2$, and GTP?

Solution

Note: The appendix contains the answer.

Oxaloacetate is a component of gluconeogenesis. Consequently, molecules that enter the citric acid cycle and eventually become oxaloacetate are **glucogenic** (ie, they can be used as a starting material to produce glucose). Acetyl-CoA, however, is *not* classified as glucogenic. The two carbons contributed to citrate by acetyl-CoA *are* incorporated into oxaloacetate, but two *other* carbons are lost along the way; therefore the *net* number of carbons available for glucose synthesis is *not* increased by running acetyl-CoA through the cycle.

Another way to view this is that acetyl-CoA enters the cycle by *reacting* with oxaloacetate, and ultimately oxaloacetate is *regenerated*. Therefore, adding acetyl-CoA into the cycle does not generate any *new* oxaloacetate but instead consumes *existing* oxaloacetate before regenerating it. Figure 12.16 shows two carbons entering the cycle when acetyl-CoA reacts with oxaloacetate and two carbons exiting as CO_2 before oxaloacetate is regenerated.

Chapter 12: Aerobic Respiration

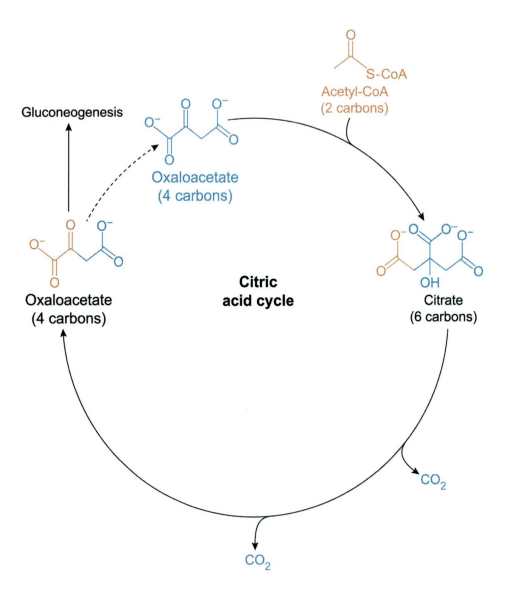

Figure 12.16 Acetyl-CoA is not glucogenic because the carbons it contributes are effectively lost during the citric acid cycle.

In contrast, if a molecule enters the cycle at *any other point* (eg, glutamate becomes α-ketoglutarate, see Lesson 13.2), these molecules do generate *new* oxaloacetate that can be converted to glucose.

Lesson 12.2

The Electron Transport Chain

Introduction

The **electron transport chain** (ETC) is a series of oxidation-reduction reactions catalyzed by protein complexes in the inner mitochondrial membrane. These reactions begin with the oxidation of either NADH or $FADH_2$, which were produced by the citric acid cycle and other upstream metabolic processes. The electrons from NADH and $FADH_2$ are passed through a series of intermediates, with molecular oxygen (O_2) being the final electron acceptor. Oxygen is converted to water by this process. Figure 12.17 provides an overview of the ETC.

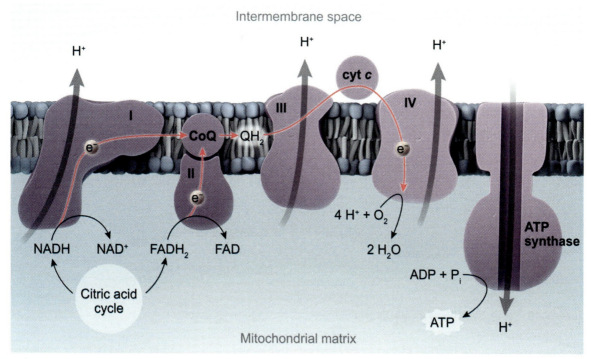

*I = Complex I, II = Complex II, etc.

Figure 12.17 Overview of the electron transport chain.

Three of the complexes in the ETC use the energy released by the reactions they catalyze to pump protons out of the mitochondrial matrix and into the intermembrane space (ie, the region between the inner and outer mitochondrial membranes). This produces a proton gradient such that the matrix has a lower proton concentration (and therefore a higher pH) than the intermembrane space. This gradient is critical for the production of ATP by ATP synthase, as discussed in Lesson 12.3.

This lesson explores the functions of each of the four complexes in the ETC.

12.2.01 Complex I

The Complex I Reaction

The first complex of the ETC (Complex I) is named NADH:ubiquinone oxidoreductase or, alternatively, NADH dehydrogenase. As these names suggest, Complex I catalyzes the **oxidation of NADH** by dehydrogenation (specifically, by removal of a hydride ion). The hydride is transferred to ubiquinone (also called coenzyme Q, UQ, or Q), which becomes ubiquinol, also called UQH_2 or QH_2. The NADH becomes NAD^+, which can then return to the citric acid cycle (or other pathways) to again be reduced to NADH. The general reaction between NADH and ubiquinone is shown in Figure 12.18.

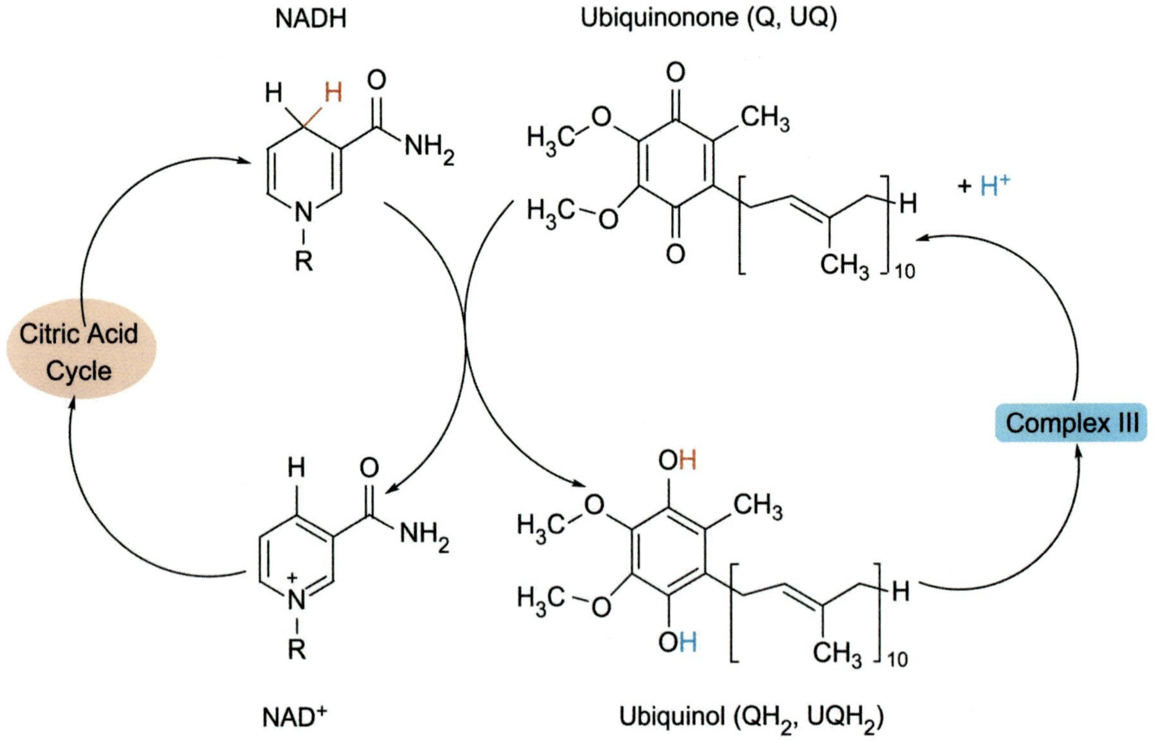

Figure 12.18 Net reaction between NADH and ubiquinone, facilitated by Complex I.

NADH is a relatively hydrophilic molecule. Therefore, this molecule is found in aqueous environments such as the cytosol or, in the context of the ETC, the mitochondrial matrix. **Ubiquinone**, on the other hand, is highly hydrophobic and is found within the **inner mitochondrial membrane**.

Accordingly, NADH interacts with the portion of Complex I that faces the aqueous mitochondrial matrix. The two electrons are passed to the transmembrane portion of the complex through a series of reactions, where they are finally passed to ubiquinone, as depicted in Figure 12.19.

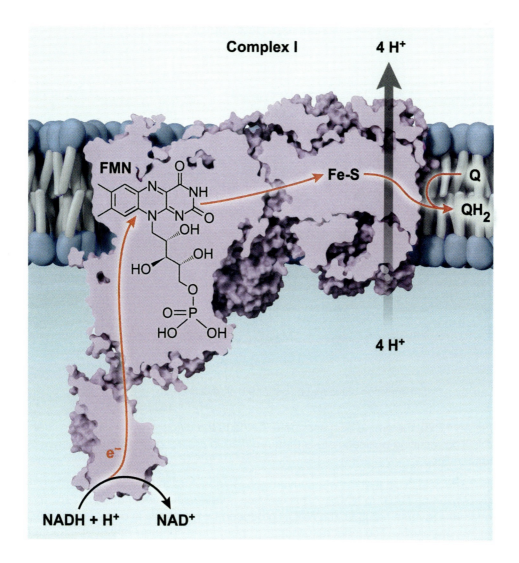

Figure 12.19 NADH passes two electrons to ubiquinone (Q) through Complex I, yielding NAD^+ and ubiquinol (QH_2).

The transfer of electrons from NADH to ubiquinone has a positive standard potential $E'^°$ of 0.365 V. Therefore, $\Delta G'^°$ is negative and the reaction is spontaneous (ie, it releases energy that can be used to do work). Complex I uses the energy released by this reaction to pump four protons (H^+) against their concentration gradient, out of the mitochondrial matrix and into the intermembrane space.

The Malate-Aspartate Shuttle

Lesson 11.1 explains that glycolysis produces NADH, which must be converted back to NAD^+ for glycolysis to continue. Under low oxygen conditions, NAD^+ is regenerated by fermentation. In contrast, when oxygen is present, NAD^+ may be regenerated by the action of Complex I of the electron transport chain.

However, the NAD^+ and NADH associated with glycolysis are found in the *cytosol* of eukaryotes, whereas the NAD^+ and NADH associated with the ETC are found in the *mitochondrial matrix*. The inner mitochondrial membrane is not permeable to either molecule, so NADH generated by glycolysis *cannot interact with Complex I directly*.

To allow regeneration of cytosolic NAD^+ without using fermentation, cells may use the **malate-aspartate shuttle**, shown in a simplified form in Figure 12.20.

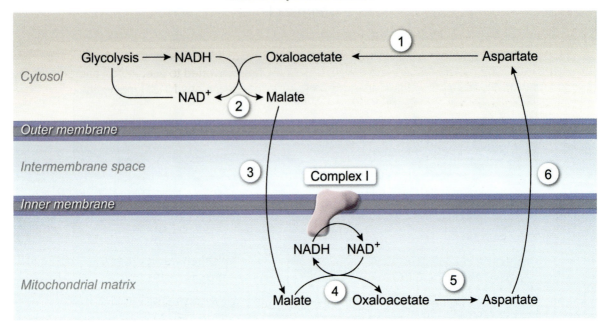

Figure 12.20 An abbreviated depiction of the malate-aspartate shuttle.

In Step 1 of Figure 12.20, the amino acid aspartate (found in the cytosol) is converted to oxaloacetate by transamination. The resulting oxaloacetate is then *reduced to malate* by a cytosolic version of **malate dehydrogenase** in Step 2.

Recall from the citric acid cycle that malate dehydrogenase converts malate to oxaloacetate while simultaneously converting NAD^+ to NADH. This reaction is *reversible*, and indeed occurs in reverse in the cytosol. Through this reverse reaction, the NADH produced by glycolysis is converted to NAD^+ to facilitate additional glycolysis. The inner mitochondrial membrane contains malate transporters that allow the resulting malate (with its new high-energy electrons) to enter the mitochondrial matrix (Step 3 of Figure 12.20).

Within the matrix, malate is converted back to oxaloacetate by *mitochondrial* malate dehydrogenase (the same enzyme used by the citric acid cycle). This reaction (Step 4 of Figure 12.20) also converts mitochondrial NAD^+ to NADH. The overall result of Steps 1–4 is that NADH in the cytosol is oxidized to NAD^+, while NAD^+ in the mitochondria is reduced to NADH. Therefore, the *electrons* from cytosolic NADH are transported into the mitochondria. The mitochondrial NADH can then be oxidized to NAD^+ by Complex I before repeating the process.

To allow the malate-aspartate shuttle to continue indefinitely, the oxaloacetate produced at the end of Step 4 is converted to aspartate in the mitochondria by a transamination reaction (Step 5 of Figure 12.20). The inner mitochondrial membrane contains aspartate transporters through which aspartate can enter the cytosol, where it may again be used to shuttle the electrons of cytosolic NADH to the mitochondrial matrix.

A second shuttle, the glycerol-3-phosphate shuttle, essentially converts NADH to $FADH_2$ and yields the same result as Complex II. This shuttle is covered in detail in Concept 12.2.02.

> **Concept Check 12.3**
>
> A molecule of glucose undergoes glycolysis. The NADH produced by this process interacts with the malate-aspartate shuttle, and the pyruvate molecules are converted to acetyl-CoA and enter the citric acid cycle. After all steps of the cycle are complete, how many NADH molecules enter the ETC from digestion of the glucose molecule?
>
> **Solution**
>
> *Note: The appendix contains the answer.*

12.2.02 Complex II

The Complex II Reaction

The citric acid cycle produces two reduced cofactors that enter the electron transport chain: NADH and $FADH_2$. Complex I is the entry point of NADH into the ETC, and Complex II is the entry point for $FADH_2$.

As discussed in Lesson 12.1, **Complex II is the succinate dehydrogenase enzyme** found in the citric acid cycle. When this enzyme oxidizes succinate to fumarate, it reduces FAD to $FADH_2$ in the process. This $FAD/FADH_2$ unit is a prosthetic group of the enzyme. $FADH_2$ passes its electrons through a series of reactions that converts $FADH_2$ back to FAD. As with Complex I, the electrons in Complex II are eventually transferred to ubiquinone (Q) in the inner membrane (see Figure 12.21).

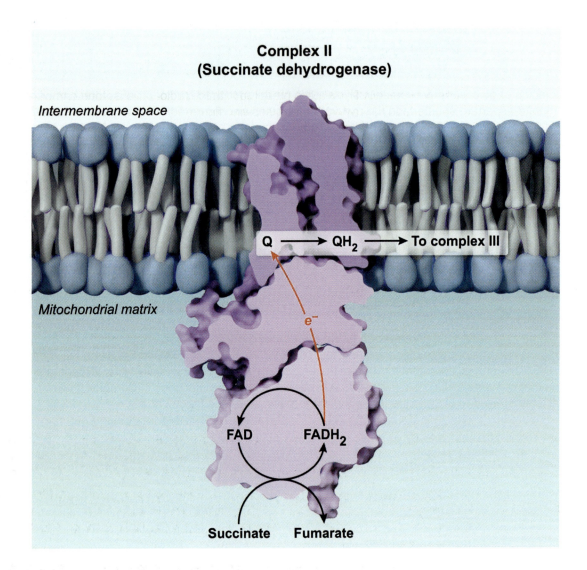

Figure 12.21 Complex II transfers electrons from succinate to FAD and then to ubiquinone.

The standard potential E'° for electron transfer from FADH$_2$ to ubiquinone is approximately +0.015 V. Although this standard potential is positive, and therefore $\Delta G'^{\circ}$ is negative, the magnitude of $\Delta G'^{\circ}$ is small (ie, close to 0). Consequently, although this reaction is spontaneous, it does not release enough energy to pump protons against their gradient. Therefore, *this reaction does not directly contribute to the pH gradient* across the mitochondrial membrane.

The Glycerol-3-Phosphate Shuttle

The **glycerol-3-phosphate shuttle** is another shuttle that allows electron transfer from cytosolic NADH to mitochondrial electron acceptors. In this shuttle, the *cytosolic* isoform of the enzyme glycerol-3-phosphate dehydrogenase removes two electrons from NADH and passes them to dihydroxyacetone phosphate (DHAP), an intermediate of glycolysis. The result is NAD$^+$ and glycerol 3-phosphate.

Glycerol 3-phosphate then transfers two electrons to an FAD prosthetic group within a different isozyme of glycerol-3-phosphate dehydrogenase, this one a transmembrane enzyme of the inner mitochondrial membrane. This process yields FADH$_2$ and restores DHAP. The resulting FADH$_2$ then passes the electrons to ubiquinone. The overall effect of this process is synthesis of ubiquinol *without* pumping any protons from the mitochondrial matrix to the intermembrane space. Therefore, although this shuttle does not use Complex II directly, it produces the same result. Figure 12.22 shows this shuttle.

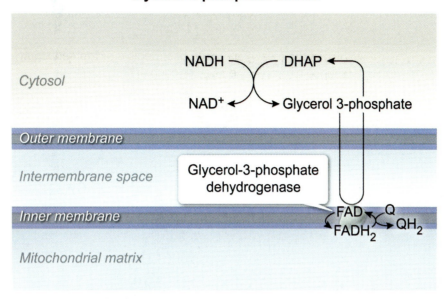

Figure 12.22 The glycerol-3-phosphate shuttle.

As discussed in Lesson 12.3, the amount of ATP produced depends on the *number of protons* pumped into the intermembrane space. The glycerol-3-phosphate shuttle consumes cytosolic NADH without pumping any protons, whereas the malate-aspartate shuttle allows cytosolic NADH to feed its electrons into Complex I, which *does* pump protons. Consequently, compared to the malate-aspartate shuttle, the glycerol-3-phosphate shuttle results in *fewer protons pumped* and therefore *less ATP produced*.

Although the conditions under which cells use the malate-aspartate shuttle as opposed to the glycerol-3-phosphate shuttle are not completely understood, the glycerol-3-phosphate shuttle requires fewer steps and is faster. Therefore, it is likely that this shuttle is used when cells need cytosolic NAD^+ quickly.

12.2.03 Complex III

The Complex III Reaction

Complexes I and II produce ubiquinol (the reduced form of ubiquinone) in their reactions. Other enzymes in other pathways also produce ubiquinol in the inner mitochondrial membrane (eg, the glycerol-3-phosphate shuttle). Once produced, ubiquinol can interact with Complex III of the ETC, also known as ubiquinone:cytochrome *c* oxidoreductase.

This complex facilitates transfer of electrons from ubiquinol to a protein called cytochrome *c* (abbreviated as cyt *c*), which is a soluble protein often found associated with the outer leaflet of the inner membrane (ie, cyt *c* is in the intermembrane space). Figure 12.23 depicts the overall reaction carried out by Complex III.

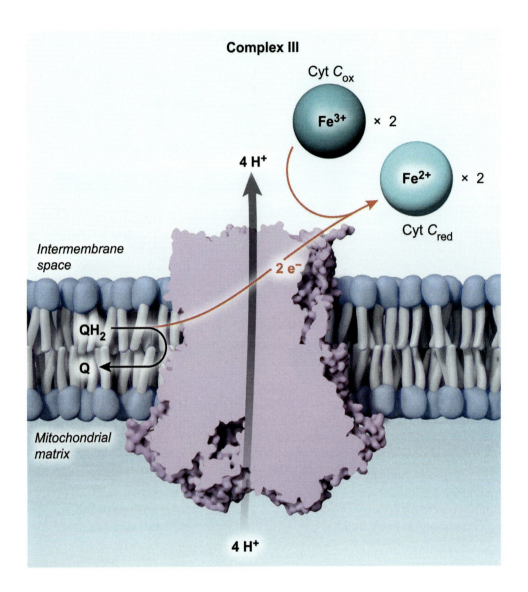

Figure 12.23 Net reaction of Complex III.

Cyt c is one of many proteins with a heme prosthetic group (a porphyrin ring with an iron center). The iron atom can exist in either the oxidized (+3) state or the reduced (+2) state. These states are represented as cyt c_{ox} and cyt c_{red}, respectively. Because iron's oxidation state only changes by one unit, cyt c_{ox} can accept one (and only one) electron from a donor such as ubiquinol. However, ubiquinol carries *two electrons* that it must donate to revert to its ubiquinone form. To accomplish this, ubiquinol and ubiquinone undergo a process called the Q cycle (Figure 12.24).

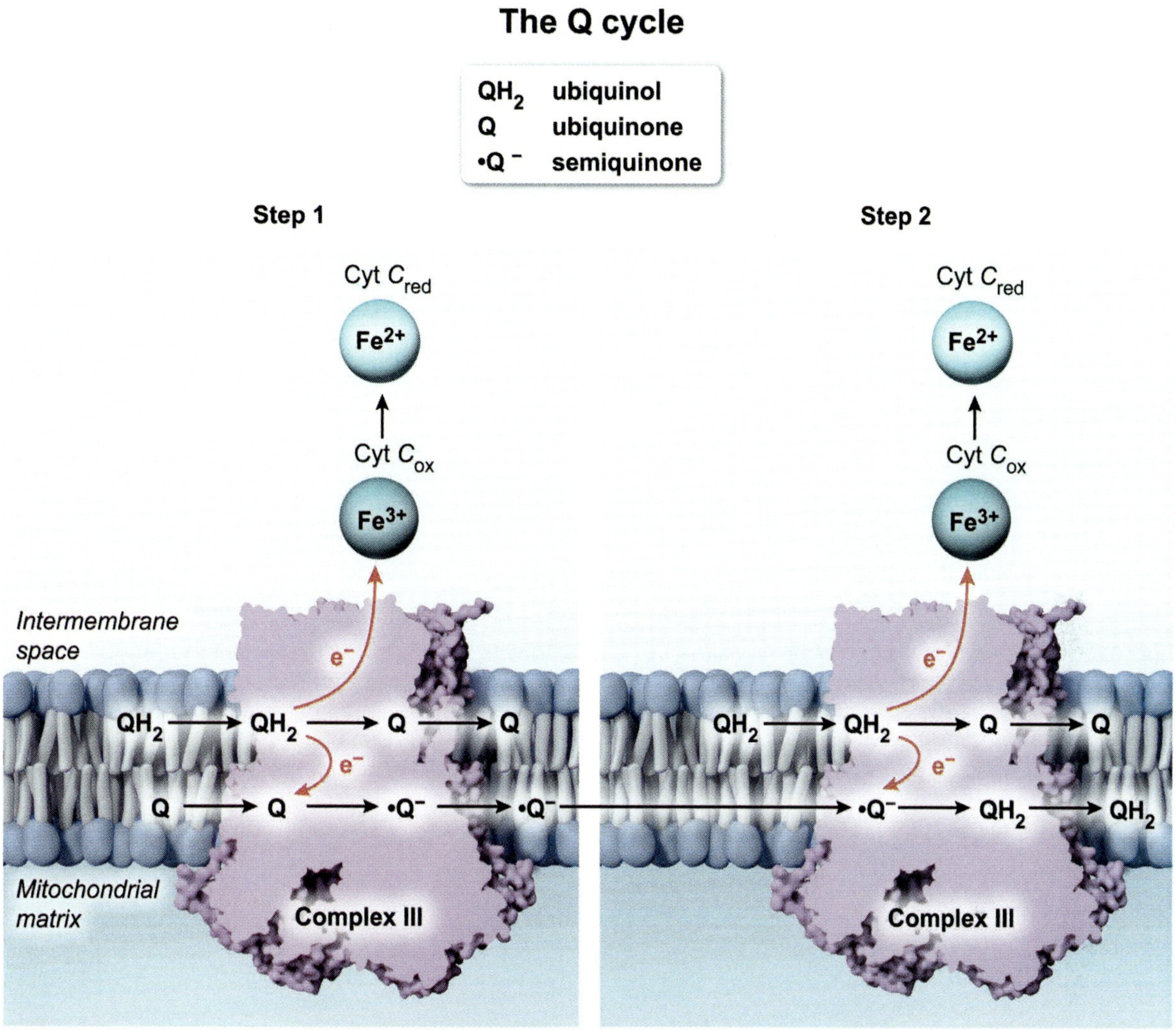

Chapter 12: Aerobic Respiration

The Q cycle

QH_2	ubiquinol
Q	ubiquinone
$•Q^-$	semiquinone

Step 1

Step 2

Figure 12.24 The Q cycle. Two versions of Complex III are shown for clarity, but all reactions take place in the same complex.

The Q cycle starts with two coenzyme Q molecules: one in its reduced form (ubiquinol, QH_2), carrying the high-energy electrons, and the other in its oxidized form (ubiquinone, Q). As shown in Step 1 of Figure 12.24, ubiquinol (QH_2) passes one electron to cyt c and another electron to ubiquinone (Q). The original QH_2 molecule has therefore become fully oxidized to Q and leaves the cycle. Meanwhile, the original Q molecule becomes a partially reduced form called semiquinone, denoted $•Q^-$. The single dot indicates that semiquinone is a radical, meaning it has an unpaired electron.

In Step 2 of Figure 12.24, another QH₂ molecule then enters the cycle and passes one electron to a cyt c molecule and the other electron to the •Q⁻. The original QH₂ from this step becomes Q, and •Q⁻ becomes fully reduced to QH₂. The *net* result from the two steps of this cycle is that one QH₂ transfers its electrons to two oxidized cyt c molecules, yielding one Q and two reduced cyt c molecules, as described by the following summed equations:

Step 1: $QH_2 + \text{cytochrome } c_{ox} + \cancel{Q} \rightarrow \cancel{Q} + \text{cytochrome } c_{red} + \cancel{•Q^-}$

Step 2: $\cancel{QH_2} + \text{cytochrome } c_{ox} + \cancel{•Q^-} \rightarrow Q + \text{cytochrome } c_{red} + \cancel{QH_2}$

Net reaction: $QH_2 + 2 \text{ cytochrome } c_{ox} \rightarrow Q + 2 \text{ cytochrome } c_{red}$

During this process, **four protons** are produced and pumped into the intermembrane space, increasing the pH gradient across the membrane.

Oxidative Stress

Oxygen gas must cross into the hydrophobic portion of the inner mitochondrial membrane before it can interact with Complex IV (see Concept 12.2.04). The various forms of coenzyme Q, including semiquinone, are found within this membrane. The semiquinone produced in Complex III is highly reactive and may react with oxygen to form reactive oxygen species (ROS) such as superoxide ($•O_2^-$).

Superoxide can lead to the condition of oxidative stress, in which components of a cell that should not be oxidized become oxidized. The resulting damage to various components of the cell can lead to apoptosis (ie, controlled cell death, see Figure 12.25). In this process, the cyt c molecules that are normally associated with the inner mitochondrial membrane are released into the cytosol, where they activate proteases such as caspase. These proteases then degrade specific proteins, and the cell dies.

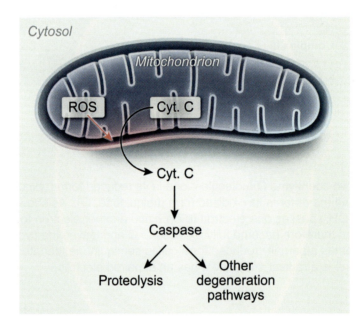

Figure 12.25 Reactive oxygen species (ROS) produced in the electron transport chain can lead to apoptosis, in which cyt c activates degradative pathways.

To protect against oxidative stress, cells use the enzyme superoxide dismutase to convert superoxide into hydrogen peroxide (H_2O_2). Catalase then reduces H_2O_2 to H_2O using glutathione as an electron donor. Glutathione is then restored to its reduced form by NADPH, which becomes $NADP^+$. Therefore, control of oxidative stress ultimately depends on the presence of NADPH, which is provided by the oxidative phase of the pentose phosphate pathway (see Concept 11.3.01).

12.2.04 Complex IV

The Complex IV Reaction

The reduced cyt c produced by Complex III migrates to Complex IV, also known as cytochrome c oxidase. Oxygen gas (O_2) binds tightly to a heme group within Complex IV, and four cyt c_{red} molecules successively pass their electrons to the bound oxygen through a series of intermediates. Each oxygen atom also reacts with two protons, and the result is two water molecules per molecular oxygen (O_2) reduced, as shown in Figure 12.26. Note that the four cyt c_{red} shown in this figure must have come from *two* molecules of either NADH or $FADH_2$.

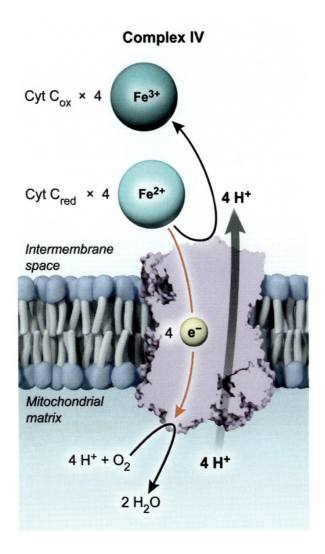

Figure 12.26 Complex IV activity.

Because oxygen is the final electron acceptor in the chain, the ETC cannot function in the absence of oxygen. In anaerobic conditions (ie, no oxygen or low oxygen), reduced cyt c cannot lose an electron and therefore remains in reduced form. This results in a buildup of cyt c_{red} and depletion of cyt c_{ox} by Complex III.

When cyt c_{ox} is depleted, ubiquinol is unable to donate its electrons and remains in reduced form. Therefore, ubiquinol builds up and ubiquinone is depleted. This in turn results in a buildup of NADH and $FADH_2$, which inhibits the citric acid cycle as no NAD^+ or FAD is available to receive electrons. For this reason, both the ETC and the citric acid cycle function only under aerobic conditions (see Figure 12.27), and fermentation is used under anaerobic conditions.

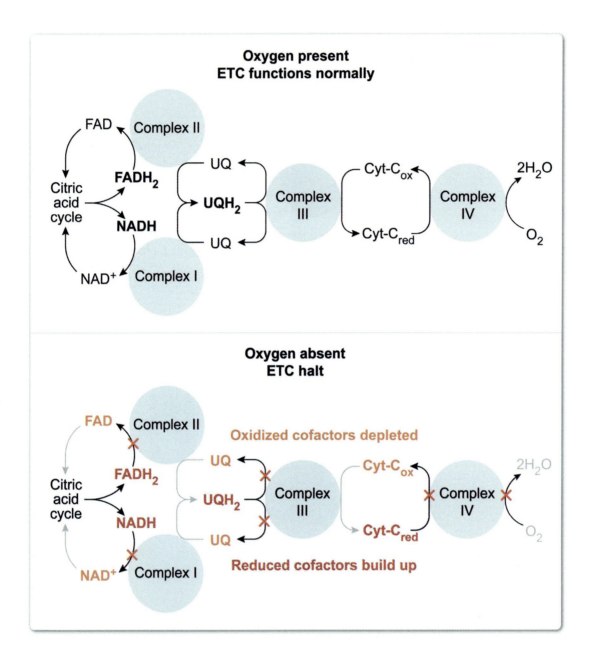

Figure 12.27 Electron transport chain and citric acid cycle in the presence and absence of oxygen.

Note that some single-cell organisms and a few invertebrates can use molecules other than oxygen as their final electron acceptors and their electron transport chains function under anaerobic conditions. However, most multicellular organisms and many single-celled organisms require aerobic conditions for their ETC to function.

Chapter 12: Aerobic Respiration

> ☑ **Concept Check 12.4**
>
> If the citric acid cycle is inhibited, what effect will this have on the ratio of ubiquinone to ubiquinol in the inner mitochondrial membrane?
>
> **Solution**
>
> *Note: The appendix contains the answer.*

Net Proton Yield of the ETC

For each oxygen molecule that is reduced to water, Complex IV pumps four protons into the intermembrane space. Note, however, that full reduction of O_2 requires *four* electrons, whereas each NADH or $FADH_2$ molecule contributes only *two* electrons. Therefore, two NADH molecules, two $FADH_2$ molecules, or one NADH molecule and one $FADH_2$ molecule are required to fully reduce a single O_2 molecule. In other words, Complex IV pumps *two* protons for every NADH or $FADH_2$ molecule that enters the ETC.

Every NADH molecule that enters the ETC causes four protons to be pumped into the intermembrane space by Complex I, another four protons by Complex III, and another two protons by Complex IV, for a total of *10 protons per NADH*. $FADH_2$ also results in four protons pumped by Complex III and two protons pumped by Complex IV, for a total of *six protons per $FADH_2$*. The number of protons pumped per reduced cofactor is shown in Figure 12.28.

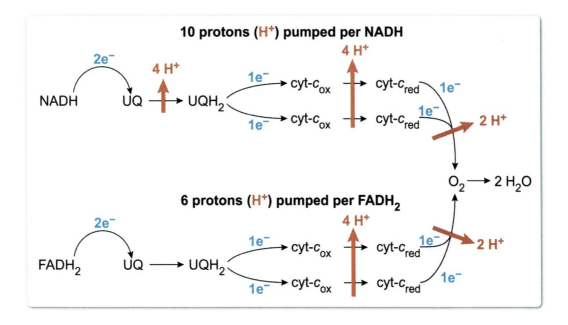

Figure 12.28 Protons pumped per reduced cofactor (NADH or $FADH_2$) that enters the ETC.

Chapter 12: Aerobic Respiration

Lesson 12.3
Oxidative Phosphorylation

Introduction

The electron transport chain produces a higher proton concentration (and a lower pH) outside the mitochondrial matrix than within it, or in other words, a pH gradient across the membrane. The separation of charge and the difference in concentration serve as a source of stored energy referred to as the **proton motive force**.

The stored energy is released when the excess protons in the intermembrane space are allowed to flow *down* their concentration gradient back into the mitochondrial matrix. The enzyme complex called ATP synthase couples this flow of protons with the synthesis of ATP through condensation of ADP with inorganic phosphate (P_i). Because this mode of ATP synthesis is powered by oxidation of NADH and $FADH_2$, it is called **oxidative phosphorylation**.

This lesson explores the proton motive force and how it facilitates oxidative phosphorylation.

12.3.01 The Proton Motive Force

Lesson 12.2 discusses the four complexes of the electron transport chain that together result in the pumping of protons from the mitochondrial matrix into the intermembrane space. Proton pumping results in a pH difference across the inner mitochondrial membrane, with the intermembrane space having a lower pH (ie, a higher H^+ concentration) than the mitochondrial matrix. Specifically, in a typical healthy mammalian cell, the intermembrane space pH is approximately 7.4 while the pH in the mitochondrial matrix is close to 7.8.

Because protons have a positive charge, the difference in proton concentration across the membrane also yields a voltage across the membrane, with the intermembrane space being more positively charged than the mitochondrial matrix. The resulting electrochemical gradient stores potential energy, called the **proton motive force** (pmf). Note that the term "force" in this case is a misnomer because the pmf does not refer to the acceleration of mass directly, but rather to the potential energy in the gradient and the work done as protons move along it.

The energy stored by the pmf is released by allowing the protons to cross the membrane back into the mitochondrial matrix (ie, by allowing protons to flow *down* their electrochemical gradient). However, as discussed in Lesson 3.3, ions cannot readily cross the hydrophobic environment of a phospholipid bilayer such as the inner mitochondrial membrane. To facilitate reentry into the matrix, a channel or carrier protein is needed. The enzyme ATP synthase facilitates diffusion (passive transport) of protons into the matrix. Active transport of protons out of the matrix through the ETC, and passive transport back into the matrix, are shown in Figure 12.29.

Chapter 12: Aerobic Respiration

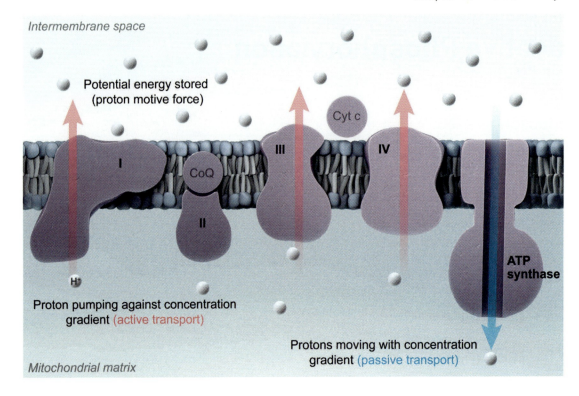

Figure 12.29 The proton motive force stores energy, which is released as protons flow through ATP synthase.

 Concept Check 12.5

The electron transport chain is primarily regulated by the availability of NADH, FADH$_2$, and oxygen. Assuming oxygen levels are in excess, what effect would an increase in citric acid cycle activity have on the proton motive force?

Solution

Note: The appendix contains the answer.

12.3.02 ATP Synthase

ATP Synthase Mechanism

ATP synthase, sometimes called Complex V of the electron transport chain, is a translocase enzyme (Concept 4.2.08). Many translocases couple the energy released by ATP hydrolysis with the transport of solutes *against* their electrochemical gradient (eg, the Na$^+$/K$^+$ pump). ATP synthase, however, facilitates the reverse process: it couples transport of solutes (H$^+$) *down* their gradient with *synthesis* of ATP. Proteins that couple solute transport to chemical reactions are said to be chemiosmotic, and therefore ATP synthase facilitates ATP production by **chemiosmosis**.

Figure 12.30 highlights several components of ATP synthase. These components do not need to be memorized, but an understanding of their interactions greatly facilitates understanding of the underlying mechanisms of ATP synthase.

Chapter 12: Aerobic Respiration

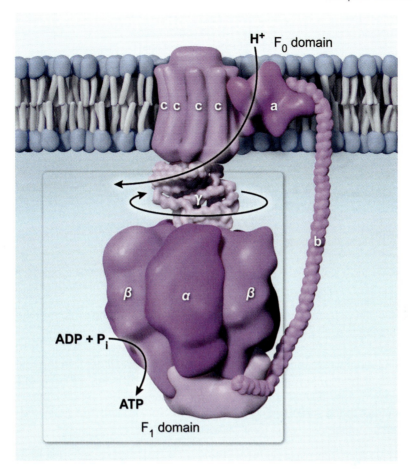

Figure 12.30 Components of ATP synthase.

ATP synthase consists of two major domains called the F_o and F_1 domains. Each domain contains multiple subunits. The F_o domain is a transmembrane complex through which protons can flow. The protons bind a set of subunits called the c subunits. Each c subunit can bind one proton, and every time this occurs, the set of subunits rotates. Once a subunit makes a complete rotation, it releases its proton into the mitochondrial matrix and is then ready to bind another proton.

The F_o domain also contains a homodimeric arm consisting of two identical subunits called the b subunits. This arm extends from the membrane into the mitochondrial matrix, where it interacts with the F_1 domain. The F_1 domain contains three sets of heterodimers, each consisting of an α subunit and a β subunit.

The αβ dimers surround the γ subunit, which is connected to the c subunits of F_o. As the c subunits rotate, so does the γ subunit. This links proton movement to γ subunit rotation. Rotation of the γ subunit causes the αβ complexes to *change conformations*. In other words, at any given time, each αβ complex is in a conformation different from the other two αβ complexes. The rotation of the γ subunit is shown in Figure 12.31.

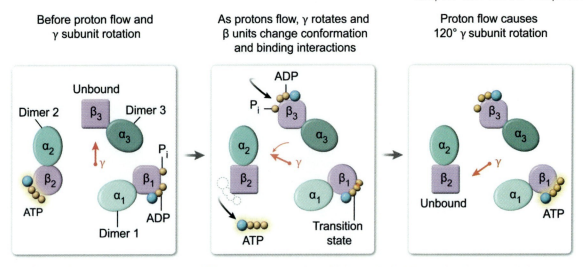

Proton flow causes the γ stalk of ATP synthase to rotate, which results in ATP synthesis at an αβ dimer.

Figure 12.31 Rotation of the γ subunit from one αβ complex to the next, causing conformational changes and different binding preferences in each β subunit.

One αβ complex is in a conformation that readily binds ADP and P_i (Dimer 1 in the first panel of Figure 12.31), another is in a conformation that binds ATP (Dimer 2), and a third is in a low-affinity conformation that binds neither (Dimer 3). As the γ unit rotates, the αβ complex that *was* bound to ADP and P_i (Dimer 1) changes to the ATP-binding conformation. In so doing, the ADP and P_i molecules are brought closely together, which catalyzes their conversion to ATP.

Simultaneously, the complex that was originally bound to ATP (Dimer 2) changes to the low-affinity conformation that binds neither substrate, and the ATP that *was* bound to it is released. Finally, the complex that *was* unbound (Dimer 3) changes to the conformation that binds ADP and P_i and binds them. This cycle continues as long as protons continue to flow through the F_o domain.

ATP Yield Per Glucose

The most common model of human ATP synthase proposes that three protons flowing through F_o cause rotation of the γ subunit from one αβ complex to the next; therefore, nine protons are required for one complete rotation of the γ unit, which produces three ATP molecules.

Once ATP is produced in the mitochondrial matrix, it must be transported to the cytosol. This is accomplished through an antiport system in which ATP exits the matrix while ADP enters. ATP can then freely diffuse from the intermembrane space into the cytosol for use as needed. To phosphorylate the ADP that has entered the matrix, inorganic phosphate must also enter. This occurs through a **symport system** in which a proton moves from the intermembrane space into the matrix through the phosphate transporter.

Therefore, although this proton does not move through ATP synthase, a *fourth* proton (in addition to the three needed to cause rotation) must enter the matrix to synthesize *cytosolic* ATP. Consequently, according to this model, producing ATP for use in the cytosol requires a total of *four protons* moving into the matrix per ATP synthesized (see Figure 12.32).

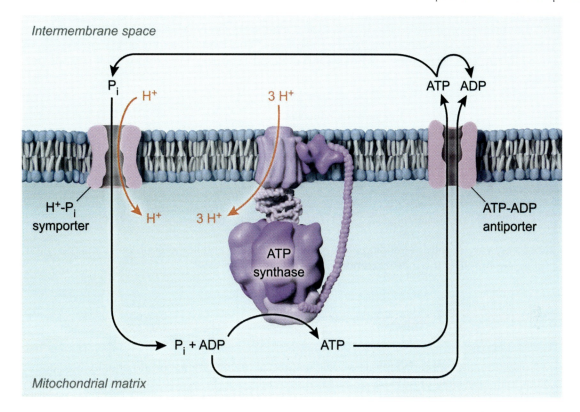

Figure 12.32 Transport of P_i into the matrix requires cotransport of one H^+ ion and indirectly facilitates export of ATP to the cytosol. For this reason, ATP synthesis is said to require four protons per ATP.

Each NADH that enters the electron transport chain pumps a total of 10 protons into the intermembrane space, as discussed in Lesson 12.2. Therefore, each NADH yields an average of 2.5 ATP:

$$\frac{10 \text{ \sout{protons}/NADH}}{4 \text{ \sout{protons}/ATP}} = 2.5 \text{ ATP/NADH}$$

Similarly, each $FADH_2$ pumps six protons, so each yields an average of 1.5 ATP per $FADH_2$ consumed:

$$\frac{6 \text{ \sout{protons}/FADH}_2}{4 \text{ \sout{protons}/ATP}} = 1.5 \text{ ATP/FADH}_2$$

Note that fractional molecules of ATP are not possible (there is no such thing as half an ATP); however, the numbers 2.5 and 1.5 represent average *ratios* of ATP produced per molecule of reducing cofactor. In other words, the electrons of one NADH molecule pump 10 protons to produce *two* molecules of ATP with two protons left over. The electrons of a second molecule of NADH pump *another* 10 protons, and the combined 12 protons produce *three* molecules of ATP. Together, two NADH molecules produce five ATP molecules, for an *average* of 2.5 ATP molecules per NADH.

Based on this model of ATP synthase activity and the ratio of ATP produced per proton, an estimate of the number of ATP molecules produced per glucose consumed can be made. Lesson 11.1 shows that glycolysis directly yields two net ATP molecules per glucose, as well as two NADH molecules. Lesson 12.1 shows that the citric acid cycle produces one GTP per pyruvate, or two per glucose. Because GTP is energetically equivalent to ATP, it is counted in the estimate. Pyruvate oxidation through the cycle also yields four NADH molecules per pyruvate (ie, eight per glucose), and one $FADH_2$ molecule per pyruvate (ie, two per glucose).

Taken together, each glucose molecule produces a total of 10 NADH molecules: two from glycolysis (assuming the malate-aspartate shuttle is active) and eight from pyruvate oxidation and the citric acid cycle. Each glucose also produces two FADH₂ molecules. This leads to a total of 112 protons pumped per glucose (10 NADH × 10 protons + 2 FADH₂ × 6 protons). Dividing 112 protons by 4 protons per ATP synthesized yields

$$\frac{112 \;\cancel{\text{protons}}/\text{glucose}}{4 \;\cancel{\text{protons}}/\text{ATP}} = 28 \text{ ATP/glucose}$$

Adding in the two ATP molecules from glycolysis and the two GTP molecules from the citric acid cycle gives

$$28 \text{ ATP/glucose}_{\text{Oxidative phosphorylation}} + 4 \text{ ATP/glucose}_{\text{Substrate-level phosphorylation}} = 32 \text{ ATP/glucose}_{\text{total}}$$

This number may vary slightly depending on ambient conditions, and some sources list as many as 38 ATP per glucose, but 32 is the most commonly accepted number.

> ### ✓ Concept Check 12.6
>
> If a cell uses the glycerol-3-phosphate shuttle instead of the malate-aspartate shuttle, how does this change the net yield of ATP per glucose consumed?
>
> #### Solution
> *Note: The appendix contains the answer.*

12.3.03 Uncoupling of ATP Synthase

Because ATP synthase relies on the proton gradient generated by the electron transport chain, ATP synthase function is said to be coupled to the ETC. However, *it is possible to uncouple these processes*. In isolated mitochondria, for example, addition of the molecule 2,4-dinitrophenol (2,4-DNP) or of the molecule FCCP allows protons to cross the membrane *without* passing through ATP synthase. Figure 12.33 shows a simplified mechanism of FCCP-mediated uncoupling. 2,4-DNP uncoupling follows a similar mechanism.

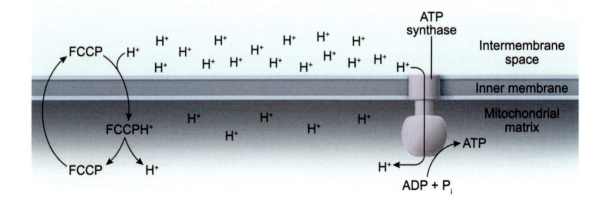

Figure 12.33 FCCP uncouples proton flow from ATP synthesis by bringing protons into the matrix without interacting with ATP synthase.

When these uncouplers are present, the electron transport chain continues to function as long as it has fuel (eg, pyruvate, acetyl-CoA, succinate). Therefore, oxygen consumption continues and may even *increase* because the smaller pH gradient makes it easier for protons to be pumped. However, ATP production *slows* because many of the pumped protons return to the matrix *without providing energy to ATP synthase*.

Many endothermic (ie, warm-blooded) organisms *deliberately* uncouple proton flow from ATP synthesis, at least partially, to help maintain body temperature. Certain cells (eg, brown adipose tissue cells) express an H^+ channel protein called thermogenin or uncoupling protein 1 (UCP1) in their inner mitochondrial membranes. In cold conditions, these channels are induced to open and allow proton flow into the matrix as shown in Figure 12.34.

When protons flow through ATP synthase, the released energy is coupled to ATP synthesis. In other words, the energy of the proton gradient is used to do work. In contrast, when protons flow through uncouplers such as thermogenin, the energy is *not* used to do work and therefore is converted to *heat* instead. This heat may then be used to help maintain a high body temperature in cold external conditions.

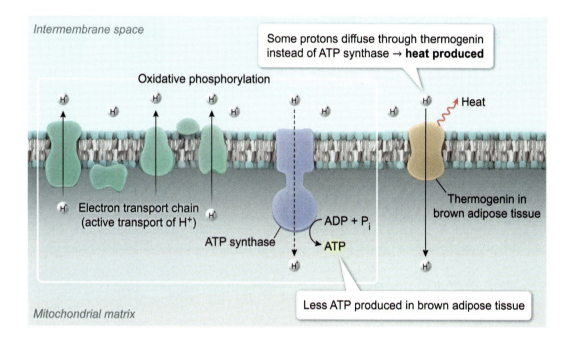

Figure 12.34 Under cold conditions, thermogenin allows protons to flow into the mitochondrial matrix and release heat.

Concept Check 12.7

Which of the following graphs best depicts oxygen consumption and ATP synthesis in the presence and absence of an uncoupler?

Graph 1

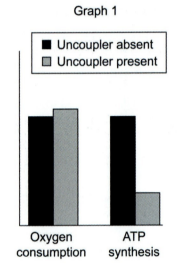

Graph 2

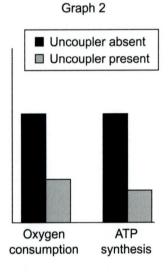

Graph 3

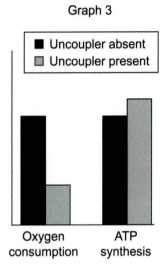

Solution

Note: The appendix contains the answer.

Lesson 13.1
Fatty Acid Metabolism

Introduction

In addition to carbohydrates, other molecules may serve as sources of energy for ATP production or as end products of anabolic pathways. Lipids in particular can store significant amounts of energy. Consequently, lipid catabolism serves as a source of energy in many cells and provides the energy necessary to carry out gluconeogenesis in the liver. When cells have sufficient ATP, they may convert acetyl-CoA molecules to fatty acids and triglycerides to store energy for later use.

This lesson describes the catabolism and anabolism of lipids, with a focus on fatty acids, and explains how lipid metabolism is regulated.

13.1.01 Lipid Transport and Catabolism

Lipids are generally hydrophobic and poorly soluble in the aqueous environments found in most parts of the body. Therefore, lipid transport through the bloodstream requires the assistance of **amphiphilic molecules** (ie, molecules with both hydrophilic and hydrophobic groups) such as transport proteins and phospholipids. The hydrophobic portions of these transport molecules interact favorably with hydrophobic lipids, while the hydrophilic portions interact favorably with the aqueous environment. After a lipid-rich meal, lipids are transported using molecules called **lipoproteins**, which consist of the components shown in Figure 13.1.

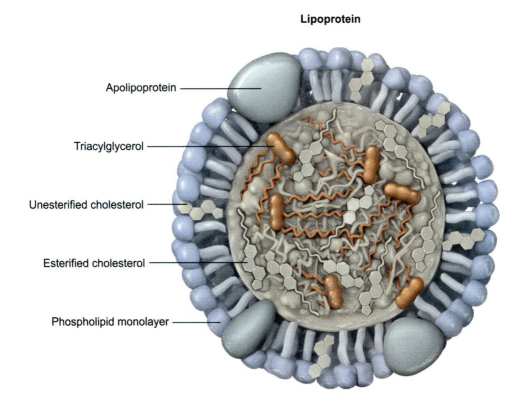

Figure 13.1 General structure of a lipoprotein.

To facilitate transport through the bloodstream and lymphatic system, dietary lipids in the intestine are first emulsified by bile salts (see Biology Chapter 15), which allows them to be accessed by intestinal hydrolytic enzymes. The hydrolyzed products can then be absorbed by intestinal cells, where free fatty acids are converted to triacylglycerides (also called triglycerides or triacylglycerols).

The triacylglycerides derived from dietary sources are then packaged into lipoproteins called chylomicrons, which also contain some cholesterol and cholesteryl esters. Lipoproteins are surrounded by a phospholipid monolayer, which helps to solubilize the lipids for transport to the tissues.

Triglycerides stored in the liver may similarly be packaged into another type of lipoprotein called very-low-density lipoproteins (VLDLs), which can then be transported to other tissues. Once chylomicrons and VLDLs deposit their triglycerides in the target tissues (eg, muscle, adipose), they become chylomicron remnants and intermediate-density lipoproteins (IDLs), respectively, which are enriched in cholesterol and cholesteryl esters compared to their precursors. IDLs and chylomicron remnants commonly return to the liver, where they deposit cholesterol and any remaining triglycerides.

IDLs may also be metabolized into low-density lipoproteins (LDLs), which are even more enriched in cholesterol and serve to deliver cholesterol to various peripheral tissues as needed. High-density lipoproteins (HDLs) may assist in delivering cholesterol from peripheral tissues back to the liver by sequestering excess cholesterol that has been excreted by the peripheral tissues. The pathways of various types of lipoproteins throughout the body are shown in Figure 13.2.

Chapter 13: Noncarbohydrate Metabolism

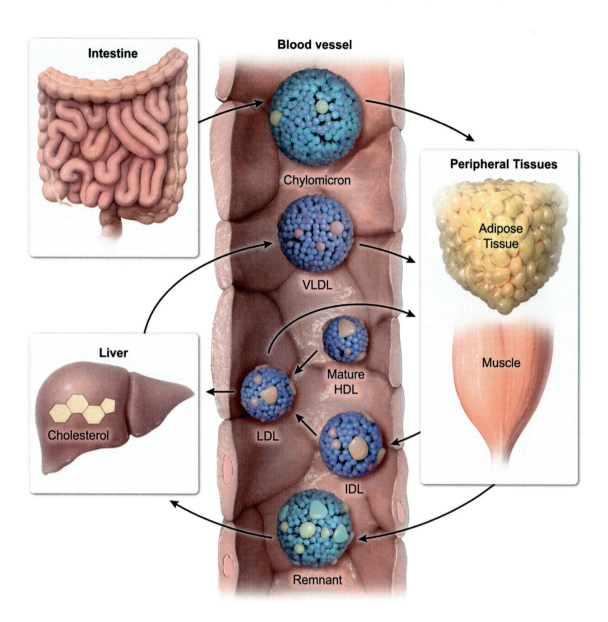

Figure 13.2 Overview of pathways for different types of lipoproteins.

Once triglycerides and fatty acids are delivered to their target tissues, they may be further metabolized for energy or for the formation of glycerophospholipids, as needed. During the further metabolism of a triglyceride, each of the ester bonds may be hydrolyzed by **lipase** enzymes.

When all three ester linkages are hydrolyzed, the result is three free fatty acids and one glycerol molecule. The fatty acids may then be digested by β-oxidation, as described in Concept 13.1.02. The glycerol molecule can be phosphorylated to become glycerol 3-phosphate, which can then be oxidized to form dihydroxyacetone phosphate (DHAP). Figure 13.3 shows degradation of a triglyceride into its components, along with the metabolic fates of those components.

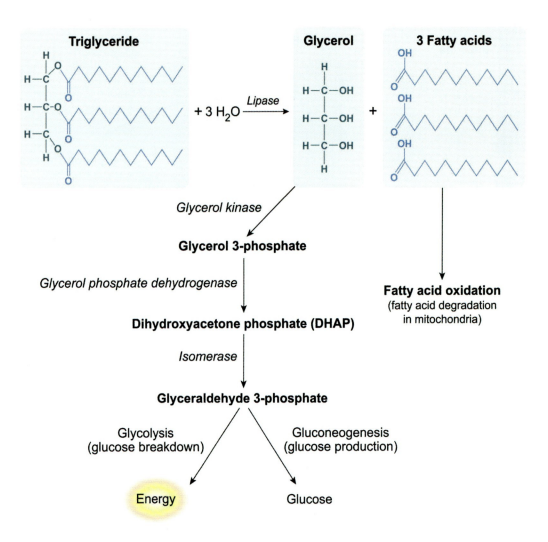

Figure 13.3 Triglyceride degradation.

13.1.02 Fatty Acid Oxidation

Fatty acid oxidation is, as the name suggests, the oxidation of fatty acids to form acetyl-CoA. Most commonly, fatty acids are digested by β-oxidation in the mitochondria, although β-oxidation can also occur in peroxisomes. β-Oxidation is so named because the oxidation occurs at the β-carbon (ie, the carbon that is two units away from the carboxylic acid group).

Before β-oxidation can begin, fatty acids must enter the **mitochondrial matrix**. Short-chain fatty acids (those with fewer than six carbons) and some medium-chain fatty acids (which have six to 12 carbons) can enter the mitochondrial matrix directly. Within the matrix, these fatty acids react with coenzyme A to become fatty acyl-CoA. β-Oxidation can then proceed.

In contrast, fatty acids with more than eight carbons typically require assistance to enter the matrix, and this assistance is largely provided by the **carnitine shuttle** shown in Figure 13.4.

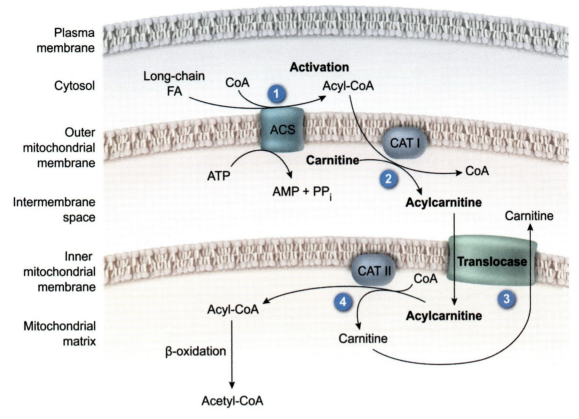

ACS = acyl-CoA synthetase; **CAT** = carnitine acyl transferase; **CoA** = coenzyme A; **FA** = fatty acid.

Figure 13.4 The carnitine shuttle.

Fatty acids that require this shuttle are first converted to **acyl-CoA molecules**. Note that an acyl-CoA molecule is any hydrocarbon chain attached to coenzyme A through a thioester linkage and should not be confused with acetyl-CoA, which is specifically a *two-carbon chain* with a thioester linkage to coenzyme A. As shown in Step 1 of Figure 13.4, acyl-CoA molecules form in the cytosol when fatty acids react with coenzyme A, as catalyzed by acyl-CoA synthetase (ACS). This reaction also requires hydrolysis of ATP to AMP and pyrophosphate (PP_i).

Acyl-CoA freely passes across the *outer* mitochondrial membrane. The inner leaflet of this membrane (ie, the side that faces the intermembrane space) contains the enzyme **carnitine acyl transferase I (CAT I)**, which catalyzes the transfer of the acyl group from coenzyme A to carnitine (Step 2 in Figure 13.4). This enzyme is also sometimes called carnitine palmitoyl transferase I (**CPT I**). The resulting molecule is **acylcarnitine**, which is transported into the mitochondrial matrix through a translocase (Step 3 in Figure 13.4). At the same time, free carnitine exits the mitochondria through the same translocase (ie, through antiport).

Within the matrix, **acyl carnitine transferase II** (**CAT II**, also called carnitine palmitoyl transferase II, **CPT II**) catalyzes the transfer of the acyl group back onto coenzyme A, reforming **acyl-CoA** and **free carnitine** (Step 4 in Figure 13.4). The free carnitine can then exit the matrix through the antiporter to react with another acyl-CoA in the intermembrane space.

Once an acyl-CoA molecule is in the mitochondrial matrix, it can undergo β-oxidation. This process consists of four repeating steps: oxidation, hydration, another oxidation, and deacetylation.

1. **Oxidation by acyl-CoA dehydrogenase**. Three different isoforms of this enzyme act on short-, medium-, and long-chain fatty acids, respectively. The reaction forms a *trans* double bond

between the α-carbon and the β-carbon (ie, carbons 2 and 3) of the chain. This converts the acyl-CoA into a *trans*-enoyl-CoA as shown in Figure 13.5.

The oxidation of these carbons is accompanied by the reduction of FAD to FADH$_2$, which is a prosthetic group in the enzyme. The enzyme then transfers electrons from FADH$_2$ to ubiquinone, forming ubiquinol and regenerating FAD. In this way acyl-CoA dehydrogenase fills a similar role to Complex II of the electron transport chain, and the resulting ubiquinol interacts with Complex III.

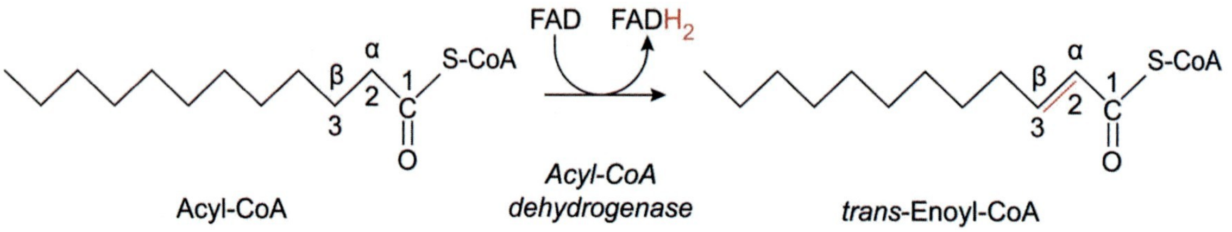

Figure 13.5 Oxidation of acyl-CoA by formation of a double bond between carbons 2 and 3 (the α- and β-carbons, respectively). FADH$_2$ forms in the process.

2. **Hydration by enoyl-CoA hydratase**. Enoyl-CoA hydratase is a lyase that adds water across the double bond of *trans*-enoyl-CoA molecules. This results in a hydroxyl group on carbon 3 (the β-carbon) of the chain, converting *trans*-enoyl-CoA to L-3-hydroxyacyl-CoA, also called L-β-hydroxyacyl-CoA. Figure 13.6 shows this reaction.

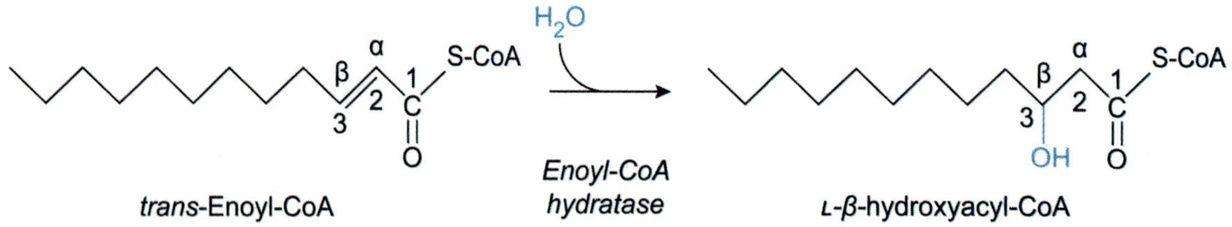

Figure 13.6 Hydration of *trans*-enoyl-CoA by enoyl-CoA hydratase.

3. **Oxidation by β-hydroxyacyl-CoA dehydrogenase**. In this reaction, the hydroxyl group that was added in Step 2 is oxidized to a ketone, forming a 3-ketoacyl-CoA, also called a β-ketoacyl-CoA, as shown in Figure 13.7. Simultaneously, NAD$^+$ is reduced to NADH. The NADH then enters the electron transport chain at Complex I.

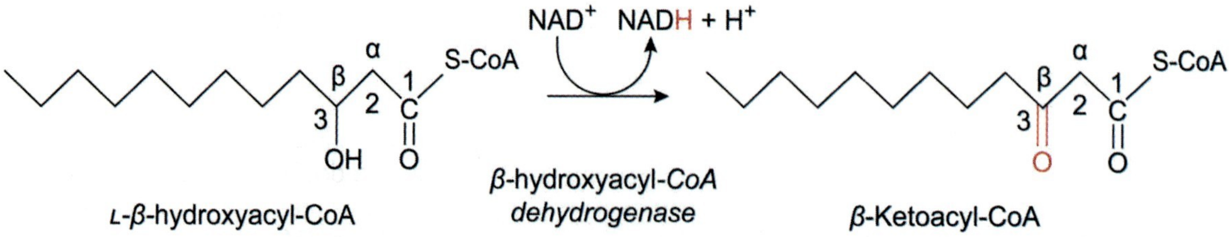

Figure 13.7 Oxidation of L-β-hydroxyacyl-CoA to β-ketoacyl-CoA using β-hydroxyacyl-CoA dehydrogenase. NADH forms as well.

4. **Deacetylation by acetyl-CoA acyltransferase**. In this reaction, coenzyme A attacks the β-carbon. The result is release of acetyl-CoA and a new acyl-CoA chain that is two carbons shorter than the original chain, as shown in Figure 13.8. Acetyl-CoA acyltransferase enzymes are also known as thiolase enzymes.

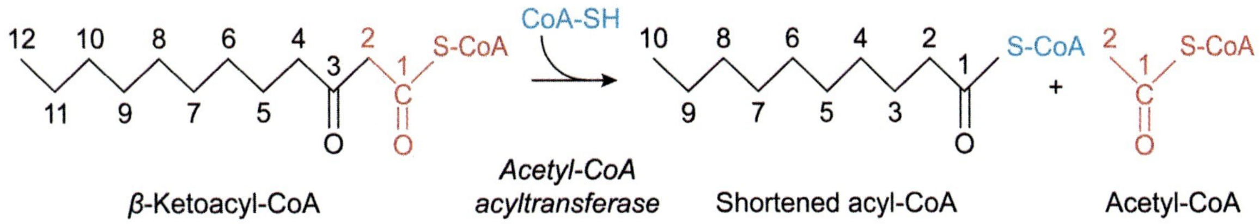

Figure 13.8 Splitting of β-ketoacyl-CoA into a shortened acyl-CoA and acetyl-CoA by acyl-CoA acyltransferase.

The new acyl-CoA chain can then repeat Steps 1–4, shortening by an additional two carbons with each round. Once a four-carbon chain is produced, it undergoes β-oxidation one more time and is split into two acetyl-CoA molecules. Therefore, a 16-carbon saturated fatty acid undergoes *seven* rounds of β-oxidation to produce *eight* acetyl-CoA molecules along with *seven* FADH$_2$ and *seven* NADH molecules. Each acetyl-CoA molecule can enter the citric acid cycle.

☑ Concept Check 13.1

How many rounds of β-oxidation can the following acyl-CoA molecule undergo?

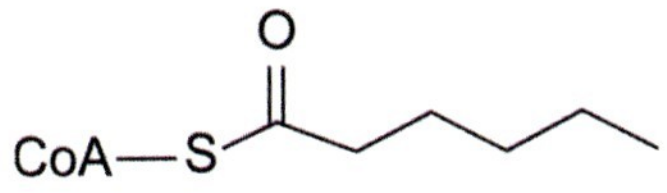

How many acetyl-CoA, NADH, and FADH$_2$ molecules are produced if it undergoes all rounds?

Solution

Note: The appendix contains the answer.

Unsaturated Fatty Acids

Unsaturated fatty acids have at least one C=C double bond in the chain. If the double bond is between an odd-numbered carbon and a higher even-numbered carbon (eg, between carbons 9 and 10), β-oxidation proceeds normally until the chain has been shortened enough that the double bond is between carbons 3 and 4. At this point, instead of forming a new double bond, the enzyme enoyl-CoA isomerase simply alters the position of the pre-existing double bond. In addition to moving the position of the double bond, enoyl-CoA isomerase also converts it from a *cis* double bond between carbons 3 and 4 to a *trans* double bond between carbons 2 and 3.

The *formation* of a double bond between carbons 2 and 3 is the first step in β-oxidation of a saturated fatty acid and produces FADH$_2$. However, in an unsaturated fatty acid the double bond needs only to have its position and configuration shifted. No oxidation-reduction reaction is required to produce the double bond because the bond is *already present*. Consequently, *one fewer FADH$_2$ molecule is produced* during β-oxidation of a fatty acid of this form, and therefore 1.5 fewer ATP molecules are produced on average per fatty acid oxidized (Figure 13.9).

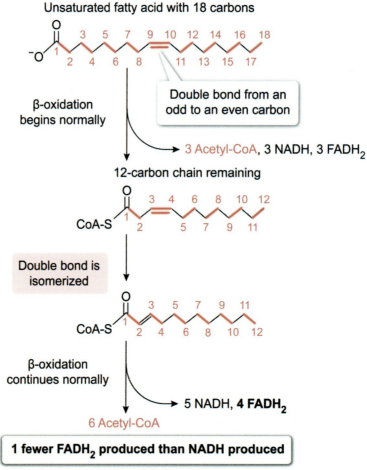

Figure 13.9 β-Oxidation of a fatty acid with a C=C double bond between an odd-numbered carbon and a higher even-numbered carbon.

In contrast, if the original fatty acid contains a double bond between an even-numbered carbon and a higher odd-numbered carbon (eg, between carbons 8 and 9), another mechanism must occur. In this case when the fatty acid is shortened to the point at which the double bond is between carbons 4 and 5, acyl-CoA dehydrogenase acts normally to introduce a double bond between carbons 2 and 3. However, the resulting fatty acid, which now has a *pair* of *conjugated* double bonds, is no longer able to fit into the active site of enoyl-CoA hydratase.

Consequently, a different enzyme called 2,4-dienoyl-CoA reductase removes *both* double bonds and creates a new one between carbons 3 and 4 (Figure 13.10). This process consumes NADPH to yield NADP$^+$. Enoyl-CoA isomerase can then act on the new double bond to place it between carbons 2 and 3, and β-oxidation again proceeds normally.

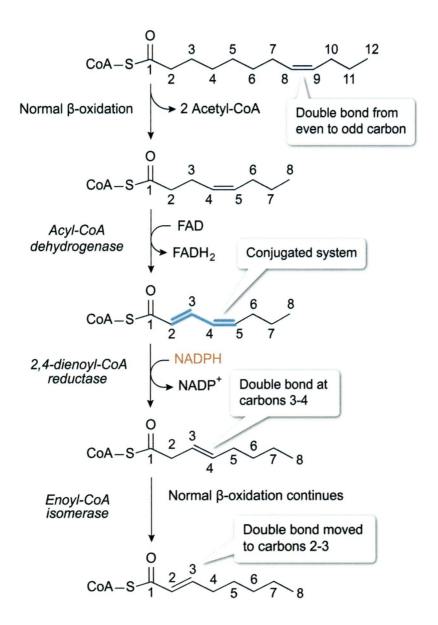

Figure 13.10 β-Oxidation of an unsaturated fatty acid with a double bond starting at an even-numbered carbon.

The consumption of NADPH in this case is energetically equivalent to consumption of NADH. This is because the enzyme nicotinamide nucleotide transhydrogenase interconverts these molecules in the mitochondria. Therefore, fatty acids of this nature effectively produce *one fewer NADH* than unsaturated fatty acids of the same length. Polyunsaturated fatty acids may contain double bonds of both types, and accordingly may produce fewer NADH molecules, fewer $FADH_2$ molecules, or both (relative to a saturated fatty acid), upon β-oxidation.

Odd-Chain Fatty Acids

Fatty acids with an odd number of carbons undergo normal β-oxidation until the last round. In fatty acids with an even number of carbons, the final round of β-oxidation involves a four-carbon molecule (butyryl-CoA) that becomes two acetyl-CoA molecules. When the number of carbons is odd, however, the final round releases a three-carbon molecule called propionyl-CoA as shown in Figure 13.11.

Carbons 1-16 are 2-carbon units incorporated into acetyl-CoA.
Carbons 17-19 are carbons incorporated into propionyl-CoA.

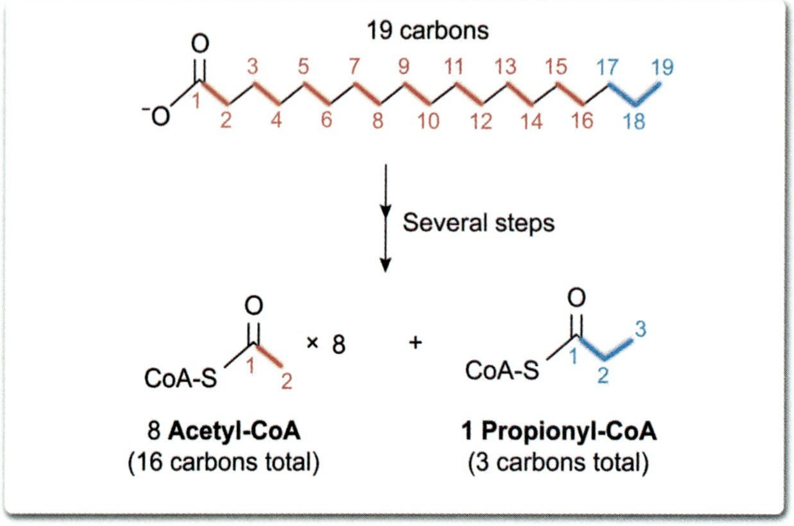

Figure 13.11 Example of β-oxidation of an odd-chain fatty acid yielding propionyl-CoA.

Propionyl-CoA is *not* a substrate for β-oxidation. Instead, this molecule becomes carboxylated by combining with bicarbonate (HCO_3^-) to form D-methylmalonyl-CoA. This reaction is catalyzed by propionyl-CoA carboxylase and requires hydrolysis of ATP to ADP and P_i. D-Methylmalonyl-CoA is converted to L-methylmalonyl-CoA by methylmalonyl-CoA epimerase, and L-methylmalonyl-CoA is then converted to succinyl-CoA by methylmalonyl-CoA mutase. This process is shown in Figure 13.12.

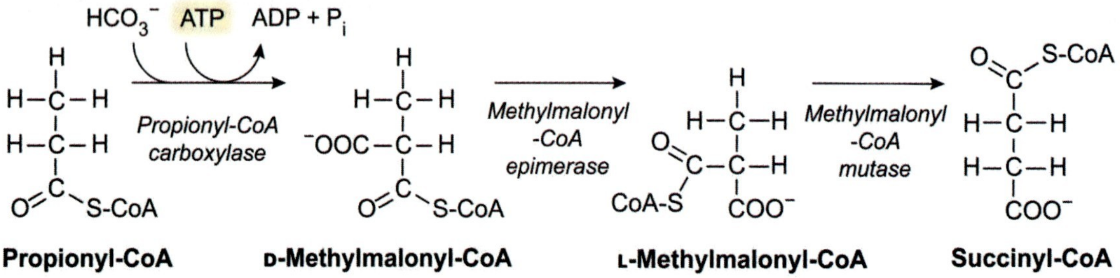

Figure 13.12 Conversion of propionyl-CoA to succinyl-CoA.

The details of this pathway are unlikely to be tested on the exam, but it is important to note the overall conversion of propionyl-CoA to **succinyl-CoA**, and the comparison to even-chain fatty acids, which only produce acetyl-CoA. Although the carbons of acetyl-CoA can be incorporated into oxaloacetate through the citric acid cycle, they must first *react* with oxaloacetate at the beginning of the citric acid cycle to do so. Therefore, no *net* oxaloacetate is produced or consumed by this pathway. Consequently, from a *net carbon standpoint*, acetyl-CoA cannot be converted to glucose.

In contrast, the propionyl-CoA produced by odd-chain fatty acids *can be used for gluconeogenesis* because succinyl-CoA is converted to oxaloacetate *without* first reacting with oxaloacetate.

> ### ✓ Concept Check 13.2
>
> Which of the following molecules produces more acetyl-CoA upon β-oxidation? Which produces more FADH$_2$?
>
> Molecule 1
>
> Molecule 2
>
> **Solution**
>
> Note: The appendix contains the answer.

13.1.03 Ketone Bodies

In the liver, β-oxidation is largely used to provide the energy needed for gluconeogenesis. This is accomplished as the NADH and FADH$_2$ from β-oxidation enter the electron transport chain and as the acetyl-CoA from β-oxidation enters the citric acid cycle to produce additional NADH, FADH$_2$, and GTP. The purpose of gluconeogenesis in the liver is to provide glucose that can be exported to other tissues, which can then use that glucose during fasting.

Often, β-oxidation produces more acetyl-CoA than is needed to power gluconeogenesis. Rather than using the excess acetyl-CoA in the citric acid cycle within liver cells, acetyl-CoA itself can be converted to other forms that can be exported into the bloodstream and sent to other tissues. However, as discussed previously, acetyl-CoA *cannot* be converted to glucose. Instead, it is converted to a class of molecules known as ketone bodies, as shown in Figure 13.13.

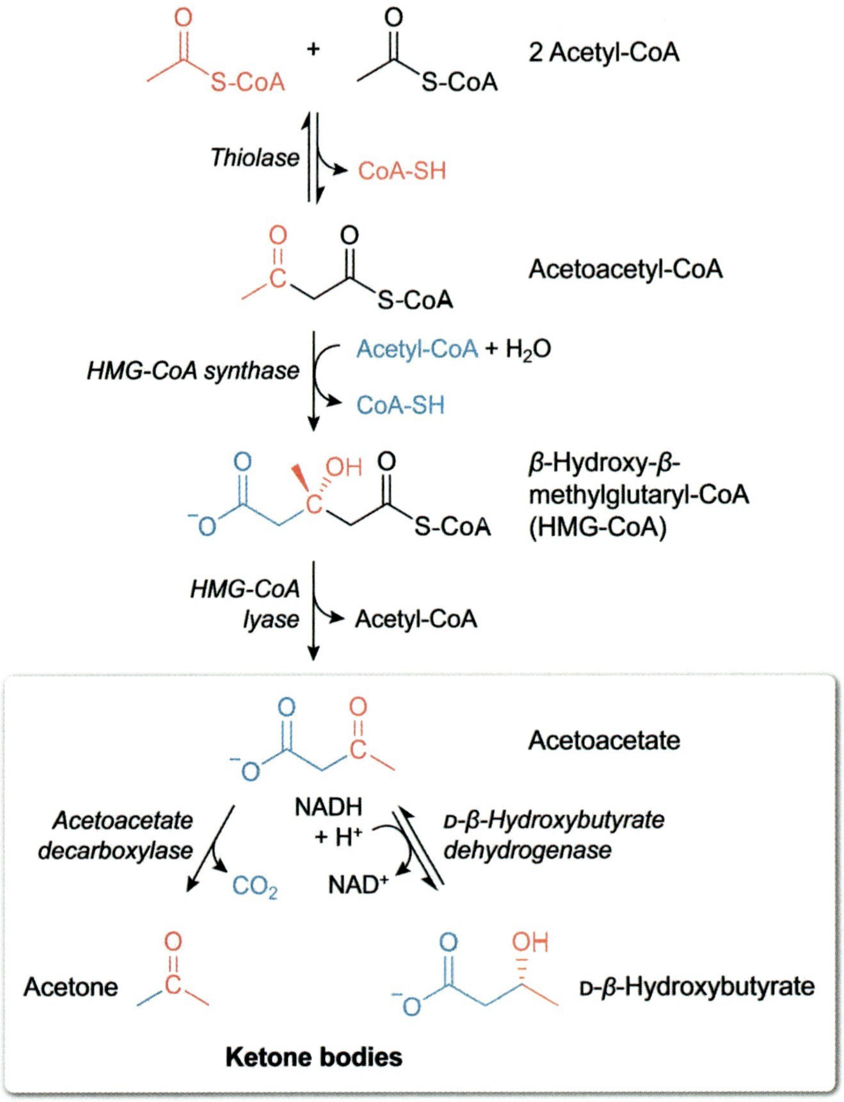

Figure 13.13 Ketone body synthesis from acetyl-CoA.

Ketone body synthesis begins when two acetyl-CoA molecules condense to form acetoacetyl-CoA as catalyzed by the final thiolase enzyme of β-oxidation (also known as acetyl-CoA acetyltransferase). Acetoacetyl-CoA then condenses with *another* acetyl-CoA molecule to form HMG-CoA as catalyzed by HMG-CoA synthase. HMG-CoA is then separated by HMG-CoA lyase, forming another acetyl-CoA molecule and the first ketone body, acetoacetate.

Acetoacetate can be exported directly, or it may be reduced to form **D-β-hydroxybutyrate**. Reduction of acetoacetate requires oxidation of NADH to NAD^+. Note that although D-β-hydroxybutyrate does not contain a ketone functional group, it is still classified as a ketone body because it is *derived* from a ketone-containing molecule.

A small percentage of acetoacetate is decarboxylated to form acetone, which does not generally provide energy to cells and is instead exhaled. High levels of acetone, which may be detected on the breath, are indicative of inability to digest glucose (and therefore a need to metabolize ketone bodies instead), due to either lack of nutrition or metabolic disorders such as diabetes.

Acetoacetate and D-β-hydroxybutyrate are transported to muscles, including the heart, as well as other tissues, as shown in Figure 13.14. In these tissues, D-β-hydroxybutyrate is oxidized back to acetoacetate, producing NADH in the process. Therefore, transport of D-β-hydroxybutyrate from the liver to other tissues effectively also transports NADH from the liver to those tissues.

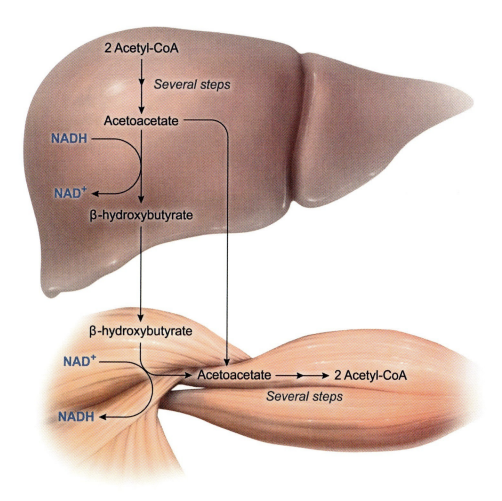

Figure 13.14 Export of β-hydroxybutyrate and acetoacetate from the liver to skeletal muscle.

Acetoacetate is converted to acetoacetyl-CoA and can then split into two molecules of acetyl-CoA via the *final step* of the β-oxidation pathway (ie, thiolase, also known as acetyl-CoA acetyltransferase). The rest of the steps are not needed to process acetoacetyl-CoA; because it is already in the fully oxidized state (ie, the β-carbon is already a ketone), no additional oxidation is needed and no NADH or $FADH_2$ is generated in this process. The resulting two acetyl-CoA molecules can enter the citric acid cycle in the target tissue, providing energy.

13.1.04 Fatty Acid Synthesis

Fatty acid oxidation primarily takes place when glycogen stores are depleted and energy is needed for gluconeogenesis. In other words, fatty acid catabolism occurs mostly during fasting. In contrast, when glucose is abundant (ie, in the well-fed state), excess acetyl-CoA from glycolysis and pyruvate decarboxylation may be *stored* as fatty acids.

Acetyl-CoA synthesis occurs in the mitochondria. In contrast, fatty acid synthesis occurs in the cytosol. Consequently, acetyl-CoA must be transported out of the mitochondria when fatty acid synthesis is active. However, acetyl-CoA cannot freely cross the inner mitochondrial membrane. Instead, acetyl-CoA in the mitochondrial matrix exits through the **citrate shuttle** shown in Figure 13.15.

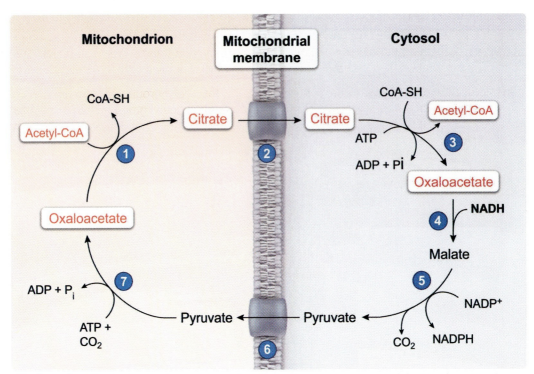

Note: Red text indicates the portion of the shuttle that transports acetyl-CoA to the cytosol.

Figure 13.15 The citrate shuttle.

The first step of the citrate shuttle is a reaction between acetyl-CoA and oxaloacetate to form citrate (ie, the first step of the citric acid cycle). Transport proteins allow citrate to exit the mitochondrial matrix (Step 2 in Figure 13.15). The outer mitochondrial membrane is porous and *also* allows citrate to cross it to enter the cytosol. Cytosolic citrate in the well-fed state can then serve both as a substrate source for fatty acid synthesis and as an allosteric signal to feedback inhibit glycolysis (see Chapter 11).

To act as a substrate for fatty acid synthesis, cytosolic citrate is split into acetyl-CoA and oxaloacetate by the enzyme citrate lyase (Step 3 in Figure 13.15). Recall that in the citric acid cycle, the condensation of acetyl-CoA and oxaloacetate is irreversible. Because of this irreversibility, citrate synthase cannot simply catalyze the reverse reaction, and instead citrate lyase couples the reaction to ATP hydrolysis. Therefore, *transport of each acetyl-CoA into the cytosol requires consumption of one ATP molecule*.

As shown in Step 4 of Figure 13.15, oxaloacetate is then converted to malate by cytosolic malate dehydrogenase (the same enzyme used in the malate-aspartate shuttle). Various pathways allow malate to reenter the mitochondrial matrix, including direct entry through a transporter; a common pathway during fatty acid synthesis is decarboxylation to form pyruvate. This step involves malic enzyme, which couples the oxidation of malate with the reduction of $NADP^+$, as shown in Step 5. *NADPH is required for subsequent steps in fatty acid synthesis*, so using this method provides needed material.

Pyruvate enters the mitochondria through the same transporter that it uses to enter the citric acid cycle (Step 6 in Figure 13.15). During well-fed states, mitochondrial pyruvate may be carboxylated by pyruvate carboxylase (the same enzyme used in gluconeogenesis). This regenerates oxaloacetate, as shown in Step 7, which can then react with more acetyl-CoA. Carboxylation of pyruvate requires hydrolysis of another ATP molecule, bringing the *energy cost per acetyl-CoA transported* to *two ATP equivalents* when this pathway is used.

Chapter 13: Noncarbohydrate Metabolism

Fatty Acid Synthesis Mechanism

Fatty acid synthesis is facilitated by a complex enzyme known as **fatty acid synthase**. The **acyl carrier protein (ACP)** is a domain on the enzyme to which thioesters may be transferred. The active site of this domain includes a prosthetic group called pantothenic acid, which contains a thiol (–SH) group.

In the first step of fatty acid synthesis, shown in Figure 13.16, the acetyl group of acetyl-CoA is transferred to the ACP thiol group, releasing free coenzyme A. The acetyl group is then transferred to the **β-ketoacyl synthase (KS) domain** by reacting with a cysteine side chain. At this point, the enzyme is primed for fatty acid synthesis.

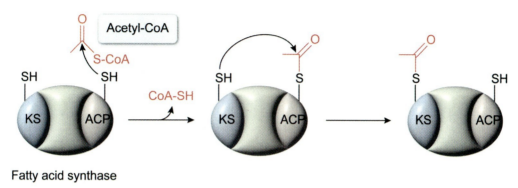

KS = β-ketoacyl synthase, ACP = acyl carrier protein

Figure 13.16 Initial step of the fatty acid synthase mechanism that links acetyl-CoA to the KS domain.

The ACP domain continues to react with thioesters, but now an *activated* form of acetyl-CoA is required. The necessary molecule is called malonyl-CoA and forms by carboxylation of acetyl-CoA. Malonyl-CoA synthesis is catalyzed by **acetyl-CoA carboxylase**, which serves as a major rate-determining enzyme of both fatty acid synthesis *and* oxidation (see Concept 13.1.05). Once malonyl-CoA is formed, it reacts with the ACP domain and releases its coenzyme A group as shown in Figure 13.17.

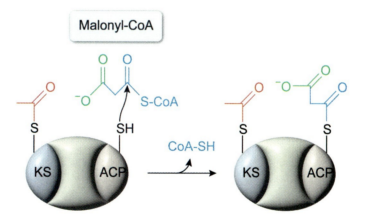

Figure 13.17 Linkage of malonyl-CoA to the ACP domain of fatty acid synthase.

At this point, a repeating series of four steps is used to synthesize fatty acids. These steps are similar to those of β-oxidation in reverse: condensation, reduction, dehydration, and another reduction.

1. **Condensation of malonyl and acetyl groups**. In this step, the acetyl group on the KS domain is transferred to the middle carbon of the malonyl group attached to the ACP domain. This process,

shown in Figure 13.18, releases CO₂ and forms a four-carbon group that contains both a thioester and a ketone, with the ketone at the β position.

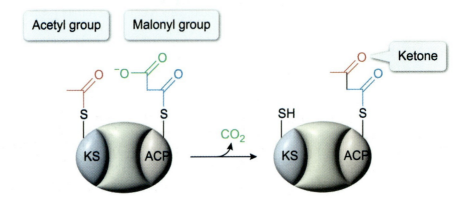

Figure 13.18 A reaction between the acetyl and malonyl groups yields a four-carbon chain and releases CO_2.

2. **Reduction of the ketone group**. The ketone at the β position is reduced to an alcohol, while **NADPH** is oxidized to NADP⁺. This reaction is shown in Figure 13.19. The four-carbon unit remains bound to the ACP domain prosthetic group throughout this process.

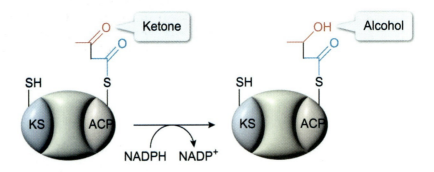

Figure 13.19 Reduction of the ketone group to an alcohol with oxidation of NADPH to NADP⁺.

3. **Dehydration**. The hydroxyl component of the alcohol is removed along with a hydrogen atom from the adjacent α carbon, forming water. A double bond also forms between the two affected carbon atoms, resulting in an alkenyl group, as shown in Figure 13.20.

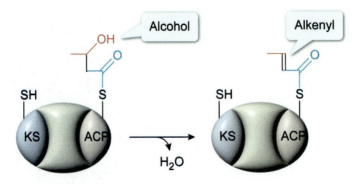

Figure 13.20 Dehydration of the alcohol to form an alkenyl group.

4. **Reduction of the alkene.** The double bond that formed between two carbons in Step 3 is reduced to a single bond (ie, an alkyl group), while NADPH is oxidized to NADP$^+$. The result is a saturated fatty acyl chain linked to the ACP domain (Figure 13.21).

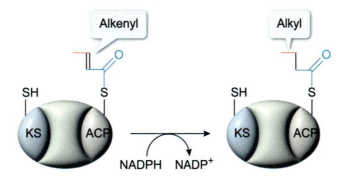

Figure 13.21 Reduction of the alkenyl group to an alkyl group, along with oxidation of NADPH.

After the four steps in the ACP domain are complete, the new four-carbon fatty acyl group is transferred back to the KS domain, and a new malonyl-CoA molecule reacts with the ACP prosthetic group. Steps 1–4 are then repeated, except instead of transferring a *two*-carbon acetyl group to malonyl-CoA, the *four*-carbon fatty acyl group is transferred to yield a six-carbon chain (Figure 13.22).

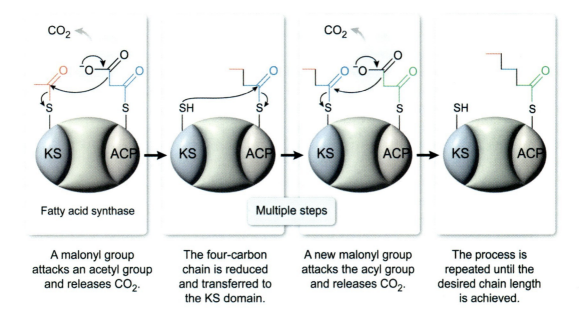

Figure 13.22 Elongation of a fatty acyl chain by repeating rounds of reactions with malonyl groups and subsequent reductions.

Once the four ACP steps are completed again, the new six-carbon acyl chain is transferred to KS, and the process is repeated to form an eight-carbon chain, a 10-carbon chain, and so on. Typically, this repeats until a **16-carbon chain** is produced. The 16-carbon chain is hydrolyzed from fatty acid synthase to yield **palmitate**. Subsequent reactions in the smooth endoplasmic reticulum may further modify the chain, such as as by extending it or by adding double bonds. Each fatty acid chain may then be esterified onto glycerol backbone for formation of a triglyceride for storage or for formation of a glycerophospholipid.

Concept Check 13.3

Glycolysis of how many glucose molecules is required to provide the carbon to synthesize one palmitate molecule?

Solution

Note: The appendix contains the answer.

Energy Cost of Fatty Acid Synthesis

Each malonyl-CoA molecule requires one ATP equivalent to produce. Seven malonyl-CoA molecules are added to one acetyl-CoA molecule to make palmitate; therefore, seven ATP equivalents are required for this process. In addition, each round of four ACP steps requires two NADPH molecules (ie, 14 NADPH molecules are needed to produce palmitate). The net equation for palmitate synthesis is

$$8 \text{ acetyl-CoA} + 7 \text{ ATP} + 14 \text{ NADPH} + 14 \text{ H}^+ \rightarrow \text{palmitate} + 8 \text{ CoA} + 7 \text{ ADP} + 7 \text{ P}_i + 14 \text{ NADP}^+ + 6 \text{ H}_2\text{O}$$

Note that the citrate shuttle also requires two ATP equivalents per acetyl-CoA transported when pyruvate is used for shuttling back into the mitochondria, and therefore this system requires an additional 16 ATP equivalents to transport the eight acetyl-CoA molecules. The overall cost of palmitate synthesis is shown in Figure 13.23.

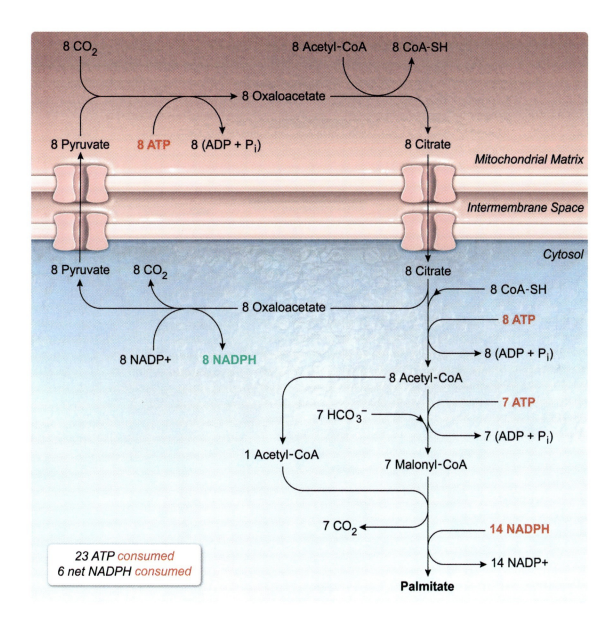

Figure 13.23 Energy cost of palmitate synthesis including the cost of acetyl-CoA transport out of the mitochondrial matrix.

As mentioned previously, the citrate shuttle provides one NADPH per acetyl-CoA transported, which can then be used in fatty acid synthesis. However, *two* NADPH molecules are required for each round of the ACP mechanism. Therefore, the citrate shuttle does not provide enough NADPH to power fatty acid synthesis by itself. The primary source of cytosolic NADPH in most eukaryotic cells is the oxidative phase of the pentose phosphate pathway (see Lesson 11.3).

In summary, fatty acid synthesis has a high energy cost. The benefit of this process is that fatty acids can store a large amount of chemical energy in a relatively small volume, and this energy can be used when the metabolic needs of the body are not being met by the diet (ie, during fasting).

13.1.05 Regulation of Fatty Acid Metabolism

Fatty acid oxidation and fatty acid synthesis are opposed metabolic processes. Assembling acetyl-CoA molecules into fatty acids costs more energy than is obtained by oxidizing fatty acids to acetyl-CoA.

Therefore, allowing both fatty acid synthesis and oxidation to occur simultaneously wastes energy. To avoid this waste, the two processes are reciprocally regulated, meaning when one process is active the other is inactive.

One of the major factors that aids in regulation of fatty acid metabolism is **compartmentalization**. Fatty acid synthesis takes place in the cytosol, whereas β-oxidation occurs in the mitochondrial matrix. Acetyl-CoA cannot freely cross the inner mitochondrial membrane and must use shuttles such as the citrate shuttle. Similarly, many fatty acids require the carnitine shuttle to enter the mitochondria. Figure 13.24 shows the compartmentalization of β-oxidation and fatty acid synthesis.

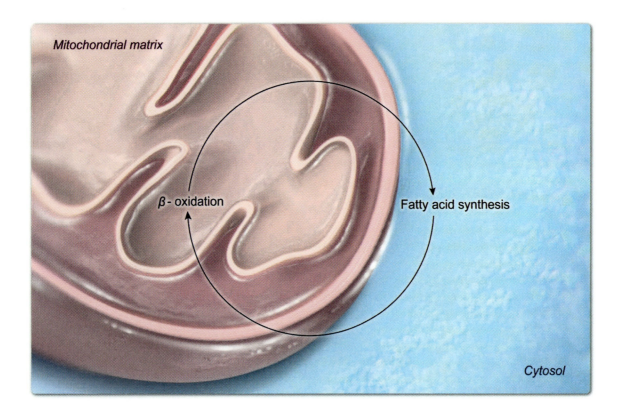

Figure 13.24 β-Oxidation and fatty acid synthesis occur in different compartments within the cell.

Therefore, control over these processes involves regulation of both the enzymes involved and in the shuttles involved. This control is also governed in part by blood glucose levels. Under high-glucose conditions (eg, after a meal), the pancreas releases **insulin** into the bloodstream. When insulin binds to its receptors, a cascade is triggered that **dephosphorylates** acetyl-CoA carboxylase (ACC), the enzyme that produces malonyl-CoA for fatty acid synthesis. Dephosphorylation *activates* this enzyme and induces malonyl-CoA production, which upregulates the *synthesis* pathway overall (see Figure 13.25).

In addition to its role as a substrate in fatty acid synthesis, malonyl-CoA *inhibits carnitine acyltransferase I* and slows the carnitine shuttle. This prevents many fatty acids from entering the mitochondria, which *prevents them from undergoing β-oxidation*. In this way, ACC activation both *stimulates fatty acid synthesis* and *inhibits fatty acid oxidation*, preventing a futile cycle.

Chapter 13: Noncarbohydrate Metabolism

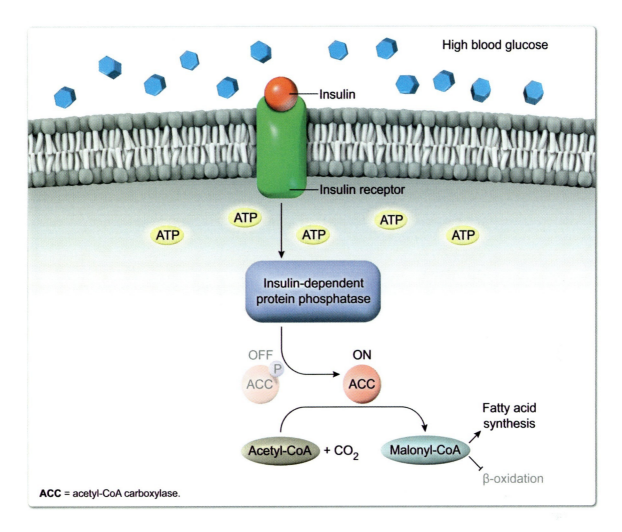

Figure 13.25 In high-glucose conditions, insulin causes activation of acetyl-CoA carboxylase (ACC), which produces malonyl-CoA to upregulate fatty acid synthesis and downregulate β-oxidation.

Conversely, when blood glucose is low (eg, after a period of fasting), the pancreas releases glucagon, which induces the reverse effect of insulin.

Glucagon binding to its receptor results in activation of AMP-activated protein kinase (AMPK), which phosphorylates ACC and inactivates it. This inactivation results in lower malonyl-CoA levels, which in turn causes less fatty acid synthesis. Simultaneously, carnitine acyltransferase I inhibition ceases as malonyl-CoA is depleted, allowing fatty acids to enter the mitochondria. By these means glucagon upregulates β-oxidation and downregulates fatty acid synthesis, as shown in Figure 13.26.

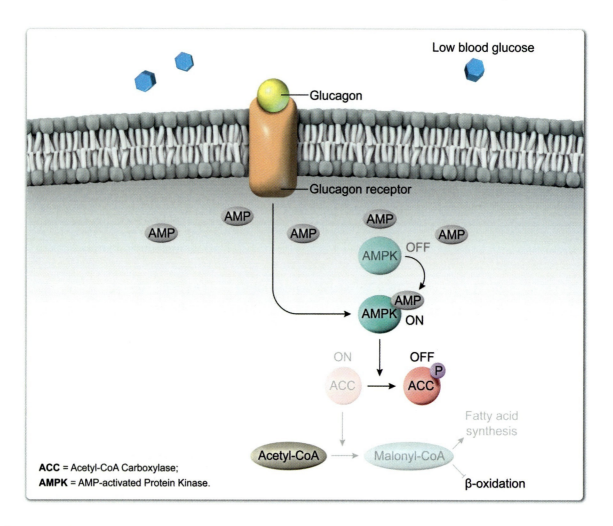

Figure 13.26 Upregulation of β-oxidation and downregulation of fatty acid synthesis under low blood glucose conditions.

Lesson 13.2
Protein Catabolism

Introduction

Proteins are not the primary source of energy for most organisms because proteins serve important biological roles as enzymes, transport systems, structural units, and more. However, under conditions of starvation an organism may digest its own proteins for energy, and even under well-fed conditions a basal level of protein digestion and recycling occurs. Proteins obtained through the diet are also broken down into their constituent amino acids to be used for synthesis of other proteins needed by the body. An abundance of protein digestion products in the blood or urine may indicate malnutrition or various disease states.

This lesson describes the means by which proteins are digested into individual amino acids, and the mechanisms by which amino acids may be digested for energy or other purposes.

13.2.01 Protein Degradation

Protein catabolism begins with degradation of proteins into short peptides or individual amino acids. Protein obtained through the diet is digested by various **proteases** found throughout the digestive tract, including pepsin, trypsin, chymotrypsin, and carboxypeptidase. The amino acids and short peptides produced in the digestive tract are absorbed by intestinal cells (Figure 13.27), and the short peptides are further digested to amino acids and allowed to enter the blood stream. The amino acids are then transported to various tissues throughout the body where they can be incorporated into the proteins needed by those tissues.

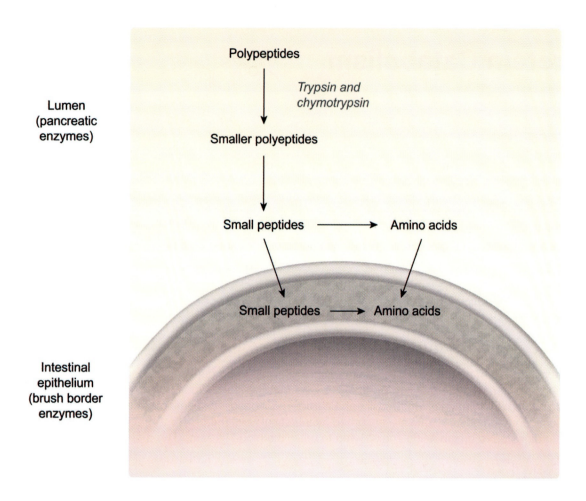

Figure 13.27 Digestion of polypeptides into smaller units within the intestine.

Within a living cell, functional proteins are frequently degraded when they are no longer needed, when they misfold, or when the cell needs the amino acids for other uses. Proteins in the cytosol may be tagged for degradation by **ubiquitin**, which targets them to the proteasome. Proteins within certain organelles such as the endoplasmic reticulum or the Golgi apparatus may be sent to the lysosome for degradation. Lysosomes also digest proteins obtained from the surrounding environment through endocytosis.

Once proteins within a cell are digested, the individual amino acids may be recycled and incorporated into other proteins, or they may be further digested for energy. The remainder of this lesson describes the digestion of amino acids and the preparation of their waste products (primarily urea) for excretion.

13.2.02 Transamination, Deamination, and Deamidation

The first step in the catabolism of many amino acids is **transamination**, in which the α-amino group of an amino acid is transferred to another molecule such as α-ketoglutarate. This process converts the amino acid into a form called an **α-keto acid**, which has a ketone group where the α-amino group previously was. α-Ketoglutarate becomes glutamate when it receives the amino group, as shown in Figure 13.28.

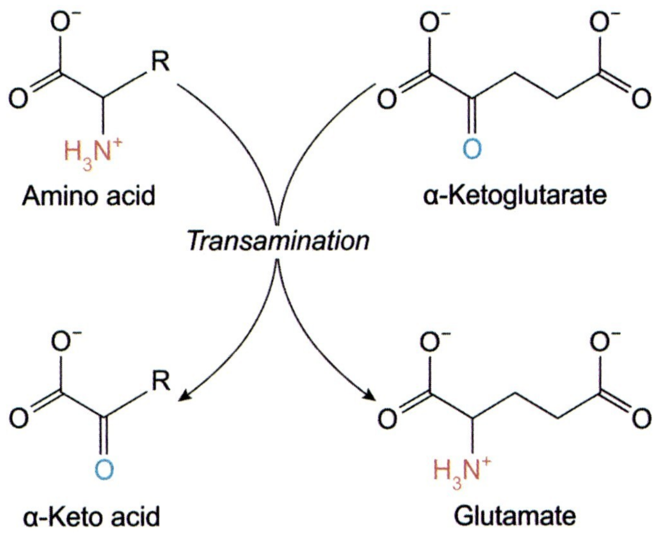

Figure 13.28 Transamination converts an amino acid to an α-keto acid while also converting α-ketoglutarate to glutamate.

Several α-keto acids are of particular importance: **pyruvate** is the corresponding α-keto acid of alanine, **oxaloacetate** is the corresponding α-keto acid of aspartate, and **α-ketoglutarate** is the corresponding α-keto acid of glutamate, as shown in Table 13.1. Each of these α-keto acids is involved in the citric acid cycle or, for pyruvate, in a metabolic step that directly precedes the cycle. Therefore, each of these amino acids can be converted into a form that can easily enter the citric acid cycle.

Table 13.1 Structural relationships of alanine, aspartate, and glutamate to their corresponding α-keto acids: pyruvate, oxaloacetate, and α-ketoglutarate, respectively.

α-Amino acid	α-Keto acid
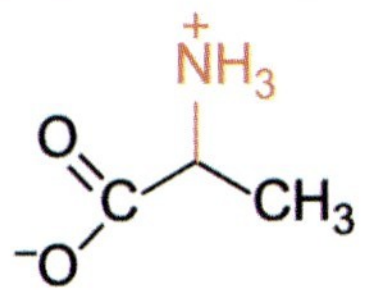 Alanine	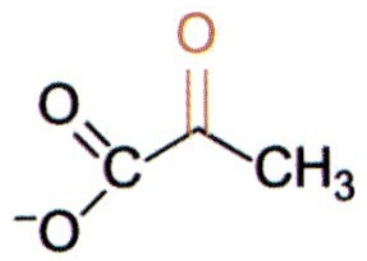 Pyruvate
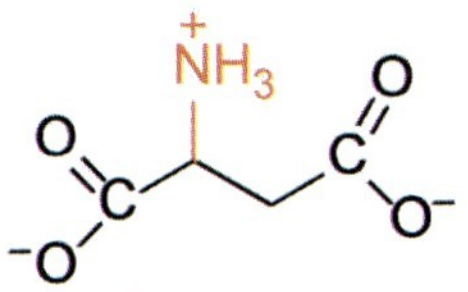 Aspartate	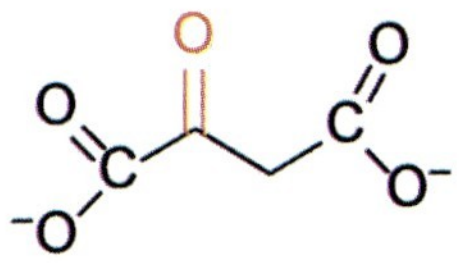 Oxaloacetate
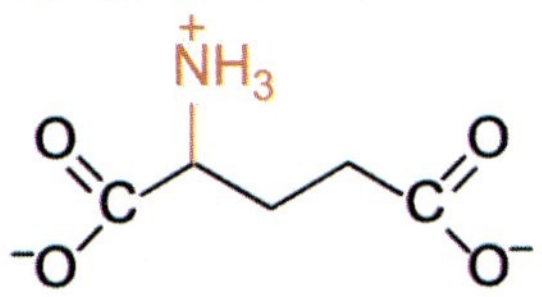 Glutamate	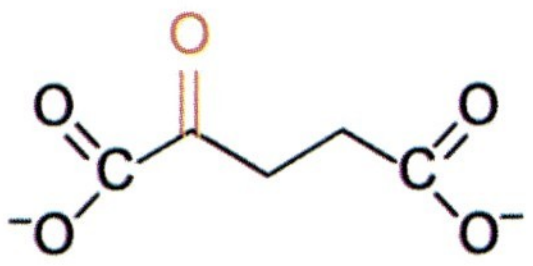 α-Ketoglutarate

Other amino acids can also react through transamination directly or can undergo other pathways that eventually result in loss of nitrogen atoms.

Chapter 13: Noncarbohydrate Metabolism

> ✓ **Concept Check 13.4**
>
> The following α-keto acid was most likely produced by a transamination reaction between α-ketoglutarate and which amino acid?
>
> **Solution**
>
> *Note: The appendix contains the answer.*

Glutamate, formed from α-ketoglutarate, serves as a repository for amino groups. In the liver, glutamate is **deaminated** in a redox reaction that requires NAD^+ or $NADP^+$ (either may be used). This process regenerates α-ketoglutarate and releases the nitrogen atom in the form of ammonium (NH_4^+). Ammonium then enters the **urea cycle** (Concept 13.2.03) to be excreted. Figure 13.29 shows an overview of this process.

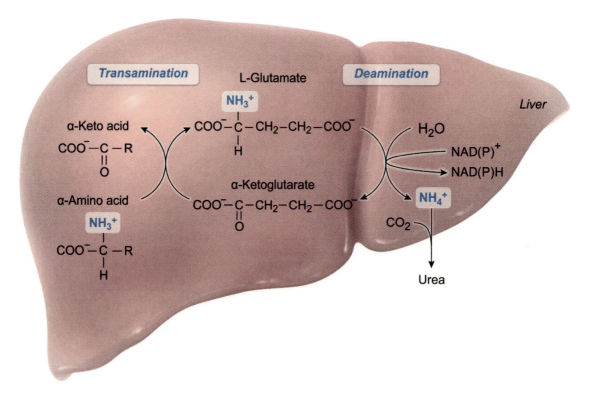

Figure 13.29 Overview of the handling of nitrogen from amino acids in the liver.

In cells that do *not* use the urea cycle (eg, muscle cells), the ammonium that would be produced by deamination would not excreted but would instead equilibrate with ammonia (NH_3), which is highly toxic. To avoid toxicity, non-liver cells use other pathways to transport ammonium groups to the liver for processing. Glutamate cannot be readily exported to the blood stream, so glutamate in muscle transfers its amino group to pyruvate, which is often abundant in muscle because it is the final product of glycolysis. This forms alanine, which is then exported into the blood and carried to the liver.

In the liver, alanine transfers its amino group back to α-ketoglutarate, which is metabolized normally through deamination and entry of NH_4^+ into the urea cycle. The pyruvate that is regenerated by this process then undergoes gluconeogenesis and is sent back to the muscles as glucose. Muscular glucose undergoes glycolysis to become pyruvate, which repeats the process. This process is called the glucose-alanine cycle and is shown in Figure 13.30.

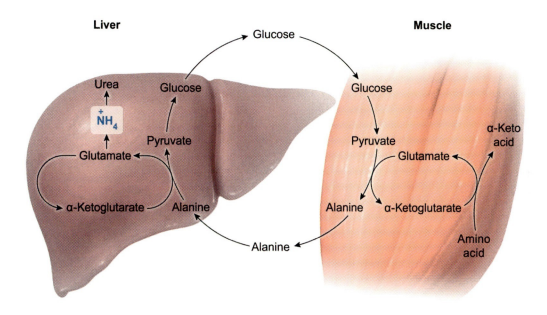

Figure 13.30 The glucose-alanine cycle.

Many non-liver cells produce free ammonia from non–amino acid sources such as nucleotide degradation. To package the ammonia in a nontoxic form, this free ammonia reacts with the side chain of glutamate to form glutamine, which can be exported to the liver along with alanine. Glutamine synthesis by this method requires ATP hydrolysis for energy.

In the liver, glutamine is **deamidated** to release ammonium from its *side chain*. This process regenerates glutamate, which can then be catabolized normally (ie, deamination to α-ketoglutarate). Note that the functional group of the glutamine side chain is an *amide*, not an amine, and therefore removal of the –NH₂ group is deamidation, *not* deamination. The released ammonium ion undergoes the urea cycle in the liver. Figure 13.31 shows the role of glutamine in transport of ammonium to the liver.

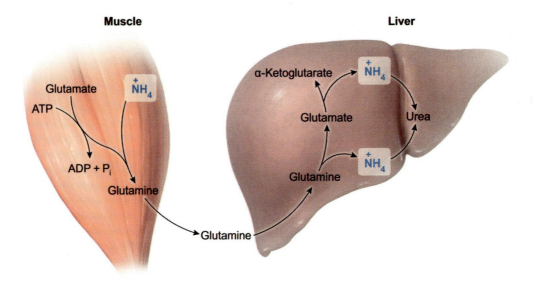

Figure 13.31 Glutamate reacts with free ammonium in the muscles and other tissues to form glutamine, which is transported to the liver for deamidation.

13.2.03 The Urea Cycle

Once in the liver, both glutamate and glutamine are transported to the mitochondria, where they are deamidated (glutamine) and deaminated (glutamate) to produce ammonium ions (which are in equilibrium with ammonia). In humans and many other terrestrial (ie, land-dwelling) animals, these ammonium ions and ammonia molecules then enter the **urea cycle**.

The details of the urea cycle (shown in Figure 13.32) are unlikely to be tested on the exam, but study of the urea cycle helps highlight the interplay between different metabolic pathways (eg, the citric acid cycle). Defects in the urea cycle can lead to severe metabolic disorders with effects ranging from chronic vomiting to impaired brain function to death.

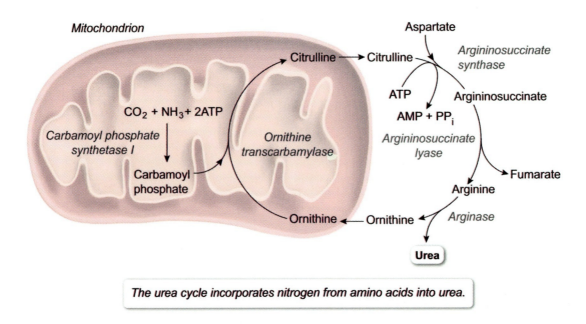

The urea cycle incorporates nitrogen from amino acids into urea.

Figure 13.32 The urea cycle.

A few chemical features are worth noting. Some parts of the urea cycle occur in the mitochondrial matrix, and others occur in the cytosol. The first step in the mitochondrial matrix converts ammonium to carbamoyl phosphate through a reaction with CO_2 and two ATP molecules. Each ATP is converted to ADP, and therefore the urea cycle up to this point requires two ATP equivalents.

Carbamoyl phosphate reacts with ornithine to form citrulline, which exits the mitochondrial matrix through a transporter. Citrulline reacts with aspartate in the cytosol to form argininosuccinate in a process that requires conversion of ATP to AMP and PP_i. This constitutes another two ATP equivalents, for a total of *four*.

Argininosuccinate, as its name implies, consists of an arginine component and a succinate component. This molecule is split into arginine and fumarate (a derivative of succinate). A portion of the arginine side chain is then hydrolyzed to form urea, which exits the liver and is transported to the kidneys for excretion. This regenerates ornithine. Note that ornithine, citrulline, and argininosuccinate are all α-amino acids, but they are not found in proteins. Therefore, they are examples of *nonproteinogenic* amino acids.

The fumarate produced by the urea cycle can enter the citric acid cycle. Oxaloacetate can be transaminated to form aspartate, which can react with citrulline. In this way, the citric acid cycle and the urea cycle in the liver are closely linked, as shown in Figure 13.33.

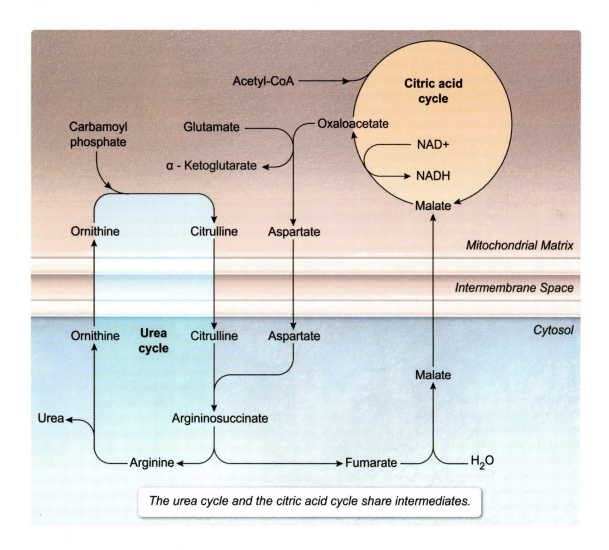

Figure 13.33 The urea cycle is linked to the citric acid cycle.

13.2.04 Glucogenic and Ketogenic Amino Acids

This lesson's discussion of amino acid catabolism has so far been limited to the handling of nitrogen. After an amino acid has lost its nitrogen atoms through deamination or deamidation, the carbon skeleton remains. Different amino acids produce different carbon skeletons, which are metabolized by distinct pathways.

The pathways of degradation for most amino acids are unlikely to be tested on the exam, but a few trends are noteworthy. Specifically, amino acids may be **glucogenic** (ie, precursors to gluconeogenesis), **ketogenic** (ie, precursors to acetyl-CoA and ketone body formation), or both. Table 13.2 summarizes these amino acid groups.

Table 13.2 Glucogenic and ketogenic classifications of the 20 proteinogenic amino acids.

Glucogenic only	Both glucogenic and ketogenic	Ketogenic only
Alanine, A	Phenylalanine, F	Leucine, L
Cysteine, C	Isoleucine, I	Lysine, K
Aspartate, D	Threonine, T	
Glutamate, E	Tryptophan, W	
Glycine, G	Tyrosine Y	
Histidine, H		
Methionine, M		
Asparagine, N		
Proline, P		
Glutamine, Q		
Arginine, R		
Serine, S		
Valine, V		

Glucogenic amino acids are those whose carbon skeletons can be converted to pyruvate or a citric acid cycle intermediate. Pyruvate can be converted directly to oxaloacetate as part of the gluconeogenesis pathway, and citric acid cycle intermediates can also be converted to oxaloacetate through the cycle. Oxaloacetate can then be converted to phosphoenolpyruvate by the enzyme PEPCK, which can continue through gluconeogenesis to generate *glucose*.

Most of the proteinogenic amino acids are *glucogenic only*. Note that although pyruvate theoretically *can* become acetyl-CoA, protein and amino acid catabolism primarily occurs during fasting. Under this condition, gluconeogenesis in the liver is generally upregulated and pyruvate dehydrogenase is downregulated. Consequently, pyruvate is primarily used for gluconeogenesis during protein catabolism, and alanine, serine, glycine, and cysteine (each of which can become pyruvate) are not considered ketogenic.

Although the specific pathways of glucogenic amino acid catabolism do not need to be memorized, an overview is provided in Figure 13.34.

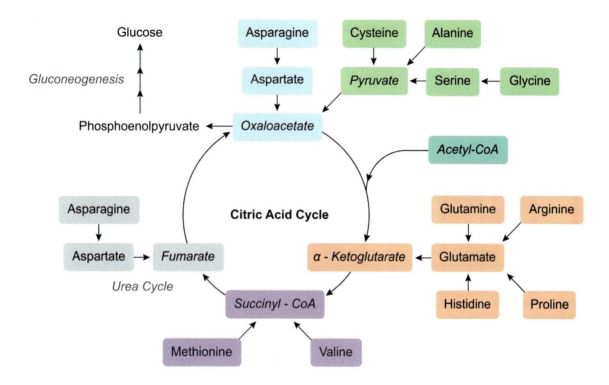

Figure 13.34 Overview of the catabolism of purely glucogenic amino acids. Asparagine and aspartate have two points of entry.

Ketogenic amino acids are those whose carbon skeletons can be converted to acetyl-CoA or to acetoacetyl-CoA. Under fasting conditions, these generate *ketone bodies*. This pathway is not glucogenic because, as described elsewhere, acetyl-CoA can be converted to oxaloacetate only by first *reacting* with oxaloacetate, and therefore no *net* increase in oxaloacetate occurs. Two amino acids, lysine and leucine, are *ketogenic only*.

The remaining amino acids are *both glucogenic and ketogenic* because their carbon skeletons are split during catabolism. Some portions of the skeleton are glucogenic while others are ketogenic. These amino acids are tryptophan, phenylalanine, tyrosine, isoleucine, and threonine. These amino acids may be remembered using the mnemonic "FITTT" because "F"-enylalanine, isoleucine, and all of the amino acids that start with T are in this category. Figure 13.35 shows an overview of ketogenic amino acid catabolism, including amino acids that are also glucogenic.

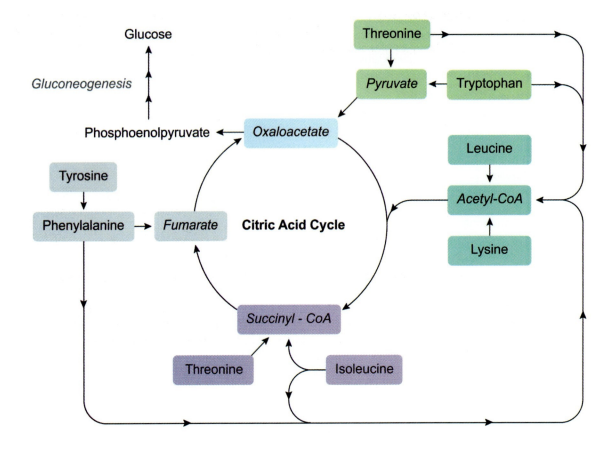

Figure 13.35 Overview of metabolism of ketogenic amino acids. Lysine and leucine are *solely* ketogenic.

Several metabolic disorders stem from an inability to catabolize amino acid carbon skeletons. Impaired catabolism of the branched-chain amino acids (see Lesson 1.2) causes maple syrup urine disease, which can be fatal. Inability to catabolize phenylalanine leads to phenylketonuria, which causes severe vomiting and impaired brain development. Other defects in amino acid digestion underlie many other congenital disorders, giving amino acid metabolism significant clinical importance.

> **Concept Check 13.5**
>
> Suppose the peptide VDLY is fully catabolized. Assume each amino acid begins its catabolic pathway by transferring its amino group to α-ketoglutarate and that each amino group enters the urea cycle through carbamoyl phosphate.
>
> 1. How many urea molecules are produced?
> 2. How many of the metabolized amino acids can contribute to glucose synthesis?
> 3. How many of the metabolized amino acids can contribute to ketone body synthesis?
>
> **Solution**
>
> *Note: The appendix contains the answer.*

END-OF-UNIT MCAT PRACTICE

Congratulations on completing **Unit 4: Metabolic Reactions**.

Now you are ready to dive into MCAT-level practice tests. At UWorld, we believe students will be fully prepared to ace the MCAT when they practice with high-quality questions in a realistic testing environment.

The UWorld Qbank will test you on questions that are fully representative of the AAMC MCAT syllabus. In addition, our MCAT-like questions are accompanied by in-depth explanations with exceptional visual aids that will help you better retain difficult MCAT concepts.

TO START YOUR MCAT PRACTICE, PROCEED AS FOLLOWS:

1) Sign up to purchase the UWorld MCAT Qbank
   IMPORTANT: You already have access if you purchased a bundled subscription.
2) Log in to your UWorld MCAT account
3) Access the MCAT Qbank section
4) Select this unit in the Qbank
5) Create a custom practice test

Unit 5 Biochemistry Lab Techniques

Chapter 14 Biomolecule Purification and Characterization

14.1 Gel Electrophoresis

- 14.1.01 Principles of Electrophoresis
- 14.1.02 Native Electrophoresis
- 14.1.03 Denaturing Electrophoresis
- 14.1.04 Reducing Gels
- 14.1.05 Isoelectric Focusing
- 14.1.06 2D Gels

14.2 Blotting Techniques

- 14.2.01 Principles of Blotting Techniques
- 14.2.02 Southern Blots
- 14.2.03 Northern Blots
- 14.2.04 Western Blots

14.3 Chromatography

- 14.3.01 Principles of Chromatography
- 14.3.02 Size Exclusion Chromatography
- 14.3.03 Ion-Exchange Chromatography
- 14.3.04 Affinity Chromatography

14.4 Additional Techniques

- 14.4.01 Dialysis
- 14.4.02 Biomolecule Quantitation
- 14.4.03 Binding Assays
- 14.4.04 Melting Temperature Assays
- 14.4.05 Fluorescence
- 14.4.06 Circular Dichroism
- 14.4.07 Structure Determination

Lesson 14.1
Gel Electrophoresis

Introduction

This book has presented several principles of biochemistry: the nature of amino acids, their formation into proteins, the ability of many proteins to act as enzymes, structural features of various biomolecules, and metabolic reactions. These principles were not always known, however, and had to be determined experimentally. This chapter presents some of the techniques used to probe biomolecules. In particular, it explores several techniques used to purify and characterize proteins and other biomolecules.

One of the most important tools in biochemistry is electrophoresis. This technique includes a broad range of methods that separate biomolecules based on their size, electric charge, and other characteristics by generating an electric field that forces molecules to migrate through a gel. This lesson discusses multiple types of electrophoresis and the principles that govern this technique.

14.1.01 Principles of Electrophoresis

Broadly speaking, **electrophoresis** involves the use of an electric field to cause charged particles to migrate through a gel or some other substance. In a biochemical context, electrophoresis is used to separate various charged particles (typically proteins, DNA, or RNA) from each other. Separation occurs because different proteins (or different DNA or RNA molecules) have different abilities to migrate through the gel.

The most common substances used to make gels for electrophoresis are agarose (a complex carbohydrate) and polyacrylamide (an organic polymer). Both substances can form a porous matrix. *Larger molecules* migrate through the matrix *more slowly* than smaller molecules because it is more difficult for larger molecules to navigate the pores, as shown in Figure 14.1.

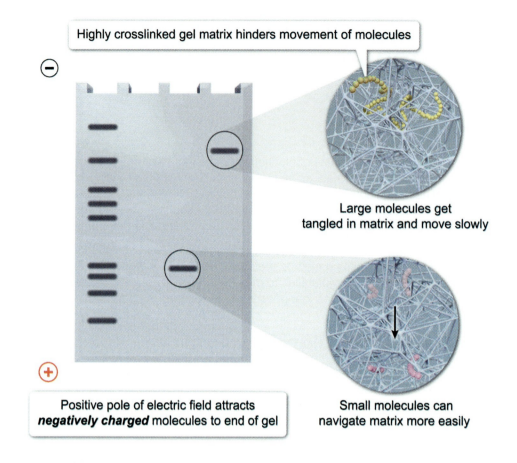

Figure 14.1 Relative motion of large and small particles through a gel in an electric field.

In addition, molecules with a higher magnitude of charge (positive or negative) experience a greater force when exposed to an electric field, so they move more quickly through the gel. Accordingly, electrophoresis may separate molecules by size, charge, or both.

A gel typically consists of a relatively small percentage of the matrix-forming agent (agarose or polyacrylamide) and a much larger percentage of aqueous solution (eg, a Tris-HCl buffer adjusted to a desired pH). Varying the percentage of agarose or polyacrylamide alters the size of the pores within the gel. For example, a 10% polyacrylamide gel forms smaller pores than a 5% polyacrylamide gel.

Typically, agarose gels form larger pores than polyacrylamide gels. Consequently, **agarose** gels are most often used to separate *larger molecules (eg, large proteins, DNA)*, and polyacrylamide gels are used to separate smaller proteins. Most proteins are small enough that they can be separated by **polyacrylamide gels**. For this reason, protein separation is often carried out by polyacrylamide gel electrophoresis (PAGE), whereas DNA and RNA separations are more commonly carried out by agarose gel electrophoresis. These gels are shown in Figure 14.2.

Chapter 14: Biomolecule Purification and Characterization

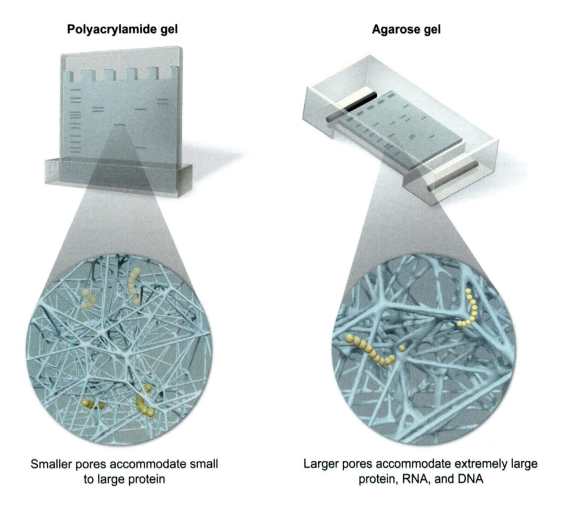

Figure 14.2 Polyacrylamide and agarose gels, and the molecules they separate by electrophoresis.

Electrophoresis occurs in an electrolytic cell. Unlike galvanic cells, electrolytic cells use a power source to drive electrons toward the *negative* electrode, where they accumulate and are eventually *gained* by electrolytes, causing a *reduction* reaction.

By definition, *reduction always occurs at the cathode* (see General Chemistry Chapter 9). Therefore, in electrolytic cells, the negatively charged electrode (ie, the electrode that the power source moves electrons to) is the cathode and the *positively charged electrode* is the **anode**.

Once the current is applied, negatively charged particles within the gel begin migrating *toward the positively charged anode*. Positively charged particles do the opposite. It may be helpful to remember that the *anode attracts anions* in these cells. The smallest molecules and those with the largest charge migrate the farthest through the gel while the current is applied.

In many electrophoresis experiments, the molecules of interest are either intrinsically negatively charged (eg, DNA, RNA) or they are given a negative charge by altering experimental conditions. For this reason, samples are usually loaded at the end of the gel nearer to the cathode (see Figure 14.3). Certain variations on electrophoresis in which some or all of the molecules are positively charged require a different setup. However, if such a setup is not specified, it can generally be assumed that the sample is loaded into the gel near the cathode.

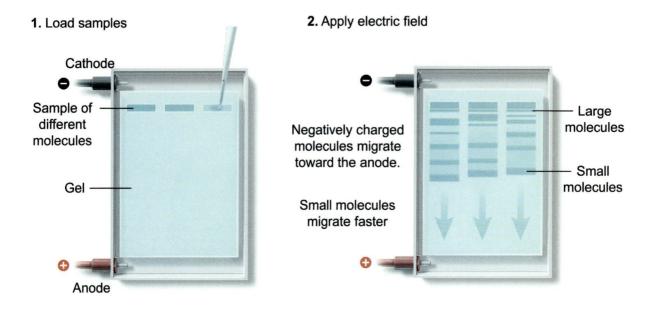

Figure 14.3 Electrophoresis of negatively charged particles: Samples are loaded near the cathode and migrate toward the anode, with small molecules migrating faster.

Once the applied electric current is turned off, the molecules in the gel stop migrating. In general, at this point the molecules are invisible and must be **visualized** for analysis. Proteins are commonly visualized by applying a staining molecule called Coomassie, which binds to proteins and colors them blue.

The gel is soaked in Coomassie stain and then rinsed, and the proteins appear in bands (Figure 14.4). The width or color density of a band corresponds to the amount of protein in that band. Note that electrophoresis and staining are only effective for detecting sufficiently large proteins (typically around 70 amino acids or larger).

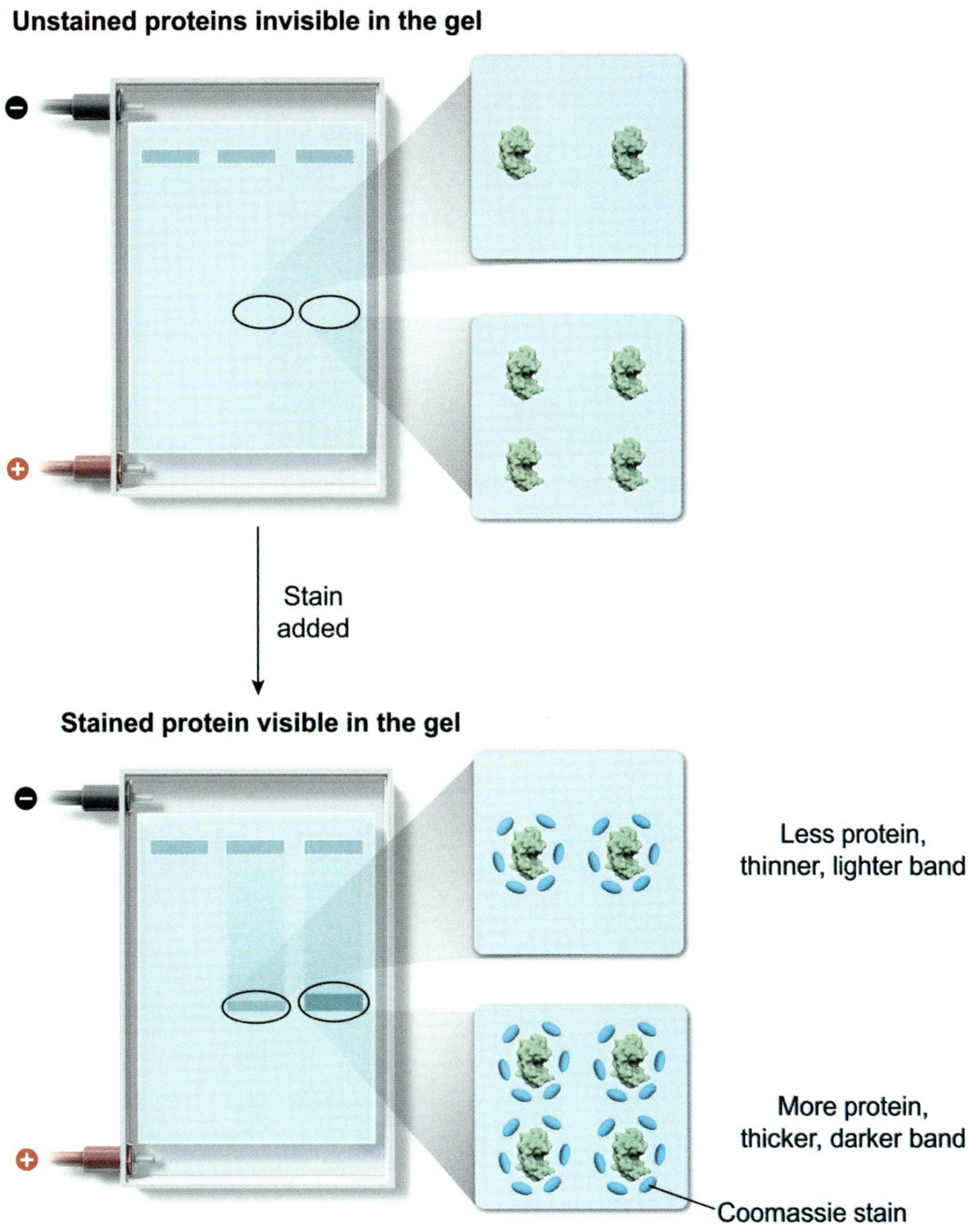

Figure 14.4 Coomassie stain makes proteins in a gel visible.

In DNA and RNA gels, a fluorescent dye such as ethidium bromide is often included. These dyes bind to nucleic acids, which significantly increases the dye's fluorescence intensity. DNA gels of this nature are generally visualized by exposing them to ultraviolet light, which causes the dye to fluoresce bright orange, as shown in Figure 14.5.

Chapter 14: Biomolecule Purification and Characterization

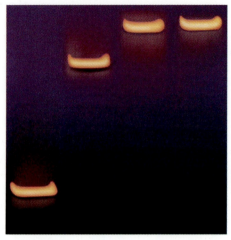

DNA fluoresces under UV light when treated with ethidium bromide

Figure 14.5 Visualization of DNA in a gel.

Whether a gel is used to analyze DNA, RNA, or proteins, it often contains multiple lanes, each of which represents a different sample. At least one lane contains a "ladder," which is a mixture of molecules of known size and charge (see Figure 14.6). The positions of the bands in the ladder can be used to estimate the sizes of the molecules in the other lanes.

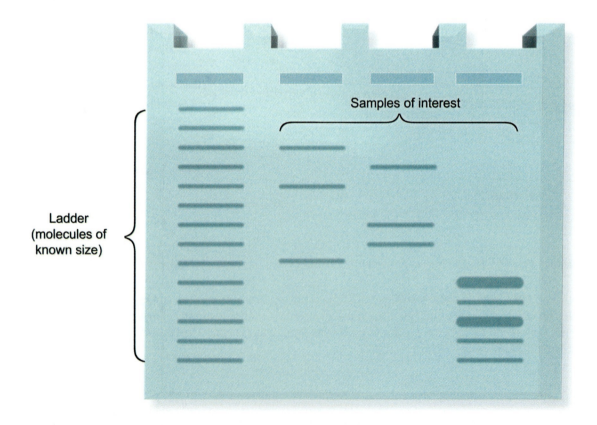

Figure 14.6 A typical gel with a ladder lane and several sample lanes.

By analyzing the intensity of bands at a given position (ie, distance of migration), the relative amounts of a specific protein or DNA strand from each sample may be compared. This is commonly used to determine how protein expression varies in different conditions. An increase in band density, for example, corresponds to an increase in gene expression. Similarly, decreases in intensity may correspond to decreased expression or increased degradation.

14.1.02 Native Electrophoresis

Native electrophoresis refers to electrophoretic techniques that preserve the functional structure of the molecule of interest. DNA and RNA electrophoresis methods are usually native. Because all DNA and RNA molecules are negatively charged due to their phosphate groups, all migrate toward the anode without any modifications. In addition, because all nucleic acids contain one phosphate group for every nitrogenous base, the *charge:mass ratio* of every nucleic acid is *approximately the same*.

Therefore, although a longer DNA strand has more negative charge and experiences more force in an electric field, it also has more mass that must migrate. Consequently, the primary factor limiting how quickly nucleic acids migrate through a gel is the size of the pores. Accordingly, native gels separate linear DNA predominantly by size, with *larger molecules migrating more slowly* (and a shorter distance) than smaller molecules, as shown in Figure 14.7.

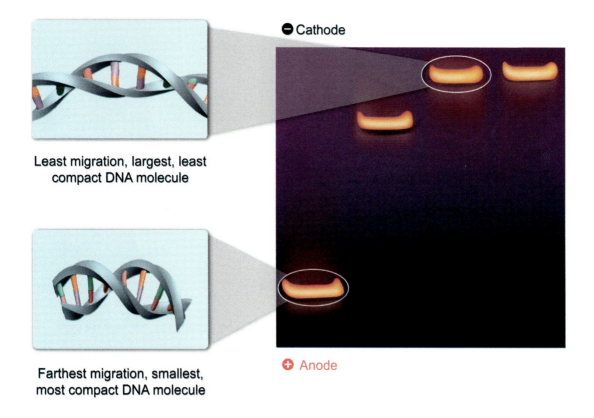

Figure 14.7 DNA strands of different sizes migrate different distances through a gel.

Importantly, circularized DNA molecules and folded RNA molecules may behave differently than their mass alone would suggest. Folded RNA and circularized DNA (especially supercoiled DNA) are more compact, allowing them to navigate the pores of the gel more easily and migrate faster. Binding of nucleic acids to other molecules (eg, proteins) generally hinders migration through the gel by making the complex more massive.

When comparing individual nitrogenous bases, guanine is larger than adenine, which is larger than thymine or uracil. Cytosine is the smallest of the nitrogenous bases. Consequently, guanine alone migrates more slowly through a gel than cytosine alone. However, a base pair consisting of A and T is approximately the same molecular weight as a pair consisting of C and G. Therefore, it is convenient to measure the size of double-stranded DNA in base pairs (bp) or kilobase pairs (kbp), as shown in Figure 14.8.

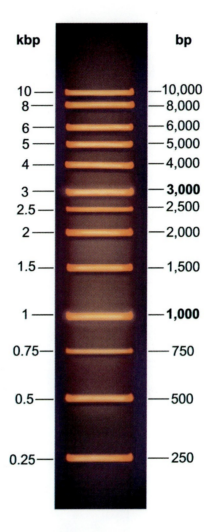

Figure 14.8 A standard DNA ladder with sizes of each band marked in both base pairs and kilobase pairs.

Concept Check 14.1

A DNA strand is digested by restriction enzymes and analyzed by agarose gel electrophoresis, yielding the following results. Each band corresponds to a distinct sequence of DNA.

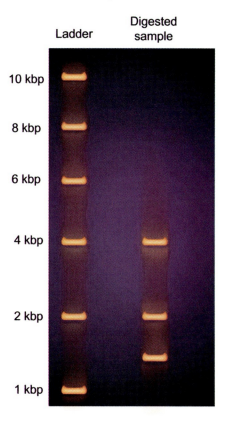

Based on these results, what was the size of the undigested DNA strand in kilobase pairs?

Solution

Note: The appendix contains the answer.

Native electrophoresis may also be used for proteins. This often involves polyacrylamide gels; therefore, this type of electrophoresis is commonly called **native PAGE**. Native PAGE for proteins is often more complicated than electrophoresis of DNA because proteins may have varied charges depending on their isoelectric points (see Chapter 2).

Because native PAGE is meant to preserve protein structure, it usually occurs at physiological pH (ie, pH 7–7.4). Proteins with a pI higher than this pH range are positively charged; those with a lower pI are negatively charged. Positively charged proteins migrate toward the cathode, whereas negatively charged proteins migrate toward the anode.

If two proteins are the same size and have the same sign but different magnitude charges, the one with the greater charge migrates faster. For example, a 50 kDa protein with a −2 charge tends to move toward the anode more quickly than a 50 kDa protein with a −1 charge. Similarly, two proteins of equal charge migrate differently if they are different sizes. Even two proteins of identical molecular weight and charge may migrate differently if one is folded to be more compact than the other, because more compact molecules navigate the pores of the gel more easily. Different factors that affect migration are shown in Figure 14.9.

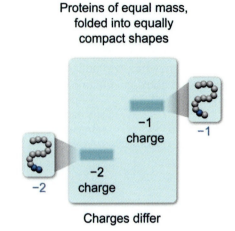

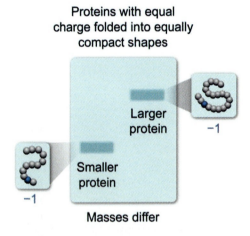

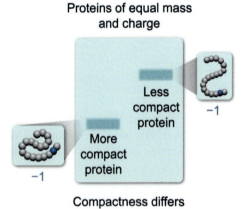

Figure 14.9 Charge, mass, and shape (ie, compactness) all affect how proteins migrate in native PAGE.

These factors together make it difficult to predict how far a protein will migrate in a native gel. However, once the protein of interest is identified, it remains intact, including any quaternary structure. Because protein structure is preserved, the protein can continue to function, even within the gel. Consequently, native gels may be used to assess binding interactions, because interactions with a ligand alter the mobility of the protein. This is called a **mobility shift** (Figure 14.10).

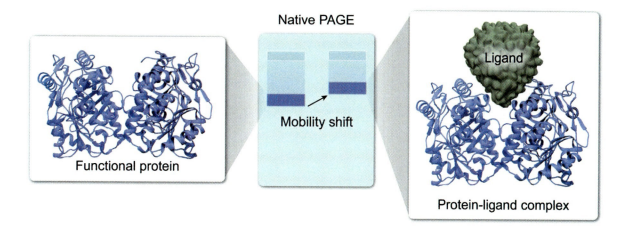

Figure 14.10 Native PAGE preserves secondary, tertiary, and quaternary protein structure. Binding to a ligand can result in a mobility shift.

14.1.03 Denaturing Electrophoresis

Denaturing electrophoresis is most commonly used to separate proteins. Denaturing electrophoresis is typically carried out by adding a denaturing agent, usually **sodium dodecyl sulfate (SDS)**, to the protein mixture. This is generally used with polyacrylamide gel electrophoresis, and the technique is often referred to as SDS-PAGE. Other denaturing agents are occasionally used, but SDS illustrates the overall principles of denaturing electrophoresis.

SDS is a detergent. It consists of a sodium ion and a dodecyl sulfate molecule. Dodecyl sulfate consists of a *negatively charged* head group (the sulfate portion), which interacts with water, and a hydrophobic 12-carbon chain. The hydrophobic tail interacts favorably with hydrophobic side chains in a protein, which disrupts the hydrophobic effect (Lesson 2.3) and causes the protein to *unfold*. The head group confers a *uniform negative charge* on the proteins in the sample. Figure 14.11 shows interactions between SDS and proteins.

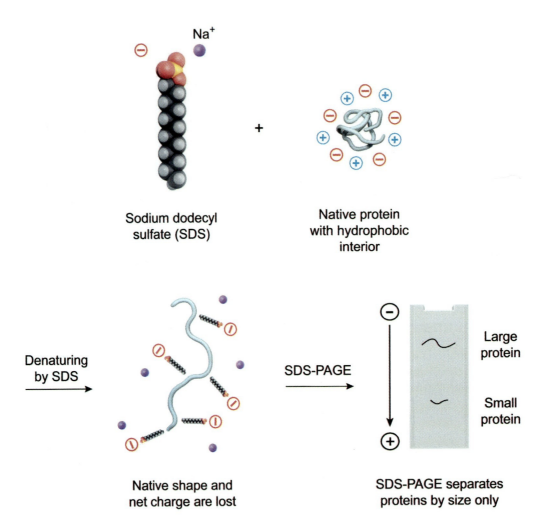

Figure 14.11 Sodium dodecyl sulfate (SDS) denatures proteins and confers a uniform negative charge, allowing separation of proteins by size only.

As a denaturing agent, SDS disrupts secondary, tertiary, and quaternary structure and renders proteins nonfunctional. Note, however, that SDS by itself does *not* break disulfide bonds; additional reagents, described in Concept 14.1.04, can be used together with SDS to break disulfide bonds. Because SDS is a denaturing agent, proteins separated by SDS-PAGE cannot be used to assess protein activity unless the proteins are first allowed to renature.

Concept Check 14.2

Protein A binds Protein B without forming any disulfide bonds. Native PAGE is performed with one lane containing Protein A only, one containing Protein B only, and one lane containing both. The process is repeated using SDS-PAGE. Both gels are stained, and the following results are obtained.

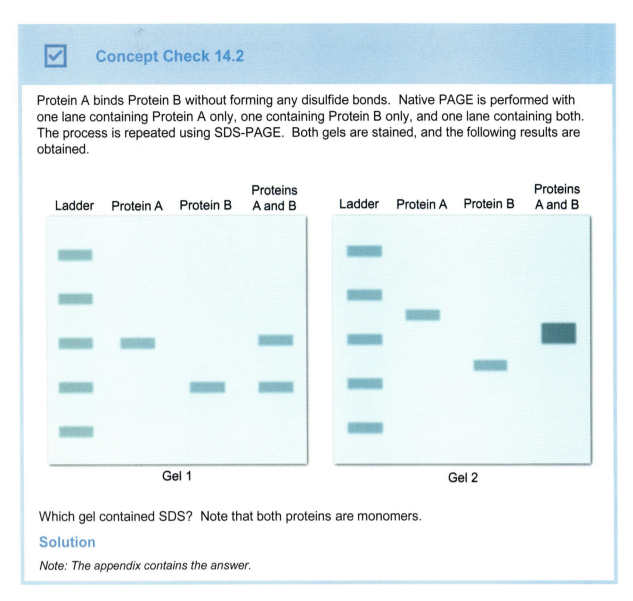

Which gel contained SDS? Note that both proteins are monomers.

Solution

Note: The appendix contains the answer.

The advantage of SDS-PAGE is that because SDS coats the proteins in a solution with negative charges, all proteins, regardless of their isoelectric point, have a *net negative charge*. Therefore, all proteins in the mixture migrate toward the anode. Because the charge distribution is approximately uniform and denaturation causes the proteins to adopt noncompact unfolded shapes, the proteins in the mixture have *similar charge:mass ratios* when SDS is used. These features allow separation of proteins largely based on molecular weight (molecular mass) alone, with larger, massive proteins migrating more slowly than smaller, less massive proteins.

As discussed in Lesson 2.2, the average molecular weight of an amino acid in a protein is 110 daltons. Consequently, the molecular weight of a protein can be estimated from the number of amino acids it contains and vice versa. Accordingly, the size of a protein in an SDS gel is commonly given in kilodaltons (kDa), as shown in Figure 14.12. Note that a dalton is an alternate name for an atomic mass unit and therefore is numerically equal to the molar mass in g/mol.

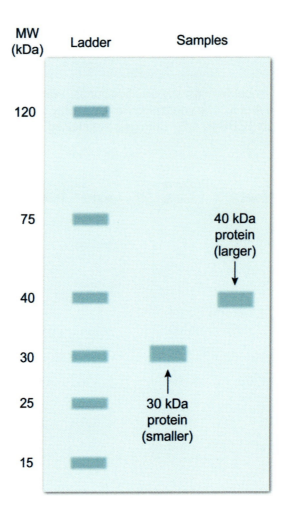

Figure 14.12 An SDS-PAGE gel with approximate molecular weights of the proteins marked.

Because SDS breaks quaternary structure, a protein that consists of multiple subunits dissociates into its individual subunits when subjected to SDS-PAGE if no disulfide bonds were involved in the interactions between subunits. Under these conditions, if all the subunits are identical (eg, in a homodimer), they all migrate equally quickly through the gel and appear in a *single band*. In addition, they migrate more quickly than they would have if they were held together as a single unit; therefore, they move farther down the gel.

In contrast, if the subunits in a protein are *not* identical (eg, in a heterodimer), they will migrate through an SDS-PAGE gel at different speeds (again assuming they are not linked by disulfide bonds). Therefore, such a protein would appear as a single band on a native gel but appears as *multiple distinct bands* in an SDS-PAGE gel. The larger subunits move more slowly than the smaller subunits and migrate a shorter distance. Figure 14.13 shows homodimers and heterodimers in native and SDS-PAGE gels.

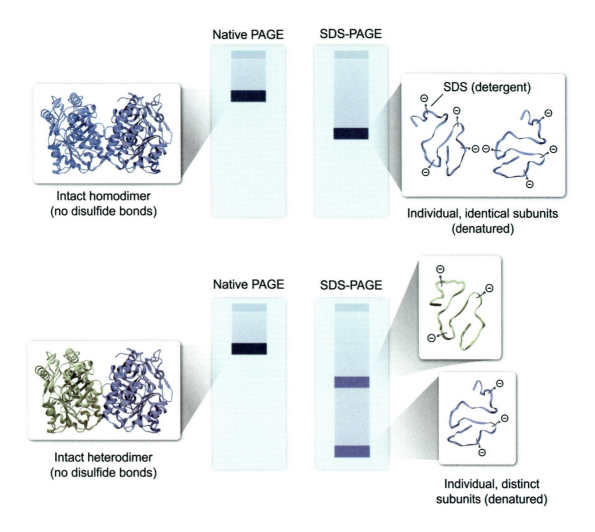

Figure 14.13 Homodimers and heterodimers on native and SDS-PAGE gels.

SDS-PAGE is often used to assess the presence of a protein in a mixture or to compare amounts of a protein in different mixtures. It can also be used to detect certain mobility shifts. For instance, certain post-translational modifications add enough mass or cause enough of a shape change (even under denaturing conditions) that protein migration slows by a detectable amount (Figure 14.14). Small, electrically neutral modifications such as monosaccharides may be too small to detect by electrophoresis, but larger or charged modifications such as glycans, ubiquitin, or phosphate groups can also cause a significant change in protein mobility.

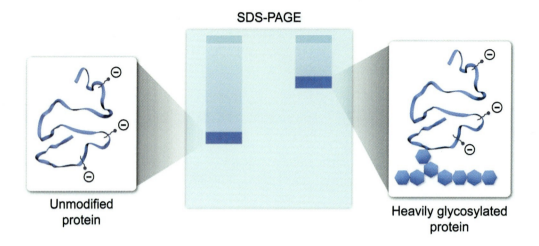

Figure 14.14 Post-translational modifications may cause gel mobility shifts.

14.1.04 Reducing Gels

The previous concept described how denaturing agents such as SDS disrupt secondary, tertiary, and quaternary structure but noted that SDS does *not* break disulfide bonds. **Disulfide bonds** form in proteins when the thiol (–SH) groups of two cysteine residues undergo oxidation to form an –S–S– bond. This converts the *cysteine* residues to *cystine*. Disulfide bonds can be broken by adding a **reducing agent** to the SDS-PAGE experiment.

The most common reducing agents for this purpose are dithiothreitol (DTT) and β-mercaptoethanol (BME). Each of these agents reduces the disulfide bonds in proteins by becoming oxidized and forming disulfide bonds themselves (Figure 14.15). The resulting reaction converts cystine (oxidized cysteine) back to two separate cysteine residues.

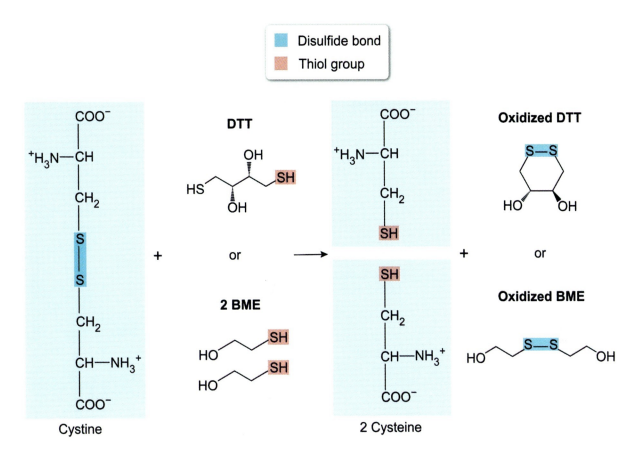

Figure 14.15 Reduction of disulfide bonds converts cystine to two cysteine side chains while oxidizing either DTT or BME.

SDS-PAGE experiments that use DTT or BME are typically designated specifically as **reducing SDS-PAGE**. When DTT and BME are absent, it is often designated as *nonreducing* SDS-PAGE. Reducing and nonreducing gels yield different results when the proteins of interest contain disulfide bonds. A protein with subunits held together by one or more disulfide bonds appears as a single band in nonreducing SDS-PAGE because the disulfide bond keeps the subunits together, although they are denatured.

In contrast, the *subunits separate* when a *reducing agent is added*. If the protein is a homodimer (ie, identical subunits), the gel will still show a single band, but it will appear further down the gel (closer to the anode) because each individual subunit is smaller than the combined subunits. If the protein is a heterodimer, two distinct bands will appear, with the band closer to the cathode corresponding to the larger subunit. Figure 14.16 shows the separation of disulfide-linked homo- and heterodimers by addition of a reducing agent.

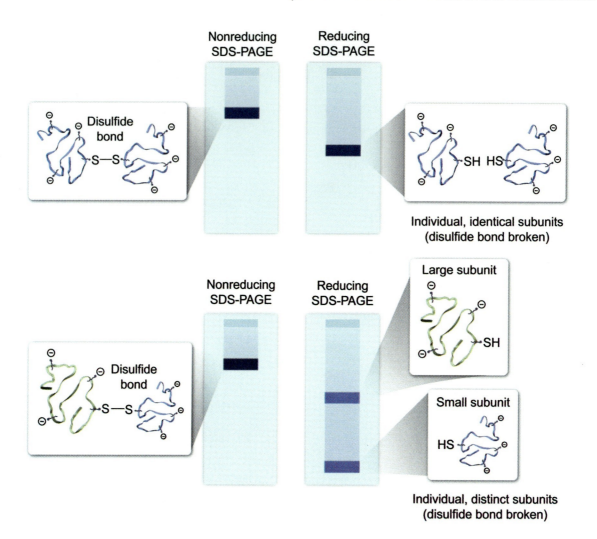

Figure 14.16 A reducing agent separates disulfide-linked homodimers to produce a single, faster-migrating band and separates disulfide-linked heterodimers to produce two distinct bands.

✅ Concept Check 14.3

Two solutions are prepared. Solution 1 contains a heterodimeric protein in which the subunits are held together by only noncovalent interactions. Solution 2 contains a separate heterodimeric protein, in which the subunits are held together by both noncovalent interactions and disulfide bonds. Which type(s) of gel can separate the subunits in each solution?

a) Native PAGE
b) Nonreducing SDS-PAGE
c) Reducing SDS-PAGE

Solution

Note: The appendix contains the answer.

14.1.05 Isoelectric Focusing

In addition to separation based on size or charge, proteins can be separated based on their isoelectric point (pI) by performing an isoelectric focusing experiment. To accomplish this, a polyacrylamide gel is prepared with a stable pH gradient. In other words, one end of the gel contains a low pH, the other end contains a high pH, and the pH gradually increases from the low end to the high end. Typically, these gels do not contain SDS or other denaturing agents and allow the proteins to maintain their native structures.

The end of the gel with a high pH (ie, low H^+ concentration) is placed near the cathode (negative charge), and the end with a low pH (ie, high H^+ concentration) is placed near the anode (positive charge). Proteins may then be loaded onto the gel at any position (near the anode, near the cathode, or in the middle). In other words, isoelectric focusing is a case in which the sample does not have to be loaded near the cathode.

Proteins at a pH that is less than their pI will pick up protons from the environment and gain a positive charge, whereas those at positions where the pH is greater than their pI will lose protons to their environment and become negatively charged. This is shown in Figure 14.17.

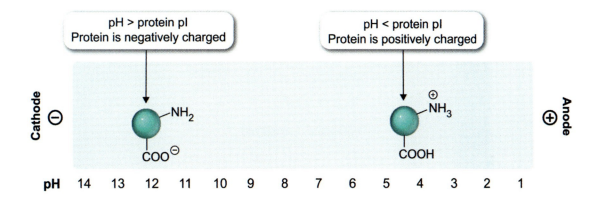

Figure 14.17 Proteins placed near the cathode in an isoelectric focusing gel tend to be negatively charged, while those near the anode tend to be positively charged.

When a current is applied and an electric field generated, positively charged proteins migrate closer to the negatively charged cathode, where the pH of the gel is higher. As the gel pH increases, these proteins *lose protons* and therefore become *less positively charged*. Eventually they reach a position in the gel where pH is equal to pI and the net charge of the protein drops to 0. Particles with 0 net charge *do not migrate* in an electric field because they experience no net force. Therefore, at this point migration stops.

Similarly, proteins with a negative charge migrate toward the positively charged anode, where the pH is lower. These proteins pick up protons and become less negatively charged. When the pH of the gel matches the isoelectric point, the net charge becomes 0 and migration stops. Therefore, regardless of where a given protein is loaded onto an IEF gel, it migrates (or focuses) toward the point where the gel's pH equals its isoelectric point.

This technique facilitates empirical determination of a protein's pI. As explained in Lesson 2.1, calculation of a protein's isoelectric point becomes difficult in complex, folded proteins because the pK_a values of the side chains may change due to interactions with each other. Therefore, a calculation using the pK_a values of free amino acid side chains may be inaccurate. Isoelectric focusing, however, directly measures pI: whichever pH stops the protein from migrating is the pI of that protein. Figure 14.18 shows a protein's migration toward its isoelectric point.

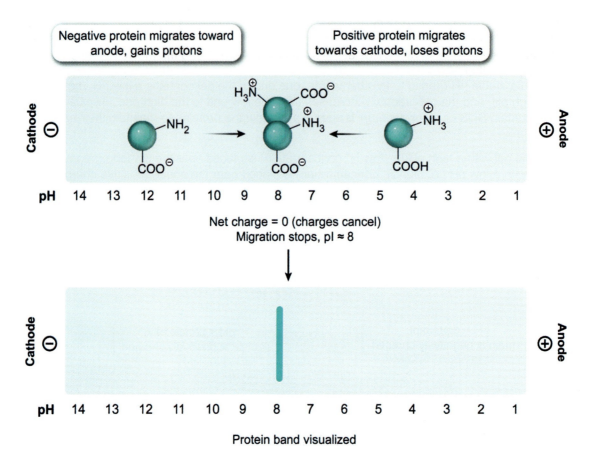

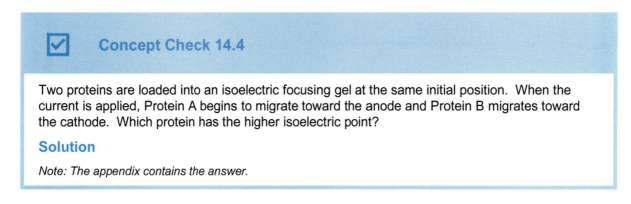

Figure 14.18 Migration through the isoelectric focusing gel occurs until the protein's net charge becomes 0.

> ☑ **Concept Check 14.4**
>
> Two proteins are loaded into an isoelectric focusing gel at the same initial position. When the current is applied, Protein A begins to migrate toward the anode and Protein B migrates toward the cathode. Which protein has the higher isoelectric point?
>
> **Solution**
>
> *Note: The appendix contains the answer.*

14.1.06 2D Gels

SDS-PAGE separates proteins by size, and isoelectric focusing separates them by isoelectric point. However, in a mixture of proteins, it is possible that two different proteins will be of similar size and migrate to the same position in SDS-PAGE, or they may have the same (or nearly the same) isoelectric point and migrate to the same position in isoelectric focusing. To further separate these proteins, both techniques may be applied, one after the other. Because this technique involves two parameters, or dimensions, it is called **two-dimensional (2D) electrophoresis**.

Typically, a sample of proteins is first subjected to isoelectric focusing. The resulting isoelectric focusing gel is treated with SDS to make all proteins within it negatively charged, and the treated gel is then aligned with the sample-loading edge of an SDS-PAGE gel. When a current is applied, the proteins migrate from the IEF gel into the SDS-PAGE gel, where they separate based on size. Unlike traditional gels, the result does not have evenly distributed lanes. Instead, 2D gels generally have multiple spots at various locations throughout the gel, each of which corresponds to a specific protein (Figure 14.19). The final location of a spot represents the isoelectric point along one axis and the molecular weight in kilodaltons on the other.

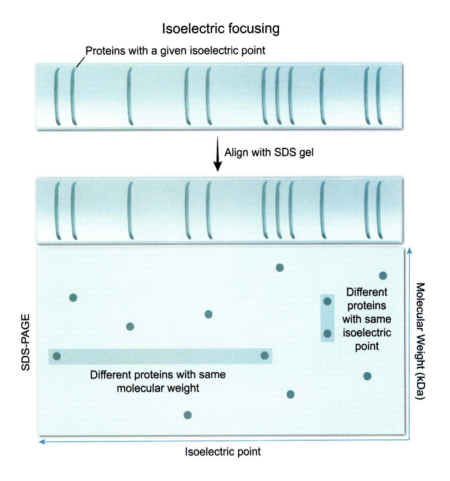

Figure 14.19 Example of a 2D gel using isoelectric focusing in one dimension and SDS-PAGE in the other.

Other variations of 2D gels are also possible. For instance, isoelectric focusing may be followed by native PAGE instead of SDS-PAGE, or native PAGE may occur first, followed by SDS-PAGE. Therefore, the positions of each protein would show their isoelectric point and native size:charge ratio, or their native size:charge ratio and denatured size, respectively.

Lesson 14.2
Blotting Techniques

Introduction

Lesson 14.1 discusses gel electrophoresis as a way of separating a mixture of biomolecules based on some physical or chemical property, such as size or isoelectric point. The stains that detect these molecules, however, are generally *nonspecific*. In other words, Coomassie stains both the protein of interest and *all other* proteins in the sample. Similarly, ethidium bromide stains both the RNA or DNA of interest and *all other* nucleic acids in the sample. Although these stains can be useful when working with purified samples or monitoring the progress of a purification experiment, the nonspecific nature of the dyes makes it difficult to interpret data with impure samples, such as crude lysates.

To visualize and analyze only a *specific* protein or nucleic acid of interest, **blotting techniques** have been developed that use stains that target only particular biomolecules, as shown in Figure 14.20. This lesson provides an overview of the principles of these blotting techniques and discusses in detail three blotting techniques that are likely to appear on the exam.

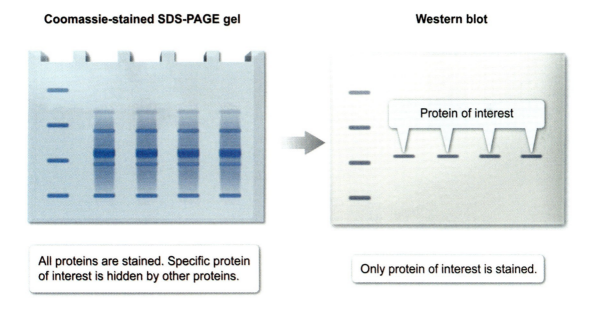

Figure 14.20 Blotting techniques allow for the visualization of specific biomolecules within a mixed sample.

14.2.01 Principles of Blotting Techniques

The generalized process of blotting involves the **transfer** of biomolecules from the sample to a **membrane**, followed by **visualization** of specific target molecules. Often, the sample had been previously processed through gel electrophoresis to allow for the separation of biomolecules by size (eg, Southern, northern, and western blots); however, blots can also be performed without electrophoresis, as discussed at the end of this concept.

Transfer of Biomolecules to Membranes

The membranes used with blotting techniques are "sticky" for their respective biomolecules. This means that once a biomolecule adheres to the membrane, it is immobilized and unlikely to diffuse or be washed away during the subsequent staining, blocking, and washing steps of the visualization procedure, discussed later in this concept.

Blots for analyzing DNA or RNA often use a positively charged nylon membrane, which binds to the negatively charged sugar phosphate backbone of nucleic acids. In contrast, blots used to analyze proteins often use a nitrocellulose or polyvinylidene fluoride (PVDF) membrane, which binds proteins through a combination of hydrophobic and electrostatic interactions.

Two common ways to transfer samples from an electrophoretic gel to a membrane are capillary transfer and electroblotting. These methods are illustrated in Figure 14.21. Capillary transfer is discussed in more detail with Southern blots in Concept 14.2.02, and electroblotting is discussed in more detail with western blots in Concept 14.2.04.

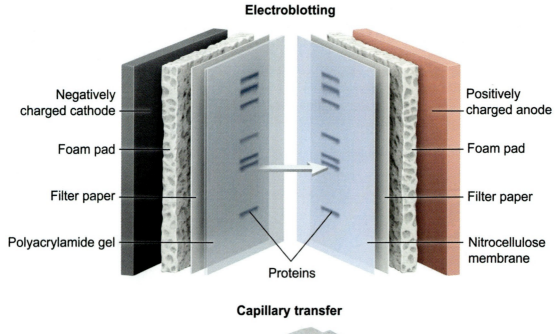

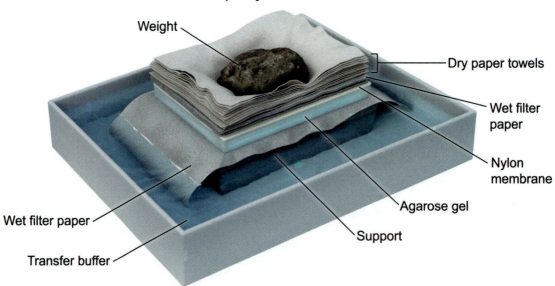

Figure 14.21 Capillary transfer and electroblotting are two methods of transferring biomolecules to a membrane.

Probes and Blocking

To specifically stain a biomolecule of interest, the membrane must be incubated with **probes** that specifically bind to the target molecule, but *not* to any other molecules. Blots intended to visualize nucleic acids often use single-stranded DNA primers that **hybridize** with the target sequence. Blots intended to visualize proteins often use antibodies that specifically recognize the target protein.

Note that these probes are the same type of molecule that the membrane is "sticky" to. In other words, DNA primers are nucleic acids and can nonspecifically stick to nylon; antibodies are proteins and can nonspecifically stick to nitrocellulose or PVDF. To prevent this, the membrane must be **blocked** prior to incubation with probes.

Blocking involves incubating the membrane with a solution that contains nucleic acid or protein to saturate the nonspecific binding sites. By performing blocking prior to adding any probe, experimenters can ensure that the probe binds *only to the target molecule*, rather than binding the membrane itself. Ultimately, this helps minimize background noise and improves signal clarity. The effect of blocking for a western blot is shown in Figure 14.22. A common blocking reagent for Southern and northern blots is salmon sperm DNA; two common blocking reagents for western blots are purified solutions of bovine serum albumin (BSA) and fat-free cow's milk.

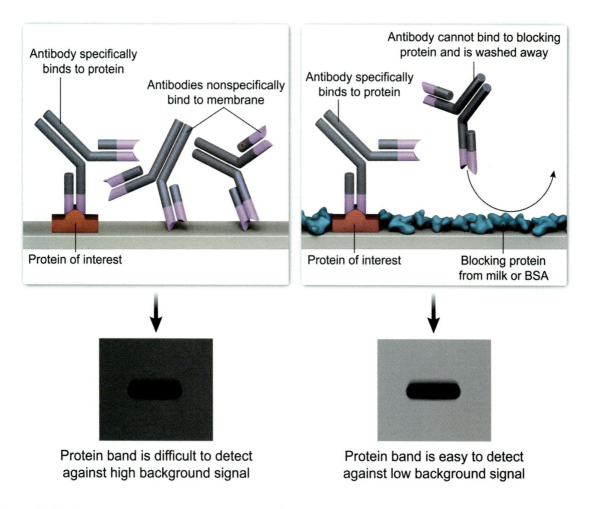

Figure 14.22 Blocking reagents bind nonspecifically to blotting membranes, preventing nonspecific binding of the probe (eg, an antibody) to the membrane. This reduces background noise and improves signal detection.

Visualization

To determine both the *location* of the target molecule on the membrane *and* the relative *amount* of it, the probe molecule that binds to it must be **labeled**. This label can be incorporated either *directly* (ie, the label is covalently incorporated onto the primary probe molecule), or it can be *indirect* (ie, the label is covalently incorporated in a *secondary* probe that then binds the first probe). Indirect methods are discussed in more detail with western blots in Concept 14.2.04.

Several types of labels are common in modern biology and biochemistry, including radioactive, chemiluminescent, and fluorescent labels. In **radioactive** labeling, certain atoms in the probe molecule are replaced with radioactive isotopes of that element. For example, the ^{31}P atom in the sugar-phosphate backbone of DNA probes can be replaced with a radioactive isotope such as ^{32}P. When the ^{32}P isotope undergoes β^- decay (becoming ^{32}S), it emits a high-energy particle that reacts with radiographic film (sometimes called x-ray film).

Like photographs, the film can be *developed* to allow for visualization of the location and position of the radioactive labels on the blotting membrane; the developed signals appear as **bands** on the film. The detection of radioactive labels using radiographic film is known as **autoradiography**. An example of an autoradiograph is shown in Figure 14.23.

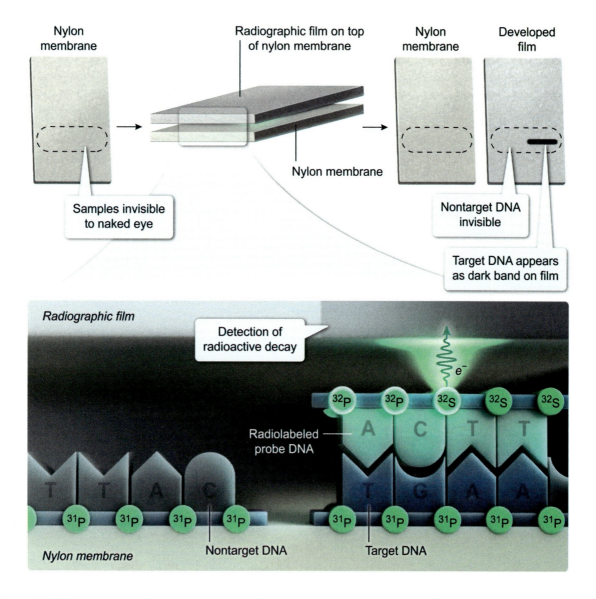

Figure 14.23 Autoradiography uses radiographic film to capture radioactive decay of isotopically labeled probes. Once the film is developed, locations where decay was detected appear as dark bands.

Unlike other labeling methods, radioactive labels have the benefit of causing minimal disruption to the chemical properties of the probe molecule. This greatly reduces the risk of any effect on binding interactions. However, radioactive labels do pose a danger to the experimenter due to radioactive emissions. To avoid these risks, chemiluminescent or fluorescent molecules may instead be linked to the probe molecule. These molecules can then be detected by film or specialized digital cameras, yielding similar results to autoradiography without releasing high-energy particles or photons.

Loading Controls

Blotting is considered a *semiquantitative* technique. Although blots are not generally used to calculate precise quantities or percent changes, they *can* indicate whether the expression of the target molecule has increased or decreased. This can be done by comparing the size and intensity of the bands between two experimental conditions, as in gel electrophoresis (Lesson 14.1).

When using blots in a semiquantitative way, it is critically important to ensure that samples are loaded in a consistent manner. Differences in sample collection may result in *inconsistent* concentrations of total protein or total nucleic acid. If these differences are not accounted for, it is impossible to determine if an increased band size is due to an increase in expression or due to the experimenter loading a higher concentration of cell lysate material, as shown in Figure 14.24.

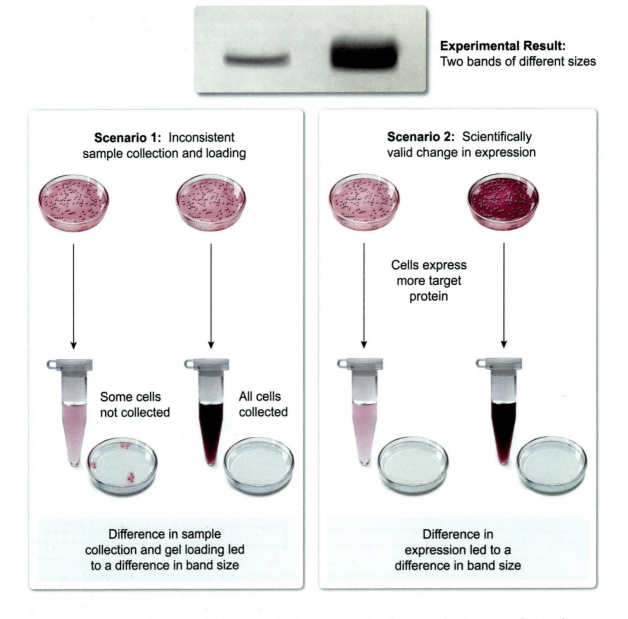

Figure 14.24 Differences in blot band size may be due to an expression change or due to an error in sample collection and loading.

To prevent this ambiguity, the contents of the cell lysate must be *quantified* prior to loading (biomolecule quantitation is discussed in Concept 14.4.02). Based on the results of the biomolecule quantitation, equal masses of the analyte (ie, protein, nucleic acid) can be loaded onto the gel or blot.

Proper loading can be verified through the probing of particular molecules *in addition to* the target of interest. Tubulin, β-actin, and glyceraldehyde-3-phosphate dehydrogenase (GAPDH) are some commonly probed additional targets because they are expressed by all cells and because expression levels are typically unaffected by experimental conditions. Consequently, the genes that express tubulin, β-actin, GAPDH, and other similar proteins are often called **housekeeping genes**.

When interpreting a change in band size for a target of interest, it is common practice to compare it to an *unchanged* band size for the housekeeping gene (or gene product). This serves as a **loading control** and ensures any change in the target protein is due to changes in the experimental conditions and *not* due to sample collection errors. The use of loading controls is shown in Figure 14.25.

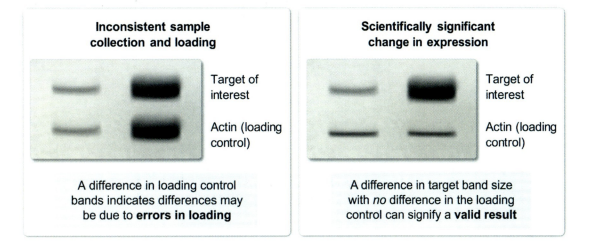

Figure 14.25 The use of loading controls helps ensure interpretation of a scientifically valid result.

 Concept Check 14.5

An experimenter is about to load an SDS-PAGE gel in preparation for a western blot. She is preparing her samples and has 10 µL of crude lysate available per gel lane. She quantifies her two samples of interest and finds Sample A has a total protein concentration of 2 mg/mL and Sample B has a total protein concentration of 4 mg/mL.

If the experimenter wants to maximize the total amount of protein loaded per well, how many microliters of Sample A should she add to the Sample A gel lane? How many microliters of Sample B should she add to the Sample B gel lane?

Solution

Note: The appendix contains the answer.

Nonelectrophoretic Blots: The Dot Blot

The following concepts in this lesson discuss the most commonly used blotting methods, which all follow a gel electrophoresis procedure. However, blots can also be performed *without* first running an electrophoretic gel. These blots are known as dot blots. Although dot blots are less informative because of the lack of electrophoresis data (eg, molecular size), the presence of a signal and the relative signal intensities can still provide valuable information in a quicker and less labor-intensive experiment.

Because dot blots do not involve gels, the transfer process simply involves spotting dissolved sample onto the blotting membrane, similar to spotting thin-layer chromatography (TLC) plates (see Organic Chemistry Lesson 13.3). The spots can air dry, or the solvent can be removed through vacuum filtration. Either way, the target molecule adheres to the dry membrane, leaving "dots" that can be visualized through one of the methods described previously (Figure 14.26).

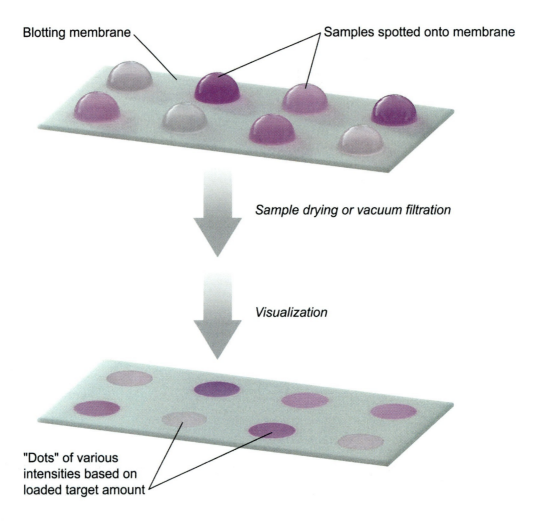

Figure 14.26 A dot blot is an example of a nonelectrophoretic blotting technique.

14.2.02 Southern Blots

The first of the electrophoretic blotting methods is the Southern blot. The Southern blot is named after its developer, Edwin Southern, and is used to detect DNA. The Southern blotting method begins with agarose gel electrophoresis (Lesson 14.1) to separate DNA fragments based on their size.

DNA agarose gels are typically run in native conditions, which keeps the DNA double stranded. This double-stranded DNA must be denatured before it is transferred to a membrane so the probes can base pair with the denatured single strands. Typically, denaturation is done by incubating the gel in an alkaline (ie, high-pH) solution.

Once the DNA has been denatured into single strands, it must then be transferred from the gel to the blotting membrane. The classical way to perform transfers from agarose gels is through upward capillary transfer.

In upward capillary transfer, shown in Figure 14.27, the membrane is placed on top of the gel, and the gel-membrane pair is sandwiched between two sheets of wet filter paper to prevent either from drying out. The bottom piece of filter paper is connected to a reservoir tank filled with buffer; the upper piece of filter paper is overlaid with dry paper towels. A weight is placed upon the whole stack to add pressure, which maintains contact between the membrane and the gel.

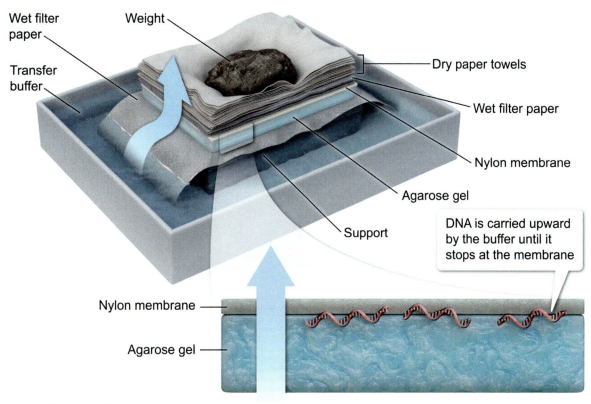

*Arrows indicate buffer movement

Figure 14.27 In upward capillary transfer, single-stranded nucleic acids are carried upward by buffer, transferring them from an agarose gel to a nylon membrane.

The dry paper towels near the top of the stack draw buffer up toward them, resulting in buffer flow from the tank to the lower filter paper and then through the entire setup. As the buffer moves upward, it also carries the DNA in the gel upward. The DNA stops migrating once it adheres to the "sticky" nylon membrane.

Once the DNA is immobilized on the membrane, the membrane must be blocked (see Concept 14.2.01) before being incubated with probe. The oligonucleotide probe has a sequence that is the *reverse complement* of a specific region in the target sequence, allowing the probe to base pair with the target DNA. This annealing of probe to target DNA is called **hybridization**.

After washing off unbound probe, the hybridized target-probe pair can then be visualized as described in Concept 14.2.01. An overview of the Southern blot procedure is shown in Figure 14.28.

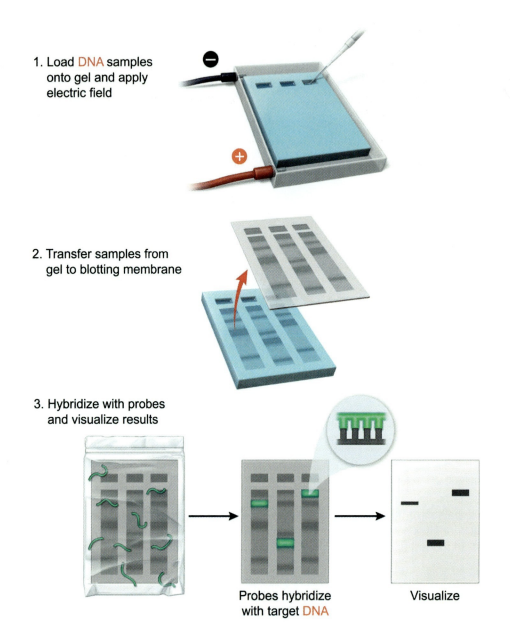

Figure 14.28 Overview of the Southern blot procedure.

Southern blots can be used in genomic analysis, such as testing for gene duplication, deletion, or rearrangement events. Southern blots can also be used to examine smaller point mutations (eg, insertions, deletions, substitutions); however, PCR-based methods (see Biology Lesson 4.1) are much more common for this purpose in modern biochemistry.

14.2.03 Northern Blots

Unlike the Southern blot, which targets DNA, the second electrophoretic blotting method targets *RNA*. This blot is not named after a scientist; however, the direction-based naming system was kept by naming the RNA blot a **northern blot**.

The process of northern blotting is very similar to the process of Southern blotting because both blots target nucleic acids. However, there are some important differences. Although RNA is not typically double stranded, it can still have local regions of secondary structure due to internal base pairing, which means that RNA still needs to be denatured. Because RNA is much more reactive than DNA, alkaline conditions (which would hydrolyze RNA) cannot be used.

Instead, regions of RNA base pairing are denatured with a chemical denaturant such as formaldehyde. To prevent these structured and folded regions from affecting electrophoretic migration, formaldehyde is also included in the agarose gel and buffer during electrophoresis, in which it acts similarly to the denaturing effect of SDS in SDS-PAGE gels.

The electrophoresed RNA is then transferred, hybridized, and visualized in a manner similar to that of Southern blots (Concept 14.2.02). An overview of the northern blot procedure is shown in Figure 14.29.

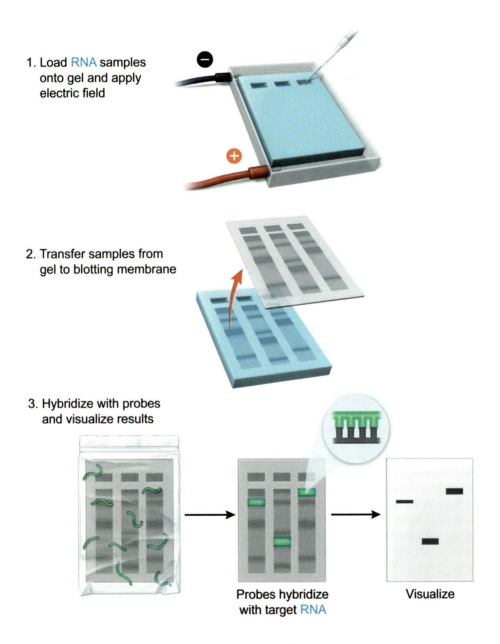

Figure 14.29 Overview of the northern blot procedure.

Northern blots can be used to analyze expression and sequence various RNA subtypes (eg, mRNA, tRNA, miRNA) and can be used in combination with PCR-based methods.

14.2.04 Western Blots

The third electrophoretic blotting method targets *proteins*. To maintain the directional theme, but to distinguish itself from the north-south nucleic acid axis, protein blots are called western blots.

Like Southern and northern blots, western blots begin with running an electrophoretic gel. In the case of western blots, this is typically a polyacrylamide gel. Instead of transferring to the membrane via upward capillary motion, protein gels are typically transferred using an electric field (**electroblotting**).

As with gel electrophoresis (Lesson 14.1), negatively charged proteins travel toward the *positively charged anode* of an electrolytic cell. If the protein gel was run with SDS (as in SDS-PAGE), typically enough SDS remains bound to protein to maintain this directional transfer. Proteins in native gels require additional treatment, either during or after electrophoresis, to ensure the proteins migrate toward the anode during transfer to the membrane. A nitrocellulose or PVDF membrane is placed on the anode-facing surface of the gel, and proteins bind to this membrane as they leave the gel, stopping their migration. A typical electroblotting setup is shown in Figure 14.30.

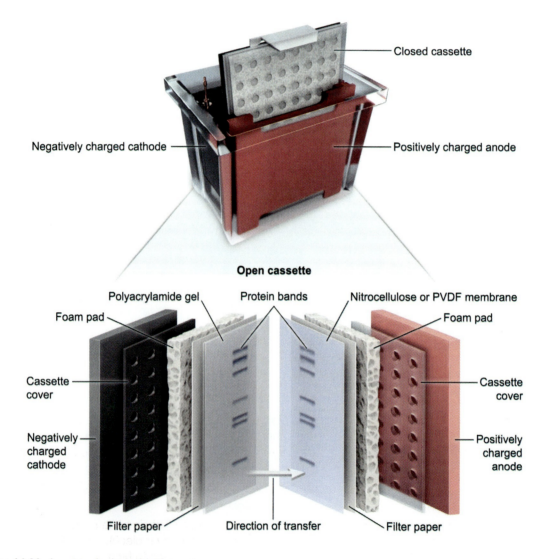

Figure 14.30 A wet tank electroblotting setup.

Like the upward capillary transfer method described in Concept 14.2.02, contact between the gel and the membrane is maintained, in this case helped by the pressure from the foam pads (or sponges) in a closed cassette. Unlike upward transfer, electroblotting requires electric power to drive molecular movement. This process is faster than capillary transfer but also generates much more heat. Because polyacrylamide gels are much thinner than agarose gels, heat buildup is less of an issue; however, transfers are still usually performed with ice blocks or in a refrigerated room to minimize heating.

Once protein transfer to the membrane is complete, the membrane must be blocked (typically with a solution of bovine serum albumin [BSA] or with fat-free cow's milk) before being incubated with probe. In western blots, the probe is usually an **antibody** that recognizes the target protein as its antigen.

The antibody that *directly binds the target protein* is known as the **primary antibody**. The primary antibody may be directly (ie, covalently) linked to a label for visualization, or another antibody (a **secondary antibody**) may be used. The secondary antibody recognizes and binds to the primary antibody. Typically, a secondary antibody binds *any* primary antibody produced by a specific organism (eg, mouse, goat). If a secondary antibody is used, it serves as the molecule that the label is linked to. Primary and secondary antibodies are shown in Figure 14.31.

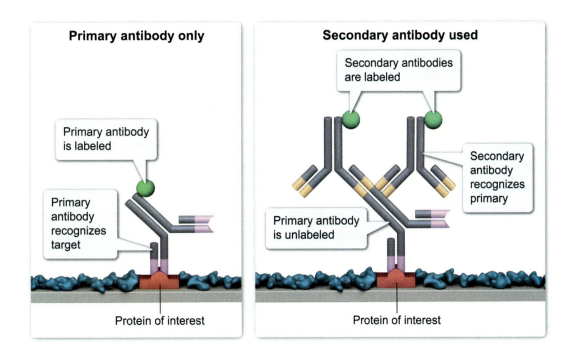

Figure 14.31 Primary and secondary antibodies.

The use of secondary antibodies brings many advantages, the first of which is cost reduction. Chemically linking target-specific antibodies to labels is costly, labor intensive and may produce low yields. By using mass-produced secondary antibodies, different specialized primary antibodies produced by the same species can be used in a variety of experiments (including immunohistochemistry, immunofluorescence, immunoprecipitations, and ELISAs, in addition to western blots), and each can be detected by the same secondary antibody.

Second, the use of secondary antibodies can improve signal detection. This is because it is possible for multiple secondary antibody molecules to bind to a single primary antibody molecule, resulting in *multiple* labels per primary antibody. The potential for multiple secondary antibodies to bind is due partly to the dimeric nature of antibodies and partly due to the use of polyclonal secondary antibodies (ie, different antibodies in the same mixture that bind different epitopes on the target molecule).

Visualization of Antibody Labels

For visualization, the relevant antibody is typically linked either to an enzyme or to a fluorescent molecule (a fluorophore). The enzyme (typically horseradish peroxidase) can be used to catalyze either a chemiluminescent or a chromogenic reaction, whereas the fluorophore can be visualized spectroscopically. An overview of the western blotting process is illustrated in Figure 14.32.

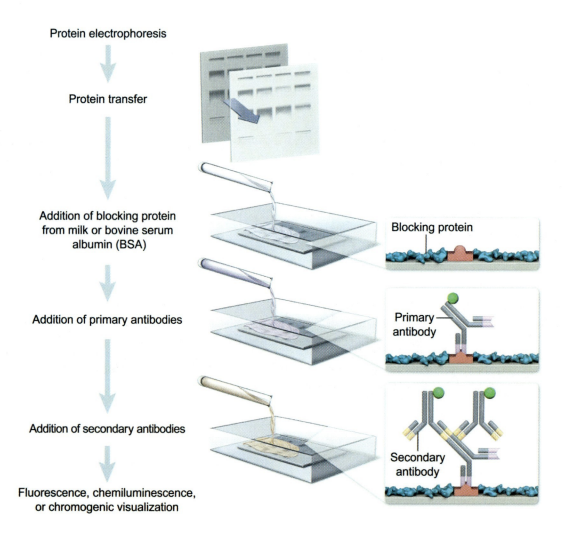

Figure 14.32 Overview of the western blot procedure.

Lesson 14.3
Chromatography

Introduction

Many of the laboratory techniques in biochemistry separate proteins or other biomolecules from each other for analysis or purification. The techniques discussed so far in this chapter—gel electrophoresis and blotting—are primarily used for analysis of a protein's properties but are not well suited for purifying *large* amounts of protein.

To study many aspects of a protein, the protein must first be expressed in large amounts in an organism or in cell culture. The cells expressing the protein must then be lysed (ie, broken open) to release the protein of interest. However, these cells also contain many other proteins and biomolecules that are not of interest to the study. These impurities must be removed while retaining the protein of interest.

This lesson discusses several chromatography techniques, each of which separates mixtures of proteins and other biomolecules by various physical properties. Chromatography techniques that are commonly used to separate small molecules are discussed in Organic Chemistry Lesson 13.3.

14.3.01 Principles of Chromatography

Like electrophoresis, chromatography is a set of techniques that separate molecules by one or more physical properties. Every chromatography technique includes a **stationary phase** and a **mobile phase**. The stationary phase typically consists of small beads or a gel packed within a column. The mobile phase consists of a solution that carries the molecules of interest. The mobile phase is allowed to flow through the stationary phase, and the molecules in the mobile phase are separated based on how strongly they interact with the stationary phase (Figure 14.33).

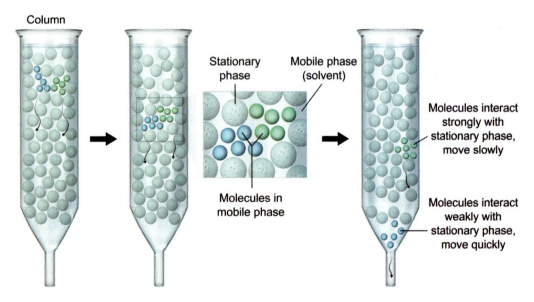

Figure 14.33 Column chromatography separates molecules in the mobile phase.

Molecules that interact *strongly* with the stationary phase move through the column *more slowly* than molecules that interact poorly with the stationary phase. Therefore, the molecules that interact poorly exit the column (ie, elute) sooner.

Some biomolecules have color that can be detected by the human eye, but most are colorless and must be detected by other means. The side chains of tryptophan and tyrosine strongly absorb 280 nm light. Therefore, to detect proteins eluting from a column, a spectrophotometer that detects absorbance of 280 nm light is commonly included as part of the experimental setup. The light source and detector are positioned at the end of the column where the buffer exits, and absorbance can be plotted as a function of time (Figure 14.34).

Small volumes (aliquots) of the buffer exiting the column are collected, and those with strong **absorbance at 280 nm (A_{280})** most likely contain protein. The measurement of A_{280} does *not* show whether the protein in a given aliquot is the specific protein of interest, however. Each aliquot that contains protein can subsequently be tested for the presence of the protein of interest by assays that detect specific protein activity (eg, enzymatic function, ligand binding).

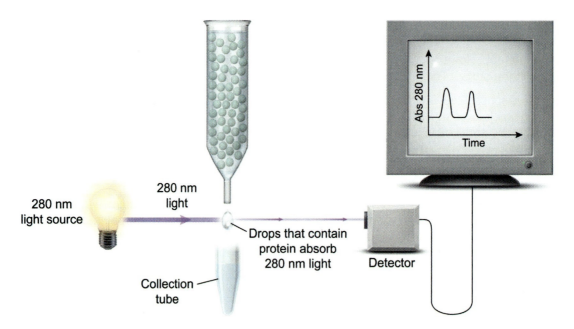

Figure 14.34 Detection of protein in several aliquots collected from a column using 280 nm light.

Importantly, none of the chromatography techniques described in this lesson are perfect. All allow some impurities to be included with the protein of interest. To increase the purity of a sample, several chromatography techniques may be used in succession to separate proteins based on different properties.

14.3.02 Size Exclusion Chromatography

Size-exclusion chromatography (SEC), also called gel filtration chromatography, uses beads that contain pores through which some of the proteins may migrate. In contrast to other forms of chromatography, none of the proteins in a mixture bind to SEC beads. Instead, small proteins enter the pores, whereas large proteins cannot and are *excluded* from the pores. In this way, small proteins have more interactions with the beads. Different beads have different pore sizes, or cutoffs, that accommodate different sizes of proteins.

Because proteins move through an SEC column at different rates depending on whether and to what extent they can enter the pores, SEC separates proteins by their physical size. In this way SEC is similar to SDS-PAGE (see Lesson 14.1), but it differs in several important ways.

First and most importantly, the correlation between size and rate of migration in SEC is the *opposite* of the correlation in SDS-PAGE. Large proteins move through SEC columns more quickly than small proteins, so *large proteins elute first*. In contrast, large proteins move more slowly in electrophoresis.

The reason for this difference is that whereas electrophoresis forces *all* proteins to move through the pores, in SEC only sufficiently small proteins will do so. Proteins that are too large to fit inside the pores simply do not enter them, and instead traverse only the portions of the column not occupied by beads, known as the void volume, and so travel a shorter path. Small proteins that can enter all the pores must cross more volume (the total volume) and travel a longer path. Because large proteins travel a shorter path, they elute more quickly. Separation of large and small molecules by SEC is shown in Figure 14.35.

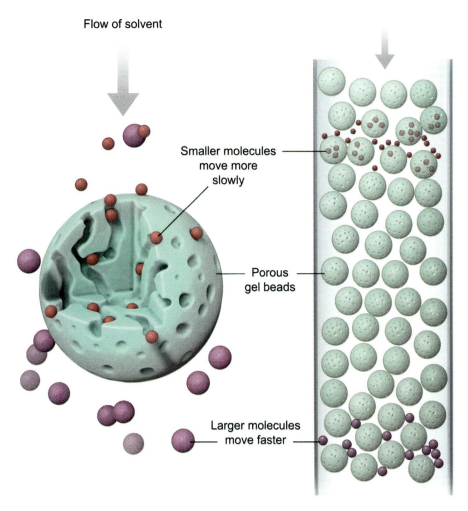

Figure 14.35 Large proteins do not fit in the pores in a size-exclusion chromatography column and travel a shorter path than small proteins.

Most proteins resolved by an SEC column are of intermediate size: small enough to enter some (but not all) pores, and large enough that they only partially enter them. These proteins travel at an intermediate rate depending on their size, which allows SEC to provide resolution of a range of proteins from small to medium to large, as shown in Figure 14.36. However, proteins that differ in size by only a few amino acids are unlikely to be separated well by SEC.

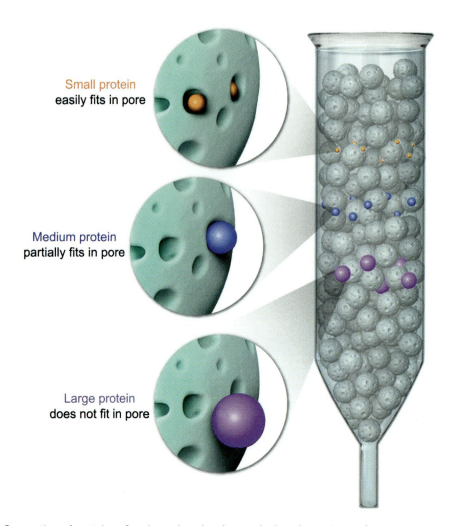

Figure 14.36 Separation of proteins of various sizes by size-exclusion chromatography.

The second important way in which SEC and electrophoresis differ is the force that causes proteins to migrate. Electrophoresis uses an electric current to force proteins through a gel matrix. In contrast, SEC relies only on gravity or, in some cases, the action of a pump to move the mobile phase carrying the proteins through the column. All proteins in an SEC column migrate in the same direction, regardless of their charge. Therefore, proteins of the same size move through the column at the same overall rate, even if their charges differ.

Finally, SEC does not typically denature the proteins in the mixture. Denaturants and reducing agents *may* be added for specialized purposes, but they are generally not necessary. Therefore, the proteins in the column usually retain their secondary, tertiary, and quaternary structures. This allows proteins collected by SEC to remain functional, which allows for further study.

Concept Check 14.6

A mixture of three proteins is separated by size-exclusion chromatography. The proteins have molecular weights of 20 kDa, 50 kDa, and 75 kDa. Assign each of these proteins to the peaks in the following chromatogram:

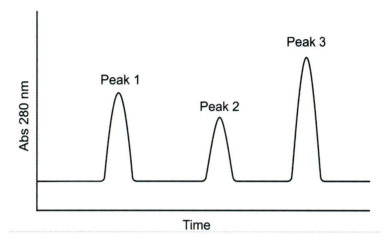

Solution

Note: The appendix contains the answer.

14.3.03 Ion-Exchange Chromatography

Most proteins are electrically charged (ie, they are ions) at physiological pH. **Ion-exchange chromatography** exploits this fact by using electrically charged beads as a stationary phase. Positively charged proteins bind to negatively charged beads, and vice versa. Bound proteins can then be made to unbind by the gradual addition of salt to the mobile phase.

When cationic (positive) proteins are bound to negatively charged beads, the addition of salt provides new cations that compete with the protein for binding. When the salt concentration is sufficiently high, the cations in the salt outcompete the protein and the protein elutes. Therefore, the column *exchanges* cationic proteins for cations from the added salt, and this form of chromatography is called **cation-exchange chromatography**. The same principle applies to **anion-exchange chromatography**, but *anionic* proteins bind to *positively* charged beads and exchange with *anions* from the salt.

Any protein with an isoelectric point (pI) *below* the pH of the buffer has a net negative charge, and any protein with a pI *above* the pH of the buffer has a net positive charge. A larger difference between pH and pI yields a greater charge magnitude. For example, at pH 7 a protein with a pI of 2 has a greater negative charge than a protein with a pI of 4, and a protein with a pI of 10 has a larger positive charge than a protein with a pI of 8.

A protein with a large magnitude charge can bind to an ion-exchange column more tightly than a protein with a smaller magnitude charge but of a similar size. Therefore, a greater salt concentration is required to compete with highly charged proteins. Accordingly, proteins with small charges elute at lower salt concentrations than proteins with large charges, and the order of elution can be predicted from the isoelectric points of each protein in a mixture. Figure 14.37 depicts both cation- and anion-exchange chromatography.

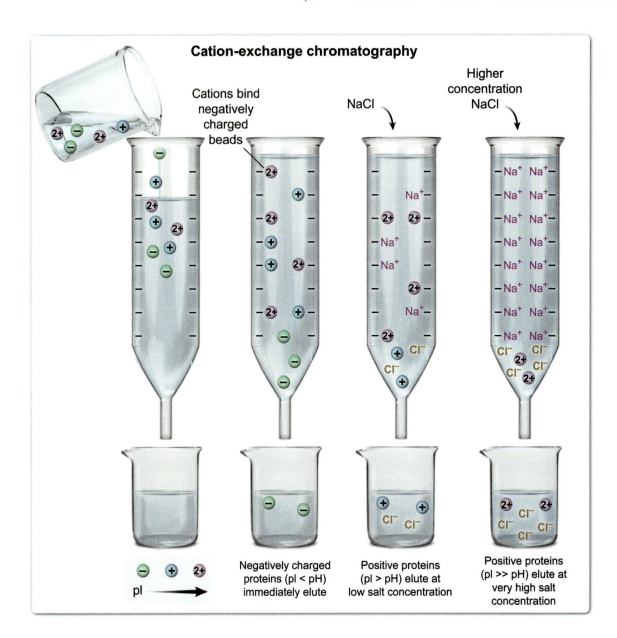

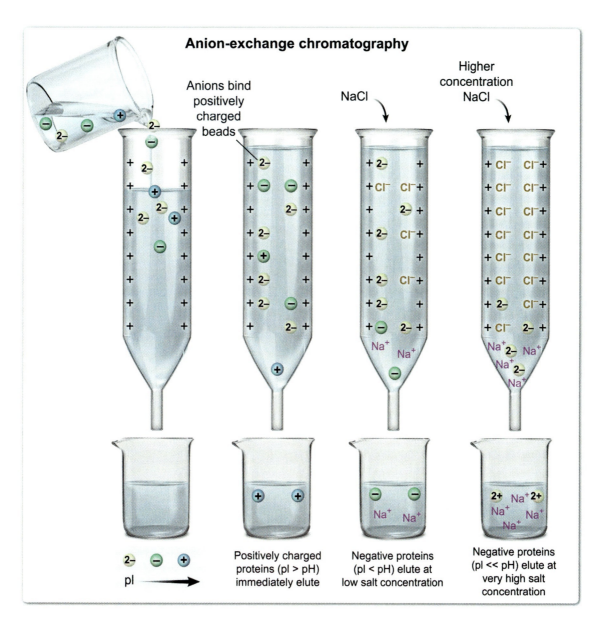

Figure 14.37 General procedures for cation- and anion-exchange chromatography.

Uncharged proteins do not bind the column and elute even with no salt added. Proteins with the same charge as the beads are repelled by the beads and exit the column even more quickly than uncharged proteins. These proteins are said to be "sped on" by the column.

Concept Check 14.7

The following table displays several proteins and their isoelectric points. For each protein, determine whether it would bind a cation-exchange column or an anion-exchange column at pH 7, and determine which would require the highest NaCl concentration to elute.

Protein A	pI = 9
Protein B	pI = 3
Protein C	pI = 10
Protein D	pI = 6

Note: Assume all four proteins are approximately the same size. Also assume that both columns bind ions of the same magnitude with the same affinity.

Solution

Note: The appendix contains the answer.

In addition to increasing salt concentration, *changes in pH* may be used to elute proteins from ion-exchange columns (Figure 14.38). If a protein is positively charged, raising the pH in the column deprotonates some of the side chains and decreases positive charge. A sufficiently high pH brings the charge to zero, or even negative, causing the protein to elute from a cation-exchange column. In anion-exchange columns, a gradual *decrease* in pH helps elute bound proteins as negatively charged side chains are neutralized by protonation.

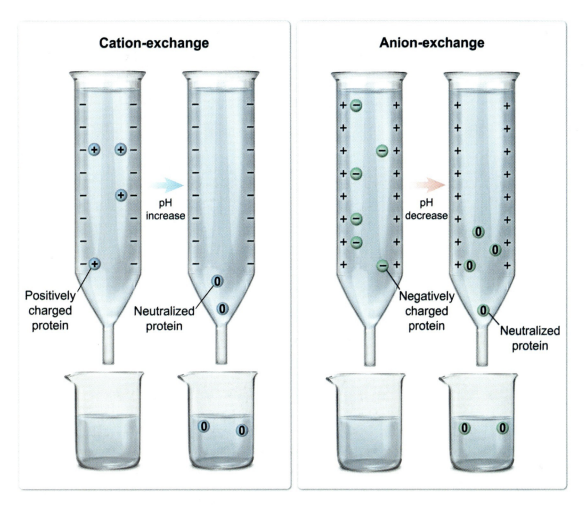

Figure 14.38 Proteins bound to ion-exchange columns may be eluted by changes in pH, which neutralize the proteins.

Combinations of changes in pH and salt concentration may also be used. For example, a positively charged protein may be eluted from a cation-exchange column by slightly increasing the pH of the mobile phase while also increasing the salt concentration.

The pH of the buffer can also be altered to *cause* binding. For instance, a researcher may wish to bind all proteins in solution to a cation-exchange column and then elute them one at a time. If some of the proteins are negatively charged while others are positively charged at pH 7, the researcher may decrease the pH so that *all* proteins in the mixture become positively charged.

After all proteins are bound to a column, they may be selectively eluted by different pH and salt combinations. Figure 14.39 shows the effect of pH on the net charges of various proteins and their ability to bind different ion-exchange columns.

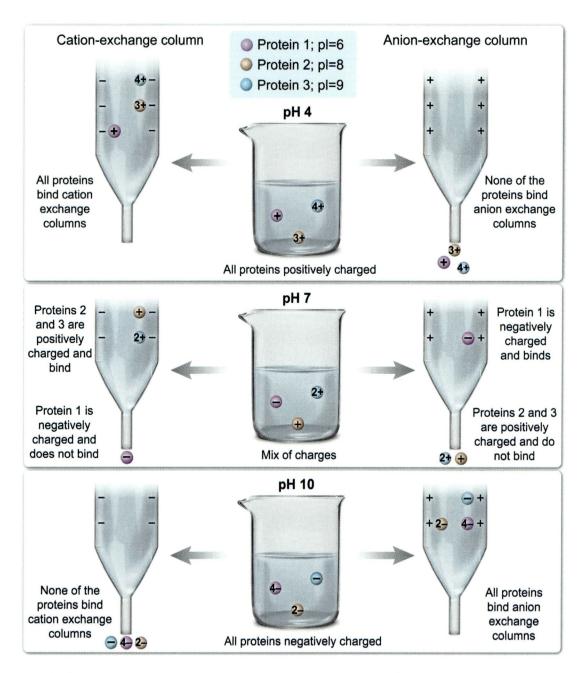

Figure 14.39 Example of protein net charges at various pH levels, allowing for binding to different ion-exchange columns.

14.3.04 Affinity Chromatography

In affinity chromatography, the beads in the column are linked to chemical groups that bind only certain proteins (or other molecules of interest) with high specificity. In many commonly used setups, the molecules attached to the beads are antibodies. Other molecules may also be used. Any molecule that does not specifically bind is washed out of the column, while molecules that *do* bind the beads are retained.

Various methods may be used to elute the protein of interest. Each method involves adding a new buffer to the column that alters the environment and, consequently, the binding interactions. Changes in salt concentration or pH may alter binding interactions between an antibody and a protein, for example, allowing the protein to unbind and exit the column. Alternatively, another ligand that acts as a competitive inhibitor of the binding interaction may be added to the mobile phase to disrupt the protein-bead interaction. The steps of affinity chromatography are shown in Figure 14.40.

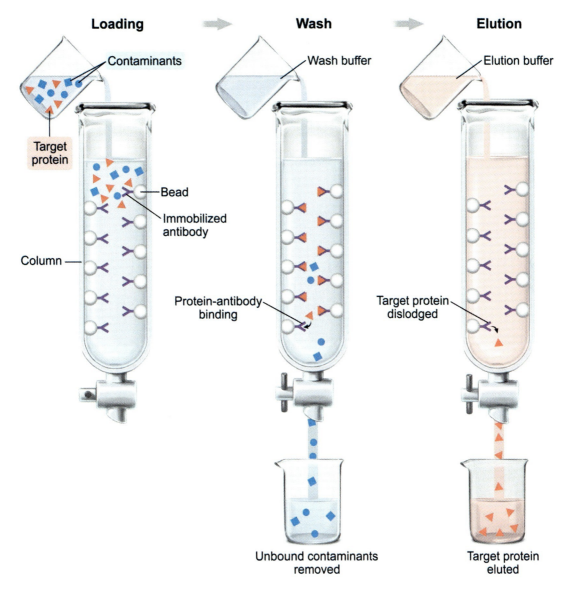

Figure 14.40 The steps of affinity chromatography.

Affinity chromatography is highly advantageous because it can select for a specific protein of interest while eliminating other proteins in a single step. In contrast, size-exclusion chromatography cannot separate the protein of interest from other proteins of similar size, and ion-exchange chromatography cannot separate the protein of interest from other proteins with the same charge. Therefore, when possible, affinity chromatography is desirable.

However, each specific protein that a researcher may wish to purify requires a column with unique beads. Whereas a single SEC column or a single ion-exchange chromatography column can be used to help isolate *any* protein of interest, an affinity chromatography column with beads that bind a certain protein can only be used to purify that protein. Producing different affinity columns for every possible protein of interest is not practical.

To solve this problem, many proteins of interest are expressed recombinantly (see Biology Chapter 4) with extra amino acids added to either the N- or C-terminus. These additions are typically referred to as "tags." Common tags include the the Myc tag and polyhistidine tags. Myc tags (and other similar tags) are each targets of specific antibodies (Figure 14.41), and beads with antibodies that bind these tags are sold commercially. Myc-tagged proteins can be eluted by the addition of short peptides expressing the myc sequence. Similarly, the *imidazole* side chain of histidine binds tightly to nickel, and columns with beads that contain nickel are also sold commercially. Polyhistidine tags are eluted by addition of free imidazole.

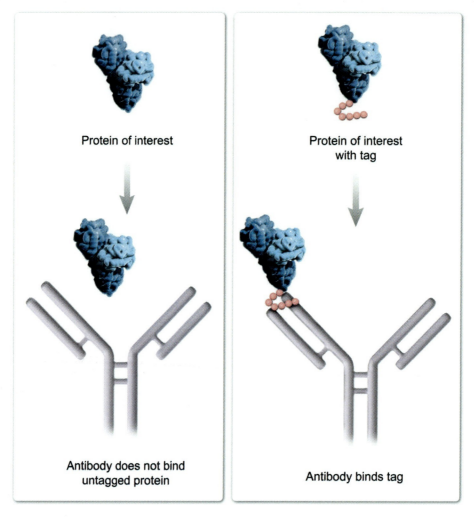

Figure 14.41 A protein may be tagged with an extra amino acid sequence that can be bound by specific antibodies.

These affinity columns can separate any protein that has the appropriate tag from all proteins that do not. However, the addition of the tag may alter the properties of the protein itself. Consequently, proteins purified in this way must have their tags removed after purification or control experiments must be performed to show that the tag does *not* interfere with or alter protein function.

Another important aspect of affinity chromatography is that a protein of interest may bind to other proteins. If the protein that binds to the beads *also* binds to another protein, *both* proteins are retained by the column until elution occurs. This fact may be used to determine which proteins in a system interact with each other.

To look for binding interactions, a variation of affinity chromatography called **co-immunoprecipitation (coIP)** may be used. In this method, the beads are commonly *not* packed into a column but are instead placed in the bottom of a test tube. The protein mixture is allowed to interact with the beads, during which time antigenic proteins interact and bind with the antibodies on the beads. By binding to the beads, the antigenic proteins (ie, proteins that bind the antibody) fall out of the bulk solution (immunoprecipitation).

Proteins that bind the protein of interest but do not *directly* bind the antibody are said to have *co-immunoprecipitated*.

The overlaying solution (supernatant) is removed, the beads are washed several times to remove impurities, and the remaining, bound proteins are then eluted. Any proteins that bound the tagged, antigenic protein remain in the tube and can be detected by western blot or other techniques (Figure 14.42). Co-immunoprecipitation often helps identify various proteins that participate in a complex.

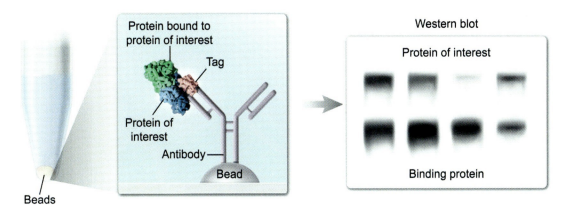

Figure 14.42 Co-immunoprecipitation uses beads linked to antibodies to detect protein-protein interactions.

✓ Concept Check 14.8

Cyclins are proteins that bind to other proteins called cyclin-dependent kinases (CDKs). Different cyclins bind to different CDKs and are expressed during different phases of the cell cycle. A certain CDK was modified with a Myc tag at its C-terminus and purified from different samples of culture using co-immunoprecipitation. Each sample was collected in a different phase of the cell cycle. The proteins that were bound to the beads were then analyzed by western blot. Based on the following results, during which phase is this cyclin expressed?

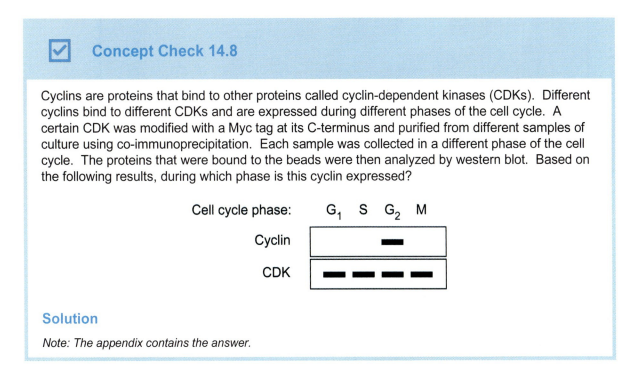

Solution

Note: The appendix contains the answer.

Chapter 14: Biomolecule Purification and Characterization

Lesson 14.4
Additional Techniques

Introduction

In addition to electrophoresis (Lesson 14.1), blotting (Lesson 14.2), and chromatography (Lesson 14.3), biochemistry employs a variety of other techniques to study the chemistry of biomolecules and living systems.

This lesson provides a broad, though not comprehensive, overview of commonly used biochemical techniques, with an emphasis on methods that may be referenced on the exam. A deep understanding of the technical aspects of each method is unlikely to be required; however, knowledge of the underlying principles of each method is still helpful in understanding passages and in interpreting data derived by a specific method.

Research methods in any natural science, but especially in biochemistry, often draw upon basic science principles covered in other courses (eg, general and organic chemistry, physics, biology). Although this lesson strives to be accessible to any student of biochemistry, referring to the relevant lessons of other subjects as needed may be helpful.

14.4.01 Dialysis

Lesson 14.3 discusses chromatography as a means of protein purification. Methods such as ion-exchange chromatography and affinity chromatography can separate a protein of interest from contaminating proteins; however, removal from the column yields a protein of interest collected in an elution buffer. Depending on the method, the elution buffer may include contaminating ligands and other contaminants, or a pH or salt concentration that is incompatible with later experiments. The purified protein must therefore be restored to conditions that *are* compatible.

One method of exchanging the elution buffer with a more compatible buffer is **dialysis**. In dialysis, the sample of interest is loaded into a container with a **porous membrane**. The pores must be *large* enough for salt, water, and small labels to pass through but *small* enough that the protein of interest *cannot* pass through.

The other side of the porous membrane is exposed to the desired final buffer (ie, the **dialysate**). Undesired small molecules or ions diffuse *out* of the sample and into the dialysate fluid, and desired solutes diffuse *in* (Figure 14.43). Water also travels across the membrane by osmosis, equalizing osmotic pressure across the membrane. Eventually, the small, *permeable* solutes reach a diffusive equilibrium wherein they are in **equal concentrations** across the membrane.

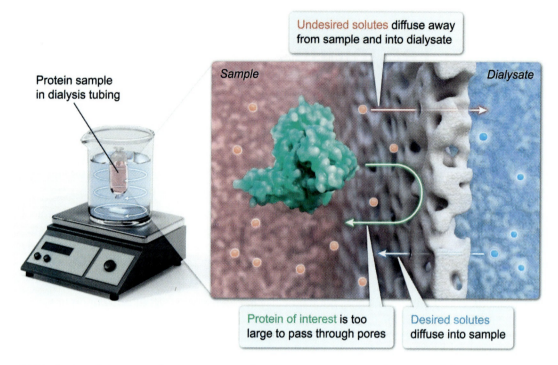

Figure 14.43 During dialysis, small solutes (eg, small molecules, ions, oligopeptide tags) diffuse across the membrane, whereas large proteins are retained on one side of the membrane.

When dialyzing a protein for purification purposes, several rounds of dialysis are usually employed. In each round, sufficient time is given for the sample and dialysate buffers to equilibrate, which "cleans" the sample and "dirties" the dialysate. Every new round replaces the dialysate fluid with fresh fluid, allowing the sample to become even cleaner, as shown in Figure 14.44.

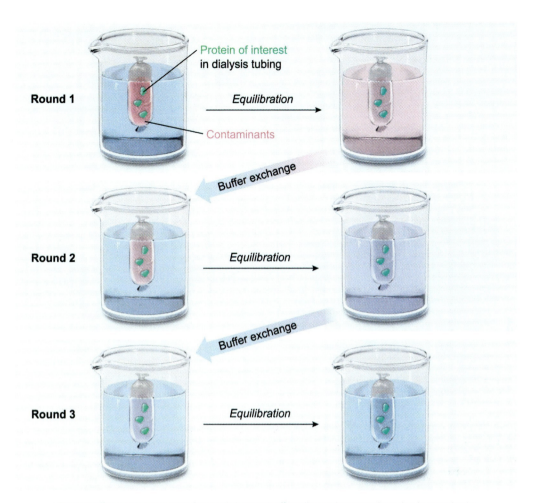

Each buffer exchange with fresh dialysate buffer allows the sample to be increasingly cleaner with each round of dialysis.

Figure 14.44 Several rounds of dialysis are often needed to fully purify a sample of small molecule or small ion contaminants.

The principle of dialysis also has clinical relevance. For example, hemodialysis is a treatment for patients whose kidneys cannot sufficiently filter their blood. In this case, patient blood is diverted from the body and into a dialysis machine, where it is exposed to dialysate fluid across a porous membrane. Excess waste products diffuse out of the blood and into the dialysate fluid, and nourishing electrolytes and other small molecules diffuse from the dialysate fluid and back to the blood (Figure 14.45). Cleaned blood exits the dialysis machine and is returned to the patient.

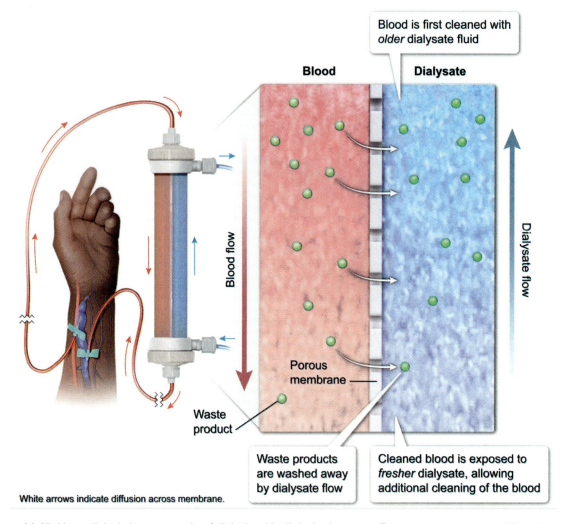

Figure 14.45 Hemodialysis is an example of dialysis with clinical relevance. Excess waste products diffuse through a porous membrane, which cleans and purifies patient blood.

Concept Check 14.9

A His-tagged protein is purified using an affinity chromatography column. An elution buffer containing 300 mM imidazole is used to elute the protein from the nickel-linked chromatography column before the sample is dialyzed with phosphate-buffered saline. At the completion of several rounds of dialysis, the concentration of imidazole in the final dialysate is measured to be 2 mM. What is the concentration of imidazole still in the sample?

Solution
Note: The appendix contains the answer.

14.4.02 Biomolecule Quantitation

After a protein or any other biomolecule has been collected, most downstream applications require knowledge of sample concentration. In other words, the amount of the specific biomolecule of interest in the sample needs to be *quantified*.

The **absolute quantitation** methods discussed in this lesson provide concentration values in terms of molarity (eg, mM) or in mass per volume (eg, mg/mL). Importantly, absolute quantitation often must be used *prior* to electrophoresis and blotting to ensure that changes in band density are due to changes in expression level and not due to loading different amounts of total protein or nucleic acid into different lanes.

Review of UV-Vis Spectroscopy Principles

Various methods exist to quantify biomolecules, many of which rely on ultraviolet-visual (UV-vis) absorption spectroscopy. The principles of UV-vis spectroscopy are covered in detail in Organic Chemistry Lesson 14.5. In brief, different molecules absorb specific energies and wavelengths of ultraviolet and visible light. The wavelength that a molecule absorbs most strongly, its lambda max (λ_{max}), is determined empirically. The amount of absorbance A is related to concentration of the molecule (c) by the equation

$$A = \varepsilon c \ell$$

in which ε represents the absorption coefficient (ie, the absorptivity) of the analyte and ℓ is the pathlength (ie, the length of the sample through which the light passes). This equation demonstrates that *absorbance is directly proportional to sample concentration* (Figure 14.46).

Therefore, biomolecule concentration can be determined from sample absorbance. For purified molecules that strongly absorb UV or visible light with known ε values, calculation of concentration is straightforward. For *impure* samples, samples with poor absorptivity at their λ_{max}, or samples with poorly defined ε values, however, additional measures must be taken.

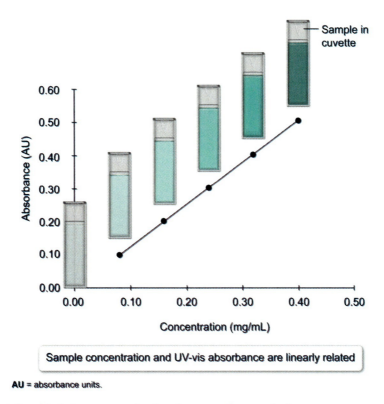

AU = absorbance units.

Figure 14.46 The relationship between sample absorbance and concentration.

Absorbance and Quantitation of Purified Biopolymers

Nucleic acids (ie, DNA, RNA) and most proteins can be directly detected by ultraviolet spectroscopy. Because tryptophan residues have a strong λ_{max} at 280 nm, many proteins can be quantified by measuring absorbance at 280 nm (ie, A_{280}). In contrast, nucleotides absorb strongly at 260 nm, so DNA and RNA can be quantified by measuring absorbance at their λ_{max} of 260 nm (A_{260}).

The values of the absorption coefficients of some proteins are known; however, when measuring the absorbance of a protein with an unknown absorption coefficient, it is common practice to use an *average literature value*. Alternatively, the absorption coefficient of a protein of interest can be estimated based on the amount of tryptophan residues (and, to a lesser extent, the amount of tyrosine and cysteine residues) in that protein. However, estimates of a protein's absorptivity based on these assumptions come with two caveats when interpreting data.

The first caveat is that concentrations determined using average coefficient values may be inaccurate if the protein has a greater-than-average or less-than-average percentage of tryptophan residues.

The second caveat is that the use of an average literature absorption coefficient value results in data that correlate more closely with the *mass* of protein in a sample than it does to the *moles* of protein in a sample. A large protein with a typical percentage of tryptophan residues would absorb 280 nm light more strongly than a smaller protein that has the same tryptophan percentage, because the large protein contains more *total* tryptophan residues. In other words, these measurements facilitate calculation of the protein concentration in mg/mL This concept is illustrated in Figure 14.47.

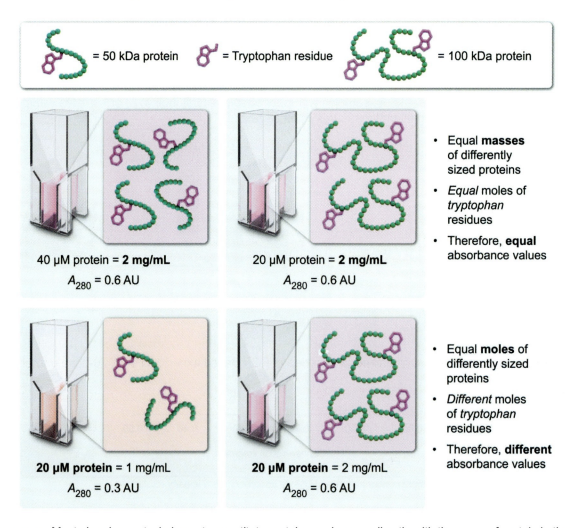

Figure 14.47 Most absorbance techniques to quantitate proteins scale more directly with the mass of protein in the sample than with the number of moles.

When needed, the mass concentration (eg, mg/mL) of a sample of purified protein can be converted to molar concentration by dividing by the molar mass (Recall that the molar mass of a protein in g/mol is numerically equal to its molecular mass in Da. A protein with a mass of 50 kDa has a molar mass of 50 kg/mol).

Typical absorption spectra for purified protein and purified DNA are given in Figure 14.48. Although both types of biopolymers have clearly different λ_{max} values, the spectra do overlap (eg, tryptophan absorbs 260 nm light, just not as strongly as 280 nm light). Consequently, protein contamination of DNA samples can affect DNA concentration measurements and vice versa.

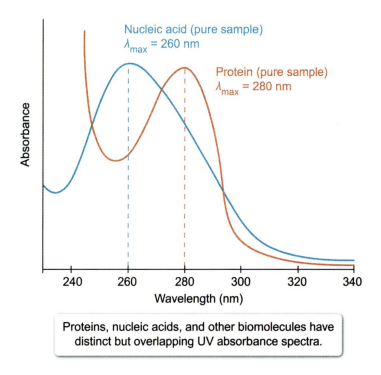

Proteins, nucleic acids, and other biomolecules have distinct but overlapping UV absorbance spectra.

Figure 14.48 Typical UV absorbance spectra for proteins and nucleic acids.

Consequently, although A_{280} and A_{260} measurements are quick and simple to set up, they are mainly used for relatively pure samples. When quantifying impure samples (eg, crude lysates), different protocols are used that typically involve staining the biopolymer of interest.

✓ Concept Check 14.10

A set of DNA primers are synthesized for a series of PCR-based experiments. A 20-base primer is synthesized for a qPCR experiment, and a 40-base primer is synthesized for a site-directed mutagenesis experiment. 10 μM solutions of each primer are prepared. How will the A_{260} values compare between the two equimolar preparations?

Solution
Note: The appendix contains the answer.

Absorbance and Quantitation of Impure Samples

Various methods of protein quantitation for impure samples exist (eg, Bradford assay, BCA assay, Lowry assay). Each method differs slightly in their preparation, the resultant λ_{max}, and their compatible contaminating reagents; however, the basic principle of each technique is similar. Proteins within a sample react with reagents to produce an intense visible color upon interaction. These proteins are said to be stained.

The intensity of many of these protein stains may vary with time or may be altered based on the environmental conditions of the experiment or the lab. As such, literature values for the absorption coefficient are unreliable. Instead, it is common practice to produce a new **standard curve** for each experiment.

Standard curves involve measuring the absorbance of samples with a known protein concentration. For example, purified bovine serum albumin (BSA) can be obtained as a powder that can be accurately weighed out and diluted in the same buffer used for the protein of interest. The absorbance measurements are plotted against the mass concentrations of the known standard, and a line of best fit (ie, the standard curve) is calculated. From the standard curve, the measured absorbance of the sample containing the protein of interest can then be correlated to a mass concentration (Figure 14.49).

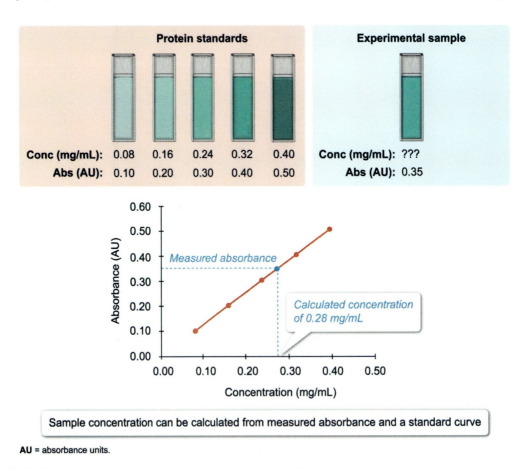

AU = absorbance units.

Figure 14.49 The use of a standard curve to calculate concentration from absorbance.

Instrument limitations often restrict the linear relationship between absorbance and concentration to a small range. Therefore, it is common practice to choose concentrations of the protein standard that give absorbance measurements between 0.1 and 1 absorbance units (AU). If the sample of interest gives an absorbance measurement beyond that range, it can be diluted by a known factor. The resulting absorbance and concentration value can be multiplied by the dilution factor to give the original concentration.

For example, a 10-fold dilution is prepared by mixing 1 mL of sample with 9 mL of dilution buffer (10 mL total volume). If the diluted sample yields an absorbance corresponding to 0.45 mg/mL of protein, then the undiluted sample has a concentration of 0.45 × 10 = 4.5 mg/mL.

In *very* concentrated samples, a diluted sample can be diluted again, multiple times if needed, in a process known as a serial dilution. In this case the overall dilution factor is the product of the individual dilution factors. For instance, mixing 1 mL of a protein with 9 mL of buffer, and then mixing 1 mL of that diluted solution with 9 more mL of buffer results in a 100-fold dilution.

Figure 14.50 depicts an example of a concentration calculation using serial dilution.

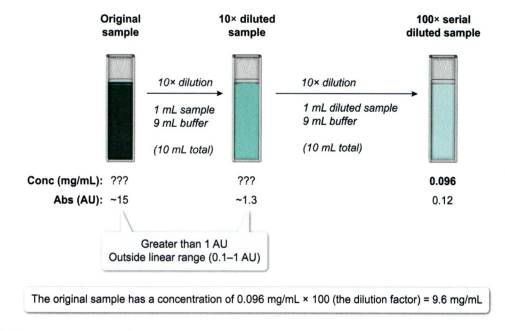

Figure 14.50 Concentrations of diluted samples can be used to calculate the original concentration by multiplying by the dilution factor.

☑ Concept Check 14.11

Various samples of cells are lysed to prepare crude lysates. Prior to running a gel, the researcher wants to quantify each crude lysate sample to ensure consistent loading. The Bradford assay, which uses a Coomassie stain, was used to produce a standard curve that yielded a line of best fit described by the equation

$$A = 0.25c$$

in which A is the measured absorbance and c is the protein concentration in mg/mL.

If a 3-fold dilution of a crude lysate results in an absorbance value of 0.50, what is the concentration of the crude lysate?

Solution

Note: The appendix contains the answer.

14.4.03 Binding Assays

One of the most fundamental types of protein study is a **binding assay**, which analyzes the binding interaction between a protein and a ligand. Binding interactions are also discussed in Lesson 3.1. In brief, the strength of a binding interaction between a protein and its ligand can be described by the K_d value (ie, the equilibrium dissociation constant) of the interaction. The K_d value is defined as:

$$K_d = \frac{[P][L]}{[PL]}$$

in which [P] represents the concentration of free (ie, unbound) protein, [L] represents the concentration of free ligand, and [PL] represents the concentration of the protein-ligand complex. From this relation, an equation describing the fraction of protein bound by ligand (θ) can be derived.

$$\text{Fraction bound, } \theta = \frac{[PL]}{[P]+[PL]}$$

$$= \frac{[L]}{K_d + [L]}$$

Binding Assays That Assume the Free Ligand Approximation

Importantly, the equations describing protein binding are defined using [L], which is the concentration of *free, unbound* ligand. However, the experimenter usually only controls [L_{tot}] the *total* concentration of ligand, which is equal to [L] + [PL]. The free ligand approximation can be used to assume [L] ≈ [L_{tot}], but this assumption is only valid if the ligand concentration is much higher than the total protein concentration.

When this approximation is valid, various methods can be used to measure protein binding. For example, if the ligand-bound protein has a different absorption spectrum relative to unbound protein and ligand, then protein binding can be assessed using UV-vis spectroscopy. Figure 14.51 shows how UV-vis spectroscopy can be used to measure oxygen binding by hemoglobin.

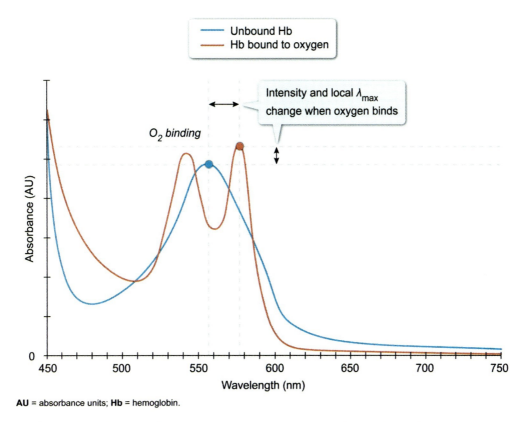

Figure 14.51 The binding of oxygen changes the absorbance spectrum of hemoglobin. Oxygen saturation (ie, the percent of hemoglobin bound to oxygen) can be determined by monitoring this change.

AU = absorbance units; Hb = hemoglobin.

✓ Concept Check 14.12

A binding assay is performed to measure the interaction between a protein and its ligand. Like hemoglobin, the protein is colored and undergoes a shift in λ_{max} depending on the binding state. When unbound to ligand, the protein has a λ_{max} of 490 nm. When ligand is added, the λ_{max} changes until it reaches 420 nm when saturated. The following plot shows the binding curve.

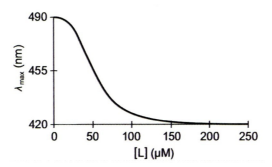

Based on these data, what is the K_d for the binding interaction? What else can be stated about the interaction between the protein and its ligand?

Solution

Note: The appendix contains the answer.

Chapter 14: Biomolecule Purification and Characterization

Isothermal Titration Calorimetry

Binding affinity between a protein and a ligand can also be determined by measuring the thermodynamics of the binding reaction. Specifically, the enthalpy of the binding reaction (ΔH) can be measured by a technique called isothermal titration calorimetry (ITC).

ITC measures the release of heat that occurs as aliquots of ligand are gradually injected (or titrated) into a sample of purified protein and bind. In an isothermal titration calorimetry experiment, the small reaction vessel is maintained at a constant temperature (ie, isothermal conditions), and the power needed to maintain that temperature is monitored over time.

If the addition of ligand results in an exothermic binding reaction, the ITC machine records a decrease in machine power needed to maintain temperature because the *heat released by the reaction* helps maintain it. Each time an injection occurs, the change in power needed is plotted as a deflection (ie, either a peak or an inverted peak) on a graph. For any given injection, the area between the peak and the baseline represents the enthalpy change ΔH that results from the ligand injection (Figure 14.52).

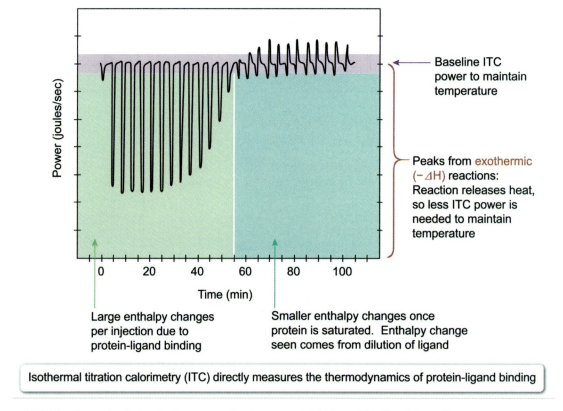

Isothermal titration calorimetry (ITC) directly measures the thermodynamics of protein-ligand binding

Figure 14.52 Isothermal calorimetry is one way to measure protein-ligand binding interactions.

As with other binding assays, this signal (the ΔH) eventually stops changing between injections, indicating that the protein is saturated and therefore no more binding can occur. Because reaction thermodynamics are directly measured by this technique, rather than indirectly monitored through a reporter (eg, radioactivity, fluorescence), ITC determination of K_d does *not* require use of the free ligand approximation.

14.4.04 Melting Temperature Assays

Lesson 2.3 and Lesson 8.2 introduced the **melting temperature (T_m)** of proteins and nucleic acids as a descriptor of biopolymer stability. Specifically, T_m is the temperature at which 50% of the biopolymer becomes denatured. This concept describes assays that can be used to determine the T_m.

In general chemistry and physics, "melting" is often used to describe the transition from a solid phase to a liquid phase. As the temperature of a solid rises past its melting point, the average kinetic energy of the constituent molecules eventually surpasses the energy needed to break the intermolecular forces holding it together. Therefore, the material changes from a rigidly packed solid to a fluid, freely diffusible liquid.

However, with biochemical melting temperatures, the polymers typically remain in an *aqueous* phase both before *and* after the transition. Therefore, the use of the term "melting" in this context does *not* refer to a transition from one state of matter to another. Instead, melting of biopolymers, like melting a solid to form a liquid, involves the *breaking of noncovalent bonds* (eg, intermolecular forces).

As the temperature of a biopolymer sample increases, the bonds maintaining secondary, tertiary, and quaternary protein structure break apart, as do the hydrogen bonds maintaining nucleic acid base-pairing (Table 14.1). Due to the cooperative nature of folding, these bond-breaking events tend to occur near the same temperature (ie, as one bond breaks, it becomes easier for other bonds to break). Therefore, the melting of biopolymers refers to the denaturing breakage of their noncovalent bonds, and the halfway point of this transition is the polymer's melting temperature, T_m.

Table 14.1 Like the melting of a solid to a liquid, the melting of a biopolymer involves the breaking of noncovalent bonds.

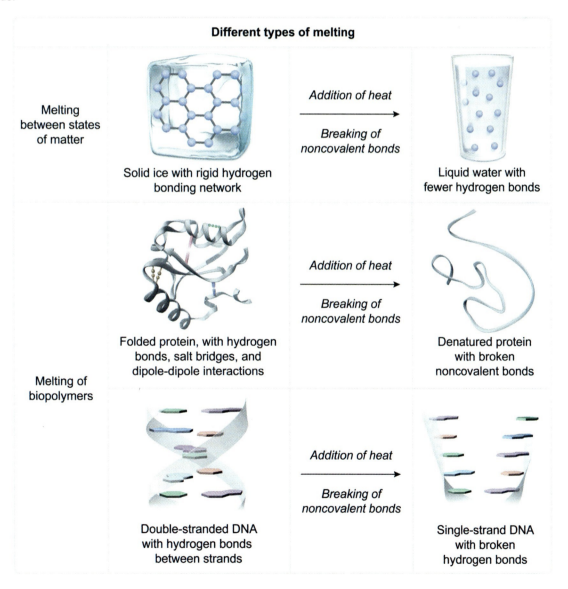

Measuring Melting Temperature through Calorimetry

Like the solid-to-liquid melting transition (in which heat added during a phase change does *not* increase the temperature), the melting transition of biopolymers is similarly marked by a *rise in heat capacity*. These heat capacity changes can be measured using differential scanning calorimetry (DSC). DSC experiments *increase* the temperature at a constant rate while monitoring the power needed to maintain that rate.

As noncovalent bonds begin to break, the power needed *increases*, because some of the energy is used to break those bonds (an endothermic process) rather than increase temperature. The point of highest power input, indicating the most energy per unit time, is taken as the T_m. At temperatures above the T_m, when most polymers have already been denatured, the heat capacity begins to drop—few noncovalent bonds remain to absorb energy, and therefore more energy can be used to increase the kinetic energy (and temperature) of the sample (Figure 14.53).

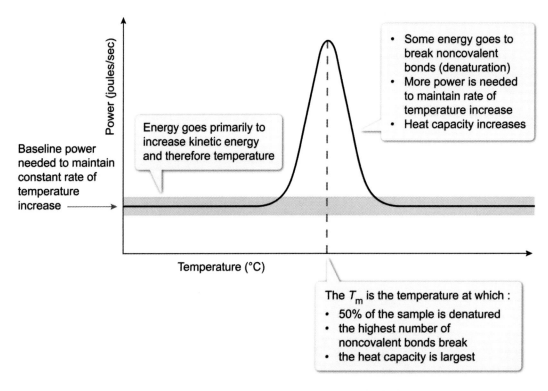

Figure 14.53 Calorimetry can be used to measure the melting temperature (T_m) of biopolymers. In calorimetry, T_m is the temperature that yields the largest heat capacity.

Concept Check 14.13

Differential scanning calorimetry data for two samples are presented in the following image.

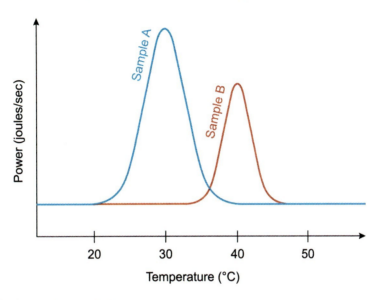

1) What is the T_m for each sample?
2) If both samples are DNA duplexes with similar GC content, which of the samples contains a longer DNA sequence?

Solution

Note: The appendix contains the answer.

Measuring Melting Temperature through Spectroscopy

Melting temperature can also be assessed with spectroscopic measurements that report on protein conformation (ie, native versus denatured). Fluorescent techniques (discussed in detail in Concept 14.4.05) can be used with both protein and nucleic acids. Certain specialized fluorescent dyes bind specifically to double-stranded regions of DNA or RNA. Therefore, intact nucleic acids provide a strong initial signal, but as temperatures increase and the nucleic acid denatures, the fluorescence signal *decreases*.

Protein folding, similarly, can be measured with fluorescence. The fluorescence signal can come either from intrinsic tryptophan residues or from external fluorophores that have been covalently linked to the protein (ie, the protein is fluorescently labeled). A fluorescence signal can be affected by the environment, which can include the conformational state of a protein that a fluorophore is attached to. Consequently, fluorescence intensity and λ_{max} changes can be monitored as temperature increases from low (native conformation) to high (denatured conformation). Circular dichroism is another spectroscopic technique that can be used to assess protein folding (see Concept 14.4.06).

The spectroscopic data can be plotted as a function of temperature to produce a curve representing protein conformation. Because folding is a positively cooperative process, the resulting graph has a sigmoidal shape, ranging from 0% denatured at low temperatures to 100% denatured at high temperatures. The temperature that results in 50% denaturation is the T_m (Figure 14.54).

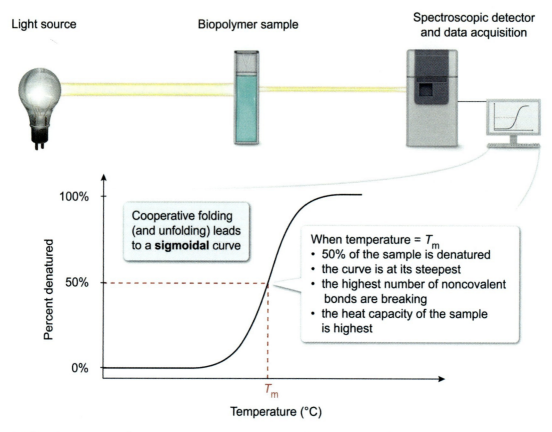

Figure 14.54 Spectroscopic data (eg, fluorescence) can be used as reporters of protein conformation to perform melting temperature assays.

14.4.05 Fluorescence

Fluorescence is a commonly utilized phenomenon in the natural sciences. Fluorescence techniques are like UV-vis spectroscopy in that they involve the absorption of photons in the UV-visual spectrum. Fluorescence differs in that the absorption of a photon by the fluorescent molecule (also called excitation) is followed by the *release* of a photon with **lower energy** (and therefore a *longer* wavelength). For example, a fluorescent molecule (ie, a fluorophore) might absorb a high-energy blue photon (~488 nm) and release a lower-energy green photon (~510 nm).

In addition to chemical fluorophores, there are also genetically encoded fluorescent tags, such as green fluorescent protein (GFP), that can be used to visualize localization and expression of the tagged proteins both in cell culture and in intact organisms *in vivo*. Various applications of fluorescence are shown in Figure 14.55.

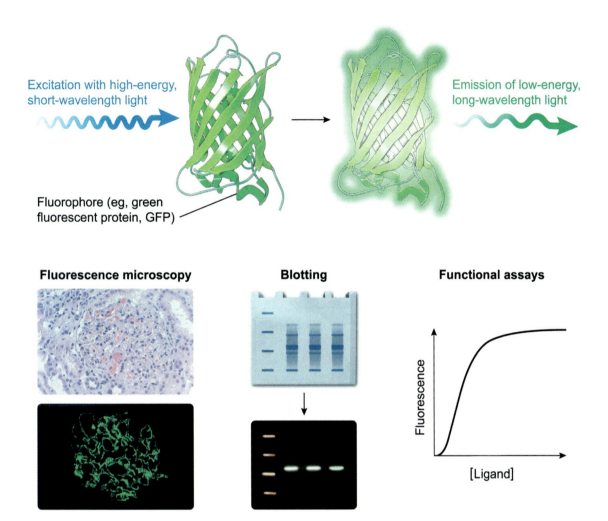

Figure 14.55 Examples of the uses of fluorescence in biochemistry.

What causes the shift in wavelength between the excitation photon and the emission photon? The absorbance of ultraviolet and visible light occurs when the photon's energy matches the energy needed to excite an electron to a higher energy state. In fluorescent molecules some—though not all—of the absorbed energy is then released as vibrational energy, or heat, to the surroundings. The *remaining* energy can then be emitted as light. Because the remaining energy is less than the initial energy, it is emitted as longer-wavelength photon. Additional heat energy is lost as heat after the photon is emitted due to vibrational relaxation (Figure 14.56).

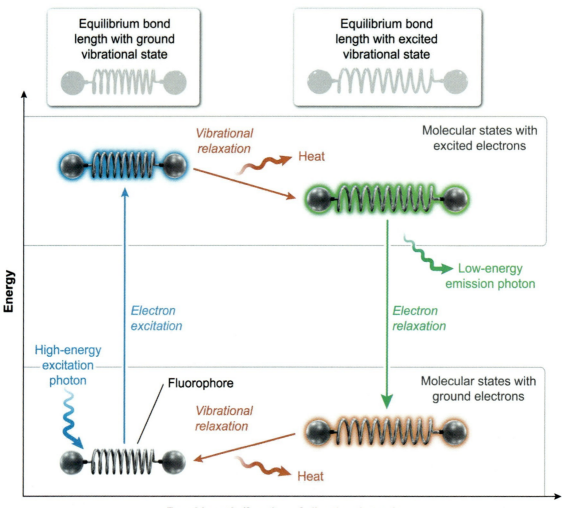

Figure 14.56 The fluorescence cycle.

14.4.06 Circular Dichroism

Circular dichroism (CD), introduced briefly in Concept 14.4.04, is a spectroscopic technique that reports the overall secondary structure of proteins. Like other UV-vis absorbance techniques, CD depends on the ability of the analyte (in this case, regions of protein secondary structure) to absorb electromagnetic radiation. CD differs from other UV-vis absorbance techniques, however, because CD relies on the absorbance of **circularly polarized light**.

Circularly polarized light differs from linearly polarized light (also known as plane-polarized light). Plane-polarized light oscillates in a two-dimensional plane, whereas circularly polarized light propagates in a spiraling, helical fashion. Circularly polarized light can be conceptually understood as the combination (ie, interference) of two rays of equal-amplitude light, in which one ray's polarization axis is both rotated 90° (eg, vertical and horizontally polarized rays) *and* phase-shifted by 90° with respect to the other ray, as illustrated in Figure 14.57. For more on circular polarization of light, see Physics Concept 4.3.05.

Chapter 14: Biomolecule Purification and Characterization

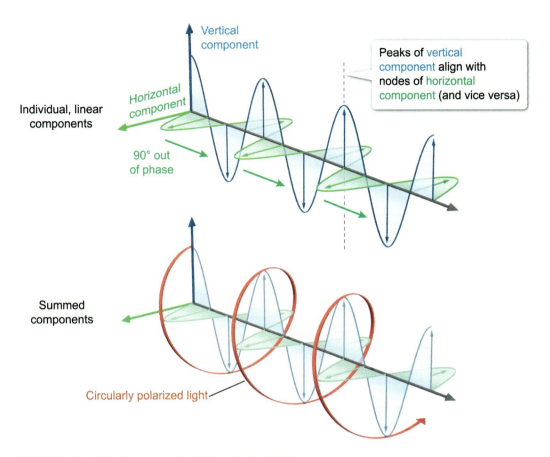

Figure 14.57 Circular dichroism uses circularly polarized light.

CD spectra of proteins uses circularly polarized light in the far-UV range (~180–250 nm). Light in this region is absorbed by the peptide bonds of the polypeptide backbone. Backbone interactions (eg, interactions that form secondary structure elements) alter the absorbance spectrum; therefore, different secondary structure elements (eg, α-helices, β-sheets) absorb right-handed and left-handed circularly polarized light to different extents.

The CD spectrum of a protein plots the difference in the individual right- and left-handed absorbance spectra. This measurement describes the *ellipticity* of the protein sample. Representative CD spectra for secondary structural elements are shown in Figure 14.58.

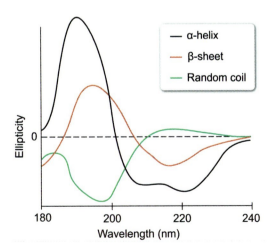

Figure 14.58 Representative circular dichroism spectra for secondary structure elements.

Most proteins have a mixture of secondary structural elements, so their CD spectra will be a linear combination of the individual spectra shown in Figure 14.58. For example, consider a protein with a structure that is 50% α-helix, 40% β-sheet, and 10% random coil. Its CD spectrum will reflect those ratios, such that at any given wavelength the ellipticity value will be equal to 0.5 times the α-helix value plus 0.4 times the β-sheet value plus 0.1 times the random coil value.

Detailed analysis of a mixed-structure spectrum is not likely to be required on the exam. Instead, the it is more likely to test identification of a structure with a single secondary structural feature given the reference spectra. In addition, the exam may test for an understanding of *how* a spectrum changes in response to conformational changes (eg, ligand binding, denaturation).

14.4.07 Structure Determination

Determination of protein structure can have great scientific and biomedical significance. For example, identification of a protein's structural features at the atomic level can allow scientists to discover the role each amino acid residue plays in protein function. Not only can this help in understanding and identifying diseases, but structural information can be used in the design of therapeutics to target those diseases (Figure 14.59).

Currently, most high-resolution structures are experimentally determined by cryogenic electron microscopy (cryo-EM), x-ray crystallography, or nuclear magnetic resonance (NMR). Recently, computational techniques have advanced and allowed artificial intelligence models to use the data from experimentally determined structures to predict structures of novel proteins with relatively high accuracy. Although it is *unlikely that the exam will ask questions about the technical details* of these methods, a brief overview of each may be instructive for *synthesizing information* throughout biochemistry and other subjects.

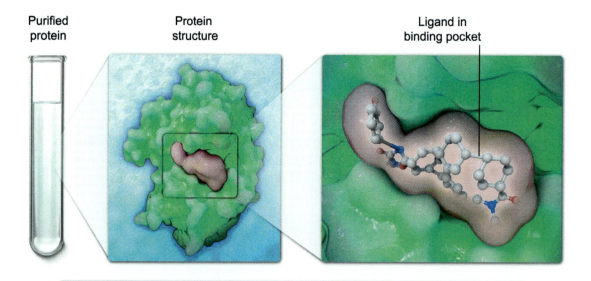

Figure 14.59 Studies of the three-dimensional structure of biomolecules are of great importance in biomedical science.

Cryogenic Electron Microscopy (Cryo-EM)

The *resolution* of a microscope (ie, its ability to "see" small objects) is dependent on the wavelength of radiation used. Because protein particles are much smaller than even the shortest wavelength of visible light (~400 nm), visible light *cannot* be used to visualize proteins.

Instead, **electron microscopy** uses high-energy *electrons* to irradiate samples. Because of the wave-particle duality of matter, these energetic electrons can have a wavelength that as short as 2 pm (2×10^{-12} m) or smaller, which is small enough to visualize proteins at atomic resolution. Unlike other high-energy (short wavelength) radiation (eg, gamma waves), for which the development of optical lenses is difficult, electron beams can be precisely focused due to their charged nature.

In cryo-EM, samples are suspended in ice and examined at low temperatures. Freezing must happen rapidly to prevent the water molecules from forming ordered, hydrogen-bonded crystal lattices, which would destroy proteins. The low temperature slows the damaging effects of the high-energy electron beam.

Microscopy generally allows a researcher to take a picture of a sample; however, this picture is flat and two-dimensional. Therefore, if a three-dimensional structure of a purified protein is needed, multiple pictures must be taken of a sample from different angles. If multiple protein particles in solution exist in random orientations, then multiple angles can be visualized from a single picture; otherwise, a sample can be physically rotated in the microscope to provide the angles needed.

Once two-dimensional images—or projections—have been obtained from various angles via many different particles, then a three-dimensional model can be reconstructed from those projections, as shown in Figure 14.60.

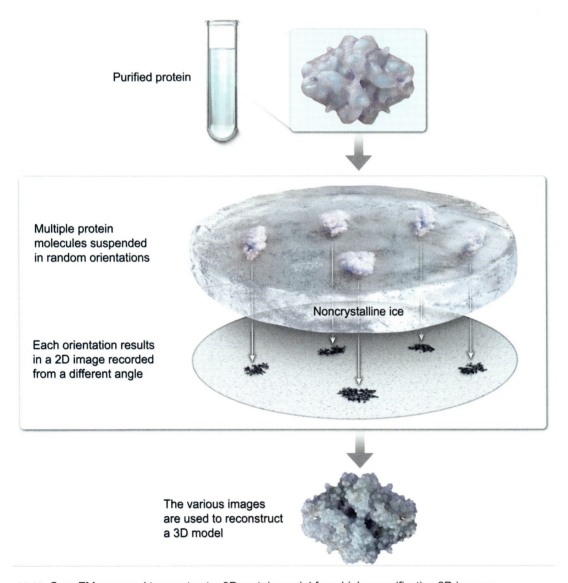

Figure 14.60 Cryo-EM can used to construct a 3D protein model from high-magnification 2D images.

X-ray Crystallography

Like high-energy electrons, x-rays have a very short wavelength and can be used to determine atomic resolution structures of proteins. As the name suggests, x-ray crystallography (XRC) involves the formation of crystals. To form the needed crystals, a protein-containing solution is purified and then slowly concentrated (often with agents to prevent denaturation) until crystals appear.

The solid crystal makes it difficult to take a direct "picture" of a molecule. Instead, crystallography relies on the scattering of x-rays as they encounter atoms. Most scattered rays interfere destructively; however, the regular spacing of biomolecules present in a crystal allows for certain scattered rays to experience *constructive* interference. This is similar to the concept of multiple-slit diffraction (see Physics Lesson 4.3), and the interference pattern seen is called a **diffraction pattern** (Figure 14.61).

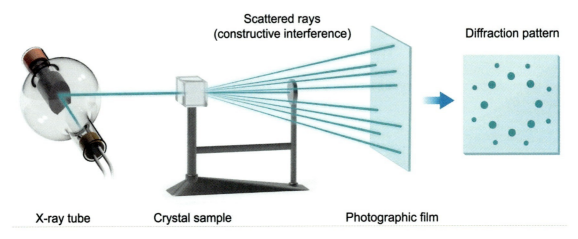

Figure 14.61 In x-ray crystallography, a diffraction pattern is formed from constructively interfering rays that have scattered after interacting with a regularly spaced crystal.

From the observed diffraction pattern, the three-dimensional shape of the protein can be determined. One complication of interpreting crystal structures is that the crystallization procedure traps molecules as a solid crystal, whereas most biomolecules exist *in vivo* as dissolved aqueous solutes. This may introduce crystallization artifacts that are not relevant for aqueous molecules.

Nuclear Magnetic Resonance (NMR)

As introduced in Organic Chemistry Lesson 14.7, nuclear magnetic resonance is a spectroscopic technique that uses very-low-energy radiofrequency waves to determine molecular structure. Unlike cryo-EM and XRC, radiofrequency waves have a wavelength that is too long to directly report on the physical position of atoms. Instead, NMR measures the absorption of the photons' energy by nuclei in a magnetic field. The amount of energy an individual nucleus can absorb varies depending on the structural features of the molecule (eg, the degree of electron shielding), allowing NMR to report on chemical structure.

To summarize the NMR method, the electronic environment surrounding a nucleus causes a shift in the frequency at the λ_{max} (known as a chemical shift) relative to some reference sample (eg, tetramethylsilane [TMS]). This shift is measured in parts per million (ppm). In addition, the interaction of nuclear spins across three σ bonds or fewer (called through-bond interactions) can be seen as coupling of the spins through spin-spin splitting.

The information gained from one-dimensional proton (^{1}H) and carbon-13 (^{13}C) NMR data can be used to help in the determination of the **primary structure** of small organic molecules. As molecules grow in complexity (eg, proteins), two-dimensional (2D) NMR and heteronuclear NMR techniques (eg, techniques that assess nuclei *other* than ^{1}H) can be used.

Two-dimensional NMR allows for the explicit identification of which peaks interact by measuring the effects of each peak in a 1D spectrum on each peak on another 1D spectrum (Figure 14.62).

Heteronuclear NMR allows the probing of interactions between different elements. In protein NMR, ^{15}N is commonly used to probe the peptide bond (Figure 14.63).

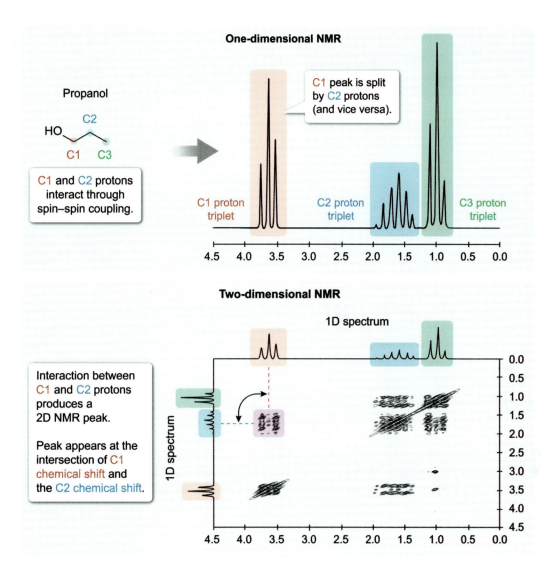

Figure 14.62 Two-dimensional NMR can be used to help identify the peaks and interactions seen in a one-dimensional NMR. This can be useful when analyzing large, complex molecules.

Chemical shift data in 2D NMR can be used to monitor ligand interactions and conformational changes. An example of this is shown in Figure 14.63. The region shown captures interactions between the amide nitrogen and protons of the peptide bond. A particular amide peak is tracked as more ligand is added to the protein sample.

At low ligand concentrations, the peak appears at the intersection of approximately 105 ppm on the ^{15}N 1D spectrum and 8.75 ppm on the ^{1}H spectrum. The peak position shifts until, at high ligand concentrations, it appears at 125 ppm on the ^{15}N spectrum and 7.25 ppm on the ^{1}H spectrum. This indicates changes in the chemical environments of these atoms, and therefore a change in the protein conformation, as the protein binds its ligand.

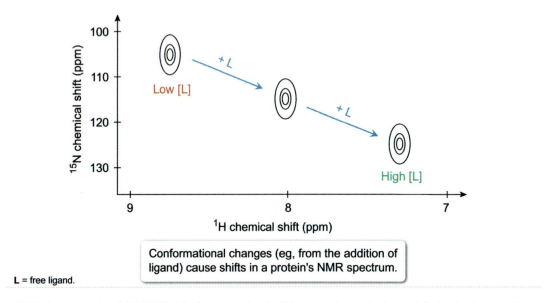

Figure 14.63 An example of 2D NMR data for a protein. In this example, a single peak is tracked across three experiments, each with different ligand concentrations.

To determine tertiary and quaternary structure, however, a specialized form of NMR must be used that can report on through-*space* interactions (ie, interactions between amino acid residues that are *not* close to each other in the primary structure but become close when the protein folds). Details of this procedure are unlikely to be tested on the exam; however, the other fundamental principles of conventional NMR still apply to NMR for protein structure.

Modeling and Artificial Intelligence

Based on the structural data obtained through cryo-EM, XRC, and NMR, a robust library of protein structures has been collected over several decades. Many proteins without experimentally determined structures can be modeled after similar proteins, based on the biochemical properties of amino acids and on analysis of sequence similarity. With the growing capabilities of artificial intelligence, predicted protein structures have increased in accuracy, as verified by subsequent experimental methods.

END-OF-UNIT MCAT PRACTICE

Congratulations on completing **Unit 5: Biochemistry Lab Techniques**.

Now you are ready to dive into MCAT-level practice tests. At UWorld, we believe students will be fully prepared to ace the MCAT when they practice with high-quality questions in a realistic testing environment.

The UWorld Qbank will test you on questions that are fully representative of the AAMC MCAT syllabus. In addition, our MCAT-like questions are accompanied by in-depth explanations with exceptional visual aids that will help you better retain difficult MCAT concepts.

TO START YOUR MCAT PRACTICE, PROCEED AS FOLLOWS:

1) Sign up to purchase the UWorld MCAT Qbank
   IMPORTANT: You already have access if you purchased a bundled subscription.
2) Log in to your UWorld MCAT account
3) Access the MCAT Qbank section
4) Select this unit in the Qbank
5) Create a custom practice test

Appendix
Concept Check Solutions

You will find detailed, illustrated, step-by-step solutions for each concept check in the digital version of this book.

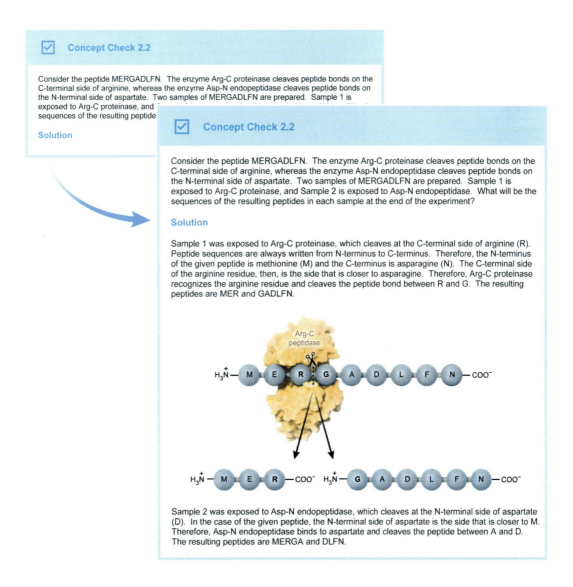

In this section of the print book, you will only find short answers to the concept checks included in each chapter. Please go online for an interactive and enhanced learning experience with visual aids.

Unit 1. Amino Acids and Proteins

Chapter 1. Amino Acids

Lesson 1.1

1.1 The amino acid that does not rotate light is glycine. The other amino acids are not glycine.

Lesson 1.3

1.2 Tyrosine is 0.099% ionized; histidine is 9.1% ionized.

1.3 3.3 (glutamate); 5.9 (alanine); 10.1 (lysine).

1.4 A) arginine; B) glutamate.

Chapter 2. Peptides and Proteins

Lesson 2.1

2.1 During peptide bond formation, multiple bonds (eg, a C–N bond) are formed and multiple bonds (eg, a C–O bond) are broken. The enthalpy of the broken bonds is greater in magnitude than the enthalpy of the formed bonds.

2.2 MER and GADLFN (Sample 1); MERGA and DLFN (Sample 2).

2.3 +1 at pH 7; +2 at pH 5.

2.4 DAES < DALS < KAES < KALS < KAHS.

Lesson 2.2

2.5 Around 545 amino acids

2.6 Peptide B

2.7 Two

2.8 On the surface of the transmembrane domain; buried in the cytosolic and extracellular domains.

Lesson 2.3

2.9 Hydrophobic residues move to the surface; hydrophilic residues become buried.

2.10 The mutation increases conformational stability.

2.11 1) Prion; 2) correctly folded protein; 3) amyloid fibril.

Lesson 2.4

2.12 S7R

2.13 Position 115

2.14 Phosphorylation activates protein A and inactivates protein B.

2.15 coenzyme; prosthetic group; apoprotein; holoprotein.

Chapter 3. Nonenzymatic Protein Activity

Lesson 3.1

3.1 The mutation increases the K_d.

3.2 4.0 µM

3.3 30 °C

3.4 Approximately 0.006 atm

3.5 The curve will be left-shifted

3.6 The protein has at least 2 binding sites.

Lesson 3.2

3.7 Compound 1 is an agonist that binds the primary binding site; Compound 2 is an antagonist that binds the primary binding site.

3.8 DAG + IP$_3$ = second messenger; Fz = GPCR; Wnt = agonist.

3.9 Molecule A

Lesson 3.3

3.10 −61.5 mV

3.11 The membrane potential becomes more positive (increases).

3.12 Channel A is ligand-gated; channel B is voltage-gated.

3.13 Bring ions into a cell: Either carriers or channels

Release ions from a cell: Either carriers or channels

Move a solute down its gradient: Either carriers or channels

Move a solute against its gradient: *Only* carriers

Allows continual flow: *Only* channels

Unit 2. Enzymes

Chapter 4. Enzymes Activity

Lesson 4.1

4.1 Conditions A and B have enzyme present. The transition state free energy is lower with Condition A; it is higher with Condition B.

4.2 There is no difference in H$_2$CO$_3$ concentration at equilibrium.

4.3 −16.7 kJ/mol

4.4 Reversible

Lesson 4.2

4.5 Enoyl-ACP reductase: NADPH is oxidized

Fumarase: No redox cofactor or coenzyme

α-Ketoglutarate dehydrogenase: NAD$^+$ is reduced

4.6 A) Hydrolase, phosphatase; B) transferase, kinase.

4.7 Enzyme A is phosphoglucose epimerase; enzyme B is phosphoglucose isomerase; enzyme C is phosphoglucomutase.

4.8 Without using; synthase; lyase.

Uses; synthetase; ligase.

Lesson 4.3

4.9 The transition state

4.10 Choice B, the K44L mutant.

4.11 Lock-and-key: A → B

Induced-fit: A → C → D → B

Conformational selection: A → D → B

4.12 B, Histidine.

4.13 Ping-pong. In the alternate scenario, no conclusion (the enzyme *either* forms a ternary complex *or* is a ping-pong enzyme that binds the given substrate second).

Lesson 4.4

4.14 The *E. coli* sample shows higher activity. Nothing can be said about specific activity.

4.15 Purity increased 7.5-fold. Percent yield is 75%.

4.16 Sample A is from the mesophile; Sample B is from the thermophile.

Chapter 5. Enzyme Kinetics

Lesson 5.1

5.1 Statement 1: Both enzymes can be described by Michaelis-Menten kinetics. The irreversibility assumption is only for *experimental* conditions.

Statement 2: [P] is not held steady.

Statement 3: Although [S] must be much higher than [E_{tot}], the enzyme does not have to be saturated.

5.2 100 μM/min

Lesson 5.2

5.3 A) III; B) II; C) IV; D) I.

Lesson 5.3

5.4 The graph shifts upward and has an increased initial slope. V_{max} is increased; K_M, k_{cat}, and catalytic efficiency are unchanged.

5.5 V_{max}, k_{cat}, and K_M are increased; catalytic efficiency is unchanged.

Lesson 5.4

5.6 Molecule 1

5.7 "Uncompetitive-like" mixed inhibitor

5.8 Molecule A is a "competitive-like" mixed inhibitor.

Chapter 6. Enzyme Regulation

Lesson 6.1

6.1 The molecule is an activator that decreases K_M and increases V_{max}.

6.2 Enzyme E

Unit 3. Carbohydrates, Nucleotides, and Lipids

Chapter 7. Carbohydrates

Lesson 7.1

7.1 D-Erythrose: Four carbons; anomeric carbon is at position 1 (aldotetrose).

D-Sedoheptulose: Seven carbons; anomeric carbon is at position 2 (ketoheptose).

7.2 Sugars B and D.

7.3

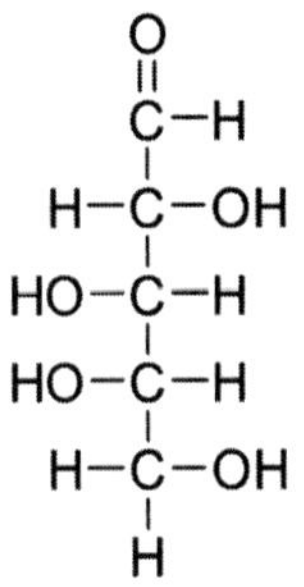

```
      O
      ‖
      C—H
      |
  H—C—OH
      |
 HO—C—H
      |
 HO—C—H
      |
  H—C—OH
      |
      H
```

7.4 D-Glucose and D-mannose.

7.5

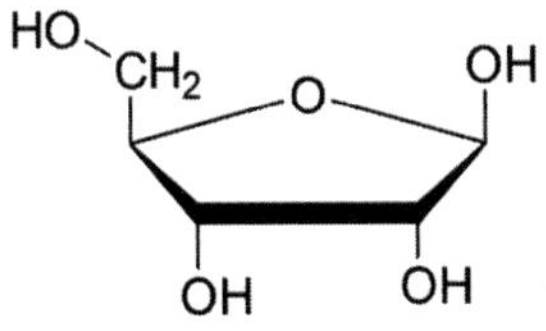

```
HO                 
   CH₂     OH
      \ O /
       ‾‾‾
      /   \
     OH    OH
```
; β anomer (β-D-ribopyranose).

7.6 Derivative 1: Ribose 5-phosphate;

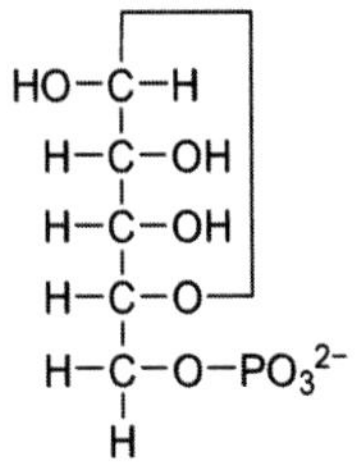

```
 HO—C—H
     |
  H—C—OH
     |
  H—C—OH
     |
  H—C—O
     |
  H—C—O—PO₃²⁻
     |
     H
```

Derivative 2: Mannose 1-phosphate;

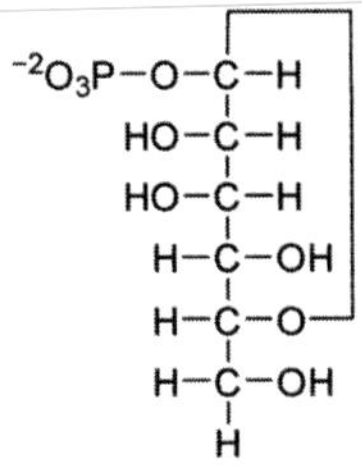

```
 ⁻²O₃P—O—C—H
        |
    HO—C—H
        |
    HO—C—H
        |
     H—C—OH
        |
     H—C—O
        |
     H—C—OH
        |
        H
```

Lesson 7.2

7.7 $C_{21}H_{36}O_{18}$

7.8   Bond 1 is an α-1,4 glycosidic linkage; bond 2 is a β-2,2 glycosidic linkage; bond 3 is a β-1,3 glycosidic linkage.

7.9   Nonreducing

7.10   The glucose unit

Chapter 8. Nucleotides and Nucleic Acids

Lesson 8.1

8.1   Position 1 = H-bond acceptor; Position 2 = H-bond donor; Position 6 = H-bond donor.

8.2   Nucleotide 1: Deoxyguanosine monophosphate (dGMP).

Nucleotide 2: Uridine diphosphate (UDP).

Lesson 8.2

8.3   To form CA, the 3′ hydroxyl group of CTP must attack the 5′ α-phosphate of ATP. To form CAU, the 3′ hydroxyl group of CA attacks the 5′ α-phosphate of UTP.

8.4   103

8.5   Sequence 1

Chapter 9. Lipids

Lesson 9.2

9.1   A) both; B) sphingolipids; C) glycerophospholipids; D) glycerophospholipids; E) sphingolipids.

9.2   1) A; 2) E.

9.3   The proportion of saturated, long-chain fatty acids will increase. The proportion of short, unsaturated fatty acids will decrease.

Lesson 9.3

9.4   Lipids can serve in *all* the given roles.

Unit 4. Metabolic Reactions

Chapter 10. Catabolism and Anabolism

Lesson 10.1

10.1   Four ATP units

Lesson 10.2

10.2   Inhibition

Chapter 11. Carbohydrate Metabolism

Lesson 11.1

11.1   1) 12 ATP molecules total; 2) 6 *net* ATP.

Lesson 11.2

11.2 4 ATP are hydrolyzed.

11.3 Liver cells: Glycolysis is inhibited, gluconeogenesis is stimulated.

Muscle cells: Glycolysis is (indirectly) stimulated; gluconeogenesis absent in muscle.

Lesson 11.3

11.4 1) Oxidative phase; 2) non-oxidative phase; 3) both phases.

11.5 5 Pyruvate, 3 CO_2, 5 NADH, 6 NADPH, 5 ATP.

Lesson 11.4

11.6 1) Eight; 2) 92.

11.7 1) 2 Pyruvate, 2 NADH, 1 ATP.

2) 2 Pyruvate, 2 NADH, 0 ATP.

11.8 1) All enzymes except glycogen synthase kinase (item D) are phosphorylated. All enzymes except glycogen synthase and phosphofructokinase-2 (items C and F, respectively) are active.

2) Only glycogen synthase kinase (item D) is phosphorylated. Only glycogen synthase and phosphofructokinase-2 (items C and F, respectively) are active.

Chapter 12. Aerobic Respiration

Lesson 12.1

12.1 1) Acetyl-CoA; 2) the carbonyl carbon.

12.2 4 NADH, 2 $FADH_2$, and 2 GTP.

Lesson 12.2

12.3 10 NADH molecules

12.4 The ubiquinone to ubiquinol ratio increases (ubiquinone accumulates).

Lesson 12.3

12.5 The strength of the proton motive force would increase.

12.6 The net yield per glucose would drop by 2 ATP.

12.7 Graph 1

Chapter 13. Noncarbohydrate Metabolism

Lesson 13.1

13.1 Two rounds of β-oxidation; three acetyl-CoA; two NADH; two $FADH_2$.

13.2 Molecule 1 produces more acetyl-CoA; molecule 2 produces more $FADH_2$.

13.3 Four glucose molecules

Lesson 13.2

13.4 Isoleucine

13.5 1) Four urea molecules; 2) three are glucogenic; 3) two are ketogenic.

Unit 5. Biochemistry Lab Techniques

Chapter 14. Biomolecule Purification and Characterization

Lesson 14.1

    14.1   7.5 kbp

    14.2   Gel 1

    14.3   Solution 1) B and C; Solution 2) C only.

    14.4   Protein B

Lesson 14.2

    14.5   10 µL of Sample A to its lane; 5 µL of Sample B (and 5 µL of buffer) to the Sample B lane.

Lesson 14.3

    14.6   Peak 1 is 75 kDa; Peak 2 is 50 kDa; Peak 3 is 20 kDa.

    14.7   Protein A would bind a cation-exchange column; Protein B would bind an anion-exchange column; Protein C would bind a cation-exchange column; Protein D would bind an anion-exchange column. Protein B would require the highest NaCl concentration to elute.

    14.8 G_2 phase

Lesson 14.4

    14.9   2 mM

    14.10   The 40-base primer has a higher A_{260} absorbance value (double to 20-base primer).

    14.11   6.0 mg/mL

    14.12   50 µM

    14.13   The T_m of Sample A is 30 °C; the T_m of Sample B is 40 °C. Sample B has a longer DNA sequence.

Index

A
α-1,6-glucosidase, 474
ABO blood types, 301, 386
absorbance, 590, 607–13, 618–21, 624
α-carbon, 4–13, 40, 157, 347, 522
ACC. *See* acetyl-CoA carboxylase
acetal, 296–97
acetaldehyde, 433–34
acetate, 434
acetoacetate, 528–29
acetoacetyl-CoA, 528–29, 547
acetylation, 85–86, 177
acetylcholine, 117–18
acetyl-CoA, 349, 361, 440, 444, 482–94, 515–36, 546–47
acetyl-CoA acetyltransferase, 528–29. *See also* acetyl-CoA acyltransferase
acetyl-CoA acyltransferase, 522–23
acetyl-CoA carboxylase (ACC), 531, 536–37
acetyl group, 85–86, 293, 482, 486, 531
acid-base chemistry, 20–27, 31–35, 71, 75, 185, 310–11, 317, 335–40
acidosis, 197
aconitase, 487
ACP. *See* acyl carrier protein
ACS. *See* acyl-CoA synthetase
activation energy (E_a), 38, 74, 127–29, 133–34, 167–68, 175, 180, 182, 195, 215
active site, 19, 25, 34, 175–83, 193, 233–37, 253–55, 524, 531
  microenvironment, 185, 193
acylcarnitine, 521
acyl carrier protein (ACP), 531–35
acyl-CoA, 520–23
acyl-CoA dehydrogenase, 521–22
acyl-CoA synthetase (ACS), 521
acyl group, 146, 349, 351, 354, 357, 376, 482, 521, 533
acyl-lipoate, 482
acyl phosphates, 402, 409, 423
acyltransferases, 145–46
adenine, 310–17, 337, 339, 342, 560
adenosine, 317–22
adenosine nucleotide, 320–21
adenosine triphosphate, 320, 324, 399. *See also* adenosine nucleotide
adenylyl cyclase, 107, 154–55, 475–76

ADP, 132, 146, 326–27, 400–402, 406–7, 409–10, 412, 414, 424–25, 436, 444–45, 487–89, 509, 512
ADP-ribosyl group, 143
adrenal cortex, 388
adrenaline, 447, 477
aerobic respiration, 400–401, 413, 432, 442–43, 481, 484
affinity (*see also* equilibrium constant, dissociation (K_d)), 92, 96–100, 104, 106, 209–11, 215, 221, 238–44
agarose, 553–54. *See also* gel electrophoresis
aggregation, 75–76, 346, 364
  of lipids, 346, 364
  of proteins, 75–76
agonist, 103–9, 16–66, 476
α-keto acids, 148, 540–41
α-ketoglutarate, 180, 487–88, 492, 494, 540–43, 548
α-ketoglutarate dehydrogenase, 488
alanine, 5–6, 8, 10, 16, 31, 41, 440, 541, 543, 546–47
albumin, 350–51, 389
alcohol dehydrogenase, 433–34
alcohols, 6, 19–20, 25, 143, 273, 281, 292–93, 350, 532
aldehyde, 143, 273, 276, 278, 292–93, 298–99, 423
aldolase, 136, 153–54, 422, 429, 438, 457
aldonic acids, 292
aldoses, 156, 276–78, 285, 292–93, 297–99, 428, 455, 458–60
aldosterone, 387
allosteric site, 97–98, 104–5, 233, 237, 244, 253, 256
allostery, 97–98, 237, 254–57, 266, 416, 444, 475, 477–78
  activation, 263, 382, 446
  inhibition, 444
alpha-carbon. *See* α-carbon
amides, 8–9, 19, 38, 85, 183, 313, 358, 361, 543, 625
amine, 293, 357, 543
amino acids,
  aliphatic, 16–17
  aromatic, 10–11, 16–18, 20

backbone, 4, 23–24, 27, 37, 43
catabolism, 546
charged, 21, 46, 49, 71, 170
   See also amino acids, ionizable
chiral, 13
digestion of, 540, 548
glucogenic, 493–94, 546–47
hydrophilic, 56, 58, 64, 66
hydrophobic, 56–66, 72, 75, 77
ionizable, 23–25, 27–29, 31–34, 43, 46, 193, 406–9
ketogenic, 546–48
metabolism, 548
neutral, 18–21
nonpolar, 15–17
nonproteinogenic, 4, 545
nucleophilic, 34, 137, 176
one-letter code, 5–11, 79
pK_a values, 28, 193
proteinogenic, 3–4, 7, 12, 15, 42, 546
side chains, 4–8, 10–11, 13, 15–21, 24–25, 28, 32–35, 43, 51, 56–57, 60, 65, 82, 85, 176–77, 571
stereochemistry, 12–13
three-letter codes, 5–11
amino acid titration, 30–33
aminoacyl-tRNA, 401
amino sugars, 293
ammonia, 542–45
AMP, 255, 320, 326, 405, 444–45, 477, 521, 545
AMP-activated protein kinase. See AMPK
amphiphiles, 347, 353, 361, 366–67, 517
AMPK (AMP-activated protein kinase), 447, 537
amyloid fibrils, 76
amylopectin, 305–6
amylose, 304–6
anabolic processes, 399, 413–14, 435, 517
anion-exchange chromatography. See chromatography, ion-exchange
annealing (of nucleic acids), 334–35, 583
anode, 555–56, 559, 561, 565, 569, 571–72, 586
anomeric carbon, 158, 276–79, 281, 283, 285, 287–89, 292, 295–99, 301–3, 306, 317
anomers, 158, 287–88, 304
antagonist, 104–5
antibodies, 90–92, 386, 393, 577, 587–88, 598, 600–601
   primary, 587
   secondary, 587
antigen, 91, 587, 600
antiport, 120, 512, 521
apoprotein, 87–88
apoptosis, 371, 504
aquaporins, 118
arabinose, 283
arachidonic acid, 384–85
arginine, 10, 17, 21–22, 24–25, 28, 41, 43, 46, 71–72, 80, 545–46
argininosuccinate, 545
Arrhenius equation, 127
asparagine, 8–9, 19, 35, 84, 546–47
aspartate, 8–9, 21, 24–25, 28, 43, 46, 72, 83, 89, 176, 498, 541, 545–47
ATP, 82, 131–32, 146–47, 150, 159–62, 255–56, 324, 326–27, 399–409, 411–13, 416, 419–20, 422–27, 436–38, 443–45, 468–69, 487–89, 509–10, 512–14, 545
   condensation. See ATP, synthesis
   equivalents, 403, 413, 435, 438, 468, 475, 530, 534, 545
   hydrolysis, 120, 132, 159, 162, 400–401, 406–12, 420–21, 510, 521, 526, 530
   synthesis, 324, 346, 402, 405, 415, 424–25, 442–43, 445, 481, 509–10, 512–17
ATP synthase, 162, 495, 509–15
autophosphorylation, 164–65, 478
autoradiography, 578–79

B

backbone, 4, 7, 15, 23, 30, 32, 39–40, 43, 49–52, 85, 167
   amino acid, 43
   nucleic acids, 85, 331–32, 341, 576, 578
   protein, 49, 85
   sphingolipids, 357
   sphingosine, 356
   steroid, 389
   sugar phosphate. See backbone, nucleic acids
β-actin, 580
base (nucleic acids), 310–11, 317, 335–36, 338–40. See also base pairs
base pairs, 312, 334–40, 560, 582–83, 615
BCAAs. See branched-chain amino acids
BCA assay. See biomolecule quantitation
β-hydroxyacyl-CoA dehydrogenase, 522
β-hydroxybutyrate, 528–29
bicarbonate, 526
bilayers. See membrane (cell), bilayers

bile salts, 362–65, 518
binding, 89–91, 93–94, 97–104, 162, 165, 168, 171, 200–201, 203, 230, 233–34, 239–40, 597–98, 600, 612–14
   affinity, 90, 92, 94, 209, 614
      See also equilibrium constant, dissociation (K_d)
   assays, 612–14
   cooperativity, 98, 100–101, 201, 230
      See also cooperativity
   curves, 95–96, 98, 100, 102, 217–18, 221, 613
   energy, 169, 175, 180–81
   equation, 95, 206
   interactions, 91, 98, 129, 562, 579, 598–600, 612–14
   pocket, 89, 91, 97–98, 172
   rate, 93–94, 200
1,3-bisphosphoglycerate, 409, 423–24, 438
bisubstrate enzymes, 141, 145, 179–80, 230
β-ketoacyl-CoA, 522–23
β-ketoacyl synthase, 531, 533
blood glucose, 444, 447–48, 536–37
bloodstream, 109, 350–51, 473, 475, 517–18, 527, 536
blotting, 575–77, 579–81, 584, 586–87, 589, 603, 607
   blocking, 577
   nonelectrophoretic (dot), 581–82
   northern blots, 577, 584–86
   Southern blots, 576, 582–85
   western blots, 575–78, 581, 586–88, 601
β-mercaptoethanol (BME). See reducing agents
bonds
   breaking, 406, 410
   forming, 406
   high-energy, 413, 489
bovine serum albumin (BSA), 577, 587, 610
β-oxidation, 349, 351, 417, 519–21, 523–28, 531, 536, 538
β-phosphate of ADP, 406–7, 409
Bradford assay. See biomolecule quantitation
branched-chain amino acids (BCAAs), 17, 548
branch points (carbohydrates), 304–6, 471, 475
brown adipose tissue, 515
BSA. See bovine serum albumin

buffer, 31–33, 378, 582–83, 585, 590, 593, 597, 603, 606, 610–11
bypass reactions, 131, 134, 415, 435–40

C

Cahn-Ingold-Prelog priority, 13, 281
calorimetry, 616
   differential scanning (DSC), 616–17
   isothermal titration (ITC), 614
cAMP (cyclic AMP), 154, 326–27, 451, 475
capillary transfer, 576, 582–83, 587
carbamoyl phosphate, 545, 548
carbohydrates, 84, 273–76, 278, 281, 285, 287, 295–301, 304, 307, 309, 316–18, 356, 358, 386, 399
   deoxy, 316
carbon-13, 624
carbon dioxide, 413, 432–33, 437, 442, 454–56, 482–83, 485, 487–89, 491, 493, 532
carbonyl, 38, 40, 85–86, 143, 273, 278, 311–13, 315, 347
carboxylate, 347
carboxylation, 436, 530–31
carboxylic acids, 4, 8–9, 23, 25, 27–28, 32, 37, 40–43, 143, 292, 347, 349, 436–37, 486–88
carboxy terminus. See C-terminus
carnitine acyltransferase I (carnitine palmitoyl transferase I), 521, 536–37
carnitine acyltransferase II (carnitine palmitoyl transferase II), 521
carnitine shuttle, 520–21, 536
carrier proteins, 109, 113, 119–21, 350, 369, 389, 509
catabolism, 389, 399–400, 413–14, 416, 419, 427, 435, 482, 517, 540, 547
catalase, 504
catalysts and catalysis, 127–29, 136, 161–63, 167, 173, 176, 178, 180, 185–87, 193, 200–203, 205, 253–54
catalytic efficiency, 209, 213–15, 220, 222–23, 225, 227–29, 241, 246–48, 265, 267
catalytic mechanisms, 35, 137, 140–41, 167, 185, 193, 197, 203, 235, 240–41, 324
catalytic perfection, 214. See also catalytic efficiency
catalytic triad, 176–77, 183, 185
cathode, 555–56, 561, 569, 571–72
CAT II. See carnitine acyltransferase II

cation-exchange chromatography. *See* chromatography, ion-exchange
CD. *See* circular dichroism
cellobiose, 302–3, 306
cells (biological), 103, 113, 116–21, 187–88, 195–96, 224, 346, 367–69, 373, 378–79, 399–400, 419–21, 432–33, 435, 442–44, 446, 462, 464–66, 504, 514–15
cells (electrolytic), 555, 586
cell signaling, 165. *See also* signal transduction
cellulose, 306–7
central dogma of molecular biology, 323–24
cGMP (cyclic GMP), 165
channels, 113–21, 509, 515
    gated, 118
    leak, 117
    ligand-gated, 117–18, 162
    mechanically gated, 118
    ungated, 117
    voltage-gated, 117–18
chaperones, 74, 76–77
Chargaff's rules, 337–40
charge-charge repulsion, 25, 406
chemical shift. *See* NMR
chemiluminescence, 578–79, 588
chemiosmosis, 510
chirality, 12–13, 157–58, 173, 274, 281, 285, 287–88, 350–51, 490
cholate, 363–64
cholesterol, 360–63, 371, 376, 378, 387, 389, 518
    cholesteryl esters, 361, 518
    metabolism, 361–62
chromatography, 192, 589–91, 593, 595, 598–601, 603
    affinity, 598–600, 603, 606
    column, 589–93, 595–600, 603
    gel filtration. *See* chromatography, size-exclusion
    ion-exchange, 593–99, 603
    mobile phase, 589, 592–93, 597, 599
    nickel-linked, 606
    size-exclusion, 191, 590–93, 599
    thin-layer (TLC), 581
circular dichroism, 55, 617, 620–22
circulatory system, 266, 351, 389
citrate, 255, 445, 449, 485–87, 489, 493, 530
citrate lyase, 530
citrate shuttle, 529–30, 534–36

citrate synthase, 486–88, 530
citric acid cycle, 160, 440, 481, 483, 485–87, 489–96, 498–99, 506–7, 510, 513–14, 523, 526–27, 529–30, 541, 544–46
citrulline, 545
clotting cascade, 263, 392
CO_2. *See* carbon dioxide
coenzyme A, 160, 185, 482–83, 486, 489
cofactors and coenzymes, 85, 87–88, 143, 159–60, 185–86, 197–98, 454, 482–83, 486, 488, 490–91, 503–4, 520–22
coIP (co-immunoprecipitation), 600–601
collision theory, 182
complex carbohydrates, 84, 289, 295–96, 298–300, 302, 330, 553
complexes
    of the electron transport chain, 162, 391, 403, 490, 495–99, 501–5, 507, 509–10, 522
    enzyme-inhibitor (EI), 238
    enzyme-product (EP), 168, 200
    enzyme-substrate (ES), 239, 241, 246
    enzyme-substrate-inhibitor (ESI), 239
    protein-ligand (PL), 92–93, 612
condensation reaction, 37–38, 130–32, 295, 300, 329–30, 486, 509, 530–31
conformational changes, 67–69, 91, 105–6, 116, 120, 170–72, 174, 255, 265, 267, 411–12, 622, 625
    binding-induced. *See* induced-fit model
conformational selection, 171–72, 174–75, 255
conformations, 39, 49, 56, 63, 66–69, 73–75, 171, 278, 511–12, 617–18, 625
Coomassie, 556–57, 575, 611
cooperativity, 97–101, 200, 206, 217, 230
Cori cycle, 433, 442–43
corticosteroids, 387–88
cortisol, 110, 387
coupled reactions, 131–32, 160, 413, 420, 444
covalent catalysis, 137, 183–84, 193, 234
covalent modifications, 263–64, 266, 445
    dynamic, 264, 266–67
    irreversible, 233
covalent regulation, 266–67, 445, 447, 449. *See also* covalent modifications
CPT II. *See* carnitine acyltransferase II
Creutzfeldt-Jakob disease. *See* prions
crude lysate, 188–89, 191–92, 575, 581, 609, 611
cryo-EM, 622–24, 626

C-terminus, 40–43, 56, 330, 599, 601
cyclization (of carbohydrates), 278–79, 287, 362
cysteine, 6, 13, 18–20, 22, 24–25, 28, 34, 43, 156, 546–47, 568–69
cystine, 568–69
cyt c (cytochrome c), 501–6
cytidine, 317
cytochrome P450 oxidase, 144
cytosine, 310–13, 315–17, 335, 337, 343, 560
cytoskeleton, 373
cytosol, 57–58, 117, 369–70, 432, 435, 438, 441–42, 445, 481, 496–98, 512–13, 529–30, 545

D

DAG. See diacylglycerol
deacetylation, 521
deamidation, 540, 543–44, 546
deamination, 315, 333, 440, 540, 542–43, 546
debranching (glycogen catabolism), 473–74
decarboxylation, 433, 482, 487–88, 530
dehydratases, 150, 154
dehydration reactions, 154, 425, 487
dehydrogenases, 144
dehydrogenation reactions. See redox reactions
deletions (mutations), 584
denaturation, 69–70, 72–75, 192–94, 342–43, 565, 582, 592, 615, 617, 622
 of nucleic acids, 343–44, 617
 of protein, 71, 73, 192
denaturing agents, 70, 72–74, 367, 563–64, 568, 571, 585, 592
deoxynucleotides (see also DNA), 291, 301, 309, 316–18, 320, 329, 331, 333, 456
2,3-dideoxyribose, 291
depolarization (membrane potential), 117–18
detergents, 366–67, 563

DHAP. See dihydroxyacetone phosphate
diacylglycerol (DAG), 109, 382
dialysis, 603–6
diastereomers, 13, 158, 283 dideoxysugars, 291
diffraction, 624
diffusion, 165, 214, 370, 385, 420, 509
digestion, 259, 263, 302, 363–65, 499, 539
dihydrolipoyl dehydrogenase, 482. See also pyruvate dehydrogenase complex
dihydrolipoyl transacetylase, 482. See also pyruvate dehydrogenase complex
 dihydroxyacetone phosphate (DHAP), 136, 153, 274–76, 422, 429–30, 438, 442, 500, 519
dilution, 610–11
dipole, 20
disaccharides, 295–96, 302
dissociation constant K_d. See equilibrium constant, dissociation (K_d)
disulfide bonds, 22, 43, 56–57, 60, 90, 156–57, 564–66, 568–70
dithiothreitol (DTT). See reducing agents
DNA, 85–86, 91, 310, 315–16, 318, 323, 331–32, 334–44, 553–55, 558–59, 561, 576, 582–85
 circularized, 559
 concentration measurements, 609
DNA polymerase, 332–33
double displacement enzyme reactions, 178–80. See also bisubstrate enzymes
double reciprocal plot. See Lineweaver-Burk plot
DSC. See calorimetry, differential scanning

E

E_a. See activation energy
eicosanoids, 384–85
electroblotting. See blotting
electrode, 555
electrolytes, 555
electronegativity, 15, 19, 357
electron microscopy, 623

electrons, 38, 141–42, 144, 391, 481–82, 488, 490–91, 495–507, 513, 619, 623
  delocalization, 38
  density, 34
  high-energy, 403–4, 424, 503, 623–24
electron shielding. See NMR (nuclear magnetic resonance)
electron transport chain, 403–4, 481, 483, 487, 490, 495, 497, 499, 504, 506, 509–10, 514–15, 522
electrophile, 34, 482
electrophoresis. See gel electrophoresis
elimination reactions, 151–52
ELISAs, 587
emulsifying agents, 363, 365
enantiomers, 157–58, 281
endergonic reactions, 130–31, 411–13
endocytosis, 540
endonuclease, 333
endoplasmic reticulum, 57, 84, 533, 540
energy currency molecules, 400–401, 403, 413–14, 424–25
enol, 409. See also tautomerization
enolase, 425
enoyl-CoA (intermediate of beta-oxidation), 522–24
enoyl-CoA isomerase, 523
enteropeptidase, 263
enthalpy, 130–31, 167, 175, 181, 614
entropy, 63–64, 168, 180, 345
enzymes,
  activators, 257
  active site, 175
  activity, 139, 187–95, 197, 199, 233, 236, 251–52, 255, 263, 265, 445
  bifunctional, 179
  binding, 170–72
  bypass, 440, 448
  catalysis, 127, 137
  classification, 139–41, 160–61
  classifying, 159
  complex, 531
  cooperative, 179, 210, 230–31
    See also cooperativity
  denaturation, 193
    See also denaturation
  digestive, 259
  expression, 251
  function, 197–98, 411, 590
  inhibitors, 233, 242
    See also inhibition
  isoform, 214
  kinetics, 199–200, 203, 217, 221, 225, 227
  multisubstrate. See bisubstrate enzymes
  mutation, 215
  rate, 197, 251
  regulation, 252, 260, 266, 326, 361, 456, 478, 480
  restriction, 561
  reversible, 253, 439
  stability, 193
  transmembrane, 500
  two-substrate. See bisubstrate enzymes
enzyme-substrate interactions, 167, 172, 193, 200
epimerases, 158
epimers, 158, 283–85, 287, 428, 430, 455
epinephrine, 447, 450, 477–79
epitope, 91–92, 587
equilibrium, 93, 113, 115, 127, 129, 134, 157, 167–68, 201–2, 241, 251–52
  acid-base, 193
  chemical, 130
  diffusive, 603
  electrochemical, 115
  substrate-binding (rapid), 238
equilibrium constant (K_{eq}), 92–94, 134, 346
  acid dissociation (K_a). See pK_a
  dissociation (K_d), 92–96, 98, 104, 206, 209, 234, 242, 612–14
  inhibitor dissociation (K_i), 242–43
exergonic reactions, 127, 129–31, 160, 402, 406, 408–10, 415, 420–22, 427, 467, 485
exonucleases, 333
exothermic reactions, 406
extracellular matrix, 373

F

FAD (flavin adenine dinucleotide), 88, 143, 185–86, 322–23, 325–26, 481–82, 490–93, 495, 498–500, 505–7, 509–10, 513–14, 522–23, 525, 527
Faraday's constant, 115
fasting, 413–14, 435, 442, 447, 450, 527, 529, 535, 537, 546–47
fatty acids, 347–50, 354, 357–58, 360–61, 366–67, 376, 383–85, 413–14, 417, 517, 519–21, 523–25, 529, 531, 535–37
  catabolism, 417, 450, 520, 529, 533, 536
  composition, 145, 347–49, 354, 376–77, 379, 384, 520, 525–26, 533

long-chain, 348, 354, 365, 444, 521
medium-chain, 348, 520
metabolism, 266, 517, 535–37
odd-chain, 348, 440, 525–26
oxidation. *See* fatty acids, catabolism
saturated, 348, 376–78, 523, 525
synthesis, 154, 179, 414, 417, 445, 449, 456, 463, 529–31, 534–38
unsaturated, 348, 377, 523, 525
fatty acid synthase, 179, 531, 533
FBPase-1. *See* fructose-1,6-bisphosphatase
FBPase-2. *See* fructose-2,6-bisphosphatase
FCCP. *See* ionophores
feedback, 251–52, 255, 258–60, 449, 530
   activation, 259, 445
   allosteric, 266
   inhibition, 252–54, 258–59, 263, 267, 444–45, 483, 487
   positive, 259–60
feedforward activation, 260–61
fermentation, 419, 432–33, 442–43, 481, 497, 506
   alcoholic, 433–34, 442
   lactic acid, 432–33, 440
FFA. *See* free fatty acids
fibrous proteins, 65–66
fight-or-flight, 477. *See also* epinephrine
final electron acceptor (oxygen), 495, 500, 505–6
Fischer projections, 274, 281, 283–86, 288, 291
flavin adenine dinucleotide. *See* FAD
flavin mononucleotide (FMN), 322–23, 326
fluid mosaic model, 370–74, 376
fluorescence, 343, 579, 588, 614, 617–19
FMN. *See* flavin mononucleotide
F_o domain, 511–12. *See also* ATP synthase
folding
   energetics, 63–64, 66
   funnel, 66–67
   protein, 54, 63–67, 69, 72, 74–75, 79, 84, 335, 345, 617, 626
   proteins, 49, 57, 59, 63, 66–67, 70, 73–77, 87, 336, 615, 617
free energy, 63, 66, 73, 94, 127, 129–35, 157, 160, 168–69, 180–81, 184, 406, 410–11
free energy change, 131, 134
free fatty acids (FFA), 350–51, 367, 518–19
free ligand approximation, 200, 203–4, 206, 612, 614
fructokinase, 429

fructose, 274–76, 279–81, 285–86, 290, 296, 302–3, 427–30
fructose 1-phosphate, 429
fructose 6-phosphate, 421, 428–29, 438–39, 445–46, 450, 458–59, 461
fructose-1,6-bisphosphatase (FBPase-1), 438, 445, 448–49
fructose 1,6-bisphosphate, 153, 260, 290, 421–22, 438, 445
fructose-2,6-bisphosphatase (FBPase-2), 445, 449–50, 480
fructose 2,6-bisphosphate, 255, 445–46
fructose catabolism, 429–30
fumarase, 154, 490
fumarate, 490, 499, 545
functional groups, 18, 20, 23, 49, 51, 145–46, 149, 151–53, 176, 182–83, 185, 281, 287, 345, 347
furanose, 278–80, 286, 302, 304, 316
futile cycles, 131, 416, 438, 444, 448–49, 475, 478–79, 536

G

GABA (γ-aminobutyric acid), 4
galactokinase, 430
galactose, 274–76, 279, 283–84, 303, 427, 430
galactose 1-phosphate, 430–31
galactose catabolism, 430–32, 467
galactose-1-phosphate uridylyltransferase (GALT), 430
GALE. *See* UDP-galactose 4-epimerase
GalNAc. *See* *N*-acetylgalactosamine
GALT. *See* galactose-1-phosphate uridylyltransferase
gangliosides, 358
GAPDH. *See* glyceraldehyde-3-phosphate dehydrogenase
gate (ion channels), 116–17
GC content, 343, 617
GDP, 106, 131, 160, 401, 405, 475, 491
gel electrophoresis, 553–56, 561, 563, 567, 571–72, 575–76, 579, 581, 585–86, 589–90, 592, 603, 607
   agarose, 555, 582–83, 585, 587
   bands, 556, 558–61, 569–70, 578–79
   isoelectric focusing, 571–72
   labels, 485, 578–79, 587, 614
   ladder, 558
   lanes, 558, 565, 607
   native gels, 559, 561–63, 565–66, 570, 573, 586

nonreducing gels, 569–70
polyacrylamide, 553–55, 561, 563, 571, 586–87
SDS-PAGE, 563–69, 572–73, 585–86, 590
genes, 48, 79, 85, 109, 391, 450, 580
GFP (green fluorescent protein), 618
GlcNAc. See N-acetylglucosamine
globular proteins, 65–66
glucagon, 447–50, 475–80, 537
glucagon signaling, 447, 449, 475, 477
glucocorticoids, 387
glucokinase, 224, 444
gluconeogenesis, 151, 153, 415–16, 435–45, 447–51, 465, 473, 476, 493, 526–27, 529–30, 543, 546, 548
precursors, 440, 442
glucosamine, 293
glucose, 132, 224, 273–76, 279, 281, 283–86, 288–90, 302–7, 399–401, 415–16, 419–20, 426–30, 435–36, 438–40, 442–44, 464–75, 477, 491, 512–14, 526–27
metabolism, 110, 400, 419, 435, 440, 444
glucose 1-phosphate, 147, 430–31, 466–68, 472–75
glucose 6-phosphate, 132, 289–90, 420–21, 428, 431, 444, 446, 454–55, 466, 468–69, 473, 477, 480
glucose-alanine cycle, 543
glucose-6-phosphatase, 439–40, 473
glucose-6-phosphate dehydrogenase, 454
glucose-6-phosphate isomerase, 421
glucose transporters, 120, 224, 420, 439
glucuronic acid, 292
GLUT2. See glucose transporters
glutamate, 8–9, 21, 24–25, 28, 31, 41, 43, 71–72, 80, 83, 89, 229, 540–44, 546
glutamine, 9, 19, 56, 79, 543–44, 546
glutathione, 504
glycans, 296–97, 308, 567
glyceraldehyde, 12–13, 274–76, 281, 284
glyceraldehyde 3-phosphate, 153, 422–23, 426, 429–30, 438, 457–59, 461
glyceraldehyde-3-phosphate dehydrogenase (GAPDH), 423–24, 438, 441, 580
glycerol, 293, 347, 350–51, 354, 356–57, 365, 382, 442, 519, 533
glycerol 3-phosphate, 442, 500, 519
glycerol-3-phosphate dehydrogenase, 500
glycerol-3-phosphate shuttle, 498, 500–501, 514
glycerophospholipids, 353–58, 360–61, 376–77, 381–82, 386, 519, 533
glycine, 5, 9, 12, 16, 51, 53, 546–47
glycogen, 306, 346, 413–14, 465–75
glycogen-branching enzyme, 471
glycogenesis, 258, 266, 306, 324, 431, 449, 465–67, 469, 473, 475, 478–80
glycogenin, 306, 469–71
glycogenolysis, 306, 450, 465, 472–79
regulation, 475
glycogen phosphorylase, 139, 147–48, 151, 265, 472–77, 479–80
glycogen synthase, 191–92, 265, 467–72, 475, 478–80
glycogen synthase kinase, 479–80
glycogen synthesis. See glycogenesis
glycolipids, 301, 307, 356, 358, 386, 427
glycolysis, 153, 156, 255–56, 258, 260, 266–67, 415–16, 419–32, 435–40, 442–51, 453–57, 462, 466, 473–76, 481, 497–500, 513–14, 529–30, 543
energy-investment phase, 419, 421–24, 426
energy-payoff phase, 423–24, 426, 430
irreversible enzymes, 436, 438, 440
regulation, 255, 260, 421–22, 430, 444–46
glycoproteins, 292, 301, 307, 427
glycosides, 301
glycosidic bonds, 146, 295–302, 306–7, 317, 322, 333, 356, 358, 371, 466–67, 474
α-1,4, 297–98, 302, 304, 306, 467, 472
α-1,6-, 305–6, 474–75
α-1,β-2-, 302
β-1,4-, 302–3, 306
β-2,6-, 297–98
formation, 295–96, 300
glycosphingolipids, 358, 386
glycosylase, 333
glycosylation, 35, 84, 146, 301
glycosylphosphatidylinositol (GPI). See lipid anchors
glycosyltransferase, 84
γ-phosphate (ATP), 264, 400–401
G protein–coupled receptors (GPCRs), 106–7, 109, 162, 311, 326–27, 447, 475–78
gradients, 46, 119–20, 162, 405, 412, 495, 500, 504, 509–10, 515, 571
concentration, 114–15, 497, 509

electrochemical, 115, 119–21, 162, 404–5, 411–13, 509–10
green fluorescent protein (GFP), 618
GTP, 106, 131, 160, 165, 324, 327, 401, 405, 437–38, 489–93, 513–14
guanidinium group, 10
guanidinyl phosphate group, 402
guanine, 310–15, 317, 337–38, 342, 560
guanosine, 317
guanosine nucleotides, 106. See also GTP
guanylyl cyclases, 162, 165

H

Haworth projection, 285–88, 298–99, 302, 306–7
head groups (lipids), 354–59, 383, 386, 563
heat, 71, 167, 406, 515, 587, 614, 616, 619
heat capacity, 616
heat shock proteins, 76
heme, 87, 502, 505
hemiacetal, 278, 295–96
hemodialysis, 605–6
hemoglobin, 60, 90, 99, 230, 612–13
hemolytic anemia, 456
Henderson-Hasselbalch equation, 28–29
hexokinase, 82, 224–25, 258, 420–21, 428–29, 435, 439–40, 444, 466, 468–69, 474
high-energy electron beam, 623
high-energy molecules, 402, 409, 491
Hill coefficient, 99–100, 102, 230. See also cooperativity
His-tagged protein, 606
histidine, 11, 18–22, 24, 28–31, 33, 35, 43, 46, 185, 546, 600
HMG-CoA reductase, 361
holoprotein, 87–88
hormones, 387, 445, 447
housekeeping genes, 580–81
hydratases, 150, 154
hydration, 154, 490, 521–22
hydrocarbon tail, 86, 347–48, 376–77, 390
hydrogen bonds, 15, 19–20, 47, 49–51, 54, 56, 60, 63, 311, 314, 335–36, 341–42
  acceptors, 311, 315
  donors, 15, 311, 314–15, 350, 361
hydrolases, 146–47, 149–51, 154, 159, 161–62, 333, 383, 490
hydrolysis, 38, 129, 131–32, 146, 149–51, 154, 160, 162, 183, 333, 400–402, 406–11, 472, 474
hydrophobic core (of proteins), 63, 71–72

hydrophobic interactions, 56, 63–65, 71–72, 167, 180, 335, 345, 353, 563
hydroxyl group, 6, 17, 20, 281, 283–84, 286–89, 291, 293, 295, 329–30, 354, 357–58, 361, 487–88, 522

I

IEF. See isoelectric focusing
IMFs. See intermolecular forces
imidazole, 11, 600, 606
immune system, 386
immunofluorescence, 587
immunohistochemistry, 587
immunoprecipitations, 587, 600
indole group, 11
induced-fit model, 171–72, 174–75, 237, 255
inhibition, 108, 253, 266, 361, 417, 449, 456, 479, 483, 537
  classes, 244–45
  competitive, 237, 239, 242, 245
  irreversible, 233–37, 264
  mixed, 233, 237, 241–43, 246–48, 256
  noncompetitive, 237, 243, 246–47
  reversible, 233, 237, 245
  uncompetitive, 237, 239–42, 246–47
inhibitor binding, 239, 242–43
inhibitors, 4, 235, 237, 239, 242–45, 252, 254, 257, 264, 434, 599
  feedback, 254
  mechanism-based, 235
  noncovalent, 234
  reversible, 207, 233, 235, 237, 244, 249, 254
  suicide, 235
inner mitochondrial membrane, 391, 403–4, 481, 490, 495–501, 504, 507, 509, 515, 529, 536
inorganic phosphate group (P_i), 402, 423, 512
inositol 1,4,5-trisphosphate (IP_3), 109, 382
insulin, 108, 110, 417, 447–49, 478–80, 484, 536–37
intensive properties, 188
intermolecular forces (IMFs), 71, 167, 366, 376, 615
ion channels. See channels
ionophores, 113, 514
IP_3. See inositol 1,4,5-trisphosphate
irreversibility assumption, 201, 204
irreversible enzymes, 204, 251, 253, 258, 260, 435

irreversible reactions, 134–35, 252, 258, 415–16, 438–39, 444, 485
isocitrate, 487–88, 492
isocitrate dehydrogenase, 487–88
isoelectric focusing, 46, 571–73
isoelectric point (pI), 29–31, 44–46, 132, 160–61, 355–56, 382, 400–401, 405, 412, 414, 512–13, 561, 571–73, 593, 596
isoleucine, 7, 13, 16–17, 546–47
isomerases, 155–56, 158, 487
isomerization, 155–58, 285, 290, 390, 421–22, 425, 428–29, 487
isomers, 7, 155–57, 173, 274, 281–82
   configurational, 155, 157
   conformational, 155–56, 281
isoprenoids, 362, 387
isozymes, 199, 213, 221, 223–24, 500
ITC. *See* calorimetry, isothermal titration

K

k_{cat}, 200, 206, 209, 211–15, 220, 222, 225, 229, 241, 246, 253, 255, 265
K_d. *See* equilibrium constant, dissociation (K_d)
ketal, 296–97
ketogenic amino acids. *See* amino acids, ketogenic
ketone bodies, 143, 273, 276, 488, 491, 522, 527–29, 532, 547–48
ketoses, 156, 276, 278, 285, 293, 297, 299–300, 428
kinases, 82, 139, 146, 150–51, 162, 264–67, 410, 430
kinetics, 129, 167, 199, 223–24, 257. *See also* Michaelis-Menten kinetics
K_M (Michaelis constant), 199, 205–7, 209–11, 213–14, 219–21, 223, 225–30, 235, 238, 240–47, 255, 257
k_{off}. *See* rate constant $k_{unbinding}$
k_{on}. *See* rate constant $k_{binding}$
Krebs cycle. *See* citric acid cycle
KS (β-ketoacyl synthase), 531, 533
K_t, 207

L

lactase, 303
lactate, 432–33, 440–43, 481
lactate dehydrogenase, 432–33. *See also* fermentation, lactic acid
lactim, 313–14
lactose, 302–4

lateral diffusion. *See* fluid mosaic model
leaflets, 368, 371, 381
   inner, 370–71, 521
   outer, 370–71, 381, 386, 501
   See also membrane (cell)
le Châtelier's principle, 134, 202, 238, 240, 251
leucine, 7, 10, 16–17, 79, 177, 546–48
leukotrienes, 384
ligands, 89–105, 117, 162, 165, 167–68, 230, 562–63, 590, 612–14, 622, 625–26
   optimal, 168, 170, 174, 234
   See also transition state analog
ligases, 148, 150, 159–61, 400, 490
light, 103, 390, 590, 607–9, 619–21
   plane-polarized, 13, 620
   visible, 607, 619, 622
Lineweaver-Burk equation, 225
Lineweaver-Burk plot, 217, 225–28, 230, 245–49
lipid anchors, 356, 421
lipidation, 86, 385
lipid bilayers. *See* membrane (cell), bilayer
lipid droplets, 345–46, 350–51, 353, 364, 367
lipid hormones, 389
lipid metabolism, 361, 456, 517
lipid modifications, 385
   glycosylation, 301
lipid rafts, 376, 385
lipids, 295, 301, 345–46, 350–51, 353–57, 361, 371, 376, 381, 391–93, 399, 517
   amphiphilic, 353, 371
   dietary, 518
   emulsification, 365
   energy storage, 304, 345–47, 350, 399, 413
   hydrolyzable, 361, 381, 383
   mobilization, 351
   nonhydrolyzable, 361, 392
   signaling, 345, 371, 381–82, 388, 390
   structural, 345, 353, 365–66, 382
   transport, 518
lipid tails, 369
lipid transport, 517
lipoic acid, 186, 482
lipoproteins, 351, 361, 367, 517–19
   chylomicron remnants, 518
   chylomicrons, 351, 518
   high-density (HDL), 351, 518
   intermediate-density (IDL), 518

low-density (LDL), 351, 518
very-low-density (VLDL), 518
liver, 435, 440, 442–45, 447–50, 473, 476–77, 480, 517–18, 527, 529, 542–46
   gluconeogenesis, 465
   glycolysis, 447
λ_{max}, 607–8, 610, 613, 617, 624
loading controls, 579–81, 607
lock-and-key, 170, 172, 174
London dispersion forces, 64, 345, 376
lyases, 150–55, 161–62, 422, 425, 487, 490, 522
lymphatic system, 518
lysates (protein purification), 191
lysine, 10, 21–22, 24–25, 28–29, 31, 34, 71–72, 80, 85–86, 92, 546–48
   acetylation, 85–86
lysosomes, 43, 57, 84, 195, 540

M
macrophages, 381
mad cow disease, 76
malate, 441, 490–91, 498, 530
malate-aspartate shuttle, 441, 497–99, 501, 514, 530
malate dehydrogenase, 491, 498
malate shuttle, 441. *See also* malate-aspartate shuttle
malate transporters, 498
malonyl-CoA, 531, 533–34, 536–37
maltose, 302, 304, 306
mannose, 274–76, 279, 283–84, 427–28
mannose 6-phosphate, 428
mannose catabolism, 428–29
MAP kinase pathway, 107
maximal reaction velocity. *See* V_{max}
melting temperature (T_m), 73, 342–43, 614–17
   assays, 74, 343, 614, 616–18
membrane (blotting), 575–78, 581–83, 586–87

membrane (cell), 113–21, 161–62, 165, 353, 356, 368–74, 376, 378–79, 385, 404, 504, 509, 575–78, 582–83, 586–87, 603–4
   asymmetry, 371
   bilayer, 58, 105, 161–62, 353, 361–62, 368–72, 375–76, 381, 389, 509
   channels, 118
   cholesterol, 361
   composition, 376, 378–79
   fluidity, 354, 361, 376–78
   glycolipids, 356, 386
   lipids, 354, 360, 364–65, 371–72, 378, 381–83
   microdomains. *See* lipid rafts
   mitochondrial, 500, 521
   monolayers, 353, 361, 367–68, 518
   phospholipids, 376
   plasma, 381, 386, 389
   semipermeable, 369
membrane (dialysis), 603–4
membrane potential, 114–16, 409, 509
   resting, 115–17
membrane proteins, 113, 162, 356, 370–72
membrane receptors, 362, 389
membrane transporters, 113, 207
messenger RNA (mRNA), 323–24, 585
metabolic pathways, 134, 142, 252, 258, 260–61, 267, 320, 324–25, 399, 416, 419
   opposed, 535
metabolism, 274, 323, 399, 444, 519, 548, 553
   anabolic, 399, 413–14
   of lipids, 517
   nonglucose, 427
metabolites, 136, 258, 260, 416, 421, 427, 437, 442, 445, 483, 491
methionine, 9, 15–16, 546
methylation (of cytosine in nucleic acids), 315–16
methylmalonyl-CoA, 526
micelles, 350, 353, 366–67
Michaelis constant. *See* K_M

Michaelis-Menten kinetics, 200–204, 207, 209–10, 217, 219–20, 224–25
   kinetic parameters, 199–200, 213, 215, 217, 226
Michaelis-Menten plot, 217–27, 230, 245–48
microenvironment, 176–77, 183
minerals (dietary), 198
mitochondria, 414, 432, 437, 441–42, 481, 485, 488–89, 498, 520–21, 525, 529–30, 534, 536–37
mitochondrial intermembrane space, 495, 497, 500–501, 504, 507, 509, 512–13, 521
mitochondrial matrix, 441, 445, 481–82, 495–98, 500, 509, 511–15, 520–21, 530, 535–36, 545
mitochondrial membranes, 481, 495, 530
mobility shifts (electrophoresis), 562–63, 567–68
molecular geometry
  planar, 38
  tetrahedral, 12
monolayers. See membrane (cell), monolayers
monosaccharides, 158, 273–83, 285–86, 289–93, 295–97, 302, 306, 358, 371, 421, 427
muscles, 117–18, 412, 432, 442, 444–46, 450–51, 465, 473, 475, 477, 542–44
mutarotases, 158
mutarotation, 158, 287–88, 296
mutase, 156
mutations, 48, 69, 74, 79–81, 83, 89–90, 92, 170, 215, 313, 316
Myc tags, 599–601
myelin sheaths, 358

N

N-acetylglucosamine (GlcNAc), 293
NAD^+ and NADH, 142–43, 185–86, 321–22, 325–26, 403, 416, 419, 423–24, 426–27, 432–36, 438, 441–43, 481–83, 487–88, 491–93, 495–501, 505–7, 513–14, 522–23, 527–29
$NADP^+$ and NADPH, 197, 321–22, 325–26, 361, 453–56, 462–64, 504, 524–25, 530, 532–35, 542
NDP. See nucleoside diphosphate
Nernst equation, 115
nickel (protein purification), 600

nicotinamide adenine dinucleotide. See NAD^+ and NADH
nitrocellulose. See membrane (blotting)
nitrogenous bases, 295, 301, 309–11, 313–17, 321–22, 333, 335, 341, 401, 559–60
N-linked glycosylation, 84
NMP. See nucleoside monophosphate
NMR (nuclear magnetic resonance), 622, 624–26
nonalcoholic fatty liver disease (NAFLD), 430
nonreducing sugars, 298–300, 302–3, 306, 330, 466–68, 471–73, 569
NSAIDs, 384
N-terminus, 40–43, 56–57, 85
NTP. See nucleoside triphosphate
nuclear decay, 578–79
nuclear magnetic resonance. See NMR
nucleic acids, 310, 329–34, 337, 340, 345, 557, 559, 575–77, 580, 584, 607–9, 614, 617
  synthesis, 329, 332, 413
nucleophiles, 34–35, 82, 85, 176–77, 183, 289, 301
nucleoside diphosphate (NDP), 160, 318, 320, 402
nucleoside monophosphate (NMP), 154, 160, 318, 320, 402
nucleosides, 295, 316–18
nucleoside triphosphate (NTP), 130–32, 146, 148, 150–51, 159–60, 318, 320, 324, 327, 329, 331, 400–402, 406, 413–14
nucleotides, 148, 274, 309–10, 313, 316, 318–27, 329–34, 371, 401, 406, 453
  function, 323
nucleotidyl cyclases, 154

O

oligonucleotides, 331
oligosaccharide, 296, 356, 471
O-linked glycosylation, 84
omega-carbon, 347
ornithine, 545
osmolarity, 346, 465
ovaries, 387
oxaloacetate, 180, 437, 440–41, 485–86, 489, 491–94, 498, 526, 530, 541, 545–47
oxidase, 144, 505

oxidative damage. *See* reactive oxygen species (ROS)
oxidative decarboxylation, 454–55, 487–88
oxidative phosphorylation, 404, 423, 509
oxidative stress, 504
oxidizing agent, 141–42
oxidoreductases, 141–44, 151, 156, 162, 197, 454, 488, 490–91, 501
oxygen, 15, 18–20, 276, 279, 301, 311, 432–33, 481, 495, 497, 504–7, 510, 515–16

P

palmitate, 385, 533–34
  synthesis, 534–35
palmitoylation, 385
pancreas, 224, 536–37
paratope, 91
partial pressure, 95
PCR (polymerase chain reaction), 343, 584–85
PDC. *See* pyruvate dehydrogenase complex
PDK. *See* pyruvate dehydrogenase kinase
pentose phosphate pathway (PPP), 149, 252, 258, 274, 276, 292, 420, 453–57, 461–64, 466, 504
  nonoxidative phase, 455–64
  oxidative phase, 149, 454–56, 462–64, 504, 535
pentose phosphates, 453–59, 461–64
PEPCK. *See* phosphoenolpyruvate carboxykinase
pepsin, 149, 259, 263, 539
pepsinogen, 263
peptide bond, 37–42, 48–49, 53, 79, 85–86, 127, 129–31, 149, 157, 183, 621, 625
  formation, 37–38, 48, 130–31, 177, 295, 329, 401, 405, 413
  hydrolysis, 38, 42–43, 127, 131, 138, 150, 172
  isopeptide, 85, 177
peptide bonds, double bond character, 38–39, 157
peptidyl-prolyl isomerase, 157
peptidyl-tRNA, 405
PFK1. *See* phosphofructokinase-1
PFK-2. *See* phosphofructokinase-2
phenols, 11, 20
phenylalanine, 10–11, 17–18, 546–47
phosphatases, 139, 150–51, 179, 264–67, 479–80

phosphate groups, 82–83, 139–40, 147, 150–51, 264, 289–91, 318, 320–21, 327, 329–31, 341–42, 354–55, 358, 402, 424–25, 436–39, 466, 468, 479–80, 489
  transfer reactions, 35, 438
phosphatidic acid (PA), 355
phosphatidylcholine (PC), 355, 371
phosphatidylglycerol (PG), 355
phosphatidylinositol (PI), 355–56, 382
phosphatidylinositol 3-kinase (PI3K), 382
phosphatidylinositol 3,4,5-trisphosphate (PIP$_3$), 117, 119, 382, 412
phosphatidylinositol 4,5-bisphosphate (PIP$_2$), 382–83
phosphatidylserine (PS), 355, 371, 381
phosphoanhydride bond, 318, 406–7, 466
phosphocholine, 358, 489
phosphodiesterase (PDE), 151
phosphoenolpyruvate (PEP), 409, 415, 425–26, 436, 438, 440–41, 546
phosphoenolpyruvate carboxykinase (PEPCK), 108, 110, 415–16, 437–38, 440–42, 546
phosphofructokinase-1 (PFK1), 255–56, 258, 260, 326, 421–22, 427, 430, 435, 438–40, 444–46, 448–49
phosphofructokinase-2 (PFK-2), 108, 267, 445–46, 448–50, 480
phosphoglucomutase, 159, 431, 466, 473
6-phosphogluconate dehydrogenase, 454–55
phosphoglucose epimerase, 159
phosphoglucose isomerase, 159, 421, 428, 439, 446
2-phosphoglycerate, 425
3-phosphoglycerate, 292, 424, 438
phosphoglycerate kinase, 147, 415, 424, 426–27, 438
phosphoglycerate mutase, 156, 425
phospholipase C (PLC), 382–83
phospholipid bilayers. *See* membrane (cell), bilayer
phospholipid monolayer. *See* membrane (cell), monolayers
phospholipids, 58, 293, 353–55, 358, 361, 367–68, 370–71, 375–76, 378, 381, 517
phosphomannose isomerase, 428
phosphoprotein phosphatase-1 (PP1), 447, 478
phosphorolysis, 147, 472, 475

phosphorylase. *See* glycogen phosphorylase
phosphorylase kinase, 476–77, 479–80
phosphorylation, 82–83, 107–8, 146–47, 264–67, 289–90, 327, 410–11, 420–21, 445, 447–50, 475–80, 483, 536
phosphotyrosine, 82, 163–64, 478
photons, 579, 618–19, 624
physiological conditions, 8, 18–22, 25, 38, 49, 187, 195, 224, 421, 426, 466
pI. *See* isoelectric point (pI)
P_i. *See* inorganic phosphate group
PI3K. *See* phosphatidylinositol 3-kinase
pi bond, 38–39, 154, 490
ping-pong. *See* double displacement
pi orbitals, 336
PIP_2. *See* phosphatidylinositol 4,5-bisphosphate
PIP_3. *See* phosphatidylinositol 3,4,5-trisphosphate
pi stacking, 336
pK_a, 27–33, 43–46, 176, 266, 447, 449, 451, 475–79, 571
  values for free amino acids, 176, 571
PKC. *See* protein kinase C
PKG. *See* protein kinase G
pmf. *See* proton motive force
polarimetry, 13
polarized light, 620–21
polyacrylamide gel electrophoresis (PAGE). *See* gel electrophoresis, polyacrylamide
polymerase chain reaction. *See* PCR
polymerases, 148–49
polypeptides, 37, 42–43, 47, 55, 59, 90, 127, 295, 540
polysaccharides, 145, 289, 293, 295–96, 304, 307, 331, 371, 465, 470
polyubiquitin tags, 85. *See also* ubiquitin
porphyrin, 87, 502
post-translational modifications (PTMs), 35, 69, 82, 84, 86–87, 91, 96, 251, 258, 263–64, 567–68
PP1. *See* phosphoprotein phosphatase-1
PP_i. *See* pyrophosphate
PPP. *See* pentose phosphate pathway
prions, 76
probes, 577–79, 582–83, 587, 625
proline, 7–8, 10, 16, 39–40, 51, 53, 66, 157, 546
propionyl-CoA, 525–26
propionyl-CoA carboxylase, 526

proproteins, 86, 263
prostaglandins, 384
prosthetic groups, 87–89, 197, 490, 499, 522, 531
proteases, 38, 129, 138, 149–51, 172, 504, 539
proteasome, 38, 85, 540
protein activity, 564, 590
protein catabolism, 85, 233, 539, 546
protein digestion. *See* protein catabolism
protein disulfide isomerase, 156
protein domains, 57, 59
protein kinase A (PKA), 266, 447, 449, 451, 475–79
protein kinase C (PKC), 382
protein kinase G (PKG), 165
protein kinases, 163, 266
  cAMP-dependent. *See* protein kinase A (PKA)
protein purification, 188–89, 191–92, 553, 589–91, 595–99, 603–6, 609, 614, 623
protein quantitation, 610
proteins
  regulatory, 476, 478
  transmembrane, 110, 113, 116, 119
protein stains, 610
protein structure, 47, 49, 65, 170, 340, 561–62, 622, 626
  α-helices, 49–56, 66, 621–22
  atomic resolution, 624
  β-sheets, 49–50, 52–57, 621–22
  β-turns, 53, 55
  primary, 48–49, 53–54, 56, 58, 65, 69, 334, 624, 626
  quaternary, 47, 59–61, 65–66, 75, 90–91, 98, 192, 562–64, 568, 592, 626
  random coil, 622
  secondary, 49, 52, 55–56, 621–22
  tertiary, 49, 55–57, 59–60, 63–66, 71, 98, 176, 336
protein synthesis, 324, 405
proteolysis, 86, 264
proton gradient, 404, 414, 495, 514–15
proton motive force (pmf), 404, 509–10
proton pumping, 403, 509
protons, 4, 8, 21–23, 25, 28, 32, 176, 313, 497, 500–501, 504–5, 507, 509, 511–15, 571
PTMs. *See* post-translational modifications
purines, 310–11, 317, 335, 337
PVDF (blotting membrane), 576–77, 586
pyranose, 278–80, 302, 304

pyrimidines, 310–13, 317, 335, 337
pyrophosphate (PPi), 148, 154, 160, 165, 329, 402, 405, 466, 468, 475, 521
pyruvate, 409–10, 415–16, 419, 426–27, 432–41, 443, 464, 481–83, 485, 488–89, 491–92, 513–15, 541, 543, 546–47
pyruvate carboxylase, 415–16, 436–37, 440, 530
pyruvate decarboxylase, 433–34
pyruvate decarboxylation, 482, 529
pyruvate dehydrogenase (PDH), 186. *See also* pyruvate dehydrogenase complex (PDC)
pyruvate dehydrogenase complex (PDC), 185–86, 482–85, 491, 546
pyruvate dehydrogenase kinase (PDK), 483–84
pyruvate dehydrogenase phosphatase, 484
pyruvate kinase, 147, 258, 260, 267, 415–16, 425–27, 435–38, 440, 444

Q

Q and QH_2. *See* ubiquinone and ubiquinol (UQ) (UQH_2)
quantitation (of biomolecules), 190, 580, 607–8, 610–11

R

racemases, 157
rapid equilibrium, 201–2, 240
rate constants, 94, 127, 200–201, 212–13
  catalytic, 212
  $k_{binding}$, 93–94
  $k_{unbinding}$, 93–94
rate-determining step (RDS), 200–201, 205, 454, 475–76
rate law, 205–6, 212, 241
reaction quotient, 134, 136, 433
reactions
  multistep, 130
  reversible, 134–35, 421, 438, 487, 490
  spontaneous, 127
  uncatalyzed, 168, 184–85
  unfavorable, 150, 309
reaction velocity. *See* V_0
reactive oxygen species (ROS), 419, 456, 504
receptor enzymes, 162–63, 165
  guanylyl cyclases, 165–66
  tyrosine kinases (RTKs), 163–65, 478
receptors, 91, 103–9, 117, 162, 165, 389, 391, 478, 536
  β-adrenergic, 477
  cytosolic, 109–11
  nonmembrane, 163
  nuclear, 391
  transmembrane, 105, 109–10, 371
red blood cells, 224, 432, 442–44, 456
redox cofactors and coenzymes, 57, 141–43, 145, 159, 162, 391–93, 453, 456, 499, 501, 504, 507, 568–70
redox reactions, 57, 141, 143–44, 162, 291–93, 299–300, 325–26, 399–400, 403–4, 432–33, 456, 490, 495, 500–501, 520–23, 530–33, 555
reducing agents, 57, 568–69
reducing and nonreducing gels, 513, 569–70
reducing and nonreducing sugars, 298–99
reduction, 144, 156, 488
regulation, 233, 251–52, 258–60, 266, 416, 421–22, 427, 444, 475, 477, 484–85
  allosteric, 162, 254, 263, 484
  covalent modifications, 263
  of DNA expression, 315
  of fatty acid metabolism, 535–36
  feedforward, 260
  of glucose metabolism, 435, 444, 447
  of glycogen metabolism, 475, 478
  hormonal, 267
  lipid modification, 385
  metabolic, 416
  of osmotic pressure, 118
  of proteins, 327
  reciprocal, 416
regulatory mechanisms, 237, 427, 444, 446–47
resonance, 20, 38, 157, 406, 408–9
rhodopsin, 390
ribitol, 293, 322
ribonucleotide reductase, 144
ribose, 274–76, 279, 281, 284, 288, 291, 296, 301, 309, 316–18, 322, 331, 333
ribose 5-phosphate, 455, 460
ribose-5-phosphate isomerase, 455
ribosomes, 264, 323–24, 401, 405
ribulose, 274–76
ribulose 5-phosphate (Ru5P), 455
RNA, 309–10, 313, 315, 317–18, 321, 323–24, 329, 331–32, 336, 339, 553, 558–59, 575–76, 584–85, 608
  blot, 584
  folded, 559
  gels, 557

RNA polymerases, 332, 339
ROS. *See* reactive oxygen species
RTKs. *See* receptor enzymes, tyrosine kinases
Ru5P. *See* ribulose 5-phosphate

S

salt bridges, 56, 60, 72, 192–93
salt concentration, 69–71, 192–93, 195, 340, 593, 597–98, 603
saponification, 367
SDH. *See* succinate dehydrogenase
SDS (sodium dodecyl sulfate), 72–73, 367, 563–65, 568, 571–72, 585–86
SDS-PAGE. *See* gel electrophoresis, SDS-PAGE
secondary structure of proteins, 50–51, 57, 65–66, 69, 98, 334, 336, 340, 584, 620–21
second messengers, 105–7, 109, 165, 346, 382
self-phosphorylation. *See* autophosphorylation
semiquinone, 503–4
separation of biomolecules, 564, 575
serine, 6, 19, 25, 34, 56, 81–84, 86, 176, 183, 264, 546–47
serine proteases, 176, 183
sex steroids, 387
sigma bond, 149–50, 154, 490
signal cascades, 107–8, 110, 266, 476
signal transduction, 103–5, 107, 109, 111, 116–17, 163–65, 381, 383–87, 446–50, 476, 478
sodium dodecyl sulfate. *See* SDS
sodium-potassium pump, 120, 162, 411
solvation layer, 63, 345, 368
spectroscopy, 617–18
sphingomyelins, 358
sphingophospholipids, 353, 358
sphingosine, 356–58
sphingosine 1-phosphate, 386
standard curve, 610–11
standard state, 135
state function, 66, 130–33, 184, 410–11
steady state approximation, 200, 202–5
stereochemistry, 13, 157–58, 173–74, 273, 275, 281, 284–85, 288–89, 297, 428, 431
steric hindrance, 39, 51, 157
steroid hormones, 109–10, 362–63, 387–91
stoichiometry, 456

structures
  atomic resolution, 624
  molecular, 173, 624
  three-dimensional, 337, 622–23
substrate analogs, 237, 244
substrate-binding site, 170, 174–75. *See also* active site
substrate cycle. *See* futile cycles
substrate-level phosphorylation, 402, 409, 424–25
substrates, 139, 141–47, 149–52, 154–55, 157–60, 165, 167–76, 178, 180–85, 197, 199–200, 203, 205–6, 209–14, 217–19, 224, 230–31, 235, 237–39, 241, 243–44, 265–66, 420, 427–28, 440, 454
  high-energy, 402–3, 410
  multiple, 178–79
  phosphorylated, 150, 410, 420
succinate, 160, 489–90, 500, 515, 545
  enzyme oxidizes, 499
succinate dehydrogenase (SDH), 88, 490, 499. *See also* complexes of the electron tranport chain
succinyl-CoA, 160, 409, 487–90, 493, 526
Succinyl-CoA synthetase, 160, 489–90
sucrose, 302–3
sugar, 84, 274, 279–81, 283, 289–92, 302, 309, 317, 331, 333, 430
sugar alcohols, 293, 322
sugar-phosphate backbone. *See* backbone, nucleic acids
symport, 120, 512
synthases, 155, 160–61
synthetase, 160–61

T

tags (protein purification), 599–600
tautomerization, 299–300, 313–15, 340, 406, 409
Tay-Sachs disease, 301
TCA (tricarboxylic acid) cycle. *See* citric acid cycle
temperature, 69–70, 73, 94, 192, 194–97, 340, 342–44, 376, 378, 614–17, 623
template (nucleic acids), 336, 339
ternary complex, 178–80, 230. *See also* bisubstrate enzymes
terpenoids, 362, 387, 390–91
testes, 387
testosterone, 387
tetramethylsilane (TMS), 624

thermodynamics, 69, 129, 133, 167–68, 264, 410, 614
thermogenin, 515
thiamine pyrophosphate (TPP), 186, 482
thioesters, 143, 402, 409, 488–89, 521, 531–32
thioether, 9, 15
thiolase, 522, 528. *See also* acetyl-CoA acyltransferase
threonine, 6, 11, 13, 17, 19, 25, 34, 56, 81–84, 163, 546–47
thymine, 310–13, 315–18, 335, 337–39, 342, 560
tissues, 432–33, 440, 442–44, 447–48, 450, 473, 477–78, 518, 527–29, 539, 544
  adipose, 350–51, 392
  muscle, 477
  nongluconeogenic, 445, 447
titration. *See* amino acid titration
TLC. *See* chromatography, thin-layer (TLC)
T_m. *See* melting temperature
TMS. *See* tetramethylsilane
Tollens test, 298–300
TPP. *See* thiamine pyrophosphate
transaldolase, 457–59
transaminases, 148, 180, 229 transamination, 498, 540–42
transcription factors, 90, 109
transferase, 146, 148–49, 457
transfer reactions, 150, 156, 421, 423, 430, 456, 466–67
transfer RNA. *See* tRNA
transition state, 127, 129, 133, 168–70, 174–75, 180–81, 184–85, 234 transition state analog, 234
transketolase, 457–58, 460
translocases, 120, 161–62, 400, 411, 510, 521
transmembrane domains, 58, 106, 166, 511
transport
  passive, 420, 509
  secondary active, 120, 412
transverse diffusion. *See* fluid mosaic model

triacylglycerides, 350, 518
tricarboxylic acid (TCA) cycle. *See* citric acid cycle
triglycerides, 347, 350–51, 353–54, 361, 367, 413–14, 442, 517–19, 533
triokinase, 429
triose phosphate isomerase, 156, 422, 438
trioses, 275
trisaccharides, 295–96, 474
triterpene, 362, 387
tRNA (transfer RNA), 324, 336–37, 405, 585 trypsinogen, 263
tryptophan, 11–12, 15, 17–18, 20, 546–47, 590, 608–9
tubulin, 580
turnover number. *See* k_{cat}
tyrosine, 11, 17–20, 22, 24–25, 28–29, 34, 43, 46, 546–47, 590, 608

U
ubiquinone and ubiquinol (UQ and UQH_2), 143, 185, 391–92, 496–97, 499–504, 506–7, 522
ubiquitin, 34, 84–85, 177, 540, 567
UDP, 467–68
UDP-galactose, 431
UDP-galactose 4-epimerase (GALE), 431
UDP-glucose, 290, 431, 466–69
UDP-glucose pyrophosphorylase, 466
UMP, 466, 468
unbinding, enzyme-product, 200
uncouplers (of ATP synthesis), 514–16
uniport, 120
UQ and UQH_2. *See* ubiquinone and ubiquinol
uracil, 310–13, 315, 317, 339, 342, 560
urea cycle, 542–45, 548
uridine monophosphate, 466
uridine nucleotides, 466–67
uridylyl group, 466
UTP, 160, 324, 331, 401, 466, 468–69
UV light-catalyzed reactions, 389
UV-vis spectroscopy, 607, 609, 612, 618, 620

V

V_0, 199, 201–2, 205–6, 210–12, 217–18, 224–25, 230–31, 239
vacuum filtration, 581
valine, 6–7, 16–17, 546
vasodilation, 165
vesicle, 114, 116, 375
visualization, 575, 578, 587–88
visual system, 390
vitamins, 87, 198, 388–92
  lipid-soluble, 392–93
V_{max}, 199, 206, 209, 211–12, 219–22, 224–27, 230, 235, 239–47, 251, 257
voltage, 115, 509

W

water cage. *See* solvation layer
wavelength, 390, 607, 618–19, 622–24
waxes, 361, 365–66
wobble pair, 339

X

x-ray crystallography (XRC), 622, 624, 626
xylulose, 274–76
xylulose 5-phosphate, 455, 458–60

Z

zwitterions, 4–5, 23–24, 43, 355, 358
zymogens, 263–64. *See also* proproteins